Current Topics in Microbiology and Immunology

228

Springer
Berlin
Heidelberg
New York
Barcelona
Budapest
Hong Kong
London
Milan
Paris
Santa Clara
Singapore
Tokyo

Protein Modules in Signal Transduction

Edited by A. J. Pawson

With 42 Figures and 14 Tables

Springer

Professor ANTHONY J. PAWSON, FRS
Samuel Lunenfeld Research Institute
Mount Sinai Hospital
600 University Avenue
Toronto, Ontario
Canada M5G 1X5

Cover illustration: The figure illustrates a series of modular interactions through which PTB, SH2 and SH3 domains collaborate to activate the Ras guanine nucleotide exchange factor Sos.

Cover design: design & production GmbH, Heidelberg

ISBN-13: 978-3-642-80483-0 e-ISBN-13: 978-3-642-80481-6
DOI: 10.1007/978-3-642-80481-6

Library of Congress Catalog Card Number 15-12910
Softcover reprint of the hardcover 1st edition 1998

Typesetting: Scientific Publishing Services (P) Ltd, Madras

SPIN: 10630881 27/3020 – 5 4 3 2 1 0 – Printed on acid-free paper

Preface

The behavior of eukaryotic cells, particularly those of multicellular organisms, depends on the transmission of signals from one cell to another. Such extracellular signals can take the form of hormones, antigens, cells surface molecules, or components of the extracellular matrix and exert their effects by binding to specific receptors, usually exposed on the surface of the target cell. These transmembrane receptors possess a cytoplasmic domain that allows communication with intracellular signaling pathways, providing access to the regulation of gene expression, cytoskeletal architecture, cell metabolism, survival, and the cell cycle. Defining the process through which a signal emanating from an individual receptor can influence so many aspects of cellular function is of central importance for our understanding of signal transduction.

Many polypeptide factors that regulate cellular growth and differentiation bind to receptors with cytoplasmic tyrosine kinase domains. Recent evidence has indicated that intracellular signaling from receptor tyrosine kinases proceeds through a series of modular protein–protein interactions, typified by the interaction of autophosphorylated growth factor receptors with the Src homology 2 (SH2) domains of cytoplasmic target proteins. Thus, a crucial role of tyrosine phosphorylation is to promote the formation of protein complexes through the creation of specific SH2 domain-binding sites, thereby regulating the activation of biochemical pathways within the cell. The interactions of SH2 domains with their ligands have two interesting features: first, phosphorylation of a tyrosine residue within the ligand is absolutely required for high-affinity binding and serves as a switch for recognition of the phosphorylated site by an SH2 domain; second, SH2 domains recognize specific phosphopeptide sequences in a fashion that is dictated by the ligand residues immediately C-terminal to the phosphotyrosine, providing an element of specificity. SH2-mediated interactions are important not only in signaling by transmembrane receptors, but also for the functions of cytoplasmic tyrosine kinases that act downstream of cytokine- and antigen-receptors.

SH2 domains can be viewed as the prototype for a growing number of protein modules that control protein–protein and protein–phospholipid interactions in a wide range of cellular processes. These include phosphotyrosine-binding (PTB) domains that, in some cases, also recognize phosphotyrosine-containing motifs on activated receptors, although in a quite different manner from SH2 domains, and SH3 and WW domains that bind proline-rich peptide sequences. PDZ domains recognize short peptide motifs at the C-terminal ends of receptors and ion channels and are involved in receptor clustering and subcellular organization as well as engaging in direct PDZ–PDZ interactions. The pleckstrin homology (PH) domain, although it has a fold very much like that of a PTB domain, apparently associates with specific phosphoinositides and probably functions to target proteins to the plasma membrane.

An important feature of these modules is that they are frequently found covalently linked within the same polypeptide chain and thereby allow the formation of a network of protein–protein and protein–phospholipid complexes that can, in principle, disseminate signaling information to a wide range of cellular processes. It is also apparent that these modular interactions are employed by a variety of cell-surface receptors and internal signaling pathways and are certainly not confined to the targets of tyrosine kinases.

The articles in this volume address the various mechanisms through which protein modules control signal transduction. Attention is focussed on the genetic, biochemical, and structural analysis of domains that act downstream of tyrosine kinases and their role in specific biological events during embryonic development, the response to insulin, and in organization of the cytoskeleton. However, the subjects of protein modules that control the functions of ion channels, as well as signaling by serine kinase receptors, and transmembrane proteins such as Notch, have also been addressed. Taken together, these chapters provide an over view of the molecular process by which signals emanating at the plasma membrane are transmitted to targets in the cytoplasm and within the nucleus.

Canada T. Pawson

List of Contents

List of Contributors

(Their addresses can be found at the beginning of their respective chapters.)

Functions of SH2 and SH3 Domains

B.J. Mayer and R. Gupta

1 Introduction

Cells have a remarkable ability to extract information from the extracellular environment and to respond by altering their transcriptional and replication programs, metabolism, shape, and many other aspects of their behavior. The transduction of extracellular signals is particularly crucial in multicellular organisms, where development and adult life requires that each cell precisely adjust its activities to conform to the needs of the whole organism. From an engineering standpoint the mechanisms used to transduce signals must be combinatorial in nature, because the limited number of total gene products implies that the trans-

Howard Hughes Medical Institute, Children's Hospital and Department of Microbiology and Molecular Genetics, Harvard Medical School, 320 Longwood Avenue, Boston, MA 02115, USA

ducers for each specific signal in each specific cell type cannot be unique. Our current understanding suggests that many types of extracellular signals are transduced by a relatively small number of enzymes including tyrosine kinases, GTP-binding proteins, and serine/threonine kinases, and that specificity of signaling arises through the assembly of multiprotein complexes involving such signaling proteins.

The importance of stable protein-protein interactions to signaling is now widely appreciated (to the point where the ubiquitous "two-hybrid screen" is almost universally applied to any protein of interest), and this understanding is due in large part to work on Src Homology 2 and Src Homology 3 (SH2 and SH3) domains. These two modules were originally noted as regions of sequence similarity found in proteins implicated in signaling (the term "Src Homology domain" is a vestige of early work in which the SH2 domain was first identified in several related oncogenic tyrosine kinases including Src (Sadowski et al. 1986). They have subsequently been found in many hundreds of different proteins in eukaryotes, making them among the most common structural motifs known. These two motifs are the archetypes for the rapidly growing assortment of independently folding protein modules whose primary role is to mediate high-affinity binding to other proteins.

1.1 SH2 Domains Bind Tyrosine-Phosphorylated Ligands

Before any specific role was assigned to the SH2 or SH3 domains, mutagenesis studies in nonreceptor tyrosine kinases had suggested that they served some sort of regulatory role. An important breakthrough was the cloning of members of otherwise unrelated families of signaling proteins that contained the domains, including PLC-γ, Ras-GAP, the p85 subunit of PI-3 kinase, and the *Crk* oncogene; this revealed the modular nature of SH2 and SH3 domains and suggested that they would be found in other proteins involved in signal transduction. Experiments by a number of groups soon showed that these proteins had the remarkable property of rapidly and tightly binding to tyrosine kinase growth factor receptors upon ligand stimulation. It subsequently became clear that this binding was due to the proteins' SH2 domains, that autophosphorylation of the receptors was required for binding, and that binding to tyrosine-phosphorylated proteins was a general property of SH2-containing proteins (Anderson et al. 1990; Margolis et al. 1990; Matsuda et al. 1990, 1991; Mayer and Hanafusa 1990; Moran et al. 1990). The demonstration that bacterially produced SH2 domains could bind in a phosphotyrosine-dependent fashion to denatured proteins on filters (Mayer et al. 1991) demonstrated that the binding of SH2 domains to tyrosine-phosphorylated proteins was direct and unlikely to require folded structure in the ligand proteins.

The realization that SH2 domains could mediate the phosphotyrosine-dependent association of proteins (and presumably affect the localization and activity of proteins that contain the domain) resolved a long-standing conundrum concerning receptor tyrosine kinases. While considerable effort had been expended in a search for the substrates of tyrosine kinases, including receptors, it was also known that the major tyrosine-phosphorylated protein in cells stimulated with mitogenic growth factors was often the receptor itself. This seemed incompatible with ideas in which the activated kinase served to amplify signals by phosphorylating many substrate molecules. But if the major role of the receptor was to autophosphorylate and thereby serve as an inducible binding site for downstream SH2-containing signaling molecules, it then was unnecessary to invoke phosphorylation of heterologous substrates to construct models of signal transmission. Activation of kinase activity could be viewed as a means of changing the binding activity of the intracellular aspect of the receptor, analogous to differences in the binding activity of G proteins depending on whether they are bound to GDP or GTP.

1.2 SH3 Domains Bind Proline-Rich Peptides

SH2 domains and tyrosine kinases seem to have evolved hand-in-hand during the transition from unicellular to multicellular organisms. The SH3 domain, on the other hand, is found in all eukaryotes including the yeasts, suggesting a longer evolutionary history and a more general function. While often found in conjunction with SH2 domains in signaling proteins, SH3 domains are also found in many proteins lacking SH2 domains, even in complex eukaryotes. Because several of the initial proteins in which SH3 domains were found were associated with the cytoskeleton (alpha spectrin, myosin isoforms, etc.), there was initially some suspicion that these domains might have a specific cytoskeletal function, but it is now clear that the role of SH3 domains is much more general.

The first SH3-binding proteins were isolated by screening bacteriophage expression libraries with labeled SH3 domains, and the binding sites were mapped to short proline-rich peptides (Cicchetti et al. 1992; Ren et al. 1993). In subsequent years numerous SH3 ligands have been identified by affinity chromatography, yeast two-hybrid screens, degenerate peptide library screens, and phage display strategies. All of the SH3-binding sites identified share the property of being proline-rich, and structural studies revealed that ligands adopt the left-handed polyproline-2 (PP-II) helix conformation. Because binding is independent of any covalent modification of the binding site (such as phosphorylation, as in the case of the SH2 domain), SH3-ligand interactions are usually constitutive and not inducible, though there are interesting exceptions, some of which will be discussed in greater detail below. But conceptually, SH3 domains are less likely to act as switches than as a more general means of assembling protein complexes via moderate-affinity interactions.

2 Domain-Ligand Interactions

Few protein-protein interactions are as well characterized biochemically and structurally as SH2- and SH3-mediated complexes. These topics have been extensively reviewed in the past several years (e.g., Cohen and Baltimore 1995; Pawson 1995), so only the highlights will be mentioned below. Although SH2 and SH3 domains differ in almost every aspect of their structure, ligands, inducibility, specificity, and so on, they do share one important property: they bind to stereotypical ligands that are defined by short stretches of contiguous peptide sequence, and in large measure the entire affinity of the domain for its protein ligand can be recapitulated by free peptide. As recently discussed by HARRISON (1996), this is a departure from earlier examples of protein-protein interactions in which whole surfaces of two interacting proteins, composed of residues from disparate regions of the primary sequence, are involved in binding.

The interaction of modular domains with short peptide ligands has obvious implications for the evolution of interactions, because it is easy to envision how binding sites and domains could be rapidly shuffled in and out of proteins in the course of evolution. There are also important implications for the experimental analysis of signaling pathways. SH2 or SH3 domains can be easily identified in a novel protein from the primary structure, and potential SH2- or SH3-binding sites can also be tentatively identified by simple sequence inspection. This allows us to anticipate the types of interactions that may be involved in the function of a novel protein and suggests specific mutations that can illuminate the significance of potential interactions. While it still might be misleading to assume that proteins behave as modular "beads on a string," identification of binding modules has proven a valuable tool for molecular analysis.

2.1 SH2 Domains

SH2 domains bind to extended peptides, with specificity largely but not exclusively determined by the three residues C-terminal to the phosphotyrosine (SONGYANG et al. 1993). The binding of an isolated SH2 domain to a specific peptide target is quite high – most reported dissociation constants are lower than 10^{-7} M (LADBURY et al. 1995), in some cases considerably lower. By contrast affinity for unphosphorylated peptides is negligible, meaning that the interaction is very tightly controlled by signals that change the phosphorylation state of potential binding sites. Discrimination between different phosphorylated sites can also be quite high. Affinities for phosphotyrosine itself are in the low millimolar range, meaning that at least four logs of affinity are provided by the peptide residues that follow the phosphotyrosine. Pioneering studies by Cantley's group using a degenerate peptide library approach have proven very useful in identifying those sites bound with highest affinity by specific SH2 domains, allowing a priori predictions about the SH2-containing proteins that might bind a specific phosphorylated site in a protein (SONGYANG et al. 1993, 1994).

2.2 SH3 Domains

SH3 domains also bind to extended peptide sequences, in this case invariably in the left-handed PP-II helix conformation (reviewed in MAYER and ECK 1995). Structural studies have shown that SH3 domains possess a shallow, extended binding groove composed of three pockets which interdigitate with the residues on the face of three turns of the helix (the PP-II helix has three residues per turn). Surprisingly, it was found that SH3 domains could bind to some ligands in an N-to-C orientation and others in a C-to-N orientation, due to the pseudosymmetrical nature of the PP-II helix (FENG et al. 1994). Two of the pockets of the SH3 domain each bind to two adjacent hydrophobic residues of the ligand (one of which is a proline) while the third pocket binds to "specificity-determining" residues, which are often but not always basic, giving two consensus binding sites: for "class 1" sites, + x Φ P x Φ P, and for "class 2" sites, Φ P x Φ P x + (where "+" denotes a basic residue, "Φ" denotes a hydrophobic residue, and "x" can be any amino acid). Because known binding sites almost invariably contain the sequence PxxP, this is often considered the core binding motif. Proline plays a critical role by favoring the adoption of the PP-II helix conformation, and in a more general sense by presenting hydrophobic patches on the typically hydrophilic surface of proteins.

Specificity of an SH3 domain for particular ligands, at least in the case of peptide ligands, is far from absolute. Most SH3-peptide ligand interactions have dissociation constants in the neighborhood of $10^{-6} - 10^{-5}$ *M*, and many peptides bind to different SH3 domains with quite similar apparent affinities in vitro (YU et al. 1994). Specificity in vivo, however, is probably significantly higher for several reasons. First, residues outside the minimal SH3 binding motif have been found to contribute significantly to affinity by binding to regions of the SH3 outside the three binding pockets (FENG et al. 1995; RICKLES et al. 1995). Furthermore, in many cases binding sites are reiterated several times in a ligand protein, raising the apparent affinity by increasing the avidity of binding relative to monovalent ligands. Because the dissociation constants of the domains for their targets are often in the same range as the intracellular concentrations of interacting proteins, local concentrations will have large effects on what potential partners actually bind in vivo. The modest affinities also imply that specific SH3-mediated interactions will be relatively short-lived and therefore subject to remodeling in response to changes in the local concentration of binding partners.

3 The Grb2 Paradigm

The importance of the interactions mediated by SH2 and SH3 domains is now beyond dispute, but it was the seminal discovery several years ago that such interactions were critical in signaling from tyrosine kinase growth factor receptors to Ras that conclusively put to rest doubts about the relevance of in vitro binding to in vivo biological activities. Because this pathway still serves (for better or for worse) as a paradigm for SH2 and SH3 function it is worth briefly considering it here. This is also a clear example of how two widely divergent experimental approaches, classical genetics and biochemistry, contributed equally to an important insight unattainable by either approach in isolation.

Genetic screens for genes involved in signaling initiated by receptor tyrosine kinases in *C. elegans* and later in *D. melanogaster* identified a gene encoding a small protein (termed Sem-5 in *C. elegans* and Drk in *Drosophila*) that appeared to function downstream of the receptor and upstream of the GTP-binding protein Ras (Clark et al. 1992; Olivier et al. 1993; Simon et al. 1993). While Ras was known to be important for signaling, especially in regulation of proliferation and differentiation, and was known to be activated by mitogenic growth factors, the mechanism of activation remained obscure. Sem-5/Drk (now more commonly referred to as Grb2, the name for the human homolog, Lowenstein et al. 1992), consists of a central SH2 domain flanked by two SH3 domains in the total absence of any other functional domains. It was clear that the SH2 domain had the potential to bind to autophosphorylated growth factor receptors after stimulation, but what might its downstream, SH3-binding effectors be? A critical insight was provided when it became apparent that SH3 domains were likely to bind proline-rich ligands (Ren et al. 1993), because a putative guanine nucleotide exchanger (GEF) for *Drosophila* Ras, termed Sos, had already been cloned and was known to have a proline-rich C-terminal tail (Bonfini et al. 1992). Given these pieces of information, it proved straightforward to demonstrate that Grb2 mediated the recruitment of Sos to the membrane after growth factor binding, leading to activation of Ras, which is confined to the membrane by virtue of its lipid modification (Buday and Downward 1993; Chardin et al. 1993; Olivier et al. 1993; Simon et al. 1993).

3.1 Advantages of Networks

The two critical interactions for this signaling pathway are the SH2-mediated interaction with liganded growth factor receptor and SH3-mediated interaction with the effector, Sos. The net effect is to greatly increase the local concentration of Sos in the vicinity of its substrate, Ras. At first glance this might seem like a needlessly complex and arcane method to activate Ras – it would be much more efficient to have activated growth factor receptor modify Ras directly, or to provide Sos with its own SH2 domain thereby eliminating the need for the Grb2 adaptor "middleman." But the system as it actually evolved provides a significant advantage: multiple branch-points for signals to diverge to multiple effectors or be integrated among many inputs. It is a mistake to consider the receptor-Grb2-Sos-Ras complex as an isolated, linear pathway. Tyrosine phosphorylated sites on activated growth factor receptors can recruit many different SH2-containing proteins other than Grb2; multiple proteins can compete for binding to each site based on abundance and local concentration and their binding can be modulated by dephosphorylation of sites. In turn, many different SH3-binding effectors can bind to the SH3 domains of Grb2, allowing a host of proteins to be recruited to the vicinity of activated receptors, and again different binding proteins can compete among themselves for binding before and after relocalization.

The multiple branch-points in such a system allow for a single biochemical event [e.g., binding of platelet-derived growth factor (PDGF) to its receptor on the cell surface] to have completely different intracellular consequences based on the specific cocktail of binding partners available in each cell. It can readily be seen how a relatively limited number of such binding partners can give virtually infinite combinations of responses to a single signal. Conversely, because the Grb2 SH2 domain can bind to sites on many different tyrosine-phosphorylated proteins, many different signal inputs can generate the same intracellular response.

While the straightforward model for Ras activation outlined above is useful in thinking about other roles for protein-interaction modules in signaling, the work of the past several years has highlighted the complexity of actual signaling pathways. For example, it is apparent that many activated receptors do not directly bind the Grb2 SH2 domain, and it can also be shown that the proline-rich tail of Sos is not required for activation of Ras in some cases (KARLOVICH et al. 1995). The focus of much future work will be on efforts to understand the behavior of signaling networks, as opposed to the more easily understood but less accurate models based on linear pathways. In the remainder of this review, we will consider some of the more recent data on new and surprising roles for Src homology domains and on the dynamic nature of the complexes they mediate in the course of signaling.

4 Unconventional Ligands for SH2 and SH3 Domains

While the great majority of the well-characterized ligands for SH2 and SH3 domains have the "conventional" properties outlined above, there are a growing number of exceptions that suggest we should keep an open mind when considering possible unexpected binding activities for these domains. At the most basic level, we can consider that SH2 domains have evolved to bind a concentrated negative charge (phosphotyrosine) in a peptide context, while SH3 domains have evolved to bind hydrophobic patches on protein surfaces, and it should not be surprising that other ligands with these general properties can be accommodated with reasonable affinity.

4.1 Phosphotyrosine-Independent SH2 Binding

Several proteins have been shown to bind to SH2 domains in the absence of detectable phosphotyrosine, including an N-terminal region of BCR to the SH2 domain of Abl (PENDERGAST et al. 1991; MULLER et al. 1992), and at least two serine/threonine phosphorylated proteins to the SH2 domain of the Src-family kinase Blk (MALEK and DESIDERIO 1994; MALEK et al. 1996). In the case of the Blk ligands, the binding was shown to be blocked by a phosphotyrosine-containing peptide and by phosphotyrosine, and a Blk SH2 domain in which a conserved serine predicted from structural studies to interact with phosphotyrosine was mutated failed to bind, strongly suggesting that the same site that interacts with conventional ligands is involved in the phosphotyrosine-independent binding. Binding was found to depend on phosphorylation, and the ability of a fragment completely lacking tyrosine residues to bind suggested that phosphoserine and/or phosphothreonine were essential.

Specificity for binding phosphotyrosine (as opposed to other phosphoamino acids) is thought to be largely due to the distance between the surface of the SH2 domain and the deeply-buried terminal amino groups of an essential basic residue, the so-called "FLVRES" arginine; only the lengthy tyrosine side chain allows the phosphate to interact productively with this arginine, whereas serine and threonine are too short (Waksman et al. 1992). In addition, in many SH2 domains other conserved basic residues make amino-aromatic interactions with the ring of tyrosine, further stabilizing the binding of phosphotyrosine (Waksman et al. 1992). How can the apparently specific binding of proteins phosphorylated on serine or threonine be explained? Probably a combination of factors are involved, including lower actual affinity than for tyrosine-phosphorylated ligands (the Blk ligands were isolated by affinity chromatography, which could allow detection of relatively low-affinity interactions), and favorable peptide context. In fact, the SH2-binding regions of the Blk ligand are quite acidic (S/T E E being common), and it is known that Src-family SH2 domains such as that of Blk bind most tightly to motifs with the consensus Y(P) E E I; it seems possible that the high concentration of negative charge in the phosphate-glutamate-glutamate region is sufficient for reasonable binding due to largely electrostatic interactions. Until such phosphotyrosine-independent interactions are rigorously shown to be important in vivo, however, they will be viewed with some caution.

4.2 SH2-Inositol Lipid Interactions

A very different class of ligand with a concentrated negative charge has also been shown to bind tightly to some SH2 domains – phosphorylated inositol lipids. Cantley's group has shown that the SH2 domains of the p85 subunit of PI-3 kinase and of Src bind with reasonably high affinity to phosphatidylinositol (3, 4, 5) trisphosphate (PI [3, 4, 5] P_3) in lipid micelles (Rameh et al. 1995). This binding was partially competed by phenyl phosphate or phosphorylated peptide ligands, suggesting that the binding site at least overlaps with that for tyrosine-phosphorylated ligands; mutation of the "FLVRES" arginine of the SH2 (which is essential for binding phosphotyrosine) did not abolish phospholipid binding, however, suggesting that the specific mode of binding was different. As suggested above, in the case of the Src SH2 it is perhaps not surprising that the multiply-phosphorylated inositol head group might bind to a site that normally binds Y(P) E E I; however, the p85 SH2 domains have been shown to bind best to Y(P) M × M sites (Songyang et al. 1993), obviously much more hydrophobic in character.

One could easily imagine how it might be useful for some SH2 domains to bind to products of PI-3 kinase such as PI (3, 4, 5) P_3, because these lipids are rapidly synthesized following growth factor receptor activation. PI-3 kinase is known to translocate to the membrane via binding of the SH2 domains of its p85 subunit to activated growth factor receptors, where it presumably begins to generate a pool of 3'-phosphorylated inositol lipids in the immediate vicinity of the receptor-PI kinase complex. The resulting very high local concentrations of phosphorylated lipid would likely lead to dissociation of the p85 SH2 domains from the receptor and association with the lipid, even if the absolute affinity for lipid were lower than for sites on the receptor. Dissociation of the PI kinase from the receptor might be important for it to diffuse freely in the membrane to access substrate, to prevent its endocytosis along with the receptor, or to allow other SH2-binding proteins to bind to the receptor.

It is likely that high-affinity binding to inositol lipids is not a general property of SH2 domains, but might be a property that evolved out of existing SH2 domains to fit the needs of particular proteins. This is reminiscent of the situation of Pleckstrin Homology (PH) domains, another modular unit of protein structure (MUSACCHIO et al. 1993). While the specific ligands for PH domains are not clear in all cases, the PH domain structures determined to date show a large, positively charged pocket (FERGUSON et al. 1995). Several PH domains have been shown to bind to phosphorylated lipids with varying affinities (HARLAN et al. 1994), and in one specific case, the PH domain of PLC-δ, quite high affinities for that enzyme's substrate (PI (4, 5) P_2) are observed (LEMMON et al. 1995). In an ironic twist, PTB domains, which are structurally indistinguishable from PH domains although there is little if any primary sequence similarity, bind very tightly to some tyrosine-phosphorylated peptides (ZHOU et al. 1995, 1996; Eck et al. 1996). What we are seeing probably reflects the working of evolution on successful structural modules, where binding specificity for one negatively charged ligand might be broadened or altered completely to accommodate other ligands.

4.3 Unusual SH3 Interactions

More information is also becoming available about less conventional ligands for SH3 domains. As outlined briefly above, until quite recently all SH3 ligands could be modeled on short, proline-rich PP-II helical peptides containing a PxxP core motif. Recent structures have revealed how regions outside the PxxP core of proteins can increase the affinity and discrimination of SH3 binding and revealed the first example of a high-affinity SH3-mediated interaction that does not involve a proline-rich helical binding peptide.

The product of the *nef* gene of HIV and SIV, which is important for efficient viral replication, has been shown to bind to the SH3 domains of a subset of Src-family kinases including Hck (SAKSELA et al. 1995). Two features of this interaction suggested that it was somewhat different from known SH3-ligand interactions. First, peptides corresponding to the known PxxP region of Nef bound extremely poorly to SH3 domains (91 μM K_d), while the intact protein bound with very high affinity (0.25 μM K_d; LEE et al. 1995). Second, changing a single amino acid in the "RT loop" of the Fyn SH3 (Arg 96 to Ile), which would not be predicted to interact directly with the PP-II helical binding peptide, converted it from a low-affinity binder of *nef* to a high-affinity binder (a greater than 50-fold increase in affinity; LEE et al. 1995). The recent structure of HIV *nef* complexed to this mutant form of Fyn SH3 reveals fairly typical binding to a *nef* proline-rich helix in the "class 2" orientation; however, an additional hydrophobic interaction between the specificity-determining isoleucine in the RT loop of the SH3 and a region of *nef* outside the proline-rich helical core was also revealed (LEE et al. 1996). Presumably it is this interaction, in the context of the whole *nef* protein, that confers high-affinity binding and specificity for particular SH3 domains. Because mutations in the PxxP SH3-binding core of *nef* render it unable to stimulate HIV replication in primary T-cell cultures (SAKSELA et al. 1995), this interaction is a promising target for antiviral drug design.

A much more unusual interaction was seen in the recent structure of a fragment of the tumor-suppressor p53 in complex with an SH3-containing protein, termed p53BP2. p53BP2 was originally isolated from a yeast two-hybrid screen using the DNA-binding core of p53 as the "bait" (IWABUCHI et al. 1994), and it binds p53 with moderate affinity ($\sim 30\ \mu M\ K_d$; GORINA and PAVLETICH 1996).

Surprisingly, the structure of the complex revealed that although the surface of the p53BP2 SH3 domain that is involved in binding is essentially the same as that used in more typical SH3-ligand interactions, the region of p53 that it binds looks nothing like a PP-II helix; instead it is assembled from two regions of an extended loop (GORINA and PAVLETICH 1996). Here we have an example of a binding surface adapted to binding one type of ligand (extended, hydrophobic helical peptides) evolving to interact with a completely different type of ligand. Like other SH3 domains the p53BP2 SH3 can bind with high affinity to PxxP ligands, as suggested by the ability to isolate such ligands from phage-display libraries (SPARKS et al. 1996), but it has also evolved a novel ability to interact with p53. Although the p53 interaction is of relatively low affinity, and significant contributions to the affinity are probably made by regions of p53BP2 outside the SH3 domain, mutational data suggest that this novel SH3-mediated interaction might play a critical role in the biological activity of p53 (GORINA and PAVLETICH 1996), highlighting the potential of other SH3 domains to participate in biologically relevant unconventional binding interactions.

5 Complex and Dynamic Networks of Interactions

It is well established that SH2- and SH3-mediated interactions are important for transmitting signals from tyrosine kinases, and series of interactions such as outlined above for the receptor-Grb2-Sos-Ras pathway are thought to be critical for signal transmission. There is very little concrete data, however, about the range of specific complexes actually induced in cells by signals, the relative importance of each specific multiprotein complex to the signal output, and the kinetics of formation and subcellular distribution of specific pools of complexed proteins. The problem lies in the very many possible interactions that can occur, and the ease with which it is possible to demonstrate interactions in vitro using purified proteins. If hundreds of SH2- and SH3-mediated interactions are possible based on in vitro binding studies, what percentage of these actually occur in vivo? If, as all indications suggest, this percentage is large, how can we understand such a complicated network of interactions? We will use the example of complexes involving the Cbl proto-oncogene to illustrate both the wealth of information available and the difficulties in interpreting this information.

5.1 Cbl-Adaptor Complexes

Cbl was originally isolated as a retroviral oncogene, and its cellular homolog contains a RING finger motif, a highly basic amino terminus, and a proline-rich carboxyl terminus (LANGDON et al. 1989). The cellular proto-oncogene can be rendered oncogenic by deletion of the entire C-terminus (leaving just the amino-terminal basic domain; BLAKE et al. 1991) or by a smaller deletion disrupting the RING finger (ANDONIOU et al. 1994), and transforming forms Cbl that retain the carboxyl terminus are highly tyrosine phosphorylated (ANDONIOU et al. 1994). Stimulation of B cells or T cells via their receptors, transformation of hematopoietic cells with BCR-Abl, or stimulation of other cells types with mitogenic growth factors induces the tyrosine phosphorylation of Cbl (BOWTELL and LANGDON 1995; DE JONG et al. 1995; ODAI et al. 1995; PANCHAMOORTHY et al. 1996; REEDQUIST et al. 1996). A homolog of Cbl also emerged from *C. elegans* screens for extragenic suppressors of a weakly inactivating mutation in the *Let-23* receptor tyrosine kinase (YOON et al. 1995). Because loss-of-function mutations in the Cbl homolog, termed *Sli-1*, restored normal levels of signaling from the crippled receptor (YOON et al. 1995), it is likely that the wild-type Cbl protein normally antagonizes receptor signaling. Because Cbl contains proline-rich potential SH3-binding sequences, and would also be predicted to bind SH2 domains in its tyrosine-phosphorylated state, a host of possible interactions are possible and in fact many signaling proteins have been shown to bind either constitutively or inducibly to Cbl.

Among these Cbl-binding proteins are the SH2/SH3 adaptor proteins Grb2, Crk, and Nck (RIVERO-LEZCANO et al. 1994 and refs. below). Like Grb2, Nck and Crk both contain a single SH2 domain and at least one SH3 domain. Unlike Grb2, however, these two adaptors can themselves induce malignant transformation if mutated or inappropriately expressed (MAYER et al. 1988; CHOU et al. 1992; LI et al. 1992). The interactions between Cbl and the adaptors Grb2 and Crk will be discussed in greater detail. Work from a number of groups has shown that Cbl exists in a constitutive complex with Grb2, and that this complex is almost entirely mediated by the N-terminal SH3 domain of Grb2 and the proline-rich carboxyl terminus of Cbl (MEISNER and CZECH 1995; ODAI et al. 1995; BUDAY et al. 1996; PANCHAMOORTHY et al. 1996; SMIT et al. 1996a, b). Therefore a plausible model suggests that Cbl is a Grb2 effector (analogous to Sos) and that receptor autophosphorylation induces relocalization of Grb2 and its associated Cbl, whereupon Cbl can be phosphorylated by the receptor. In some cases the adaptor Shc (which contains an SH2 domain, a phosphotyrosine-binding PTB domain, and tyrosine-phosphorylated sites that bind tightly to the Grb2 SH2) might serve as an intermediary between Grb2 and phosphorylated receptor.

The interaction between Cbl and the Crk adaptor is quite different. In this case there is little if any binding of Cbl to the Crk SH3 domains, but tyrosine-phosphorylated Cbl binds very tightly to the SH2 domain of Crk (DE JONG et al. 1995; REEDQUIST et al. 1996; SATTLER et al. 1996; SMIT et al. 1996). Because tyrosine phosphorylation is inducible, the Crk-Cbl complex is therefore also inducible and only seen after stimulation with various treatments that increase phosphotyrosine. This of course raises the question of what proteins might bind the Crk SH3 domains in the Cbl-Crk complex. Two interesting Crk SH3 ligands are C3G, a GEF for the Ras-like GTP-binding protein Rapl (TANAKA et al. 1994; GOTOH et al. 1995), and the nonreceptor tyrosine kinase Abl (FELLER et al. 1994; REN et al. 1994). Rap can be shown to inhibit the effects of activated Ras (KITAYAMA et al. 1989), so a C3G complex has the potential to positively or negatively affect Ras-mediated signaling. Abl in its mutated forms is oncogenic, whereas overexpression of unmutated c-Abl has cytostatic activity (WANG 1993). Crk has been shown to modify the kinase activity of Abl when complexed with it by allowing Abl to processively phosphorylate proteins that bind tightly to the Crk SH2 (MAYER et al. 1995). One could therefore imagine that formation of a Cbl-Crk-Abl complex might affect both the phosphorylation state of Cbl and the catalytic activity of Abl. It is remarkable that all three of these proteins (Cbl, Crk, and Abl) were first isolated as oncogenes in acutely transforming retroviruses, suggesting that such complexes could be intimately involved in regulating cell proliferation. The Abl protein itself has both an SH3 and an SH2 domain, so it is easy to see that the number of potentially important interactions in even this relatively simple example is enormous (Fig. 1).

5.2 Dynamics of Adaptor-Cbl Interactions

It must also be remembered that Cbl is by no means the only protein that interacts with the Grb2 SH3s or the Crk SH2, and that changes in the content of Cbl complexes in different cells over time might be critical for signal transmission. Several aspects of the dynamic nature of these complexes have recently been reported by Downward's group. The first of these involves the Crk-Cbl complex in human PC12 cells treated with the mitogenic growth factor EGF (KHWAJA et al. 1996). In unstimulated cells, Crk exists in an SH2-mediated complex with the focal adhesion protein $p130^{cas}$, which is basally phosphorylated on tyrosine. When cells are stimulated with EGF, the Crk-$p130^{cas}$ interaction rapidly disappears and is replaced with the Crk-Cbl complex. The authors propose that this "hand-off" could be due to a higher affinity of phosphorylated Cbl for the Crk SH2 (and/or a

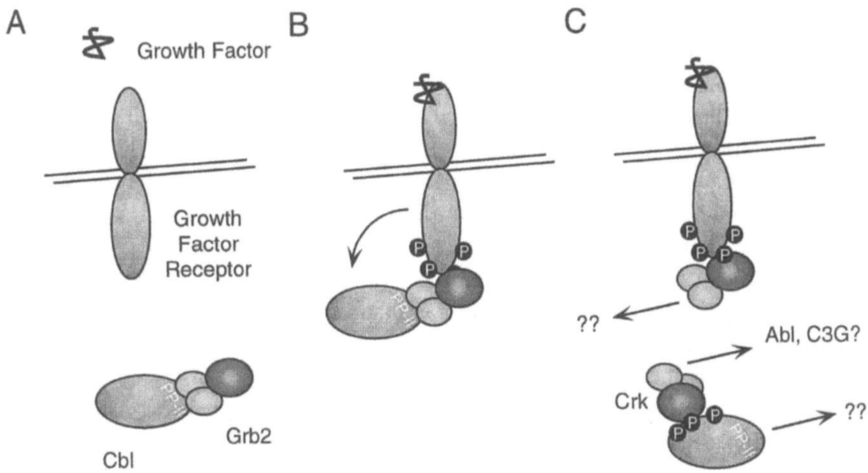

Fig. 1A–C. Cbl-adaptor complexes during receptor tyrosine kinase signaling. **A** In unstimulated cells, Cbl is complexed with the SH3 domains of the Grb2 adaptor in the cytosol. **B** Upon binding of growth factor to the receptor, the receptor autophosphorylates and Grb2-Cbl complex is recruited to the receptor via the SH2 domain of Grb2. Receptor-bound Cbl is phosphorylated on tyrosine residues. **C** After phosphorylation, Cbl is released from Grb2 and binds to the SH2 domain of Crk. The SH3 domain of Crk may bind to the potential effectors C3G or Abl. The SH3 domains of Grb2 and the proline-rich region of Cbl are also free to interact with other ligands

higher local concentration), although this has not been directly demonstrated and other mechanisms are possible.

Grb2 complexes with Cbl also change over time following stimulation. In one study, the SH3-mediated Grb2-Cbl complex in T cells rapidly diminished after activation, in parallel with increased tyrosine phosphorylation of Cbl, suggesting either that phosphorylated Cbl binds less tightly or that other Grb2 SH3 ligands compete for Grb2 binding at the membrane after activation (BUDAY et al. 1996). Because Crk binds well to the phosphorylated form of Cbl, we can envision another type of "hand-off" in which Grb2 mediates the initial translocation of Cbl to membranes resulting in Cbl phosphorylation, whereupon this complex dissociates and a new complex is formed with Crk mediated by SH2-phosphotyrosine interactions. In another cell type, EGF-treated PC12 cells, a different picture emerges; in this case the amount of Cbl complexed with Grb2 was found to increase approximately threefold after stimulation (KHWAJA et al. 1996). None of the Cbl complexed with Grb2 was tyrosine-phosphorylated either before or after stimulation, consistent with the idea that phosphorylation of Cbl is for some reason incompatible with Grb2 association. The mechanism whereby stimulation could increase the SH3-mediated Grb2 complex is completely obscure, though a similar phenomenon has been reported in some cell types for the Grb2-Sos complex so important for Ras activation (BUDAY and DOWNWARD 1993; RAVICHANDRAN et al. 1995). One possibility is global shifts in the binding equilibria due to changes in local concentrations of potential interaction partners that result from the EGF-induced relocalization of SH2-containing proteins and alterations in cell shape.

This brief review of the interactions of one potentially important signaling protein gives a sense of the difficulties encountered in analyzing these networks. Based on published work one can construct a linear pathway of information transfer via at least six SH2- or SH3-mediated interactions in response to receptor phosphorylation: receptor→Shc→Grb2→Cbl→Crk→Abl→? And this oversimplified view does not take into account dynamic changes over time or the many other potential competing interactions at each link in the chain. A major experimental hurdle is our inability to measure associations of proteins in intact cells without disturbing the system (usually by lysing cells in detergent), so we are limited to approaches that either assess what interactions can occur in vitro, or assess the effects of disrupting specific complexes or sets of complexes in vivo. A more approachable but still daunting problem is the rigorous analysis of the consequences of complex formation; for example, if Grb2-Cbl complexes are important for a specific signal is it because the complex leads to the phosphorylation of Cbl, because it brings Cbl into the vicinity of other critical interaction partners, because it induces a conformational change in Cbl, or a combination these factors plus others not yet imagined? Although there is clearly much to learn, our understanding of the general principles of SH2- and SH3-mediated interactions allows these questions to be rationally addressed.

6 Complex Networks of SH3-Mediated Interactions

In general SH3-mediated interactions are thought to be constitutive and any regulation of such interactions (as in the Cbl/Grb2 or Sos/Grb2 complexes mentioned above) is modest. There is mounting evidence, however, that in some cases networks of SH3-mediated interactions can be quite dynamic and may underlie fundamental processes such as the spatial organization of the actin cytoskeleton. The best-characterized system involving regulated assembly of multiple SH3-containing proteins is in activation of the neutrophil oxidase, and this serves as an excellent model for how other such systems might operate.

6.1 Assembly of the Neutrophil Oxidase

When presented with bacterial lipopolysaccharide or other activators, neutrophils rapidly activate a mitochondrial cytochrome oxidase and generate reactive oxygen species including superoxide which are used to kill invading microorganisms. A number of patients suffering from chronic granulomatous disease (CGD) have mutations in genes encoding critical elements in the pathway of assembly of the functional oxidase, which has facilitated the identification of the proteins involved. These include the mitochondrial oxidase itself (cytochrome b_{558}), which is composed of two chains termed $p22^{phox}$ and $gp91^{phox}$, and two cytosolic proteins termed

$p47^{phox}$ and $p67^{phox}$, each of which contains two SH3 domains and a proline-rich region that is a potential SH3-binding site. In addition to these components, the small GTP-binding protein Rac appears to be required for activation and another SH3-containing protein, $p40^{phox}$, is associated with $p47^{phox}$ and $p67^{phox}$ in cell lysates (Fig. 2a).

Binding of an N-terminal domain of $p67^{phox}$ to the membrane oxidase appears to be the critical step in activation of the enzyme, and in an in vitro reconstituted system high concentrations of recombinant $p67^{phox}$ are sufficient for full activation in the absence of $p47^{phox}$ (FREEMAN and LAMBETH 1996). The role of $p47^{phox}$ appears to be to greatly increase the affinity of $p67^{phox}$ for the oxidase upon activation, thereby triggering assembly of the functional complex (FREEMAN and LAMBETH 1996). The presence of SH3 and proline-rich regions in both $p47^{phox}$ and $p67^{phox}$ suggested that both intramolecular and intermolecular interactions might be involved in regulating the assembly of the complex. This is supported by the observation that amphophiles such as arachidonic acid or sodium dodecyl sulfate are required for maximal oxidase activity in reconstituted in vitro systems, and can unmask the SH3 domains of $p47^{phox}$ (BROMBERG and PICK 1985; SUMIMOTO et al. 1994). Because phosphorylation of $p47^{phox}$ is likely to be the trigger for assembly of the functional complex in vivo (NAUSEEF et al. 1990, 1991; TEAHAN et al. 1990), it is thought that amphophiles mimic conformational changes in $p47^{phox}$ normally induced by phosphorylation that lead to unmasking of its binding activities.

Studies of SH3-mediated interactions led to a simple model for oxidase assembly (LETO et al. 1994; SUMIMOTO et al. 1994; Fig. 2b). It was proposed that in resting cells $p47^{phox}$ exists in the cytosol in an inactive "hairpin" conformation with at least one of its SH3 domains interacting intramolecularly with its own proline-rich C-terminus. Upon phosphorylation, conformational changes would break the intramolecular interactions, leaving the proline-rich region of $p47^{phox}$ free to bind in trans to the C-terminal SH3 of $p67^{phox}$, and the N-terminal SH3 of $p47^{phox}$ to the $p22^{phox}$ component of the oxidase on the mitochondrial membrane. In this model $p47^{phox}$ serves as an inducible cross-linker whose essential function is to increase the affinity of $p67^{phox}$ for its target.

6.2 Revising the Model for Oxidase Assembly

Over the past several years, data have begun to accumulate that suggest the actual situation in vivo is much more complicated than outlined in the simple model above. For example, most of the $p47^{phox}$ and $p67^{phox}$ in resting cells appears to be already associated in the cytosol in a complex of over 200 kDa (PARK et al. 1994), suggesting that complex formation is not simply induced by phosphorylation, but is remodeled in a more subtle fashion. Additional activation-inducible binding sites between the SH3 domains of $p47^{phox}$ and the N-terminal proline-rich site of $p67^{phox}$, and between the positively charged C-terminus of $p47^{phox}$ (in the vicinity of its phosphorylation sites) and $p67^{phox}$, were also identified (DE MENDEZ et al. 1996). In addition a high-affinity binding site for $gp91^{phox}$ was identified on $p47^{phox}$,

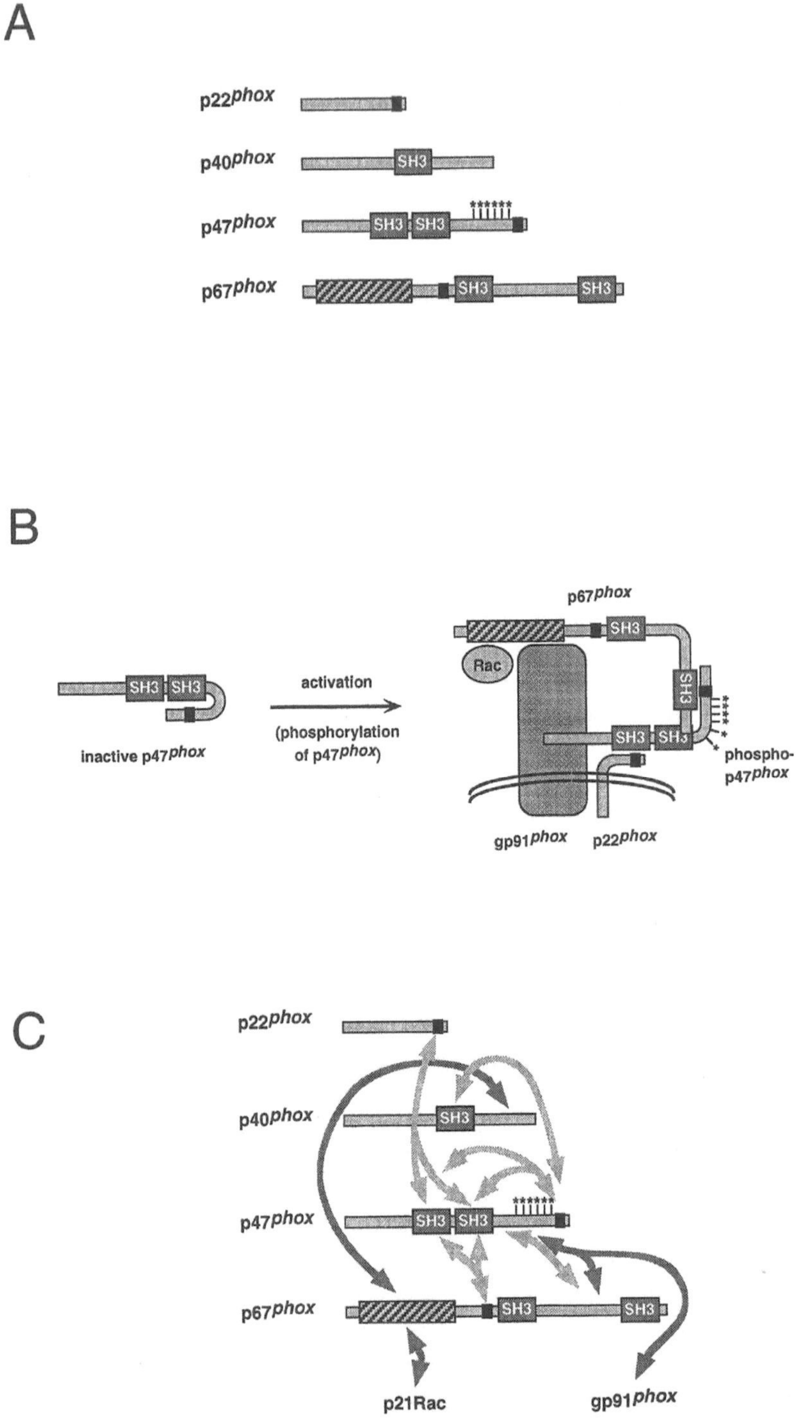
A
p22phox
p40phox
SH3
p47phox
SH3
SH3
p67phox
SH3
SH3
B
SH3
SH3
inactive p47phox
activation
(phosphorylation
of p47phox)
p67phox
SH3
Rac
SH3
SH3
phospho-
p47phox
gp91phox
p22phox
C
p22phox
p40phox
SH3
p47phox
SH3
SH3
p67phox
SH3
SH3
p21Rac
gp91phox

overlapping its phosphorylation site and the second site of interaction with $p67^{phox}$ (DE LEO et al. 1996). And if that were not complicated enough, $p40^{phox}$ (which is inessential for oxidase activation in vitro) was found to bind via its own SH3 to the C-terminal proline-rich region of $p47^{phox}$ and in an SH3-independent fashion to the N-terminus of $p67^{phox}$ (FUCHS et al. 1995, 1996; ITO et al. 1996; WIENTJES et al. 1996; Fig. 2c).

There is not yet a consensus on which interactions are most important for biological activity, due to the proliferation of possible interactions and the fact that many disparate systems (two-hybrid, GST fusions, filter overlay, surface plasmon resonance, phage display, in vivo reconstitutions, etc.) have been used to generate the binding data. The most likely model involves a preexisting complex of $p47^{phox}$, $p67^{phox}$, and $p40^{phox}$ in resting cells, or perhaps a population of different specific complexes in equilibrium with each other; this resting complex or complexes would not have high affinity for the membrane oxidase and most likely could not stimulate its activity even if present at high concentrations. Phosphorylation of $p47^{phox}$ (and perhaps $p67^{phox}$ as well) would trigger a rearrangement of inter- and intramolecular interactions in the complex such that it now has a high affinity for the membrane-associated oxidase, and once bound can stimulate the catalytic activity of the oxidase in the presence of the GTP-bound Rac. Because it is unnecessary for oxidase activity in vitro, $p40^{phox}$ is most likely to play a regulatory role in inhibiting the spontaneous activation of the complex, although this has not been directly addressed.

6.3 Other Complex SH3-Mediated Assembly Processes

This very well studied system provides a glimpse at how other complex SH3-mediated assembly processes might be regulated. One area where such regulation is almost certain to be involved is in the spatial organization of the actin cytoskeleton, which is required for processes such as bud site selection and polarized growth in yeast. Like assembly of the neutrophil oxidase, regulation of cytoarchitecture involves the inducible assembly of large multiprotein complexes on cell membranes. A number of genes encoding proteins containing SH3 domains (or potential proline-rich binding sites) have been identified in genetic screens in yeast for proteins that affect the location of bud sites or other aspects of morphology and the

◄

Fig. 2A–C. Assembly of the neutrophil oxidase. **A** Diagrammatic representation of the four components of the oxidase whose interactions are mediated by SH3-proline-rich interactions. SH3 domains are indicated, as are proline-rich SH3 binding sites (*small black boxes*). *Hatched box* denotes region of $p67^{phox}$ sufficient for activation of the oxidase in vitro. *Asterisks* indicate positions of phosphorylation sites on $p47^{phox}$. **B** Simple model for assembly of the complex. In resting cells $p47^{phox}$ is in an inactive conformation in which its SH3 domains bind to its proline-rich C-terminal tail. Upon phosphorylation, the intramolecular interactions are broken and $p47^{phox}$ can cross-link $p67^{phox}$ and $p22^{phox}$, leading to assembly of the catalytically active oxidase on the mitochondrial membrane. **C** Interactions among the components of the oxidase complex. *Light arrows* indicate SH3-proline-rich interactions; *dark arrows* indicate SH3-independent interactions. Data are assembled from references given in text

organization of the actin cytoskeleton (CHENEVERT et al. 1992; BAUER et al. 1993; HOLTZMAN et al. 1993; BENDER et al. 1996; GOODSON et al. 1996; MATSUI et al. 1996; TANG and CAI 1996). It is also intriguing that the Rho-family GTPase Cdc42 is known to be involved in regulating these processes (JOHNSON and PRINGLE 1990), reminiscent of the involvement of the Rho family GTPase Rac in assembly of the cytosolic oxidase (ABO et al. 1991; KNAUS et al. 1991). Continued genetic analysis, especially in combination with more detailed biochemical studies to elucidate critical protein-protein interactions, will surely reveal networks of SH3-mediated interactions involved in the regulated assembly of the cytoskeleton.

7 Conclusion

The realization that the controlled assembly of specific complexes of proteins lies at the heart of signal transduction has presented us with extraordinary opportunities to identify important players and to rationally modulate signaling pathways. Modular domains such as the SH3 and the SH2 are easily identified and have predictable behavior that allows the identification of potential interaction partners and suggests ways to eliminate binding to assess the importance of potential interactions. The fact that both domains recognize short peptide ligands suggests that it will ultimately be possible to design high-affinity, cell-permeable inhibitors of specific interactions that will be able to modulate the behavior of cells in a variety of experimental and clinical settings. The fact that signals emerge from networks of proteins, whose associations are mediated by SH2 and SH3 domains, also presents conceptual problems to our complete understanding of biological processes. It becomes more and more likely that responses to extracellular signals will never be simply attributable to a single protein or a single complex, but will actually be a consequence of a whole population of dynamic complexes, any one of which may not be essential but which contributes in a subtle way to the overall response. Our challenge for the future is to find new approaches to analyze such systems and ultimately to understand them to the extent we can predict the behavior of the system based on the physical properties of its components.

References

Abo A, Pick E, Hall A, Totty N, Teahan CG, Segal AW (1991) Activation of the NADPH oxidase involves the small GTP-binding protein $p21^{rac1}$. Nature 353:668–670

Anderson D, Koch CA, Grey L, Ellis C, Moran MF, Pawson T (1990) Binding of SH2 domains of phospholipase $C_{\gamma}1$, GAP, and src to activated growth factor receptors. Science 250:979–982

Andoniou CE, Thien CBF, Langdon WY (1994) Tumor induction by activated abl involves tyrosine phosphorylation of the product of the cbl oncogene. EMBO J 13:4515–4523

Bauer F, Urdaci M, Aigle M, Crouzet M (1993) Alteration of a yeast SH3 protein leads to conditional viability with defects in cytoskeletal and budding patterns. Mol Cell Biol 13:5070–5084
Bender L, Lo HS, Lee H, Kokojan V, Peterson V, Bender A (1996) Associations among PH and SH3 domain-containing proteins and Rho-type GTPases in yeast. J Cell Biol 133:879–894
Blake TJ, Shapiro M, Morse HC, Langdon WY (1991) The sequences of the human and mouse c-cbl proto-oncogenes show v-cbl was generated by a large truncation encompassing a proline-rich domain and a leucine zipper-like motif. Oncogene 6:653–657
Bonfini L, Karlovich CA, Dasgupta C, Bannerjee U (1992) The Son of sevenless gene product: a putative activator of Ras. Science 255:603–606
Bowtell DDL, Langdon WY (1995) The protein product of the c-cbl oncogene rapidly complexes with the EGF receptor and is tyrosine phosphorylated following EGF stimulation. Oncogene 11:1561–1567
Bromberg Y, Pick E (1985) Activation of NADPH-dependent superoxide production in a cell-free system by sodium dodecyl sulfate. J Biol Chem 260:13539–13545
Buday L, Downward J (1993) Epidermal growth factor regulates $p21^{ras}$ through the formation of a complex of receptor, Grb2 adapter protein, and Sos nucleotide exchange factor. Cell 73:611–620
Buday L, Khwaja A, Sipeki S, Farago A, Downward J (1996) Interactions of Cbl with two adaptor proteins, Grb2 and Crk, upon T cell activation. J Biol Chem 271:6159–6163
Chardin P, Camonis JH, Gale NW, van Aelst L, Schlessinger J, Wigler MH, Bar-Sagi D (1993) Human Sos1: a guanine nucleotide exchange factor for Ras that binds to GRB2. Science 260:1338–1343
Chenevert J, Corrado K, Bender A, Pringle J, Herskowitz I (1992) A yeast gene (BEM1) necessary for cell polarization whose product contains two SH3 domains. Nature 356:77–79
Chou MM, Fajardo JE, Hanafusa H (1992) The SH2- and SH3-containing Nck protein transforms mammalian fibroblasts in the absence of elevated phosphotyrosine levels. Mol Cell Biol 12:5834–5842
Cicchetti P, Mayer BJ, Thiel G, Baltimore D (1992) Identification of a protein that binds to the SH3 region of Abl and is similar to Bcr and GAP-rho. Science 257:803–806
Clark SG, Stern MJ, Horvitz HR (1992) C. elegans cell-signalling gene sem-5 encodes a protein with SH2 and SH3 domains. Nature 356:340–344
Cohen GB, Baltimore D (1995) Modular binding domains in signal transduction proteins. Cell 80: 237–248
de Jong R, ten Hoeve J, Heisterkamp N, Groffen J (1995) Crkl is complexed with tyrosine-phosphorylated Cbl in Ph-positive leukemia. J Biol Chem 270:21468–21471
De Leo FR, Ulman KV, Davis AR, Jutila KL, Quinn MT (1996) Assembly of the human neutrophil NADPH oxidase involves binding of $p67^{phox}$ and flovocytochrome b to a common functional domain in $p47^{phox}$. J Biol Chem 271:17013–17020
de Mendez I, Adams AG, Sokolic RA, Malech HL, Leto TL (1996) Multiple SH3 domain interactions regulate NADPH oxidase assembly in whole cells. EMBO J 15:1211–1220
Eck MJ, Dhe-Paganon S, Trub T, Nolte RT, Shoelson SE (1996) Structure of the IRS-1 PTP domain bound to the juxtamembrane region of the insulin receptor. Cell 85:695–705
Feller SM, Knudsen B, Hanafusa H (1994) c-Abl kinase regulates the protein binding activity of c-Crk. EMBO J 13:2341–2351
Feng S, Chen JK, Yu H, Simon JA, Schreiber SL (1994) Two binding orientations for peptides to the Src SH3 domain: development of a general model for SH3-ligand interactions. Science 266:1241–1247
Feng S, Kasahara C, Rickles RJ, Schreiber SL (1995) Specific interactions outside the proline-rich core of two classes of Src homology 3 ligands. Proc Natl Acad Sci USA 92:12408–12415
Ferguson KM, Lemmon MA, Sigler PB, Schlessinger J (1995) Scratching the surface with the PH domain. Nature Struct Biol 2:715–718
Freeman JL, Lambeth JD (1996) NADPH oxidase assembly is independent of p47phox in vitro. J Biol Chem 271:22578–22582
Fuchs A, Dagher M-C, Vignais PV (1995) Mapping the domains of interaction of p40phox with both p47phox and p67phox of the neutrophil oxidase complex using the two-hybrid system. J Biol Chem 270:5695–5697
Fuchs A, Dagher M-C, Faure J, Vignais PV (1996) Topological organization of the cytosolic activating complex of the superoxide-generating NADPH oxidase. Pinpointing the sites of interaction between p47phox, p67phox and p40phox using the two-hybrid system. Biochim Biophys Acta 1312:39–47
Goodson HV, Anderson BL, Warrick HM, Pon LA, Spudich JA (1996) Synthetic lethality screen identifies a novel yeast myosin I gene (MYO5): myosin I proteins are required for polarization of the actin cytoskeleton. J Cell Biol 133:1277–1291
Gorina S, Pavletich NP (1996) Structure of the p53 tumor suppressor bound to the ankyrin and SH3 domains of 53BP2. Science 274:1001–1005

Gotoh T, Hattori S, Nakamura S, Kitayama H, Noda M, Takai Y, Kaibuchi K, Matsui H, Hatase O, Takahashi H, Kurata T, Matsuda M (1995) Identification of Rapl as a target for the Crk SH3 domain-binding guanine nucleotide-releasing factor C3G. Mol Cell Biol 15: 6746–6753

Harlan JE, Hajduk PJ, Yoon HS, Fesik SW (1994) Pleckstrin homology domains bind to phosphatidylinositol 4,5-bisphosphate. Nature 371:168–170

Harrison SC (1996) Peptide-surface association: the case of PDZ and PTB domains. Cell 86:341–343

Holtzman DA, Yang S, Drubin DG (1993) Synthetic-lethal interactions identify two novel genes, SLA1 and SLA2, that control membrane cytoskelton assembly in Saccharomyces cerivisiae. J Cell Biol 122:635–644

Ito T, Nakamura R, Sumimoto H, Takeshige K, Sakaki Y (1996) An SH3 domain-mediated interaction between the phagocyte NADPH oxidase factors p40phox and p47phox. FEBS Lett 385:229–232

Iwabuchi K, Bartel PL, Li B, Marraccino R, Fields S (1994) Two cellular proteins that bind to wild-type but not mutant p53. Proc Natl Acad Sci USA 91:6098–6102

Johnson DI, Pringle JR (1990) Molecular characterization of CDC42, a Saccharomyces cerevisiae gene involved in the development of cell polarity. J Cell Biol 111:143–152

Karlovich CA, Bonfini L, McCollam L, Rogge RD, Daga A, Czech MP, Bannerjee U (1995) In vivo functional analysis of the Ras exchange factor Son of Sevenless. Science 268:576–579

Khwaja A, Hallberg B, Warne PH, Downward J (1996) Networks of interaction of p120-^{cbl} and p130^{cas} with Crk and Grb2 adaptor proteins. Oncogene 12:2491–2498

Kitayama H, Sugimoto Y, Matsuzaki T, Ikawa Y, Noda M (1989) A ras-related gene with transformation suppressor activity. Cell 56:77–84

Knaus UG, Heyworth PG, Evans T, Curnutte JT, Bokoch GM (1991) Regulation of phgocyte oxygen radical production by the GTP-binding protein Rac2. Science 254:1512–1515

Ladbury JE, Lemmon MA, Zhou M, Green J, Botfield MC, Schlessinger J (1995) Measurement of binding of tyrosyl phosphopeptides to SH2 domains: a reappraisal. Proc Natl Acad Sci USA 92:3199–3202

Langdon WY, Hartley JW, Klinken SP, Ruscetti SK, Morse HC (1989) v-cbl, an oncogene from a dual-recombinant murine retrovirus that induces early B-lineage lymphomas. Proc Natl Acad Sci USA 86:1168–1172

Lee C-H, Leung B, Lemmon MA, Sheng J, Cowburn D, Kuriyan J, Saksela K (1995) A single amino acid in the SH3 domain of Hck determines its high affinity and specificity in binding to HIV Nef protein. EMBO J 14:5006–5015

Lee C-H, Saksela K, Mirza UA, Chait BT, Kuriyan J (1996) Crystal structure of the conserved core of HIV-1 Nef complexed with a Src family SH3 domain. Cell 85.931–942

Lemmon MA, Ferguson KM, Sigler PB, Schlessinger J (1995) Specific and high-affinity binding of inositol phosphates to an isolated pleckstrin homology domain. Proc Natl Acad Sci USA 92:10472–10476

Leto TL, Adams AG, de Mendez I (1994) Assembly of the phagocyte NADPH oxidase: binding of Src homoloy 3 domains to proline-rich targets. Proc Natl Acad Sci USA 91:10650–10654

Li W, Hu E, Skolnik EY, Ullrich A, Schlessinger J (1992) The SH2 and SH3 domain-containing Nck protein is oncogenic and a common target for phosphorylation by different surface receptors. Mol Cell Biol 12:5824–5833

Lowenstein EJ, Daly RJ, Betzer AG, Li W, Margolis B, Lammers R, Ullrich A, Skolnick EY, Bar-Sagi D, Schlessinger J (1992) The SH2 and SH3 domain-containing protein GRB2 links receptor tyrosine kinases to ras signaling. Cell 70:431–442

Malek SN, Desiderio S (1994) A cyclin-dependent kinase homologue, p130^{PITSLRE}, is a phosphotyrosine-independent SH2 ligand. J Biol Chem 269:33009–33020

Malek SN, Yang CH, Earnshaw WC, Kozak CA, Desiderio S (1996) p150^{TSP}, a conserved nuclear phosphoprotein that contains multiple tetratricopeptide repeats and binds specifically to SH2 domains. J Biol Chem 271.6952–6962

Margolis B, Li N, Koch A, Mohammadi M, Hurwitz DR, Zilberstein A, Ullrich A, Pawson T, Schlessinger J (1990) The tyrosine-phosphorylated carboxyterminus of the EGF receptor is a binding site for GAP and PLC-γ. EMBO J 9:4375–4380

Matsuda M, Mayer BJ, Fukui Y, Hanafusa H (1990) Binding of transforming protein, P47$^{gag\text{-}crk}$, to a broad range of phosphotyrosine-containing proteins. Science 248:1537–1539

Matsuda M, Mayer BJ, Hanafusa H (1991) Identification of domains of the v-crk oncogene product sufficient for association with phosphotyrosine-containing proteins. Mol Cell Biol 11:1607–1613

Matsui Y, Matsui R, Akada R, Tohe A (1996) Yeast src homology 3 domain-binding proteins involved in bud formation. J Cell Biol 133:865–878

Mayer BJ, Eck MJ (1995) Minding your p's and q's. Curr Biol 5:364–367

Mayer BJ, Hanafusa H (1990) Association of the v-crk oncogene product with phosphotyrosine-containing proteins and protein kinase activity. Proc Natl Acad Sci USA 87:2638–2642
Mayer BJ, Hamaguchi M, Hanafusa H (1988) Characterization of $p47^{gag\text{-}crk}$, a novel oncogene product with sequence similarity to a putative modulatory domain of protein-tyrosine kinases and phospholipase C. Cold Spring Harbor Symp Quant Biol 53:907–914
Mayer BJ, Jackson PK, Baltimore D (1991) The noncatalytic src homology region 2 segment of abl tyrosine kinase binds to tyrosine-phosphorylated cellular proteins with high affinity. Proc Natl Acad Sci USA 88:627–631
Mayer BJ, Hirai H, Sakai R (1995) Evidence that SH2 domains promote processive phosphorylation by protein-tyrosine kinases. Curr Biol 5:296–305
Meisner H, Czech MP (1995) Coupling of the proto-oncogene product c-Cbl to the epidermal growth factor receptor. J Biol Chem 270 (43):25332–25335
Moran MF, Koch CA, Anderson D, Ellis C, England L, Martin GS, Pawson T (1990) Src homology region 2 domains direct protein-protein interactions in signal transduction. Proc Natl Acad Sci USA 87.8622–8626
Muller AJ, Pendergast AM, Havlik MH, Puil L, Pawson T, Witte ON (1992) A limited set of SH2 domains binds BCR through a high-affinity phosphotyrosine-independent interaction Mol Cell Biol 12:5087–5093
Musacchio A, Gibson T, Rice P, Thompson J, Saraste M (1993) The PH domain: a common piece in the structural patchwork of signalling proteins. Trends Biochem Sci 18:343–348
Nauseef WM, Volpp BD, Clark RA (1990) Immunochemical and electrophoretic analyses of phosphorylated native and recombinant neutrophil oxidase component of p47-phox. Blood 76:2622–2629
Nauseef WM, Volpp BD, McCormick S, Leidal KG, Clark RA (1991) Assembly of the neutrophil respiratory burst oxidase: protein kinase C promotes cytoskeletal and membrane association of cytosolic oxidase components. J Biol Chem 266:5911–5917
Odai H, Sasaki K, Iwamatsu A, Hanazono Y, Tanaka T, Mitani K, Yazaki Y, Hirai H (1995) The proto-oncogene product c-Cbl becomes tyrosine phosphorylated by stimulation with GM-CSF or Epo and constitutively binds to the SH3 domain of Grb2/Ash in human hematopoietic cells. J Biol Chem 270:10800–10805
Olivier JP, Raabe T, Henkemeyer M, Dickson B, Mbamalu G, Margolis B, Schlessinger J, Hafen E, Pawson T (1993) A Drosophila SH2-SH3 adaptor protein implicated in coupling the sevenless tyrosine kinase to an activator of Ras guanine nucleotide exchange, Sos. Cell 73:179–191
Panchamoorthy G, Fukazawa T, Miyake S, Soltoff S, Reedquist K, Druker B, Shoelson S, Cantley L, Band H (1996) $p120^{cbl}$ is a major substrate of tyrosine phosphorylation upon B cell antigen receptor stimulation and interacts in vivo with Fyn and Syk tyrosine kinases, Grb2 and Shc adaptors, and the p85 subunit of phosphatidylinositol 3-kinase. J Biol Chem 271:3187–3194
Park JW, Benna JE, Scott KE, Christensen BL, Chanock SJ, Babior BM (1994) Isolation of a complex of respiratory burst oxidase components from resting neutrophil cytosol. Biochemistry 33:2907–2911
Pawson T (1995) Protein modules and signalling networks. Nature 373:573–579
Pendergast AM, Muller AJ, Havlik MH, Maru Y, Witte ON (1991) BCR sequences essential for transformation by the BCR-ABL oncogene bind to the ABL SH2 regulatory domain in a non-phosphotyrosine-dependent manner. Cell 66:161–171
Rameh LE, Chen C-S, Cantley LC (1995) Phosphatidylinositol (3, 4, 5)P_3 interacts with SH2 domains and modulates PI 3-kinase association with tyrosine-phosphorylated proteins. Cell 83:821–830
Ravichandran KS, Lorenz U, Shoelson SE, Burakoff SJ (1995) Interaction of Shc with Grb2 regulates association of Grb2 with mSos. Mol Cell Biol 15:593–600
Reedquist KA, Fukazawa T, Panchamoorthy G, Langdon WY, Shoelson SE, Druker BJ, Band H (1996) Stimulation through the T cell receptor induces Cbl association with Crk proteins and the guanine nucleotide exchange protein C3G. J Biol Chem 271:8435–8442
Ren R, Mayer BJ, Cicchetti P, Baltimore D (1993) Identification of a 10-amino acid proline-rich SH3 binding site. Science 259:1157–1161
Ren R, Ye Z-S, Baltimore D (1994) Abl protein-tyrosine kinase selects the Crk adapter as a substrate using SH3-binding sites. Genes Dev 8:783–795
Rickles RJ, Botfield MC, Zhou X-M, Henry PA, Brugge JS, Zoller MJ (1995) Phage display selection of ligand residues important for Src homology 3 domain binding specificity. Proc Natl Acad Sci USA 92:10909–10913
Rivero-Lezcano OM, Sameshima JH, Marcilla A, Robbins KC (1994) Physical association between Src Homology 3 element and the protein product of the c-cbl proto-oncogene. J Biol Chem 269:17363–17366

Sadowski I, Stone JC, Pawson T (1986) A noncatalytic domain conserved among cytoplasmic protein-tyrosine kinases modifies the kinase function and transforming activity of fujinami sarcoma virus $P130^{gag\text{-}fps}$. Mol Cell Biol 6:4396–4408

Saksela K, Cheng G, Baltimore D (1995) Proline-rich (PxxP) motifs in HIV-1 Nef bind to SH3 domains of a subset of Src kinases and are required for the enhanced growth of Nef^{+} viruses but not for downregulation of CD4. EMBO J 14:484–491

Sattler M, Salgia R, Okuda K, Uemura N, Durstin MA, Pisick E, Xu G, Li J-L, Prasad KV, Griffin JD (1996) The proto-oncogene product $p120^{CBL}$ and the adaptor proteins CRKL and c-CRK link c-ABL, $p190^{BCR/ABL}$ and $p210^{BCR/ABL}$ to the phosphatidylinositol-3' kinase pathway. Oncogene 12:839–846

Simon MA, Dodson GS, Rubin GM (1993) An SH3-SH2-SH3 protein is required for p21Ras1 activation and binds to sevenless and Sos proteins in vitro. Cell 73:169–177

Smit L, van der Horst G, Borst J (1996a) Formation of Shc/Grb2- and Crk adaptor complexes containing tyrosine phosphorylated Cbl upon stimulation of the B-cell antigen receptor. Oncogene 13:381–389

Smit L, van der Horst G, Borst J (1996b) Sos, Vav, and C3G participate in B cell receptor-induced signaling pathways and differentially associate with Shc-Grb2, Crk, and Crk-L adaptors. J Biol Chem 271(15):8564–8569

Songyang Z, Shoelson SE, Chaudhuri M, Gish G, Pawson T, Haser WG, King F, Roberts T, Ratnofsky S, Lechleider RJ, Neel BG, Birge RB, Fajardo JE, Chou MM, Hanafusa H, Shaffhausen B, Cantley LC (1993) SH2 domains recognize specific phosphopeptide sequences. Cell 72:767–778

Songyang Z, Shoelson SE, McGlade J, Olivier P, Pawson T, Bustelo XR, Barbacid M, Sabe H, Hanafusa H, Yi T, Ren R, Baltimore D, Ratnovsky S, Feldman RA, Cantley LC (1994) Specific motifs recognized by the SH2 domains of Csk, 3BP2, fps/fes, GRB-2, HCP, SHC, Syk, and Vav. Mol Cell Biol 14:2777–2785

Sparks AB, Rider JE, Hoffman NG, Fowlkes DM, Quilliam LA, Kay BK (1996) Distinct ligand preferences of Src homology 3 domains from Src, Yes, Abl, p53 bp, PLC-γ, Crk, and Grb2. Proc Natl Acad Sci USA 93.1540–1544

Sumimoto H, Kage Y, Nunoi H, Sasaki H, Nose T, Fukumaki Y, Ohno M, Minakami S, Takeshige K (1994) Role of src homology 3 domains in assembly and activation of the phagocyte NADPH oxidase. Proc Natl Acad Sci USA 91:5345–5349

Tanaka S, Morishita T, Hashimoto Y, Hattori S, Nakamura S, Shibuya M, Matuoka K, Takenawa T, Kurata T, Nagashima K, Matsuda M (1994) C3G, a guanine nucleotide-releasing protein expressed ubiquitously, binds to the Src homology 3 domains of CRK and GRB2/ASH proteins. Proc Natl Acad Sci USA 91:3443–3447

Tang HY, Cai M (1996) The EH-domain-containing protein Pan1 is required for normal organization of the actin cytoskeleton in Saccharomyces cervisiae. Mol Cell Biol 16:4897–4914

Teahan CG, Totty N, Casimir CM, Segal AW (1990) Purification of the 47 kDa phosphoprotein associated with the NADPH oxidase of human neutrophils. Biochem J 267:485–489

Waksman G, Kominos D, Robertson SC, Pant N, Baltimore D, Birge RB, Cowburn D, Hanafusa H, Mayer BJ, Overduin M, Resh MD, Rios CB, Silverman L, Kuriyan J (1992) Crystal structure of the phosphotyrosine recognition domain (SH2) of the v-src tyrosine kinase complexed with tyrosine phosphorylated peptides. Nature 358:646–653

Wang JYJ (1993) Abl tyrosine kinase in signal transduction and cell-cycle regulation. Curr Opin Genet Dev 3:35–43

Wientjes FB, Panayotou G, Reeves E, Segal AW (1996) Interactions between cytosolic components of the NADPH oxidase: p40phox interacts with both p67phox and p47phox. Biochem J 317:919–924

Yoon CH, Lee J, Jongeward GD, Sternberg PW (1995) Similarity of sli-1, a regulator of vulval development in C. elegans, to the mammalian proto-oncogene c-cbl. Science 269:1102–1105

Yu H, Chen JK, Feng S, Dalgarno DC, Brauer AW, Schreiber SL (1994) Structural basis for the binding of proline-rich peptides to SH3 domains. Cell 76:933–945

Zhou M-M, Ravichandran KS, Olejniczak ET, Petros AM, Meadows RP, Harlan JE, Wade WS, Burakoff SJ, Fesik SW (1995) Structure and ligand recognition of the phosphotyrosine binding domain of Shc. Nature 92:7784–7788

Zhou M-M, Huang B, Olejniczak ET, Meadows RP, Shuker SB, Miyazaki M, Trub T, Shoelson SE, Fesik SW (1996) Structural basis for IL-4 receptor phosphopeptide recognition by the IRS-1 PTB domain. Nature Struct Biol 3:388–393

Function of PTB Domains

J-P. Borg and B. Margolis

1 Introduction

The Phosphotyrosine Binding (PTB) domain is one of the recently described domains involved in protein-protein interactions. This domain was initially identified in the amino-terminus of the Shc protein (Blaikie et al. 1994; Kavanaugh and Williams 1994; Gustafson et al. 1995) and was subsequently identified in the insulin receptor substrate-1 (IRS-1) and insulin receptor substrate-2 (IRS-2) proteins (Gustafson et al. 1995; Sun et al. 1995). Through database searching it was realized that several proteins contain domains related in primary sequence to the Shc PTB domain (Bork and Margolis 1995). The structure of the PTB domains is described elsewhere in this volume. This review will focus on the role of the PTB domain in Shc and IRS-1/IRS-2 signaling and in mediating interaction between these domains and phosphotyrosine-containing proteins. We will then discuss the function of PTB domains in other proteins. In the past our group has referred to these domains as the Phosphotyrosine Interaction Domain (PID) but for simplicity we will refer to them as PTB domains in this review. In any the event the names may be misnomers because phosphotyrosine binding does not appear to be a universal function for these domains.

Howard Hughes Medical Institute, Department of Internal Medicine and Biological Chemistry, University of Michigan Medical School, 1150 West Medical Center Drive, Ann Arbor, MI 48109-0650, USA

2 Shc PTB Domain

The Shc protein was originally cloned using degenerate PCR targeted to identify proteins with SH2 domains (Fig. 1). Shc was found to exist as three proteins of 46, 52 and 66 kDa with an SH2 domain at the carboxy-terminus of all three forms (PELICCI et al. 1992). The different forms of the Shc protein were found to arise from both differential splicing and alternative translational start sites. In addition there are three different genes that encode ShcA, ShcB and ShcC proteins (KAVANAUGH and WILLIAMS 1994; PELICCI et al. 1996; O'BRYAN et al. 1996). Shc is tyrosine phosphorylated by a large number of tyrosine kinases and once phosphorylated binds the Grb2/Sos complex with high affinity (ROZAKIS-ADCOCK et al. 1992). The Grb2/Sos complex then proceeds to activate Ras, triggering multiple signaling cascades. Several groups have determined that the interaction of Shc with tyrosine phosphorylated receptors and other signaling proteins is crucial for Shc mediated signal transduction. This was perhaps best demonstrated with nerve growth factor receptor (NGF-receptor; also known as TrkA) and polyoma middle

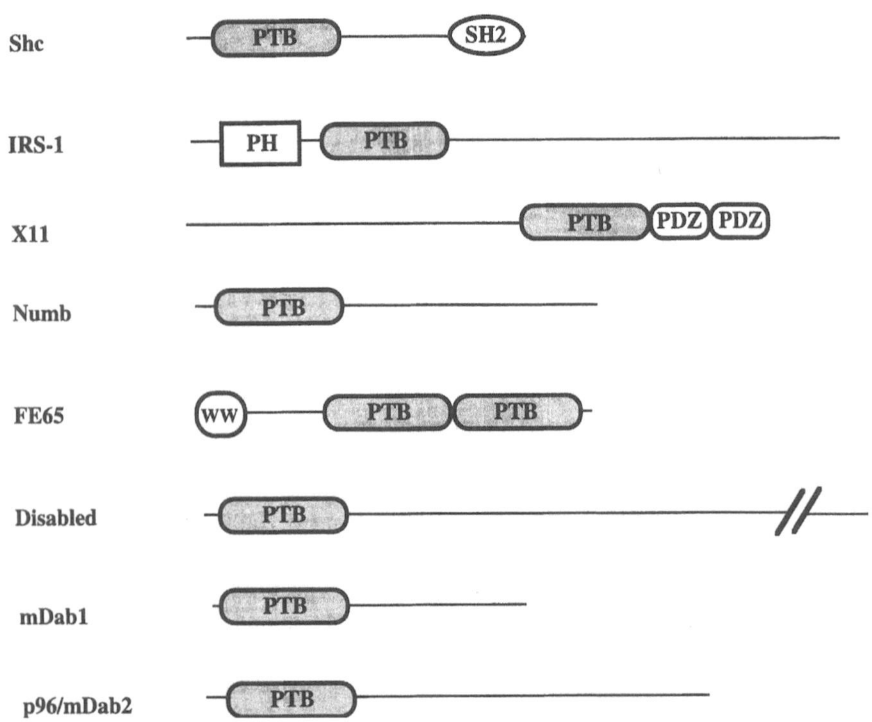

Fig. 1. Schematic representation of Phosphotyrosine Binding (*PTB*) domain containing proteins. This figure displays the major structural domains found in the PTB domain proteins discussed in this paper. Many of these proteins have several isoforms but these are not shown in this figure. For specifics on the individual domains see text and related chapters in this volume. *PH*, *P*leckstrin *H*omology domain; *SH2*, *S*rc *H*omology *2* domain; *PDZ*, *P*SD95, *D*isks Large, *Z*01 domain

T-antigen (MTAg). The binding site for Shc on NGF-receptor has been identified as a sequence motif surrounding tyrosine 490 of NGF-receptor (Stephens et al. 1994; Obermeier et al. 1993). Mutation of this site blocked Shc tyrosine phosphorylation in response to NGF and impaired the ability of NGF to induce neurite extension (Stephens et al. 1994; Obermeier et al. 1994). Polyoma MTAg transforms cells by binding to Src family tyrosine kinases, undergoing tyrosine phosphorylation and binding to downstream signaling molecules (Dilworth 1995). Studies have demonstrated an important role for Shc binding to polyoma MTAg at a motif surrounding the phosphotyrosine at residue 250. Like Y490 of NGF-receptor, mutation of Y250 blocks the ability of MTAg to bind and induce tyrosine phosphorylation of Shc as well as limiting the transforming potential of MTAg (Dilworth et al. 1994; Campbell et al. 1994). These studies demonstrate the importance of these Shc binding sites in Shc tyrosine phosphorylation and downstream signaling events. The Shc binding sites on both NGF-receptor and MTAg are composed of NPXpY motifs (where N is asparagine, P is proline, X is any amino acid and pY is phosphotyrosine; Table 1). Similar tyrosine phosphorylation sites can also be found on a variety of other proteins that bind Shc (Campbell et al. 1994).

It had been assumed that the SH2 domain of Shc mediated the binding between Shc and growth factor receptors or polyoma MTAg (Dilworth et al. 1994; Okabayashi et al. 1994; Batzer et al. 1994). However, it was also well established

Table 1. Binding sites for PTB domains

Shc PTB binding sites	
NGFR-TrkA (pY490)	QGHI ME NPQpYF SD
Polyoma MTAg (pY250)	LPSLL SNPTpYS VM
TrkB (pY515)	KI PVI ENPQpYF GI
TrkC (pY516)	RI PVI ENPQpYF RQ
EGFR (pY1148)	HQI SL DNPDpYQ QD
ErbB2 (pY122)	FSPAF DNLYpY WDQ
ErbB3 (pY1309)	T DS AF DNPDpY WHS
Flt4 (pY1337)	GGQVF YNSEpY GEL
Relation to phosphotyrosine	−9 −7 −5 −3 −1 0 +1 +3
IRS-1/2 PTB binding sites	
Insulin receptor (pY960)	PLYAS SNPEpYL SA
IGF-1 receptor (pY950)	VLYAS VNPEpYF SA
IL-4 receptor (pY475)	PLVLA DNPApYR SF
Relation to phosphotyrosine	−9 −7 −5 −3 −1 0 +1 +3
X11 PTB binding site	
bAPP$_{695}$ (Y687)	QQNGY ENPT YKFF
Other NPXY motifs	
LDL-receptor (Y828)	NSI NF DNPV YQKT
Integrin B_1 (Y795)	AVTTV VNPK YE GK

The binding sites for Shc, IRS-1/IRS-2 and X11 PTB domains are shown. The letters in bold indicate residues that are crucial for binding. The relation to the phosphotyrosine indicates the numbering of residues amino-terminal (−) and carboxy-terminal (+) to the phosphotyrosine. No phosphotyrosine is involved in the binding of X11 PTB to the NPXY peptide on bAPP. Examples of other NPXY motifs that could be potential binding partners for PTB or related domains are also shown. The numbered tyrosine beside the protein name refers to the tyrosine of the NPXY motif in the human protein.

that SH2 domain binding specificity was determined by residues that are carboxy-terminal to the phosphotyrosine and not residues amino-terminal to the phosphotyrosine (SONGYANG et al. 1993). This discrepancy was resolved when it was realized that Shc had a second domain, the PTB domain, that could interact with phosphotyrosine and recognize the NPXpY motif as its binding site. Our group identified this domain by screening a mouse fibroblast bacterial expression library with the tyrosine phosphorylated epidermal growth factor (EGF) receptor (BLAIKIE et al. 1994). The domain was simultaneously recognized by Kavanaugh and Williams, who identified the binding of a 145-kDa tyrosine phosphorylated protein to the PTB domain of Shc (KAVANAUGH and WILLIAMS 1994). Additional work by GUSTAFSON and colleagues also identified this domain in Shc using the insulin receptor in the yeast two hybrid system (GUSTAFSON et al. 1995; see below). The PTB domain is conserved in all forms of Shc including the three mammalian genes and their isoforms as well as *Drosophila* Shc (LAI et al. 1995b; BONFINI et al. 1996). Subsequent studies by several groups using peptide competition, library and random peptide screening as well as phosphatase protection assays identified the binding specificity of the Shc PTB domain (VAN DER GEER et al. 1995; KAVANAUGH et al. 1995; GUSTAFSON et al. 1995; DIKIC et al. 1995; BATZER et al. 1995; TRUB et al. 1995; PRIGENT et al. 1995; FOURNIER et al. 1996). These studies indicated that the sequence binding specificity for the Shc PTB domain was ΨXNPXpY (where Ψ is a hydrophobic residue). The role of the proline in this interaction is at best small and the binding site is also sometimes represented as ΨXNXXpY (LAMINET et al. 1996). The binding affinity between the Shc PTB domain and a peptide from NGF-receptor containing the Y490 site has been measured at approximately 50 n*M* (LAMINET et al. 1996; ZHOU et al. 1995a).

In order to determine the function of the Shc PTB domain it was important to identify mutations within the domain that reduced binding affinity. Several approaches to this problem were undertaken. VAN DER GEER and coworkers (1996) mutated all the arginines within the Shc PTB domain, reasoning that basic residues must be critical for complexing with phosphotyrosine. This group identified the arginine at residue 175 (R175) of p52 Shc as being important for the interaction with phosphopeptide. Our laboratory used random mutagenesis to identify phenylalanine 198 (F198) of p52 Shc as being important for peptide interactions (YAJNIK et al. 1996). Structural analysis of the Shc PTB domain also demonstrated that R175 and F198 were crucial for the interaction with the NPXY peptide. This analysis revealed that phosphotyrosine interacts with four residues of the PTB domain: R175, R169, S151 and R67 (ZHOU et al. 1995b). Of these, R175 appears to be the most important but mutation of S151, R169 or R67 also reduces the interaction to a variable extent (ZHOU et al. 1995b; YAJNIK et al. 1996; VAN DER GEER et al. 1996). Structural analysis also indicated an important role for F198. This residue lies in the carboxy-terminal α-helix of the Shc PTB domain and interacts with both the minus 3 asparagine and minus five hydrophobic residue of the b-turn ΨXNPXpY peptide (ZHOU et al. 1995b).

Mutations of F198 or R175 have been used to analyze the role of the PTB domain in Shc signaling. Our group has demonstrated an important role for the

PTB domain of Shc and its ability to signal downstream of insulin receptor, NGF-receptor and polyoma MTAg (ISAKOFF et al. 1996; Blaikie et al. 1997). Other have demonstrated the importance of the PTB domain in GM-CSF and IL-2 signaling using both mutations and deletions of the PTB domain (RAVICHANDRAN et al. 1996; PRATT et al. 1996). The role of the SH2 domain in Shc function is not clearly understood. It has been presumed that the combination of the SH2 and PTB domains promote tyrosine phosphorylation of Shc by many different tyrosine kinases. In the case of the EGF-receptor, both the SH2 and PTB domains can bind to the receptor. However, our data have shown that the PTB is the major domain responsible for binding full length Shc to the EGF-receptor (YAJNIK et al. 1996). Expression of the PTB domain alone can act as a dominant negative protein to impair signaling. In our hands it blocks transformation by polyoma MTAg (Blaikie et al. 1997) and has been reported to block ZAP-70 signaling (MILIA et al. 1996). The SH2 domain has been shown to be involved in mediating the binding of Shc to the ζ-chain of the T-cell receptor (RAVICHANDRAN et al. 1993). However, in general there are very little data supporting the role of the SH2 domain in mediating Shc phosphorylation. In contrast several groups have shown that overexpression or microinjection of the SH2 domain of Shc blocks EGF mediated cell division (GOTOH et al. 1995; RICKETTS et al. 1996). At least one of these studies has shown that the SH2 domain is likely to have an effect distinct from its role in Shc phosphorylation (RICKETTS et al. 1996).

Although the specific downstream signaling molecules that interact with the Shc SH2 domain are unknown, it has been demonstrated that the Shc PTB domain does contact other potential signaling molecules. One of these molecules is Ship, a phosphatase that hydrolyzes the 5′ phosphate from polyphosphoinositides (LIOUBIN et al. 1996; DAMEN et al. 1996; KAVANAUGH et al. 1996). Ship appears to be one of a family of proteins that had been previously identified as Shc binding proteins (LIU et al. 1994; SAXTON et al. 1994; CROWLEY et al. 1996; KAVANAUGH and WILLIAMS 1994). The Ship protein has an SH2 domain in addition to two ΨXNPXY motifs that are likely to be responsible for interactions with Shc. Ship has been proposed to be a negative and a positive regulator of cell signaling and further work will be necessary to define its role in Shc signaling (KLIPPEL et al. 1997; ONO et al. 1996).

3 Insulin Receptor Substrate-1 PTB Domain

An NPXpY motif also plays an important role in insulin signal transduction. In the late 1980s it was realized that a phosphotyrosine containing NPEpY motif was important for many aspects of insulin receptor signal transduction. This phosphotyrosine at residue 960 of insulin receptor was essential for the phosphorylation of the IRS-1 protein, an important docking molecule that signals downstream of insulin and cytokine receptors (WHITE et al. 1988). It was subsequently realized from the work of Gustafson and colleagues that the amino-terminus of IRS-1 contained a binding site for the Y960 site and that the Shc PTB domain could also bind to this NPEpY motif (GUSTAFSON et al. 1995). Furthermore the binding of both Shc and IRS-1 was dependent not only on the phosphotyrosine residue but also on the asparagine and proline residues within this motif. The PTB domain on IRS-1 that bound to this NPEpY motif was defined by WHITE and coworkers with the identification of a homologous region in IRS-2 (SUN et al. 1995). Two factors indicated that the IRS-1 PTB domain functioned slightly different than the Shc PTB domain. First there was no significant primary sequence similarity between the Shc and IRS-1 PTB domains (although several alignments were proposed). Second the binding specificity of the IRS-1 PTB domain was different than the Shc PTB domain (HE et al. 1995; WOLF et al. 1995). While the Shc PTB domain bound to the ΨXNPXpY motif, the IRS-1 PTB domain bound to the sequence ΨΨΨXXNPXpY (Table 1). In contrast to Shc PTB domain, which utilizes a hydrophobic amino acid five residues amino-terminal to the phosphotyrosine, the IRS-1 PTB domain utilizes hydrophobic residues minus 6 to minus 8 to the phosphotyrosine for binding (ZHOU et al. 1996; ECK et al. 1996). Additionally the proline of the NPXpY motif appears more important for IRS-1 than for Shc PTB domain binding (WOLF et al. 1995). The sequence surrounding Y960 of the insulin receptor as well as similar sites in insulin-like growth factor 1 (IGF-1) receptor and interleukin-4 (IL-4) receptor are not high affinity Shc binding sites because they do not contain a hydrophobic residue minus five to the phosphotyrosine. Nonetheless insulin receptor can still phosphorylate Shc and it has been demonstrated that PTB domain binding is crucial for this phosphorylation (ISAKOFF et al. 1996). A hydrophobic residue at the plus 1 position is necessary for the interaction of insulin and IGF-1 receptor with the Shc PTB domain and explains why the IL-4 receptor cannot interact with Shc (ISAKOFF et al. 1996). It has been found that mutations around the NPXY motif of the insulin receptor can change the binding characteristics of the receptor to favor Shc or IRS-1 family members (ISAKOFF et al. 1996; VAN DER GEER et al. 1996). Despite these difference in primary sequence and binding specificity, a structural analysis of the IRS-1 PTB domain revealed it to be remarkably similar to the structure of the Shc PTB domain (ZHOU et al. 1996; ECK et al. 1996). Like the Shc PTB domain, the structure has a seven β-strand sandwich with a carboxy-terminal α-helix. However, the residues of the PTB domain that mediate the interactions with the β-turn peptide are different for IRS-1 and Shc. In particular, the amino acids that interact with the phosphotyrosine are highly divergent.

The role of the PTB domain in IRS-1 function in living cells has not been studied as extensively as its role in Shc function. Studies of point mutants that eliminate IRS-1 PTB domain activity have not been published but studies have been performed in which deletions have been made in the PTB domain of IRS-1. Deletion of the PTB domain weakens but does not eliminate the phosphorylation of IRS-1 after insulin stimulation. Equally important for IRS-1 phosphorylation is the IRS-1 PH domain which sits just amino-terminal to the PTB domain (YENUSH et al. 1996; MYERS et al. 1995). Thus, as with Shc, the PTB domain appears to promote the interaction of IRS-1 with receptors leading to IRS-1 phosphorylation. This may be particularly important in cells with low levels of IRS-1/IRS-2 or insulin receptor and at low agonist levels.

4 PTB Domains in Other Proteins

Upon discovery of the Shc PTB domain, database searches indicated that this domain was not unique to Shc (BORK and MARGOLIS 1995). The domain is also found in several proteins, some of which are schematically illustrated in Fig. 1. Sequence alignment of these PTB domains revealed approximately 15%–30% sequence identity to the Shc PTB domain. The IRS-1/IRS-2 PTB domains do not align with these PTB domains. The PTB domains in the proteins shown in Fig. 1 (excluding IRS-1) are defined by two absolutely conserved residues: a glycine at residue 58 and a serine at residue 151 of p52 Shc. The glycine connects the two fragments of the first β-strand. As such it is a crucial structural component but not involved in peptide interactions (ZHOU et al. 1995b). The serine is involved in the interaction of the Shc PTB domain with the phosphotyrosine of the NPXpY peptide. Most PTB domains also contain a conserved phenylalanine at the residue analogous to F198 of p52 Shc. The only exception is the PTB domains found in the FE65 family of proteins, where the phenylalanine is replaced by a cysteine (YAJNIK et al. 1996). As discussed above, this phenylalanine is crucial for interactions with the hydrophobic residue minus five and the asparagine minus three to the phosphotyrosine. There are several potential proteins with PTB domains that are not displayed in Fig. 1. These include putative proteins identified in the expressed sequence tag database and via genomic sequencing.

The delineation of the function of these additional PTB domain containing proteins is an area of active investigation. Like SH2 domain proteins, many of these appear to be adaptor type proteins containing multiple protein interaction domains. A few of the PTB domain proteins have also been defined by genetic approaches in *Drosophila melanogaster*. These are the PTB domain proteins, Numb and Disabled. Their role in *Drosophila* development will be discussed below. No PTB domain proteins have been identified in yeast.

4.1 FE65 and X11

FE65 and X11 will be described together as their PTB domains appear to share a common target, β-amyloid precursor protein (βAPP). FE65 was first described as a protein expressed in the nervous system by Russo and coworkers (DUILIO et al. 1991). Although originally felt to function as a transcriptional activator, identification of two PTB domains as well as a WW domain has indicated that FE65 functions as a cytoplasmic protein involved in mediating protein interactions

(Fiore et al. 1995). WW domains are involved in binding to proline-rich sequences in a fashion similar to that seen with SH3 domains (Sudol et al. 1995; Staub and Rotin 1996). As with many proteins in the PTB domain family, several genes encode members of the FE65 family (Guenette et al. 1997).

X11 is a gene that was incidentally identified on chromosome 9 in the search for the gene responsible for Friedreich's ataxia (Duclos et al. 1993). Although originally described as a transmembrane protein, it is a cytoplasmic protein with a central PTB and two PDZ domains (Bork and Margolis 1995). PDZ domains are also protein-protein interaction domains that in most, but not all cases, interact with carboxy-terminal peptide motifs (Fanning and Anderson 1996).

Both FE65 and X11 have been identified as proteins that bind to βAPP (Fiore et al. 1995; Borg et al. 1996; Bressler et al. 1996; McLoughlin and Miller 1996; Guenette et al. 1997). βAPP is a protein of unknown function involved in Alzheimer's disease (Selkoe 1994). It is primarily expressed as three alternatively spliced proteins of 770, 751 and 695 amino acids. All forms can be processed to β-amyloid, a protein that is deposited in the extracellular space of patient's with Alzheimer's disease. The crucial role of βAPP in Alzheimer's disease has been confirmed by finding several mutations in the βAPP gene that lead to increased processing of βAPP to β-amyloid and early onset of Alzheimer's disease. The binding of FE65 to the intracellular domain of βAPP was first discovered by Russo and coworkers, who screened a yeast two hybrid library with the two PTB domains of FE65 (Fiore et al. 1995). They found that FE65 not only bound to βAPP but also to the related βAPP-like proteins. The binding of FE65 to the intracellular domain of βAPP has also been reported by several other groups (Bressler et al. 1996; Guenette et al. 1997; McLoughlin and Miller 1996). Our group screened several other PTB domains against βAPP. We found that the PTB domain of X11 also bound βAPP but not the PTB domain of p96, Numb or Shc (Borg et al. 1996). Additionally we and others narrowed the βAPP binding site to the carboxy-terminal PTB domain of FE65.

The binding of βAPP was intriguing because βAPP contains a YENPTY motif within its intracellular domain. This sequence fits with the consensus binding site for the Shc PTB domain, namely ΨXNPXY. Our group set out to determine if this motif was the binding site for the FE65 and X11 PTB domains (Borg et al. 1996). Mutation of the amino-terminal tyrosine or the asparagine residue of the YENPTY motif blocked the ability of βAPP to bind to X11. Furthermore a 12 mer peptide containing the YENPTY motif of βAPP blocked interaction between βAPP and X11. A crystal structure of the PTB domain of X11 bound to this 12 mer peptide has been solved, indicating that the PTB domain of X11 forms the same basic structure as the Shc and IRS-1 PTB domain and that the YENPTY peptide binds into the same site between the carboxy-terminal α-helix and the β-sandwich (Z. Zhang, C-H. Lee, J-P. Borg, B. Margolis, J. Schlessinger and J. Kuriyan, unpublished observations). The case with FE65 is not as clear (Borg et al. 1996). Here mutation of the amino-terminal tyrosine but not the asparagine blocked interactions between βAPP and FE65. Furthermore, peptides containing the YENPTY site that interfered with X11 binding did not affect FE65 binding. This suggests that the FE65 PTB domain binds to a different site on βAPP than X11 but that the binding site includes at least a portion of the YENPTY motif.

We have also identified mutations within the PTB domain of X11 and FE65 that interfere with binding to βAPP. These mutations target the residue analogous to the F198 residue that severely impairs Shc PTB domain binding. In most PTB domains (including X11) this residue is a phenylalanine and mutation of this residue reduces the binding ability of the X11 PTB domain. In FE65 this phenylalanine is a cysteine in both PTB domains (YAJNIK et al. 1996). Nonetheless mutation of this cysteine in the carboxy-terminal FE65 PTB domain also impairs its binding ability. This indicates that this residue whether it be phenylalanine or cysteine is in a crucial position to control interaction with peptides that bind to PTB domains. The alteration in this crucial residue may be in part responsible for the different binding specificity of X11 and FE65.

A distinct possibility is that FE65 and X11 modulate the processing of βAPP. The mechanism by which βAPP is processed to β-amyloid is a subject of extensive research. βAPP undergoes two types of processing. One involves an enzyme called α-secretase. This protease cuts at a region close to the membrane and excludes the production of the proamyloidgenic β-amyloid protein. The other enzyme, β-secretase, cuts further from the membrane and is responsible for β-amyloid production. It is felt that this cleavage occurs intracellularly either as βAPP is making its way to the cell surface or after internalization (PEREZ et al. 1996). This internalization is dependent, in part, on the YENPTY sequence that controls endocytosis of βAPP into clathrin pits (LAI et al. 1995a). Deletion of the YENPTY, the site of binding for X11 and part of the binding site for FE65, reduces the production of β-amyloid from βAPP (KOO and SQUAZZO 1994). Thus it is possible that FE65 or X11 may be responsible for the internalization of βAPP via the NPXY motif or may serve to modulate the internalization. The normal function of βAPP on the cell surface is unknown. It has been suggested that βAPP functions as a G-protein-coupled receptor, has a role in internalization of molecules or is involved in cell adhesion (OKAMOTO et al. 1995; NORDSTEDT et al. 1993; BEHER et al. 1996). X11 and FE65 could play a role in any of these functions. One key to identifying the functions of these adaptor proteins will be to isolate partners that bind to the WW domain of FE65 and the PDZ domains of X11.

4.2 Numb

Numb was originally described as a gene crucial for the development of the sensory organ precursor (SOP) cells in *Drosophila melanogaster* (UEMURA et al. 1989). It encodes a protein of 556 amino acids whose only recognizable motif is a PTB domain. Numb controls the fate of the cells that arise from the SOP, the precursor cell that generates the sensory bristles found on the body of the fly. It divides to form the IIa and IIb cells, which then divide to form the four cells that make up the external sense organ. Two of these cells, the socket and the hair, come from the IIa cell while the sheath and the neuron cell come from the IIb cell. During division of the SOP cell, Numb localizes to the cell membrane on one side of the cell (Rhyu et al. 1994). This causes all of the Numb to be localized to the IIb cell. In flies carrying loss of function Numb alleles, the IIb cells adopt the fate of the IIa cells, leading to a duplication of the hair and socket cells (Rhyu et al. 1994). Conversely when Numb is misexpressed in the IIa cell, this cell develops into sheath and neuron cells (Rhyu et al. 1994). It has been determined that the PTB domain is not responsible for the localization of Numb to the IIb daughter cell. However, the PTB domain is crucial in the effector role of Numb in that daughter cell (Frise et al. 1996).

It has been proposed that the major function of the Numb PTB domain is to antagonize signaling by the Notch receptor (Guo et al. 1996). Activated Notch has the same effect as lack of Numb function, causing the IIb cells to adopt the fate of IIa cells. Notch is present in both IIa and IIb cells but its signaling in IIb cells may be attenuated by the presence of Numb (Frise et al. 1996). In transfected *Drosophila* Schneider S2 cells, the protein suppressor of Hairless, Su(H), is localized to the nucleus but if it is coexpressed with Notch it localizes to the cytoplasm (Fortini and Artavanis-Tsakonas 1994). If the Notch receptor is activated by ligand, Su(H) moves to the nucleus. Overexpression of Numb antagonizes the agonist induced movement of Su(H) and this effect is dependent on an intact Numb PTB domain. In a mammalian neural crest cell line, MONC-1, mouse Numb overexpression promotes differentiation towards a neuronal cell fate while overexpression of the Numb PTB domain alone inhibits neuronal differentiation (Verdi et al. 1996).

Biochemical evidence suggests that the Numb PTB domain may mediate direct binding of *Drosophila* Numb to Notch (Guo et al. 1996). Similar findings were reported with mouse Numb and mammalian Notch (Zhong et al. 1996). However, further biochemical studies of this interaction will be necessary to define the sites of interaction. It is interesting that Notch does not contain a conserved NPXY or related motif that has been described as the binding site for PTB domains of X11 and Shc. It is also important to note that these interactions between Numb and Notch have been demonstrated in the yeast two hybrid system, suggesting that phosphotyrosine is not involved. In contrast, it has been reported that Numb PTB domain can bind phosphotyrosine-containing proteins from lysates of v-*src* transformed cells (Verdi et al. 1996).

4.3 Disabled and p96

Disabled was identified in *Drosophila* as an enhancer of the phenotype seen in flies deficient for *Drosophila abl*. The *abl* gene encodes a tyrosine kinase that when disrupted in *Drosophila* leads to lethality in adult flies (Henkemeyer 1987). Haploinsufficiency of the *disabled* gene enhances the phenotype of *abl* deficient flies leading to embryonic lethality (Gertler et al. 1989). The Abl and Disabled protein are tyrosine phosphorylated in *Drosophila* tissue culture cells but it is not clear if Disabled is a substrate of Abl. The PTB domain of Disabled is closely related to the

PTB domain of the p96 protein, a protein that is phosphorylated on serine residues following stimulation of mouse macrophages with CSF-1 (Xu et al. 1995). A fragment of p96 had been previously identified as the *DOC-2* gene, a gene expressed in normal human ovarian epithelial cells but possibly deficient in ovarian carcinoma (Mok et al. 1994). There are several splice variants of the p96 proteins and some have referred to this gene as mouse Disabled 2 (*mDab2*; Howell et al. 1997).

Recently Cooper and coworkers (Howell et al. 1997) have isolated a mouse protein that they refer to as mouse Disabled 1 (mDab1). It is highly related to the mDab2 family of proteins. The *mDab1* gene has multiple spliced forms, yielding proteins of 555, 271 and 217 amino acids in neuronal and hematopoietic cells. The PTB domain of mDab1 is 63% identical to the p96/mDab2 PTB domain and 52% related to *Drosophila* Disabled. However, both mDab1 and p96/mDab2 are much smaller proteins than *Drosophila* Disabled and it is not yet clear if they represent true homologues. Nonetheless, the mDab1 protein, like *Drosophila* Disabled, appears to have an important role in neuronal development. The protein is tyrosine phosphorylated before and during neurite extension in differentiating P19 embryonal carcinoma cells. It has been found that the PTB of mDab1 binds to phosphotyrosine proteins from mouse embryonic brain. Mutations of the arginine analogous to R67 of p52 Shc reduces the ability of this PTB domain to interact with these, as of yet unidentified, phosphotyrosine containing proteins. This suggests that, like Shc and IRS-1, the mDab1 PTB domain might play a role in phosphotyrosine dependent interactions.

5 Perspective

In summary the PTB domain represents a new protein-protein interaction domain that interacts with peptide sequences. The PTB domain is structurally highly related to PH domains. It is possible that PH domains are the evolutionary forerunners of PTB domains as PH domains, but not PTB domains, are found in yeast. The function of PH domains is still a matter of intense investigation but a consensus has emerged that these domains are involved in binding polar head groups of phosphoinositides and inositol phosphates (Lemmon et al. 1996). Although PH domains are quite divergent in primary sequence they tend to share a positively charged region at the base of their structure that binds polyphosphoinositides. The binding site for peptides on PTB domains sits on the opposite side of the structure that PH domains use to bind phosphoinositides. It might be hypothesized that the PH domain was present first in evolution and over time evolved to acquire peptide binding activity leading to the creation of PTB domains. It is not surprising then that the Shc PTB domain has the ability to bind both peptides and phosphoinositides (Zhou et al. 1995b).

One of the major problems with the PTB domains is nomenclature. While our group and others have referred to them as phosphotyrosine interaction domain, most groups have referred to them as phosphotyrosine binding domains. Despite this multiplicity of names, the basic concept that these domains bind phosphotyrosine as a common target is on uncertain ground. Clearly the Shc and IRS-1 PTB domains bind phosphotyrosine and it is possible that the PTB domain from mDab1 does the same. However, for FE65 and X11 the available data suggest that phosphotyrosine is not involved in the interaction. The data best support the hypothesis that PTB domains are involved in both phosphotyrosine dependent and independent binding. In fact there are reports of phosphotyrosine independent interactions for the Shc PTB domain (CHAREST et al. 1996). The data with the Numb PTB domain are somewhat contradictory, supporting both phosphotyrosine dependent and phosphotyrosine independent interactions. Nontheless, Numb represents one of the best models for understanding this issue as candidates that bind the PTB domain of Numb in vitro can be tested for biological activity in *Drosophila* development. While the issue of phosphotyrosine dependent binding is still unclear, many of the known ligands for PTB domains form a β-turn structure. This β-turn allows the ligand multiple interactions between the β-sandwich and carboxy-terminal α-helix of the PTB domain. Many proteins have potential β-turn motifs including members of the LDL and integrin receptor familes (CHEN et al. 1990; Table 1). Future work will determine if this β-turn motif is the common denominator in proteins that bind to PTB domains.

Note added in proof: Recent work has identified a protein downstream of F6F. Receptor with a PTB domain 5 elated to IRS-1 (KOUHARA et al., Cell 89, 693, 1997) and further studies on the binding specificity of the Numb PTB domain have been published (LIE et al. Proc Natl Acad Sci 94:7204, 1997)

References

Batzer AG, Rotin D, Urena JM, Skolnik EY, Schlessinger J (1994) Hierarchy of binding sites for Grb2 and Shc on the epidermal growth factor receptor. Mol Cell Biol 14:5192–5201

Batzer AG, Blaikie P, Nelson K, Schlessinger J, Margolis B (1995) The phosphotyrosine interaction domain of Shc binds an LXNPXY motif on the epidermal growth factor receptor. Mol Cell Biol 15:4403–4409

Beher D, Hesse L, Masters CL, Multhaup G (1996) Regulation of amyloid protein precursor (APP) binding to collagen and mapping of the binding sites on APP and collagen type I. J Biol Chem 271:1613–1620

Blaikie P, Immanuel D, Wu J, Li N, Yajnik V, Margolis B (1994) A region in Shc distinct from the SH2 domain can bind tyrosine phosphorylated growth factor receptors. J Biol Chem 269:32031–32034

Blaikie P, Fournier E, Dilworth SM, Birnbaum D, Borg JP, Margolis B The role of the Shc Phosphotyrosine interaction/phosphotyrosine binding, domain and tyrosine phosphorylation sites in polyoma middle Tantigen-mediated cell transformation

Bonfini L, Migliaccio E, Pelicci G, Lanfrancone L, Pelicci PG (1996) Not all Shc's roads lead to Ras. Trends Biochem Sci 21:257–261

Borg JP, Ooi J, Levy E, Margolis B (1996) The phosphotyrosine interaction domains of X11 and FE65 bind to distinct sites on the YENPTY motif of amyloid precursor protein. Mol Cell Biol 16:6229–6241

Bork P, Margolis B (1995) A phosphotyrosine interaction domain. Cell 80:693–694

Bressler SL, Gray MD, Sopher BL, Hu QB, Hearn MG, Pham DG, Dinulos MB, Fukuchi KI, Sisodia SS, Miller MA, Disteche CM, Martin GM (1996) cDNA cloning and chromosome mapping of the human FE65 gene – interaction of the conserved cytoplasmic domains of the human β-amyloid precursor protein and its homologues with the mouse FE65 protein. Hum Mol Genet 5:1589–1598

Campbell KS, Ogris E, Burke B, Su W, Auger KR, Druker BJ, Schaffhausen BS, Roberts TM, Pallas DC (1994) Polyoma middle tumor antigen interacts with SHC protein via the NPTY (Asn-Pro-Thr-Tyr) motif in middle tumor antigen. Proc Natl Acad Sci USA 91:6344–6348

Charest A, Wagner J, Jacob S, McGlade CJ, Tremblay ML (1996) Phosphotyrosine-independent binding of SHC to the NPLH sequence of murine protein-tyrosine phosphatase-PEST. Evidence for extended phosphotyrosine binding/phosphotyrosine interaction domain recognition specificity. J Biol Chem 271:8424–8429

Chen WJ, Goldstein JL, Brown MS (1990) NPXY, a sequence often found in cytoplasmic tails, is required for coated pit-mediated internalization of the low density lipoprotein receptor. J Biol Chem 265:3116–3123

Crowley MT, Harmer SL, DeFranco AL (1996) Activation-induced association of a 145-kDa tyrosine-phosphorylated. J Biol Chem 271:1145–1152

Damen JE, Liu L, Rosten P, Humphries RK, Jefferson AB, Majerus PW, Krystal G (1996) The 145-kDa protein induced to associate with Shc by multiple cytokines is an inositol tetraphosphate and phosphatidylinositol 3,4,5-triphosphate 5-phosphatase. Proc Natl Acad Sci USA 93:1689–1693

Dikic I, Batzer AG, Blaikie P, Obermeier A, Ullrich A, Schlessinger J, Margolis B (1995) Shc binding to nerve growth factor receptor is mediated by the phosphotyrosine interaction domain. J Biol Chem 270:15125–15129

Dilworth SM (1995) Polyoma virus middle T antigen: meddler or mimic? Trends Microbiol 3:31–35

Dilworth SM, Brewster CEP, Jones MD, Lanfrancone L, Pelicci G, Pelicci PG (1994) Transformation by polyoma virus middle T-antigen involves the binding and tyrosine phosphorylation of Shc. Nature 367:87–90

Duclos F, Boschert U, Sirugo G, Mandel JL, Hen R, Koenig M (1993) Gene in the region of the Friedreich ataxia locus encodes a putative transmembrane protein expressed in the nervous system. Proc Natl Acad Sci USA 90:109–113

Duilio A, Zambrano N, Mogavero AR, Ammendola R, Cimino F, Russo T (1991) A rat brain mRNA encoding a transcriptional activator homologous to the DNA binding domain of retroviral integrases. Nucleic Acids Res 19:5269–5274

Eck MJ, Dhe Paganon S, Trub T, Nolte RT, Shoelson SE (1996) Structure of the IRS-1 PTB domain bound to the juxtamembrane region of the insulin receptor. Cell 85:695–705

Fanning AS, Anderson JM (1996) Protein-protein interactions – PDZ domain networks. Curr Biol 6:1385–1388

Fiore F, Zambrano N, Minopoli G, Donini V, Duilio A, Russo T (1995) The regions of the Fe65 protein homologous to the phosphotyrosine interaction/phosphotyrosine binding domain of Shc bind the intracellular domain of the Alzheimer's amyloid precursor protein. J Biol Chem 270:30853–30856

Fortini ME, Artavanis-Tsakonas S (1994) The suppressor of hairless protein participates in notch receptor. Cell 79:273–282

Fournier E, Rosnet O, Marchetto S, Turck CW, Rottapel R, Pelicci PG, Birnbaum D, Borg JP (1996) Interaction with the phosphotyrosine binding domain/phosphotyrosine interacting domain of SHC is required for the transforming activity of the FLT4/VEGFR3 receptor tyrosine kinase. J Biol Chem 271:12956–12963

Frise E, Knoblich JA, Younger-Shepherd S, Jan LY, Jan YN (1996) The Drosophila Numb protein inhibits signaling of the Notch receptor during cell-cell interaction in sensory organ lineage. Proc Natl Acad Sci USA 93:11925–11932

Gertler FB, Bennett RL, Clark MJ, Hoffmann FM (1989) Drosophila abl tyrosine kinase in embryonic CNS axons: a role in axonogenesis is revealed through dosage-sensitive interactions with disabled. Cell 58:103–113

Gotoh N, Muroya K, Hattori S, Nakamura S, Chida K, Shibuya M (1995) The SH2 domain of Shc suppresses EGF-induced mitogenesis in a dominant negative manner. Oncogene 11:2525–2533

Guenette SY, Chen J, Jondro PD, Tanzi RE (1997) Association of a novel human FE65-like protein with the cytoplasmic domain of the beta-amyloid precursor protein. Proc Natl Acad Sci USA 93:10832–10837

Guo M, Jan LY, Jan YN (1996) Control of daughter cell fates during asymmetric division interaction of numb and notch. Neuron 17:27–41

Gustafson TA, He W, Craparo A, Schaub CD, O'Neill TJ (1995) Phosphotyrosine-dependent interaction of Shc and IRS-1 with the NPEY motif of the insulin receptor via a novel (non-SH2) domain. Mol Cell Biol 15:2500–2508

He W, O'Neill TJ, Gustafson TA (1995) Distinct modes of interaction of SHC and insulin receptor substrate-1 with the insulin receptor NPEY region via non-SH2 domains. J Biol Chem 270:23258–23262

Howell BW, Gertler FB, Cooper JA (1997) Mouse disabled (mDab1): a src binding protein implicated in neuronal development. EMBO J 16:121–132

Henkemeyer MJ, Gertler FB, Goodman W, Hoffman FM (1987) The Drosophila Abelson proto-oncogene homolog: identification of mutant alleles that have pleiotropic effects late in development. Cell 51:821–828

Isakoff SJ, Yu Y-P, Su Y-C, Blaikie P, Yajnik V, Rose E, Weidner KM, Sachs M, Margolis B, Skolnik EY (1996) The Shc PTB/PI domain is required for Shc tyrosine phosphorylation by the Insulin Receptor (IR) and recognizes an NPXY motif on the IR that is distinct from IRS-1 in vivo. J Biol Chem 271:3959–3962

Kavanaugh WM, Williams LT (1994) An alternative to SH2 domains for binding tyrosine-phosphorylated proteins. Science 266:1862–1865

Kavanaugh WM, Turck CW, Williams LT (1995) PTB domain binding to signaling proteins through a sequence motif containing phosphotyrosine. Science 268:1177–1179

Kavanaugh WM, Pot DA, Chin SM, Deuter-Reinhard M, Jefferson AB, Norris FA, Masiarz FR, Cousens LS, Majerus PW, Williams LT (1996) Multiple forms of an inositol polyphosphate 5-phosphatase form signaling complexes with Shc and Grb2. Curr Biol 6:438–445

Klippel A, Kavanaugh WM, Pot D, Williams LT (1997) A specific product of phosphatidylinositol 3-kinase directly activates the protein kinase Akt through its pleckstrin homology domain. Mol Cell Biol 17:338–344

Koo EH, Squazzo SL (1994) Evidence that production and release of amyloid beta-protein involves the endocytic pathway. J Biol Chem 269:17386–17389

Lai A, Sisodia SS, Trowbridge IS (1995a) Characterization of sorting signals in the beta-amyloid precursor protein cytoplasmic domain. J Biol Chem 270:3565–3573

Lai KM, Olivier JP, Gish GD, Henkemeyer M, McGlade J, Pawson T (1995b) A Drosophila shc gene product is implicated in signaling by the DER receptor tyrosine kinase. Mol Cell Biol 15:4810–4818

Laminet AA, Apell G, Conroy L, Kavanaugh WM (1996) Affinity, specificity, and kinetics of the interaction of the SHC phosphotyrosine binding domain with asparagine-X-X-phosphotyrosine motifs of growth factor receptors. J Biol Chem 271:264–269

Lemmon MA, Ferguson KM, Schlessinger J (1996) PH domains: diverse sequences with a common fold recruit signaling molecules to the cell surface. Cell 85:621–624

Lioubin MN, Algate PA, Tsai S, Carlberg K, Aebersold A, Rohrschneider LR (1996) p150Ship, a signal transduction molecule with inositol polyphosphate-5-phosphatase activity. Genes Dev 10:1084–1095

Liu L, Damen JE, Cutler RL, Krystal G (1994) Multiple cytokines stimulate the binding of a common 145-kilodalton protein to Shc at the Grb2 recognition site of Shc. Mol Cell Biol 14:6926–6935

McLoughlin DM, Miller CCJ (1996) The intracellular cytoplasmic domain of the Alzheimer's disease amyloid precursor protein interacts with phosphotyrosine-binding domain proteins in the yeast two-hybrid system. FEBS Lett 397:197–200

Milia E, Di Somma MM, Baldoni F, Chiari R, Lanfrancone L, Pelicci PG, Telford JL, Baldari CT (1996) The aminoterminal phosphotyrosine binding domain of Shc associates with ZAP-70 and mediates TCR dependent gene activation. Oncogene 13:767–775

Mok SC, Wong KK, Chan RK, Lau CC, Tsao SW, Knapp RC, Berkowitz RS (1994) Molecular cloning of differentially expressed genes in human epithelial ovarian cancer. Gynecol Oncol 52:247–252

Myers MG Jr, Grammer TC, Brooks J, Glasheen EM, Wang LM, Sun XJ, Blenis J, Pierce JH, White MF (1995) The pleckstrin homology domain in insulin receptor substrate-1 sensitizes insulin signaling. J Biol Chem 270:11715–11718

Nordstedt C, Caporaso GL, Thyberg J, Gandy SE, Greengard P (1993) Identification of the Alzheimer beta/A4 amyloid precursor protein in clathrin-coated vesicles purified from PC12 cells. J Biol Chem 268:608–612

Obermeier A, Lammers R, Wiesmuller KH, Jung G, Schlessinger J, Ullrich A (1993) Identification of Trk binding sites for SHC and phosphatidylinositol 3′-kinase and formation of a multimeric signaling complex. J Biol Chem 268:22963–22966

Obermeier A, Bradshaw RA, Seedorf K, Choidas A, Schlessinger J, Ullrich A (1994) Neuronal differentiation signals are controlled by nerve growth factor receptor/Trk binding sites for SHC and PLC gamma. EMBO J 13:1585–1590

O'Bryan JP, Songyang Z, Cantley L, Der CJ, Pawson T (1996) A mammalian adaptor protein with conserved Src homology 2 and phosphotyrosine-binding domains is related to Shc and is specifically expressed in the brain. Proc Natl Acad Sci USA 93:2729–2734
Okabayashi Y, Kido Y, Okutani T, Sugimoto Y, Sakaguchi K, Kasuga M (1994) Tyrosines 1148 and 1173 of activated human epidermal growth factor receptors are binding sites of Shc in intact cells. J Biol Chem 269:18674–18678
Okamoto T, Takeda S, Murayama Y, Ogata E, Nishimoto I (1995) Ligand-dependent G protein coupling function of amyloid transmembrane precursor. J Biol Chem 270:4205–4208
Ono M, Bolland S, Tempst P, Ravetch JV (1996) Role of the inositol phosphatase ship in negative regulation of the immune system by the receptor Fc-gamma-rIIb. Nature 383:263–266
Pelicci G, Lanfrancone L, Grignani F, McGlade J, Cavallo F, Forni G, Nicoletti I, Pawson T, Pelicci PG (1992) A novel transforming protein (SHC) with an SH2 domain is implicated in mitogenic signal transduction. Cell 70:93–104
Pelicci G, Dente L, De Giuseppe A, Verducci Galletti B, Giuli S, Mele S, Vetriani C, Giorgio M, Pandolfi PP, Cesareni G, Pelicci PG (1996) A family of Shc related proteins with conserved PTB, CH1 and SH2 regions. Oncogene 13:633–641
Perez RG, Squazzo SL, Koo EH (1996) Enhanced release of amyloid beta-protein from codon 670/671. J Biol Chem 271:9100–9107
Pratt JC, Weiss M, Sieff CA, Shoelson SE, Burakoff SJ, Ravichandran KS (1996) Evidence for a physical association between the Shc-PTB domain and the beta c chain of the granulocyte-macrophage colony-stimulating factor receptor. J Biol Chem 271:12137–12140
Prigent SA, Pillay TS, Ravichandran KS, Gullick WJ (1995) Binding of Shc to the NPXY motif is mediated by its N-terminal domain. J Biol Chem 270:22097–22100
Ravichandran KS, Lee KK, Songyang Z, Cantley LC, Burn P, Burakoff SJ (1993) Interaction of Shc with the zeta chain of the T cell receptor upon T. Science 262:902–905
Ravichandran KS, Igras V, Shoelson SE, Fesik SW, Burakoff SJ (1996) Evidence for a role for the phosphotyrosine-binding domain of Shc in interleukin 2 signaling. Proc Natl Acad Sci USA 93:5275–5280
Rhyu MS, Jan LY, Jan YN (1994) Asymmetric distribution of numb protein during division of the sensory organ precursor cell confers distinct fates to daughter cells. Cell 76:477–491
Ricketts WA, Rose DW, Shoelson S, Olefsky JM (1996) Functional roles of the Shc phosphotyrosine binding and Src homology 2 domains in insulin and epidermal growth factor signaling. J Biol Chem 271:26165–26169
Rozakis-Adcock M, McGlade J, Mbamalu G, Pelicci G, Daly R, Li W, Batzer A, Thomas S, Brugge J, Pelicci PG, Schlessinger J, Pawson T (1992) Association of the Shc and Grb2/Sem5 SH2-containing proteins is implicated in activation of the Ras pathway by tyrosine kinases. Nature 360:689–692
Saxton TM, Van Oostveen I, Bowtell D, Aebersold R, Gold MR (1994) B cell antigen receptor cross-linking induces phosphorylation. J Immunol 153:623–636
Selkoe DJ (1994) Cell biology of the amyloid beta-protein precursor and the mechanism of Alzheimer's disease. Annu Rev Cell Biol 10:373–403
Songyang Z, Shoelson SE, Chaudhuri M, Gish G, Pawson T, Haser WG, King F, Roberts T, Ratnofsky S, Lechleider RJ, Neel BG, Birge RB, Fajardo JE, Chow MM, Hanafusa H, Schaffhausen B, Cantley LC (1993) SH2 domains recognize specific phosphopeptide sequences. Cell 72:767–778
Staub O, Rotin D (1996) WW domains. Structure 4:495–499
Stephens RM, Loeb DM, Copeland TD, Pawson T, Greene LA, Kaplan DR (1994) Trk receptors use redundant signal transduction pathways involving SHC and PLC-gamma 1 to mediate NGF responses. Neuron 12:691–705
Sudol M, Chen HI, Bougeret C, Einbond A, Bork P (1995) Characterization of a novel protein-binding module – the WW domain. FEBS Lett 369:67–71
Sun XJ, Wang L-M, Zhang Y, Yenush L, Myers MG, Glasheen E, Lane WS, Pierce JH, White MF (1995) Role of IRS-2 in insulin and cytokine signalling. Nature 377:173–177
Trub T, Choi WE, Wolf G, Ottinger E, Chen Y, Weiss M, Shoelson SE (1995) Specificity of the PTB domain of Shc for beta turn-forming pentapeptide motifs amino-terminal to phosphotyrosine. J Biol Chem 270:18205–18208
Uemura T, Shepherd S, Ackerman L, Jan LY, Jan YN (1989) numb, a gene required in determination of cell fate during sensory organ formation in Drosophila embryos. Cell 58:349–360
Van der Geer P, Wiley S, Lai VK-M, Olivier JP, Gish GD, Stephens R, Kaplan D, Shoelson S, Pawson T (1995) A conserved amino-terminal Shc domain binds to phosphotyrosine motifs in activated receptors and phosphopeptides. Curr Biol 5:404–412

Van der Geer P, Wiley S, Gish GD, Lai VK, Stephens R, White MF, Kaplan D, Pawson T (1996) Identification of residues that control specific binding of the Shc phosphotyrosine-binding domain to phosphotyrosine sites. Proc Natl Acad Sci USA 93:963–968

Verdi JM, Schmandt R, Bashirullah A, Jacob S, Salvino R, Craig CC, Amgen EST Program , Lipshitz HD, McGlade CJ (1996) Mammalian Numb is an evolutionarily conserved signaling adapter protein that specifies cell fate. Curr Biol 6:1134–1145

White MF, Livingston JN, Backer JM, Lauris V, Dull TJ, Ullrich A, Kahn CR (1988) Mutation of the insulin receptor at tyrosine 960 inhibits signal transmission but does not affect its tyrosine kinase activity. Cell 54:641–649

Wolf G, Trub T, Ottinger E, Groninga L, Lynch A, White MF, Miyazaki M, Lee J, Shoelson SE (1995) PTB domains of IRS-1 and Shc have distinct but overlapping binding specificities. J Biol Chem 270:27407–27410

Xu XX, Yang W, Jackowski S, Rock CO (1995) Cloning of a novel phosphoprotein regulated by colony-stimulating factor 1 shares a domain with the Drosophila disabled gene product. J Biol Chem 270:14184–14191

Yajnik V, Blaikie P, Bork P, Margolis B (1996) Identification of residues within the Shc phosphotyrosine binding/phosphotyrosine interaction domain crucial for phosphopeptide interaction. J Biol Chem 271:1813–1816

Yenush L, Makati KJ, Smithhall J, Ishibashi O, Myers MG, White MF (1996) The pleckstrin homology domain is the principle link between insulin receptor and IRS-1. J Biol Chem 271:24300–24306

Zhong WM, Feder JN, Jiang MM, Jan LY, Jan YN (1996) Asymmetric localization of a mammalian numb homolog during mouse cortical neurogenesis. Neuron 17:43–53

Zhou MM, Harlan JE, Wade WS, Crosby S, Ravichandran KS, Burakoff SJ, Fesik SW (1995a) Binding affinities of tyrosine-phosphorylated peptides to the COOH-terminal SH2 and NH2-terminal phosphotyrosine binding domains of Shc. J Biol Chem 270:31119–31123

Zhou MM, Ravichandran KS, Olejniczak EF, Petros AM, Meadows RP, Sattler M, Harlan JE, Wade WS, Burakoff SJ, Fesik SW (1995b) Structure and ligand recognition of the phosphotyrosine binding domain of Shc. Nature 378:584–592

Zhou MM, Huang B, Olejniczak ET, Meadows RP, Shuker SB, Miyazaki M, Trub T, Shoelson SE, Fesik SW (1996) Structural basis for IL-4 receptor phosphopeptide recognition by the IRS-1 PTB domain. Nat Struct Biol 3:388–393

Pleckstrin Homology Domains

M.A. LEMMON and K.M. FERGUSON

1 Introduction

Pleckstrin homology (PH) domains are small protein modules of around 120 amino acids that are found in a large number of proteins involved in intracellular signaling and cytoskeletal organization, often occurring alongside SH2, SH3, PTB and other domains discussed in this volume. PH domains were first noted by MAYER et al. (1993) and HASLAM et al. (1993) as sequences found in a number of intracellular signaling molecules that show limited homology to a region repeated in the protein pleckstrin (TYERS et al. 1988). As a result, this 47-kDa protein, which is the major substrate of protein kinase C (PKC) in platelets, has lent its name to a domain now

---

Department of Biochemistry and Biophysics, University of Pennsylvania School of Medicine, A606 Richards Building, 3700 Hamilton Walk, Philadelphia, PA 19104-6089, USA

identified in more than 100 different proteins involved in different signaling and cytoskeletal organization processes. Soon after the identification of the PH domain, structural studies showed that it does indeed form an independent module with a characteristic β-sandwich structure. The functions of PH domains are now becoming more clear, and the current view is that they are involved in recruitment of their host proteins to cell membranes. In some cases this recruitment is achieved through direct interaction of the PH domain with specific membrane components, and can be directly signal-dependent – with the PH domain binding to a lipid second messenger. In this chapter, we will discuss the structure of PH domains, and the characteristics that make them ideally suited for binding to the membrane surface. We will also review the current state of knowledge regarding PH domain function and ligand-binding properties, and will consider how they may participate in defining the specificity of intermolecular interactions and compartmentalization required for the function of their host proteins in signaling processes.

2 PH Domains Show Considerable Diversity in Both Sequence and Location

Mayer et al. (1993) and Haslam et al. (1993) initially identified PH domains in protein serine/threonine kinases (such as Akt/Rac/PKB); regulators of small GTP-binding proteins (both GTPase activating proteins and guanine nucleotide exchange factors); cytoskeletal proteins (such as spectrin); putative signaling adapter molecules (such as Grb7); as well as proteins involved in cellular membrane transport (such as hSec7 and dynamin). Further database searches, using profiles generated with the 16 or so initially identified occurrences, identified putative PH domains in a large number of additional proteins; expanding these groups, and adding protein tyrosine kinases and phospholipase C isoforms (Musacchio et al. 1993; Shaw 1993, 1996; Parker et al. 1994; Gibson et al. 1994). The total number of putative PH domains identified now exceeds 100. It is not our intention in this chapter to survey the details of PH domain occurrence and sequence analysis, for which the reader is directed to reviews by Musacchio et al. (1993), Gibson et al. (1994) and Shaw (1996). Rather, we will focus on the structural and functional aspects of these domains, giving here only a précis of PH domain distribution and sequence analysis. Figure 1 shows the position of PH domains in the domain structure of selected proteins involved in cellular signaling. Few general statements can be made about PH domain location in their host proteins. They are often seen at the very N-terminus of the protein, as in the PLCs, Akt, Btk, IRS-1 and pleckstrin; they are sometimes present close to the C-terminus, as in βARK and pleckstrin. The Dbl homology region found in many guanine nucleotide exchange factors (GNEFs) is almost invariably followed by a PH domain, as seen in Vav and Sos. Otherwise, PH domains, like SH2, SH3 and PTB domains – alongside which they are often found – are present at a variety of positions, consistent with their

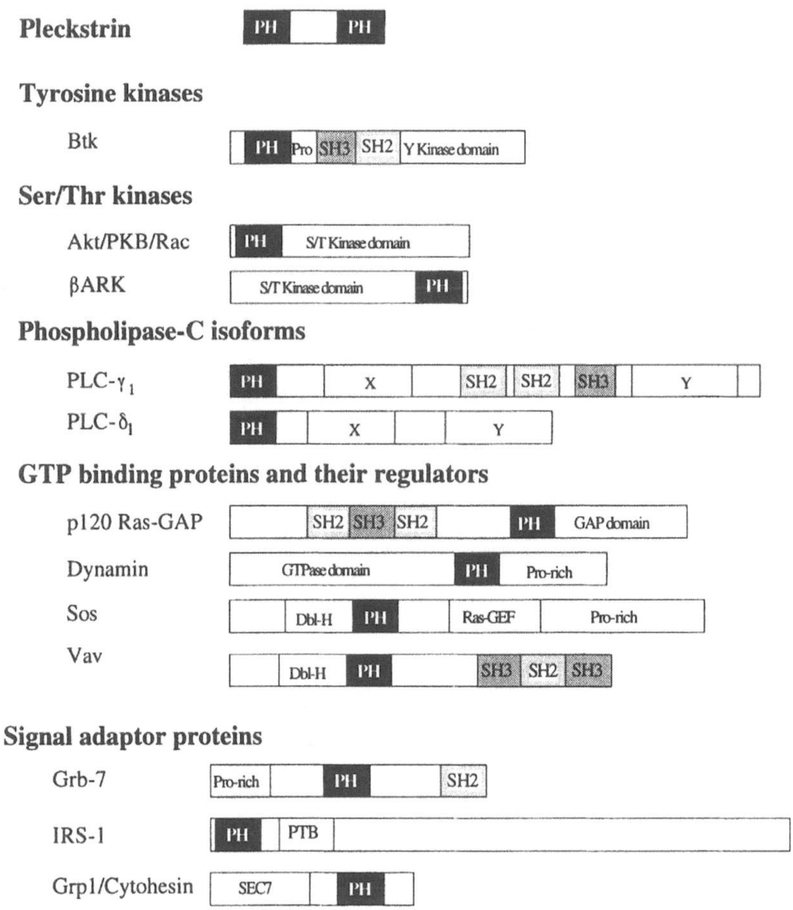

Fig. 1. Schematic representation of the domain structure of selected PH domain-containing proteins. PH domains are *black, with a white label*, and the positions of SH2, SH3, PTB and other domains are marked where they occur. *X and Y* in the phospholipase C isoforms indicate the conserved catalytic regions. *Dbl-H* indicates a region in Vav and Sos that is homologous to a portion of Dbl (the *dbl* oncogene product), and is thought to act as an exchanger for Rho-like small G-proteins. *SEC7* represents a region in Grp1 and cytohesin-1 that shares homology with a region in the yeast Sec7 gene product

being functional modules that can be inserted in a number of different environments.

The homology that defines the family of PH domains is significantly less obvious than that for SH2 or SH3 domains. A sequence alignment of 19 selected PH domains is shown in Figure 2. The overall percentage identity at the amino acid sequence level is rarely greater than 30% for PH domains from unrelated proteins, and in many cases it is less than 10%. As an example, the PH domains from dynamin-1 and PLC-δ_1, which have been shown by X-ray crystallography to represent bona fide PH domains, and to have remarkably similar structures (see below), are related by just 9.9% amino acid identity. There are only a handful of positions

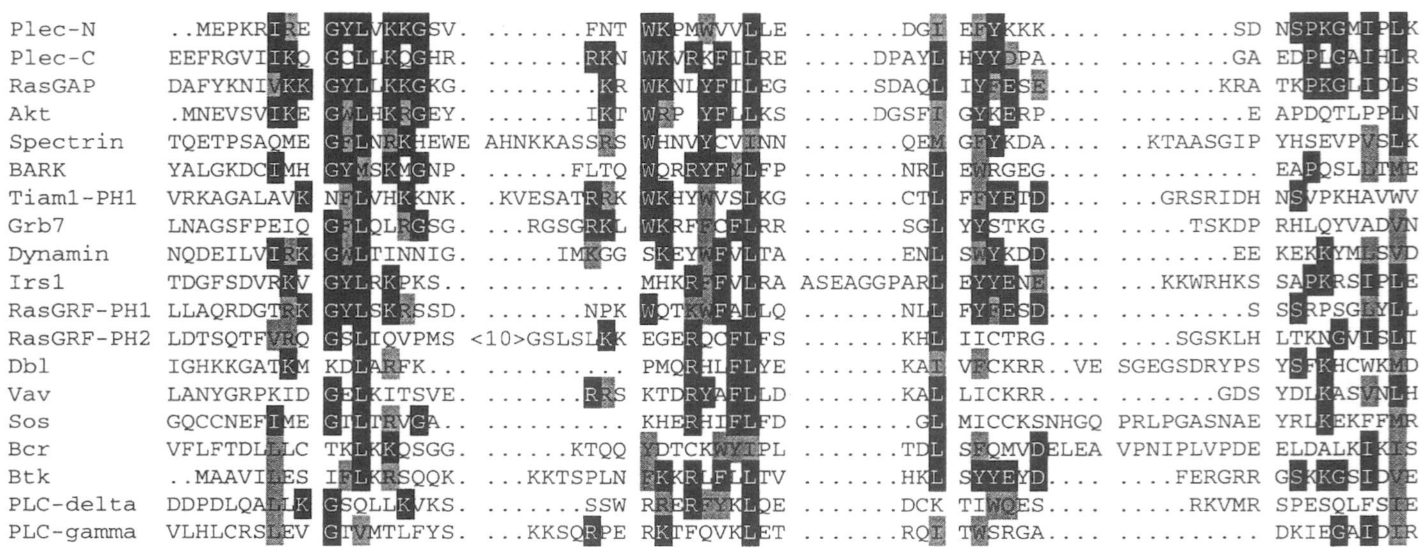
Plec-N ..MEPKRIRE GYLVKKGSV. ....FNT WKPMWVVLLE ....DGI EFYKKK.... ....SD NSPKGMIPLK
Plec-C EEFRGVIIKQ GCLLKQGHR. ....RKN WKVRKFILRE .....DPAYL HYYDPA.... ....GA EDPLGAIHLR
RasGAP DAFYKNIVKK GYLLKKGKG. ....KR WKNLYFILEG .....SDAQL IYFESE.... ....KRA TKPKGLIDLS
Akt .MNEVSVIKE GWLHKRGEY. ....IKT WRP YFLLKS .....DGSFI GYKERP.... ....E APDQTLPPLN
Spectrin TQETPSAQME GFLNRKHEWE AHNKKASSRS WHNVYCVINN ....QEM GFYKDA.... ..KTAASGIP YHSEVPVSLK
BARK YALGKDCIMH GYMSKMGNP. ......FLTQ WQRRYFYLFP ....NRL EWRGEG.... .... EAPQSLLTME
Tiam1-PH1 VRKAGALAVK NFLVHKKNK. .KVESATRRK WKHYWVSLKG ....CTL FFYETD.... ...GRSRIDH NSVPKHAVWV
Grb7 LNAGSFPEIQ GFLQLRGSG. ...RGSGRKL WKRFFCFLRR ....SGL YYSTKG.... .....TSKDP RHLQYVADVN
Dynamin NQDEILVIRK GWLTINNIG. .....IMKGG SKEYWFVLTA ....ENL SWYKDD.... ....EE KEKKYMLSVD
Irs1 TDGFSDVRKV GYLRKPKS.. .... MHKRFFVLRA ASEAGGPARL EYYENE.... ...KKWRHKS SAPKRSIPLE
RasGRF-PH1 LLAQRDGTRK GYLSKRSSD. ....NPK WQTKWFALLQ ....NLL FYFESD.... ....S SSRPSGLYLL
RasGRF-PH2 LDTSQTFVRQ GSLIQVPMS <10>GSLSLKK EGERQCFLFS ....KHL IICTRG.... ....SGSKLH LTKNGVISLI
Dbl IGHKKGATKM KDLARFK... .... PMQRHLFLYE ....KAI VFCKRR..VE SGEGSDRYPS YSFKHCWKMD
Vav LANYGRPKID GELKITSVE. ....RRS KTDRYAFLLD ....KAL LICKRR.... ......GDS YDLKASVNLH
Sos GQCCNEFIME GTLTRVGA.. .... KHERHIFLFD ....GL MICCKSNHGQ PRLPGASNAE YRLKEKFFMR
Bcr VFLFTDLLLC TKLKKQSGG. ......KTQQ YDTCKWYIPL ....TDL SFQMVDELEA VPNIPLVPDE ELDALKIKIS
Btk ..MAAVILES IFLKRSQQK. ...KKTSPLN FKKRLFLLTV ....HKL SYYEYD.... ....FERGRR GSKKGSIDVE
PLC-delta DDPDLQALLK GSQLLKVKS. ....SSW RRERFYKLQE ....DCK TIWQES.... .....RKVMR SPESQLFSIE
PLC-gamma VLHLCRSLEV GTVMTLFYS. ...KKSQRPE RKTFQVKLET ....RQI TWSRGA.... .... DKIEGAIDIR

```
Plec-N       GSTLTSPCQD FGKR...... ...MFVFKIT TTKQQ..... ..DHFFQAAF LEERDAWVRD INKAIKCIEG
Plec-C       GCVVTSVESN SNGR.....K SEEENLFEII TADEV..... ..HYFLQAAT PKERTEWIKA IQMASRTGK*
RasGAP       VCSVYVVHDS LFGR...... ...PNCFQIV VQHFS...EE HYIFYFAGET PEQAEDWMKG LQAFCNLRKS
Akt          NFSVAECQLM KTER...... .PRPNTFVIR CLQWT....T VIERTFHVDS PDEREEWMRA IQMVANSLKQ
Spectrin     EAVCEVALDY KKKK...... ....HVFKLR LNDGN..... ..EYLFQAKD DEEMNTWIQA ISSAISSDKH
BARK         EIQSVEETQI KER....... ....KCLLLK IRGGK..... ..QFILQCDS DPELVQWKKE LRDAYREAQQ
Tiam1-PH1    ENSIVQAVPE HPKK...... ...DFVFCLS NSLGD..... .AFLFQTTSQ .TELENW TA IHSACAAAVA
Grb7         ESNVYVVTQG RKLY.....G MPTDFGFCVK PNKLR..NGH KGLHIFCSED EQSRTCWLAA FRLFKYGVQL
Dynamin      NLKLRDVEKG FMSS...... ...KHIFALF NTEQRNVYKD YRQLELACET QEEVDSWKAS FLRAGVYPER
Irs1         SCFNINKRAD SKNK...... ....HLVALY TRDEH..... ...FAIAADS EAEQDSWYQA LLQLHNRAKA
RasGRF-PH1   EGSICKRMPS PKRGTSSKES DKQHHYFTVN FSNDS..... QKSLELRTDD SKDCDEWVAA IARASYKILA
RasGRF-PH2   DCTLLDDPEN MDDD..GKGQ EVDHLDFKIW VEPKD...SP PFTVILVASS RQEKAAWTSD IIQCVDNIRC
Dbl          EVGITEYVKG DN........ ....RKFEIW YGEKE..... .EVYIVQASN VDVKMTWLKE IRNILLKQQE
Vav          SFQVRDDSSG ERDN...... KKWSHMFLLI EDQGA..... .QGYELFFKT RELKKKWMEQ FEMAISNIYP
Sos          KVQINDKDDT NEYK...... ....HAFEII LKDEN..... ..SVIFSAKS AEEKNNWMAA LISLQYRSTL
Bcr          QIKSDIQREK RANK<19>LL MSPSMAFRVH SRNGK..... ..SYTFLISS DYERAEWREN IREQQKKCFR
Btk          KITCVETVVP EKNP<20>II ERFPYPFQVV YDEGP..... ...LYVFSPT EELRKRWIHQ LKNVIRYNSD
PLC-delta    DIQEVRMGHR TEGLEKFARD IPEDRCFSIV FKDQRN.... ..TLDLIAPS PRDAQHWVQG LRKIIHHSGS
PLC-gamma    EIKEIRPGKT SRDF<7>AFR PDQSHCFVIL YGMEF...RL KTLSLQATSE .DEVNMWIKG LTWLMEDTLQ
```

Fig. 2. Sequence alignment of 19 different PH domains, after the alignment by GIBSON et al. (1994). Where a given amino acid (identity) is found in five or more of these PH domains, it is marked with a *black box* (*white text*). Where a given type of amino acid (homology) is found in five or more PH domains, it is *shaded grey*. The following PH domains are shown: *Plec-N*, human pleckstrin, aa 1–105 (TYERS et al. 1988); *Plec-C*, human pleckstrin, aa 239–350 (TYERS et al. 1988); *RasGAP*, human Ras GTPase activating protein, aa 292–404 (TRAHEY et al. 1988); *Akt*, human serine/threonine kinase AKT2, aa 1–118 (CHENG et al. 1992); *Spectrin*, human β-spectrin, aa 2192–2311 (HU et al. 1992); *BARK*, human β-adrenergic receptor kinase, aa 553–656 (CHUANG et al. 1992); *Tiam1-PH1*, mouse GDP-GTP exchanger Tiam-1, aa 440–546 (HABETS et al. 1994); *Grb7*, mouse growth factor receptor bound protein 7, aa 224–345 (MARGOLIS et al. 1992); *Dynamin*, human GTPase dynamin 1, aa 515–629 (VAN DER BLIEK et al. 1993); *Irs1*, rat insulin receptor substrate 1, aa 7–119 (SUN et al. 1991); *RasGRFPH1*, rat Ras guanine nucleotide release factor, aa 17–134 (SHOU et al. 1992); *RasGRFPH2*, rat Ras guanine nucleotide release factor, aa 451–584 (SHOU et al. 1992). <10> indicates omitted aa; EKGKINKGRL; *Dbl*, human product of the *dbl* proto-oncogene, aa 704–813 (RON et al. 1988); *Vav*, mouse product of the *vav* proto-oncogene, aa 397–508 (ADAMS et al. 1992); *Sos*, human son of sevenless protein, aa 438–548 (CHARDIN et al. 1993); *Bcr*, human break point cluster region gene product, aa 728–870 (LIFSHITZ et al. 1988); <19> indicates omitted aa; GSKATERLKK KLSEQESLL; *Btk*, human Bruton's tyrosine kinase, aa 1–137 (VETRIE et al. 1993); <20> indicates omitted aa; PPERQIPRRG EESSEMEQIS; *PLC-delta*, rat phospholipase $C\delta_1$, aa 15–134 (SUH et al. 1988); *PLC-gamma*, human phospholipase $C\gamma_1$, aa 22–146 (BURGESS et al. 1990); <7> indicates omitted aa; DRYQEDP

at which the identity of the amino acid is reasonably well conserved. One of these is a tryptophan close to the C-terminus, which is found in nearly all PH domains. Another position of significant conservation is a glycine that usually precedes the first β-strand. A few additional positions exist at which a particular hydrophobic amino acid is quite well conserved, but there are no clear conserved sequence motifs that are characteristic of PH domains. Rather, the domain is defined primarily by six sequence blocks that show a conserved pattern of hydrophobic and hydrophilic amino acids (Fig. 2), which are separated by stretches of amino acids that vary widely in both length and sequence. Based upon the periodicity of amino acid type in the six blocks of amino acid homology, MUSACCHIO et al. (1993) predicted with remarkable accuracy the secondary structure content of PH domains. It was suggested that the PH domain would consist of 7 or 8 β-strands and a single C-terminal α-helix. The less obvious nature of the homology that defines PH domains, which is much more readily detected by database searches using sequence profiles rather than using particular amino acid sequences, presumably delayed the identification of this widely occurring domain compared with the other examples discussed in this volume.

3 PH Domain Structure

As is true in an increasing number of cases today, structural studies of PH domains preceded the now developing appreciation of their functional role, and have contributed substantially to the design of experiments aimed at understanding PH domain function. Structures have been reported for four distinct PH domains. Nuclear magnetic resonance (NMR) spectroscopy has been used to determine the structure of the PH domains from pleckstrin (N-terminal PH domain; YOON et al. 1994), mouse β-spectrin (MACIAS et al. 1994), *Drosophila* β-spectrin (ZHANG et al. 1995) and dynamin-1 (DOWNING et al. 1994; FUSHMAN et al. 1995). X-ray crystallography has been used to determine the structure of the dynamin-1 PH domain (FERGUSON et al. 1994; TIMM et al. 1994), as well as complexes between inositol-1,4,5-trisphosphate (Ins (1,4,5) P_3) and the PH domains from PLC-δ_1 (FERGUSON et al. 1995) and spectrin (HYVÖNEN et al. 1995). In each case, the structure of the domain is remarkably similar, and is best described as an antiparallel β-sandwich of two sheets in which the β-strands of one sheet are nearly orthogonal to those of the other sheet (Fig. 3).

Using the dynamin-1 PH domain as an example, the characteristics of the structure can be discussed (Figs. 3, 4). The strands in each β-sheet are arranged as in a "β-meander," with each strand reentering the sheet on the same side from which it left, and with strands that are adjacent in the sequence also being adjacent in the structure (RICHARDSON 1977). The N-terminal half of the PH domain forms a four-stranded antiparallel β-sheet that packs almost orthogonally against a second β-sheet of three strands (Fig. 3). Hydrophobic side-chains from each of the two

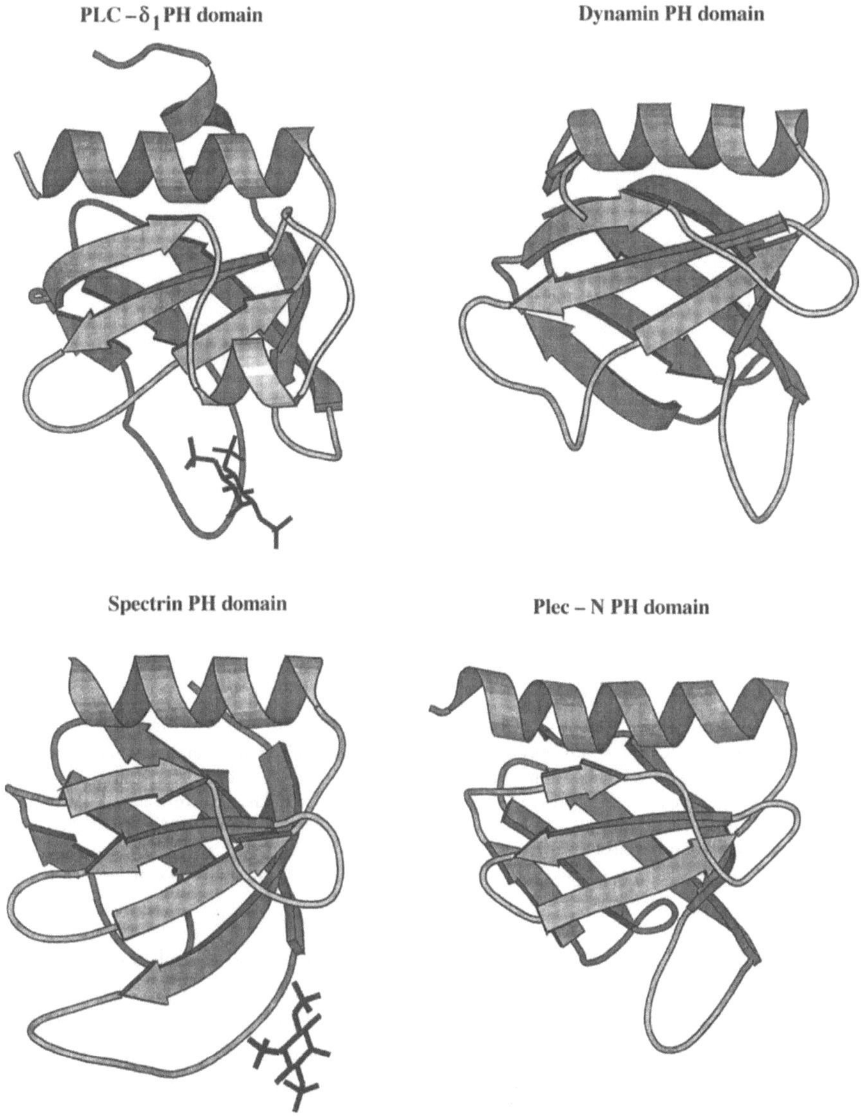

Fig. 3. MOLSCRIPT (KRAULIS 1991) representations of the PH domains from PLC-δ_1 (FERGUSON et al. 1995), dynamin-1 (FERGUSON et al. 1994), spectrin (HYVÖNEN et al. 1995) and pleckstrin (N-terminal) (YOON et al. 1994), showing the similarity in their stuctures. Each domain is shown in the same orientation, with the C-terminal amphipathic α-helix at the top (C-terminus *on left*). This places the three most variable loops at the bottom of each figure. D-*myo*-Ins (1,4,5) P_3 is shown as bound in crystal structures of the PH domains from PLC-δ_1 (K_D = 0.21 μ*M*) and spectrin (K_D = 40 μ*M*). In each case the 1-phosphate is most exposed. The PLC-δ_1 PH domain has additional α-helices at its N-terminus and in its β5/β6 loop (FERGUSON et al. 1995). Spectrin PH has an additional α-helix in its β3/β4 loop (MACIAS et al. 1994; HYVÖNEN et al. 1995)

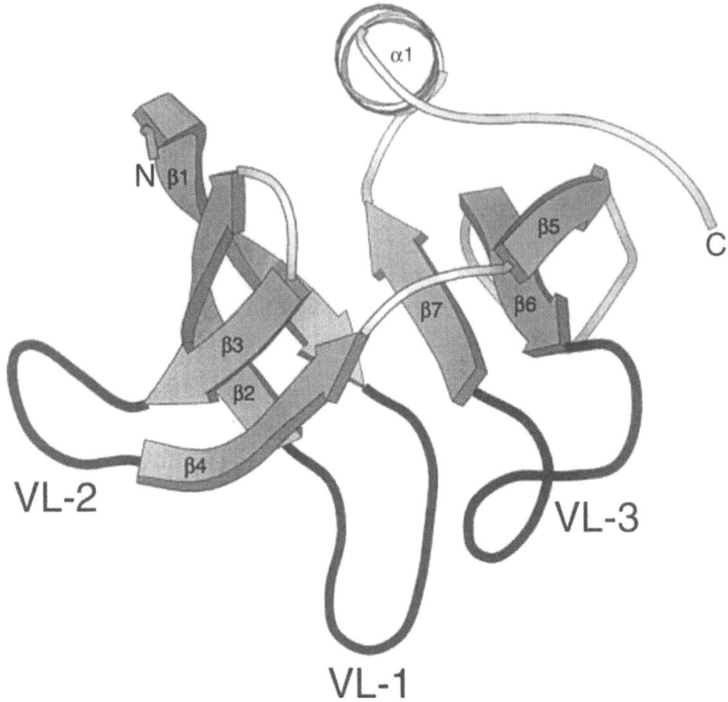

Fig. 4. MOLSCRIPT (KRAULIS 1991) representation of the PH domain from dynamin-1 (FERGUSON et al. 1994). The molecule is in an orientation rotated 90° about a vertical axis compared with that shown in Fig. 3. The N- and C-termini are marked, as are the elements of secondary structure common to all PH domains of known structure. VL-1, VL-2 and VL-3 are the variable loops (*shaded black*) that are seen in sequence alignments to be most variable in both length and sequence, and which are seen to bind to phosphoinositides in the cases of the PH domains from PLC-δ_1, spectrin and pleckstrin

β-sheets project into the center of the domain to generate the well-packed hydrophobic core of the PH domain. PH domains do not represent β-barrels, since the interstrand hydrogen bonding pattern does not continue all around the structure (WOOLFSON et al. 1993). As is typical for near-orthogonal β-sheet packing (CHOTHIA 1984), the right-handed twist of the two β-sheets results in close contact only at two (close) corners. One close corner results from strand β-1 that contributes to both β-sheets, while the other is completed by a type II β-turn between strands β4

Fig. 5. A structure-based sequence alignment of the four PH domains of known structure (Fig. 3), plus the Btk PH domain. The sequence numbering is for PLC-δ_1, and the positions of secondary structural elements are marked as *black arrows* (β-strands) *or a black rectangle* (C-terminal α-helix). The three variable loops, VL-1, VL-2 and VL-3, as marked in the dynamin-1 PH domain in Fig. 4 are marked in the sequence. For the PLC-δ_1 PH domain, residues that contact Ins (1,4,5) P_3 in the high-affinity complex are denoted by *black boxes*. Also in *black boxes* are two sites in the Btk PH domain (F25, R28) at which XLA-associated mutations have been found in Btk (MATTSSON et al. 1996) and E41, at which a gain-of-function was found in a screen of random Btk mutants (LI et al. 1995)

```
                20         30                   40         50         60              70
                 •          •                    •          •          •               •

PLC-δ1       DPDLQALLKG SQLLKVKS.. .......SSW RRERFYKLQE DCKTIWQESR KVMRS..... PESQLFSIED
Spectrin            MEG FLNRKHEWEA HNKKASSRSW H.NVYCVINN Q.EMGFYKDA KSAASGIPY. HSEVPVSLKE
Pleckstrin N  MEPKRIREG YLVKKG.... ....SVFNTW K.PMWVVLLE D.GIEFYKKK SDNS...... .PKGMIPLKG
Dynamin      QDEILVIRKG WLTINNIG.. ....IMKGGS K.EYWFVLTA E.NLSWYKDD EE.......K EKKYMLSVDN
Btk            MAAVILES IFLKRSQ... QKKKTSPLNF K.KRLFLLTV H.KLSYYEYD FERGRRGSKK GSIDVEKITC
                                           25   28             41
```

```
                   80         90        100                   110        120        130
                    •          •          •                     •          •          •

PLC-δ1       IQEVRMGHRT EGLEKFARDI PEDRCFSIVF K.......DQ RNTLDLIAPS PADAQHWVQG LRKIIHHS.
Spectrin     .AICEVALD. ........YK KKKHVFKL.R L.......SD GNEYLFQAKD DEEMNTWIQA ISSA.....
Pleckstrin N STLTSPCQD. ........FG KRMFVFKI.T T.......TK QQDHFFQAAF LEERDAWVRD INKAIKCI.
Dynamin      LKLRDVEKG. ........FM SSKHIFAL.F NTEQRNVYKD YRQLELACET QEEVDSWKAS FLRAGVYPE
Btk          VETVVPEKN. ........P# IIERFPYP.F QVV.....YD EGPLYVFSPT EELRKRWIHQ LKNVIRYND
```

and β5. The other two (splayed) corners are splayed apart. One splayed corner (Fig. 4, top) is closed off by a C-terminal amphipathic α-helix that contains the invariant tryptophan and other residues with hydrophobic side-chains that project into, and contribute to, the hydrophobic core of the domain. The other splayed corner of the β-sandwich (Fig. 4, bottom) is closed off largely by side-chains from the β1/β2 and β6/β7 connecting loops and portions of β4. It is apparent from the structures of the domains depicted in Fig. 3 that both termini are located towards the same side of the molecule that contains the C-terminal α-helix. This arrangement is consistent with the PH domains being protein modules that can be inserted at various positions in their host proteins.

A sequence alignment of the PH domains with known structure is shown in Fig. 5, together with the positions of secondary structural elements. It is clear from this alignment (as well as that in Fig. 2) that four of the interstrand loops are particularly variable in both length and sequence. These are loops β1/β2, β3/β4, β5/β6 and β6/β7. As seen in Fig. 3, the spectrin PH domain contains an additional 2-turn α-helix in the β3/β4 loop that is not seen in the other domains, and the PLC-δ_1 PH domain has an additional α-helix in its β5/β6 loop. The fact that the variable loops appear to define one face of the PH domain led to the suggestion that they might be involved in PH domain-mediated interactions. Further support for this suggestion was lent by the fact that each of the PH domains of known structure is markedly electrostatically polarized, with the positively charged face of each domain corresponding to that containing the most variable loops (Fig. 6). As will be discussed below, this positively charged, variable face has been found to interact with negatively charged lipid headgroups in some cases.

4 The Ligands of PH Domains

The identification of PH domains in many proteins involved in intracellular signaling prompted the suggestion that they may have functional similarity to SH2, SH3 and other domains discussed in this volume, promoting interactions of their host proteins with other intracellular signaling molecules. As discussed below, the cases of Bruton's tyrosine kinase (Btk) and the β-adrenergic receptor kinase (βARK) (Fig. 1) provided some immediate indication that this might be true. The PH domains of Btk and βARK replace fatty acylation sites in their homologs Src

Fig. 6. Electrostatic polarization of PH domains. The PLC-δ_1 (*left*) and dynamin-1 PH domains are shown in the same orientation as that seen for the dynamin-1 PH domain in Fig. 4. The backbone is represented as *a worm in dark gray*, and is surrounded by contours of positive electrostatic potential (+1.5 kT, *in light gray*) and negative electrostatic potential (−1.5 kT, *in dark gray*). Each domain is clearly electrostatically polarized, with the face of positive potential coinciding approximately with the face comprising the variable loops (compare Figs. 4 and 6). Ins (1,4,5) P_3 binds in the center of the positively charged face of the PLC-δ_1 PH domain. This figure was generated using the program GRASP (Nicholls et al. 1991)

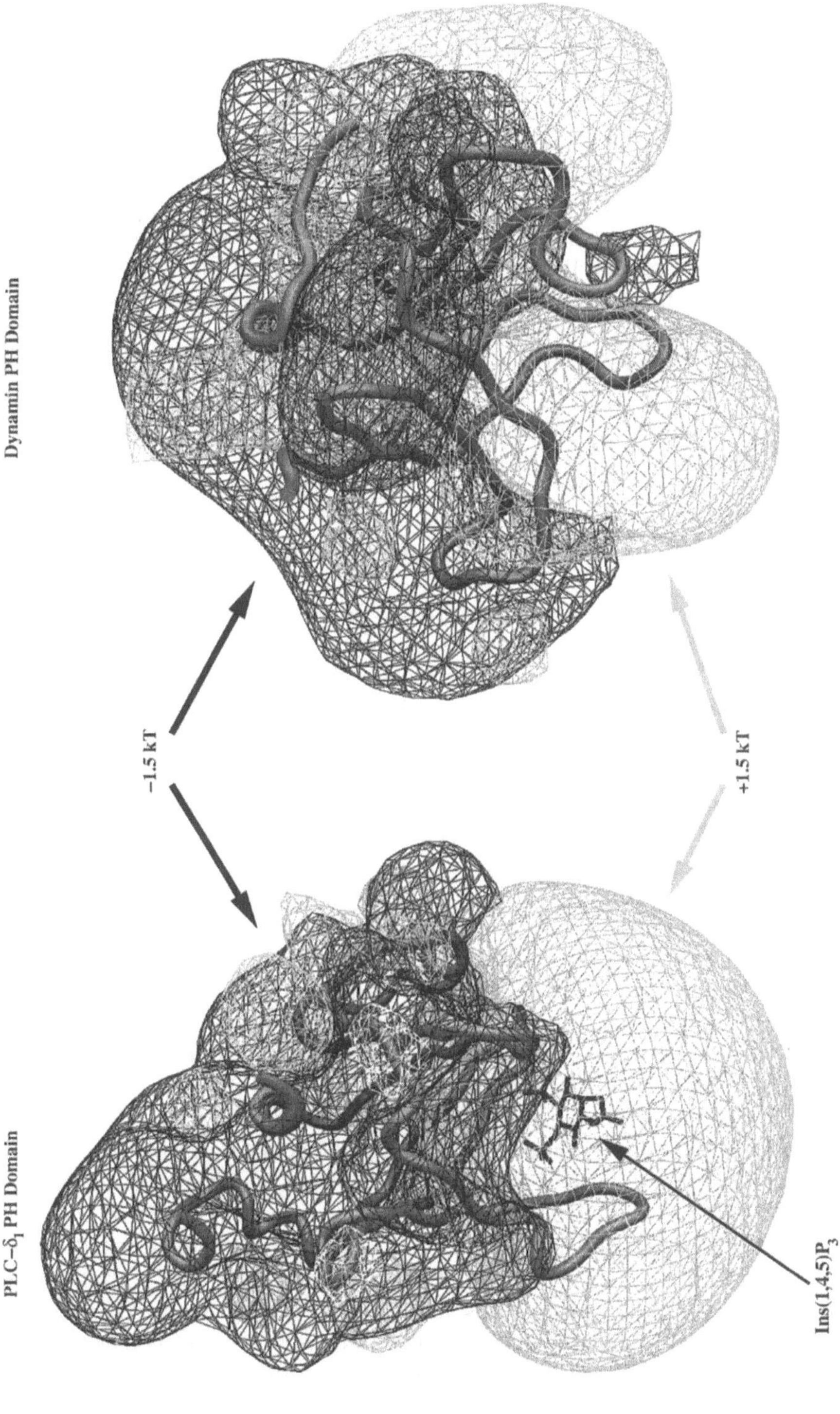
Dynamin PH Domain
PLC–δ1 PH Domain
−1.5 kT
+1.5 kT
Ins(1,4,5)P3

and rhodopsin kinase respectively, specifically implicating the PH domains of these molecules in association with the plasma membrane. Subsequent studies of a number of other signaling molecules also point to a role for PH domains in membrane association, and several isolated PH domains – most notably those from PLC-δ_1 (PATERSON et al. 1995), spectrin (WANG et al. 1996) and Sos (CHEN et al. 1997) are clearly localized to the plasma membrane when introduced into mammalian cells. However, the cellular ligands for PH domains that are responsible for this membrane association are not yet clear in most cases. The potential PH domain ligands for which there is most evidence are the βγ-subunits ($G_{\beta\gamma}$) of heterotrimeric G-proteins (INGLESE et al. 1995), protein kinase C (PKC) isoforms (YAO et al. 1994) and phosphoinositides (HARLAN et al. 1994). For the PH domains of certain proteins, all three types of ligand remain possibilities. In the following, we will summarize briefly the evidence that has led to the suggestion of these three types of PH domain ligand. In Sect. 5, the current status of studies aimed at elucidating PH domain function and ligand binding will be considered in more detail for several selected proteins.

4.1 PH Domain Binding to $G_{\beta\gamma}$-Subunits: The βARK PH Domain

The PH domain at the C-terminus of βARK (Fig. 1) is replaced in its visual homolog, rhodopsin kinase (RK), by a farnesylation sequence. Farnesylation of RK is necessary for its light-dependent translocation to the membrane where it phosphorylates, and thus desensitizes, activated rhodopsin (INGLESE et al. 1992). The analogous function of βARK requires the presence of free $G_{\beta\gamma}$-subunits at the membrane surface for βARK membrane targeting and agonist-dependent phosphorylation of the β_2-adrenergic receptor (PITCHER et al. 1992). The C-terminal portion of βARK, which includes its PH domain, is important in this $G_{\beta\gamma}$-mediated membrane targeting, with a minimal requirement for the PH domain plus an additional 14 amino acids beyond its C-terminus. This 125-amino acid C-terminal fragment of βARK can directly inhibit $G_{\beta\gamma}$-stimulated receptor phosphorylation by βARK (KOCH et al. 1993), and can also inhibit various $G_{\beta\gamma}$-mediated signaling processes (KOCH et al. 1994). These observations led to the suggestion that the βARK PH domain binds directly to $G_{\beta\gamma}$-subunits. TOUHARA et al. (1994) subsequently reported that several different PH domains, including those from spectrin, Ras-GRF, Atk, Ras-GAP, PLC-γ, IRS-1, OSBP and Racβ, could bind to $G_{\beta\gamma}$-subunits in experiments involving β-subunit precipitation from cell lysates by immobilized GST-fusion proteins of the various PH domains. A similar result was presented by others for the PH domains from Btk (TSUKUDA et al. 1994) and β-spectrin (WANG et al. 1994). Furthermore, overexpression of several different isolated PH domains (from βARK-1, PLC-γ, IRS-1, Ras-GAP and Ras-GRF) was reported to inhibit $G_{\beta\gamma}$-mediated stimulation of inositol phosphate production by the α2-C10 adrenergic receptor (LUTTRELL et al. 1995), but not by G_{α}-mediated signaling pathways. Similar inhibition of inositol phosphate production in response to activation of the M2 acetylcholine receptor was reported for the Btk PH domain, and was suggested to result from sequestration of $G_{\beta\gamma}$-subunits by the PH domain (TSUKUDA et al. 1994).

As a result of these observations, it has been suggested that most PH domains interact with the $\beta\gamma$-subunits of heterotrimeric G-proteins, and that PH domains might represent a key point of cross-talk between G-protein-coupled signaling and other pathways (INGLESE et al. 1995). Although this hypothesis is attractive, there are two major reasons to doubt it. The first is that $G_{\beta\gamma}$ binding by PH domains other than that of βARK appears to be very weak in the in vitro binding studies reported (TOUHARA et al. 1994; MAHADEVAN et al. 1995). The second is that the C-terminal extension beyond the PH domain is critical for the observed binding of βARK to $G_{\beta\gamma}$-subunits (KOCH et al. 1993; MAHADEVAN et al. 1995). In the study of $G_{\beta\gamma}$ binding by TOUHARA et al. (1994), all PH domain constructs included approximately 30 amino acids C-terminal to the consensus PH domain sequence, and as with βARK, this extension was essential for $G_{\beta\gamma}$ binding to the Ras-GRF PH domain construct. Furthermore, a 28-mer synthetic peptide containing 14 residues from the C-terminus of the βARK PH domain (beginning at the invariant tryptophan) plus 14 residues from beyond the PH domain C-terminus can inhibit $G_{\beta\gamma}$ binding to βARK (KOCH et al. 1993). This observation suggests that $G_{\beta\gamma}$ binding is most likely to be driven by a small epitope in the C-terminal extension, rather than being a property of the PH domain itself. Finally, as discussed below (Sect. 5.2), it now appears that βARK binding to the membrane surface actually requires both binding of this C-terminal extension to $G_{\beta\gamma}$ and binding of the PH domain itself to a phospholipid ligand (PITCHER et al. 1995).

4.2 Protein Kinase C Isoforms as PH Domain Ligands

Since several PH domain-containing proteins, including pleckstrin itself, are major substrates for protein kinase C (PKC), it is interesting that there have been a few reports that PH domains themselves can interact with PKC. The first of these used GST-PH fusion proteins to demonstrate that the Btk PH domain could bind in vitro to a mixture of the PKC α, β and γ isoforms from rat brain (YAO et al. 1994). Btk itself was found to co-immunoprecipitate with the βI isoform of PKC from mast cells, and to be phosphorylated in a manner that leads to a reduction in its activity. KONISHI et al. (1994, 1995) have also found that the PH domains from Akt/Rac/PKB protein Ser/Thr kinases can associate with several PKC isoforms, although the physiological relevance of these interactions has yet to be clearly demonstrated.

4.3 Phosphoinositides as PH Domain Ligands

Since all PH domain-containing proteins require membrane association for their function, an attractive hypothesis is that PH domains might bind directly to membrane lipids. Certainly, the electrostatically polarized domain would be well suited for association through its positively charged face with the negatively charged face of a lipid bilayer – in a manner reminiscent of that seen with the secretory phospholipase A_2's (Scott et al. 1994). Having determined the NMR structure of the N-terminal PH domain from pleckstrin, Fesik's laboratory showed that this PH domain could bind to phosphatidylinositol-4,5-bisphosphate (PtdIns (4,5) P_2) with a K_D in the range of 30 μM (Harlan et al. 1994). Analysis of NMR chemical shift changes in the PH domain upon addition of PtdIns (4,5) P_2 showed that the site of binding included the β1/β2 and β3/β4 variable loops on the positively charged face of the domain. PtdIns (4) P was found to bind similarly to the N-terminal pleckstrin PH domain, and the PH domains from Ras-GAP, Tsk, βARK and pleckstrin (C-terminal) showed similar PtdIns (4,5) P_2-binding properties. As will be discussed below for each individual case, high-affinity PtdIns (4,5) P_2 binding has since been reported for the PH domains from PLC-δ_1 (Lemmon et al. 1995; Garcia et al. 1995), and Sos (Kubiseski et al. 1997; Chen et al. 1997). In both of these cases binding is significantly tighter than seen with the pleckstrin PH domain, with K_D values in the 1 μM range. The PLC-δ_1 PH domain binds to both PtdIns (4,5) P_2 and its headgroup inositol-1,4,5-trisphosphate (Ins (1,4,5) P_3) with clear stereospecificity (Lemmon et al. 1995) that has been observed in a crystal structure of the PH domain/Ins (1,4,5) P_3 complex (Ferguson et al. 1995). PtdIns (4,5) P_2 binding to the PLC-δ_1 PH domain has been shown to be important for the ability of the enzyme to hydrolyze its substrate processively (Rebecchi et al. 1992; Kanematsu et al. 1992; Cifuentes et al. 1993). High affinity binding to phosphoinositides and their headgroups has also been seen for the PH domain from Akt/PKB/Rac. In this case the PH domain specifically recognizes the phosphoinositide 3-kinase (PI 3-kinase) products PtdIns (3,4) P_2 and PtdIns (3,4,5) P_3, as well as their inositol phosphate headgroups (Klippel et al. 1997; James et al. 1996; Franke et al. 1997; Frech et al. 1997). As discussed below, Akt/PKB/Rac binding to PtdIns (3,4) P_2 appears to be required for its activation in response to cellular stimulation by insulin or PDGF, and the case for physiologically important PH domain recognition of a phosphoinositide appears very convincing here. In addition to these examples, several PH domains have been reported to bind to phosphoinositides and their headgroups with lower affinities and varying degrees of specificity. These include the PH domains from spectrin (Hyvönen et al. 1995; Wang and Shaw 1995), dynamin (Salim et al. 1996; J. Zheng et al. 1996), Btk (Salim et al. 1996; Fukuda et al. 1996) and Ras-GAP1 (Fukuda and Mikoshiba 1996). Phosphoinositide binding has not been shown to be physiologically relevant for all of these PH domains, and in some cases, as discussed in Sect. 5, the contrary can be argued. Nonetheless, the volume of data concerning PH domain/phosphoinositide interactions strongly suggests that it is important in at least some of these cases, although the precise ligands (which may be related to PtdIns (4,5) P_2) may not yet have been identified.

5 Studies of PH Domain Function and its Correlation with Ligand Binding

As discussed in Sect. 4, PH domains have variously been proposed to bind to $G_{\beta\gamma}$-subunits, PKC isoforms and phosphoinositides. It is very unlikely that any of these three possibilities will provide a uniform explanation for PH domain function. It is possible that some PH domains participate in binding to more than one of these ligands (and that some will bind to none). In this section we will consider the data available regarding PH domain function in several different proteins, and, where possible, consider what is (are) likely to be the relevant PH domain ligand(s). Since the functions of the proteins are quite different, and since the extent to which their PH domains have been studied varies, each protein will be discussed separately.

5.1 The Btk PH Domain

Mutations in the PH domain of Bruton's agammablobulinemia kinase (Btk) provided the first evidence for the functional importance of PH domains in vivo. The gene encoding Btk is mutated in human X-linked agammaglobulinemia (XLA), a disease that results from defective B-cell differentiation leading to abnormally low levels of immunoglobulin production (Mattsson et al. 1996). Several XLA-linked Btk mutations occur in the N-terminal PH domain of the kinase, and similar mutations in the PH domain of the mouse Btk homologue cause the milder disease of X-linked immunodeficiency or Xid (Thomas et al. 1993; Vetrie et al. 1993). These mutations clearly interfere with Btk's role in B-cell development, yet do not inhibit its intrinsic kinase activity (as assessed by in vitro autophosphorylation assays), arguing that the PH domain participates in interactions of Btk that are important for its signaling function. Btk is one of a family of related tyrosine kinases that includes Tec (Mano et al. 1990) and Itk (Siliciano et al. 1992), and shows some similarity to the Src family of tyrosine kinases. Like the Src kinases, Btk and other Tec family tyrosine kinases contain an SH3 domain followed by an SH2 domain and the tyrosine kinase domain. Unlike the Src kinases, however, which have a myristoylation sequence N-terminal to the SH3 domain, Btk instead has an N-terminal PH domain followed by a so-called Tec homology (TH) region. As with the comparison between RK and βARK, this observation suggests a membrane association role for the PH and TH regions of Btk.

Using sequence alignments and a model of the Btk PH domain structure (based upon the dynamin PH domain), it has been argued that several XLA-linked mutations in the Btk PH domain cluster on a face of the domain opposite from that containing the C-terminal α-helix, but which includes the three variable loops of the PH domain (Fig. 4) and corresponds to the Ins (1,4,5) P_3/PtdIns (4,5) P_2-binding site of the PLC-δ_1 and pleckstrin PH domains (Vihinen et al. 1995; Mattsson et al. 1996). As will be discussed specifically below, certain of these mutations (notably at F25 and R28 of Btk) are at positions equivalent to those of residues in the PLC-δ_1

PH domain that directly contact Ins (1,4,5) P_3 (Fig. 5; FERGUSON et al. 1995). It has recently been reported that the Btk PH domain binds to both PtdIns (3,4,5) P_3 (SALIM et al. 1996) and Ins (1,3,4,5) P_4 (FUKUDA et al. 1996). The XLA-associated mutation R28Y was found to inhibit the ability of the Btk PH domain to bind to PtdIns (3,4,5) P_3, although the wild-type affinity was not reported (SALIM et al. 1996). Similarly, XLA and Xid-associated mutations were found to reduce the ability of the Btk PH domain to bind to Ins (1,3,4,5) P_4 – the wild-type giving an IC_{50} of 40 n*M* in competition assays (FUKUDA et al. 1996). Thus, these data provide some evidence in support of PtdIns (3,4,5) P_3 as a potentially relevant ligand for the Btk PH domain, although, to our knowledge, there are currently no other data clearly linking Btk stimulation to PI 3-kinase activity.

In addition to the reports of phosphoinositide binding, the Btk PH domain has been reported to interact with PKC isoforms (YAO et al. 1994), $G_{\beta\gamma}$-subunits (TSUKUDA et al. 1994) and a recently cloned 135-kDa protein named BAP-135 (YANG and DESIDERIO 1997). Btk can be co-immunoprecipitated with PKC βI from mast cells, and phosphorylation of Btk by PKC downregulates its activity in vitro. YAO et al. (1994) showed that a GST-fusion protein of the Btk PH domain could precipitate various PKC isoforms from cell lysates, and the Xid-associated R28C mutation was reported to reduce the affinity of the PH domain for PKC by three- to fivefold. Similar studies have been reported for Btk PH domain binding to $G_{\beta\gamma}$-subunits, but in this case the effect of XLA- or Xid-associated mutations on the interaction was not investigated (TSUKUDA et al. 1994). Certain $G_{\beta\gamma}$-subunits do appear to be capable of activating Btk and Tsk (a member of the same family) in cotransfection assays (LANGHANS-RAJASEKARAN et al. 1995), although analysis of Tsk activation in vitro showed that some additional unidentified plasma membrane component, possibly a phosphoinositide, was required together with $G_{\beta\gamma}$-subunits for maximum kinase activation. Another potential binding partner of Btk, named BAP-135, was recently cloned by YANG and DESIDERIO (1997) on the basis of its co-immunoprecipitation with Btk from a B-lymphoid cell line. BAP-135 appears to be phosphorylated by Btk, and analysis of the domains in Btk responsible for BAP-135 association showed that GST-fusion proteins containing the PH domain, or the PH plus TH domains, could precipitate BAP-135. While some mutations in Btk lead to a reduced level of BAP-135 phosphorylation, PH domain mutations that are associated with Xid had only a small effect (YANG and DESIDERIO 1997). Finally, LI et al. (1995) isolated a gain-of-function mutation in the Btk PH domain using a random mutagenesis approach. Mutation of Glu 41 in the Btk PH domain (predicted to be in variable loop 2; Fig. 5) to lysine resulted in both increased tyrosine phosphorylation and membrane association of Btk, and the mutated protein was able to transform NIH 3T3 cells.

As will be clear from the above description, there is no clear consensus regarding the binding partners of the Btk PH domain. In no single case is the evidence for a physiologically relevant interaction wholly convincing. We are thus left with two main possible explanations. One is that none of the potential ligands discussed above are relevant, and the true ligand has yet to be found. The other is that the Btk PH domain can interact with several of the possible ligands discussed above, and that the lack of conviction from the data so far reflects a context dependence of each reported interaction for which appropriate controls have not yet been performed.

5.2 The βARK PH Domain: Coordinate Binding to $G_{\beta\gamma}$ and Phosphoinositides

As was discussed in some detail in Sect. 4.1, the C-terminal portion of βARK can interact with $G_{\beta\gamma}$-subunits, and some experiments have suggested this to be a property of several different PH domains. The main difficulty with the view that $G_{\beta\gamma}$ binding might represent a relatively general function of PH domains stems from the finding that only the very C-terminus of the βARK PH domain and residues beyond its C-terminus are required for this interaction (KOCH et al. 1993). This view would argue that the entire β-sandwich region of the PH domain is dispensable for its function. It was recently found that $G_{\beta\gamma}$-subunits alone are not sufficient for recruitment of βARK to the membrane surface (PITCHER et al. 1995). While $G_{\beta\gamma}$-subunits present in crude lipid preparations can direct membrane association of βARK, they could not when reconstituted into pure phosphatidylcholine (PC) vesicles. However, if PtdIns (4,5) P_2 was added to the PC vesicles, the ability of $G_{\beta\gamma}$ to recruit βARK was restored. These studies argue that the C-terminus of βARK binds simultaneously to both PtdIns (4,5) P_2 (or another similar lipid) and $G_{\beta\gamma}$. The PH domain itself is likely to be responsible for PtdIns (4,5) P_2 binding, while the C-terminal extension is likely responsible for $G_{\beta\gamma}$ binding. Neither of the individual ligands alone is sufficient for membrane binding of βARK (PITCHER et al. 1995). A similar combination of ligands may explain the observations mentioned above for $G_{\beta\gamma}$-activation of the Tsk kinase, where an additional unidentified membrane component was required in addition to $G_{\beta\gamma}$ to reconstitute the activating effect in vitro (LANGHANS-RAJASEKARAN et al. 1995). These findings underscore the importance of studying weak PH domain interactions in the appropriate context, and indicate that at least some PH domains may be multifunctional with respect to ligand binding.

5.3 The PH Domain in Membrane Association of β_G-Spectrin

In what is becoming a common theme for PH domains, the PH domain of β_G-spectrin has also been reported to bind to both $G_{\beta\gamma}$-subunits (TOUHARA et al. 1994; WANG et al. 1994) and to Ins (1,4,5) P_3, the headgroup of PtdIns (4,5) P_2 (HYVÖNEN et al. 1995; WANG and SHAW 1995). β_G-Spectrin contains two domains that contribute to its direct association with the plasma membrane; one at the N-terminus of the protein, and one at the C-terminus. Both domains can inhibit binding of intact spectrin to peripheral protein-stripped bovine brain membranes, with IC_{50} values in the submicromolar range, and each of these domains appears to bind to distinct sites on the membrane (DAVIS and BENNETT 1994; LOMBARDO et al.

1994). The C-terminal membrane association domain contains a PH domain, disruption of which greatly impairs membrane binding (LOMBARDO et al. 1994). WANG et al. (1996) have shown that a fusion protein of the β_G-spectrin PH domain with green fluorescent protein (GFP) localizes to the plasma membrane when expressed in COS7 cells. They have also reported that the 109 amino acid β_G-spectrin PH domain binds strongly to bovine brain membranes (WANG and SHAW 1995). In this study it was reported that $G_{\beta\gamma}$ or other proteins are not required for membrane association, but that Ins (1,4,5) P_3 can inhibit the interaction. Binding of Ins (1,4,5) P_3 itself to the β_G-spectrin PH domain was shown by HYVÖNEN et al. (1995) to occur with a $K_D \geq 40$ μM, indicating an affinity for this molecule that is somewhat lower than suggested for PH domain binding to bovine brain membranes. Using a circular dichroism (CD)-based assay, only inositol phosphates with phosphates at the 4- and 5-positions were found to alter the spectrum (Ins (1,4,5) P_3, Ins (1,3,4,5) P_4 and 1-glycerophosphoryl-Ins (4,5) P_2). An X-ray crystal structure of the spectrin PH/Ins (1,4,5) P_3 complex showed that Ins (1,4,5) P_3 lies on the surface of the domain (Fig. 3), making hydrogen bonds with a total of five residues from the $\beta1/\beta2$ loop (four residues) and the $\beta5/\beta6$ loop (one residue) (HYVÖNEN et al. 1995). This places the Ins (1,4,5) P_3 molecule in the center of the face of this PH domain with positive electrostatic potential (Fig. 3). Most of the interactions are made by the 4- and 5-phosphates of Ins (1,4,5) P_3, with the 1-phosphate completely solvent-exposed as would be required for the PH domain to bind to this head group when presented by PtdIns (4,5) P_2 in a membrane bilayer. The Ins (1,4,5) P_3 binding affinity of the PH domain does not appear to be sufficient to account for binding of spectrin's C-terminal membrane association domain to stripped bovine brain membranes. However, it is possible that additional delocalized electrostatic attraction between the positively charged face of the PH domain and the membrane surface could cooperate with PtdIns (4,5) P_2 binding to increase the membrane binding affinity. Nonetheless, the physiological relevance of phosphoinositide binding to spectrin has not been demonstrated, and it remains possible that a distinct, as yet unidentified, ligand (or combination of ligands) is responsible for the interaction.

5.4 PH Domains in Intracellular Targeting of Guanine Nucleotide Exchange Factors (GNEFs)

One of the few statements that can be made regarding the positioning of PH domains in their host molecules is that the Dbl homology region in GNEFs for small G-proteins is always immediately followed by a PH domain (MUSACCHIO et al. 1993). The Dbl homology region itself is thought to promote guanine nucleotide exchange on Rho/Rac-like small G-proteins, resulting in their activation (CERIONE and ZHENG 1996). The PH domain of oncogenic Dbl is required for its ability to transform NIH 3T3 cells, although in vitro assays show it is not required for GNEF activity (Y. ZHENG et al. 1996). Rather, the PH domain appears to play a role in subcellular targeting of the Dbl oncogene product. Oncogenic Dbl is found in a

Triton-insoluble cell fraction, and this localization is dependent on the PH domain, which is found in a Triton-insoluble fraction when expressed alone. Co-overexpression of the Dbl PH domain inhibits the ability of oncogenic Dbl to transform cells, while overexpression of the Vav PH domain has no such effect, suggesting subcellular targeting of Dbl by its PH domain to a specific, as yet unidentified, cytoskeletal component (Y. ZHENG et al. 1996). The PH domain immediately following the Dbl homology region of both the *Ost* and *Lfc* oncogenes has also been shown to be required for the ability of these proteins to transform cells (HORII et al. 1994; WHITEHEAD et al. 1995). In the case of oncogenic Lfc, the transforming ability of a PH domain-deletion mutant can be restored by the addition of a CAAX box that promotes membrane localization of the protein through farnesylation (WHITEHEAD et al. 1995). This observation again implicates the PH domain in membrane localization of the GNEF. Since farnesylation of the PH domain-deleted oncogenic Dbl does not restore its transforming ability (Y. ZHENG et al. 1996), it can be argued that the PH domains of Dbl and Lfc are likely to target their host molecules to different specific locations in the cell. Their respective ligands are not yet clear.

Ras-GRF and Sos contain both a Ras GNEF domain (towards their C-termini) and a Dbl homology region that is followed immediately by a PH domain. Ras-GRF also has a second PH domain at its amino terminus, which was reported to bind to $G_{\beta\gamma}$-subunits in vitro (TOUHARA et al. 1994), and has been found to participate in membrane targeting of the molecule that is required for its function (BUCHSBAUM et al. 1996). In the case of Sos and its mammalian homologues, both the Dbl homology region and the PH domain appear to be required for Ras activation (MCCOLLAM et al. 1995). Sos proteins are thought to be recruited to the membrane in receptor tyrosine kinase signaling by the adaptor protein Grb2/Sem5/Drk, which binds through its SH2 domain to the activated receptor, and through its two SH3 domains to the C-terminal proline-rich tail of Sos (SCHLESSINGER 1994). Artificial membrane targeting of hSos can overcome the requirement for receptor activation (ARONHEIM et al. 1994). Moreover, overexpressed Sos mutants from which the C-terminal proline-rich tail has been deleted can activate Ras without the involvement of Grb2, but with a requirement for an intact N-terminal portion containing the Dbl homology region and PH domain (MCCOLLAM et al. 1995; KARLOVICH et al. 1995; WANG et al. 1995). CHEN et al. (1997) found that Ras activation by an overexpressed, C-terminal deleted, hSos1 is serum-dependent, indicating that the Dbl homology region and/or PH domain of Sos may also participate in its serum-dependent recruitment to the membrane. The isolated PH domain of Sos was reported to act as a weak dominant negative in the *Drosophila* Sevenless signaling pathway (KARLOVICH et al. 1995), and overexpression of the hSos1 PH domain in COS cells inhibits serum-dependent Ras activation (CHEN et al. 1997). Moreover, using indirect immunofluorescence microscopy, CHEN et al. (1997) found that the isolated Sos PH domain localizes to the plasma membrane in a serum-dependent manner when microinjected into rat embryo fibroblasts or COS cells. The membrane target responsible for this signal-dependent localization is not clear. However, KARLOVICH et al. (1995) have reported that a GST fusion protein of the Sos PH domain can precipitate Sevenless, although no evidence for a direct PH/Sevenless association was presented. PtdIns (4,5) P_2 has also been shown to bind to the mammalian Sos PH domain with relatively high-affinity (K_D = 1.1–1.8 μM) (KUBISESKI et al. 1997; CHEN et al. 1997), and clear selectivity for PtdIns (4,5) P_2 over PtdIns (4) P or PtdIns (3,4,5) P_3 was seen (KUBISESKI et al. 1997). A signal-dependent PtdIns (4,5) P_2-mediated membrane recruitment event would seem unlikely. Indeed, a Sos PH domain mutant that binds only very weakly to PtdIns (4,5) P_2 still shows serum-dependent plasma membrane localization (CHEN et al. 1997), casting significant doubt on the likelihood that PtdIns (4,5) P_2 is a physiologically relevant ligand for the Sos PH domain. Finally, it has been reported that the isolated PH domain from hSos1 induces germinal vesicle breakdown (GVBD) when injected into *Xenopus* oocytes (FONT DE MORA et al. 1996). The mechanism of this effect is far from clear: rather than acting in a dominant negative manner as discussed above, it was reported in these experiments that the isolated hSos1 PH domain actually synergizes with Ras when co-injected into oocytes. The PH domains from IRS-1, βARK or PLC-δ_1 had no effect upon GVBD when injected into oocytes, indicating differences in their specificity. Since the PLC-δ_1 PH domain binds to PtdIns (4.5) P_2 with high affinity, yet has no effect on GVBD, this observation lends further support to the existence of a distinct ligand for the Sos PH domain.

5.5 Participation of the IRS-1 PH Domain in Recruitment to the Activated Insulin Receptor

The insulin receptor substrate-1 (IRS-1) has a PH domain at its N-terminus, immediately followed by a phosphotyrosine binding (PTB) domain (see Chap. 2, this volume). Deletion of the PH domain from IRS-1 significantly reduces its insulin-stimulated tyrosine phosphorylation compared with wild-type IRS-1 (MYERS et al. 1995; VOLIOVITCH et al. 1995). In 32D cells, which lack endogenous IRS proteins, only wild-type and not the PH domain-deleted form of IRS-1 could confer insulin-dependent stimulation of PI 3-kinase activity (MYERS et al. 1995). These results indicate that the PH domain of IRS-1 is important for its interaction with the activated insulin receptor. However, the PH domain is not essential for this coupling, since the effect of its deletion is masked when the insulin receptor is expressed at high levels in 32D cells (MYERS et al. 1995). From the results of more detailed studies, YENUSH et al. (1996) argue that the PH domain is essential for insulin-dependent IRS-1 phosphorylation when insulin receptor levels are low, and that under these conditions the PTB domain is dispensable but increases the sensitivity of insulin receptor signaling. When the insulin receptor is expressed at high levels, either the PTB or PH domain is sufficient for insulin-dependent IRS-1 phosphorylation. It thus appears that the two domains are likely to cooperate under normal conditions in recruitment of IRS-1 to the activated insulin receptor. Since the PTB domain of IRS-1 binds with high affinity to phosphorylated Y960 of the activated insulin receptor (ECK et al. 1996), a ligand with similar high affinity might be anticipated for the PH domain based upon the observations of YENUSH et al. (1996). This proposed ligand does not appear to be present in the insulin receptor itself, and its identity is not at all clear.

5.6 The PH Domain of Dynamin-1 in Rapid Endocytosis

A specific role for the PH domain of the GTPase dynamin-1 in regulation of endocytosis has recently been demonstrated (ARTALEJO et al. 1997). Introduction of the dynamin-1 PH domain into adrenal chromaffin cells specifically inhibits a form of rapid membrane retrieval that follows the stimulated secretion of catecholamines by these cells and is termed rapid endocytosis. Rapid endocytosis (RE) can be inhibited by antibodies against dynamin and by nonhydrolyzable GTP analogues, but not by antibodies against clathrin (ARTALEJO et al. 1995). In studying the role of the PH domain, it was found that only the dynamin-1 PH domain, and not those from PLC-δ_1 or pleckstrin, could inhibit RE. Since the PLC-δ_1 and pleckstrin (N-terminal) PH domains both bind to PtdIns (4,5) P_2, this finding suggests that this phosphoinositide is not the relevant ligand for the dynamin-1 PH domain. That there is a highly specific ligand was suggested by the finding that the PH domain from the dynamin-2 isoform (COOK et al. 1994; SONTAG et al. 1994), despite being 81% identical in amino acid sequence to the dynamin-1 PH domain, had no effect upon RE. Mutagenesis studies showed that residues in the β1/β2 loop of the dynamin-1 PH domain were critical for its ability to inhibit RE. By altering the dynamin-2 PH domain to resemble that from dynamin-1 at just four positions in the β1/β2 and β6/β7 loops, a gain-of-function mutant was generated (ARTALEJO et al. 1997).

The nature of the ligand to which the dynamin-1 PH domain binds is not clear. However, it does not bind to $G_{\beta\gamma}$-subunits (FERGUSON et al. 1994), and reports of its binding to phosphoinositides have been quite varied. The findings of ARTALEJO et al. (1997) argue against PtdIns (4,5) P_2 as the relevant ligand. Nonetheless, while LEMMON et al. (1995) failed to detect binding of the dynamin-1 PH domain to PtdIns (4,5) P_2 in vesicles using calorimetric and equilibrium gel filtration approaches, SALIM et al. (1996) reported its selective binding to PtdIns (4,5) P_2 and PtdIns (4) P, but not PtdIns (3,4,5) P_3, using surface plasmon resonance (SPR). No indication of the strength of this binding was given. J. ZHENG et al. (1996) have reported binding of phosphoinositides to the dynamin-1 PH domain when added from an SDS solution, and measured apparent dissociation constants in the micromolar range. Binding of the dynamin-1 PH domain to Ins (1,4,5) P_3 or GPtdIns (4,5) P_2 has been studied by NMR, with the finding that the K_D is in the 1.23 μM (SALIM et al. 1996) to 4.3 mM (J. ZHENG et al. 1996) range. This very weak interaction involves the positively charged face of the PH domain, but the residues that appear to be critical for its ability to inhibit RE were not directly implicated. Since the dynamin-1 PH domain could inhibit RE when added from a 1 μM solution (ARTALEJO et al. 1997), it is expected that the affinity for its ligand will be significantly higher than those measured to date for phosphoinositides or inositol phosphates.

5.7 Specific PH Domain Binding to PtdIns (4,5) P_2: The PLC-δ_1 PH Domain

One class of PH domain-containing proteins for which membrane localization is clearly critical for function is the phospholipase C isoforms, each of which has a PH domain at its N-terminus (PARKER et al. 1994). Little is currently known about the PH domains from PLC-β or PLC-γ, but the PLC-δ_1 PH domain provides the best characterized example of PH domain recognition to date. Intact PLC-δ_1 binds with high affinity to PtdIns (4,5) P_2-containing membranes (REBECCHI et al. 1992), and proteolytic removal of the amino-terminal 60 amino acids from PLC-δ_1 abolishes this binding. The resulting large proteolytic fragment retains catalytic activity, but has a reduced capacity for processive PtdIns (4,5) P_2 hydrolysis, suggesting the presence of a noncatalytic PtdIns (4,5) P_2 binding site in the N-terminus of the enzyme (CIFUENTES et al. 1993). Ins (1,4,5) P_3, the headgroup product of PtdIns (4,5) P_2 hydrolysis by PLC, also binds to intact PLC-δ_1 (KANEMATSU et al. 1992), and inhibits its high-affinity binding to PtdIns (4,5) P_2-containing vesicles (CIFUENTES et al. 1994). PLC-δ_1 was, in fact, cloned in a search for cytoplasmic Ins (1,4,5) P_3 binding proteins, and Ins (1,4,5) P_3 inhibits PLC-δ_1 activity in an apparently noncompetitive fashion (KANEMATSU et al. 1992). Removal of the amino-terminal portion of PLC-δ_1 abolishes both high-affinity binding of the enzyme to PtdIns (4,5) P_2 and the inhibitory effect of Ins (1,4,5) P_3, indicating that PtdIns (4,5) P_2 and Ins (1,4,5) P_3 bind to the same region (YAGISAWA et al. 1994; CIFUENTES et al. 1994).

Studies employing immunofluorescence microscopy (PATERSON et al. 1995) have shown that the PH domain is both necessary and sufficient for association of PLC-δ_1 with the plasma membrane. In this case, by contrast with the observations with the Sos PH domain, membrane association is not serum-dependent (Falasca and Schlessinger, unpublished). A bacterially expressed recombinant PLC-δ_1 PH domain binds in vitro to PtdIns (4,5) P_2 in lipid vesicles with a K_D of approximately 1.7 μM and 1:1 stoichiometry (GARCIA et al. 1995; LEMMON et al. 1995). Ins (1,4,5) P_3 competes directly for this binding, and itself binds to the PH domain in a specific 1:1 interaction with a K_D of 210 nM (LEMMON et al. 1995). The specificity of inositol phosphate recognition by the isolated PH domain (LEMMON et al. 1995) closely resembles that seen for the whole protein (KANEMATSU et al. 1992; YAGISAWA et al. 1994). Phosphates at the 4- and 5-positions of the D-*myo* isomer are necessary, but not sufficient, for binding. An additional phosphate group at the 1-position is required for the highest affinity interaction. Removal, addition, or rearrangement of any phosphate groups in Ins (1,4,5) P_3 leads to a reduction in the affinity of at least 15-fold. The X-ray crystal structure of the complex between the PLC-δ_1 PH domain and Ins (1,4,5) P_3 (FERGUSON et al. 1995) provides a view of the basis for this specificity (Figs. 3, 7). The residues in the PLC-δ_1 PH domain that interact with Ins (1,4,5) P_3 are mostly in the β1/β2 and β3/β4 variable loops, placing Ins (1,4,5) P_3 in the center of the positively charged face of the PH domain. Two lysine side-chains (K30 and K57) 'clamp' the 4- and 5-phosphates of Ins (1,4,5) P_3 in the binding pocket such that they are buried beneath the surface, and each lysine forms hydrogen bonds with both phosphates. Among

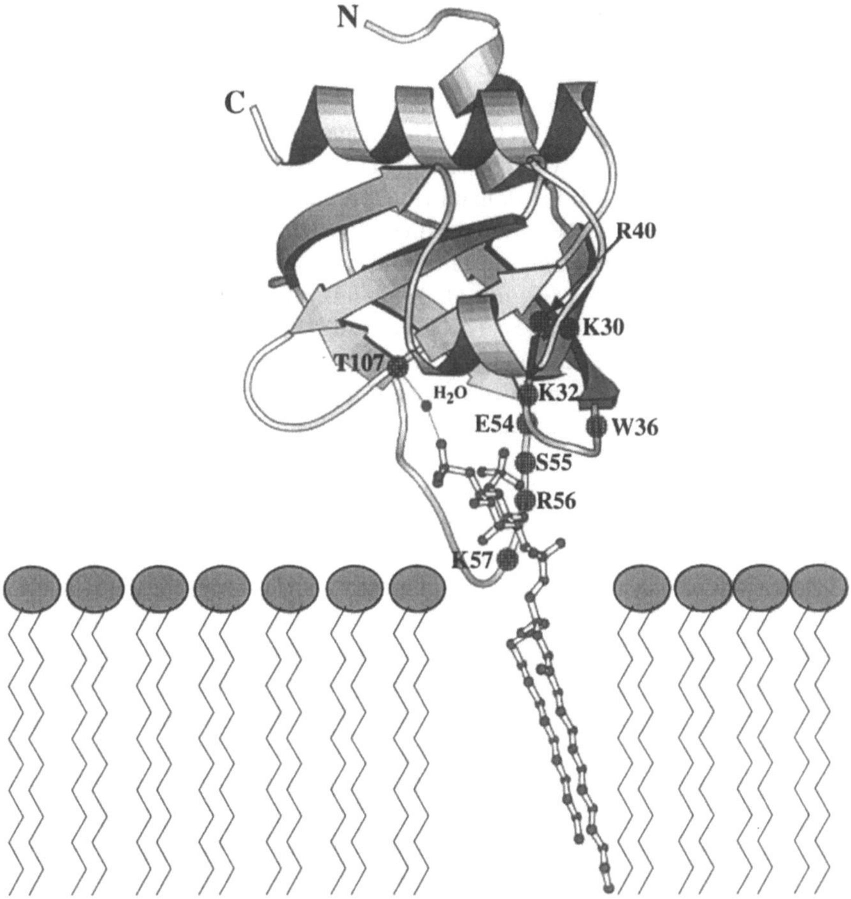

Fig. 7. Hypothetical view of how the PLC-δ_1 PH domain may associate with a membrane containing PtdIns (4,5) P_2. The PLC-δ_1 PH domain is shown in a MOLSCRIPT representation (KRAULIS 1991) with the positions of the nine amino acids that form hydrogen bonds to Ins (1,4,5) P_3 represented by dots (all H-bonds are direct except those from the backbone carbonyls of E54 and T107, which are H_2O-mediated). As discussed in the text, the side-chains K57 and K30 'clamp' the 4- and 5-phosphates into the binding site, and the W36 side-chain is hydrogen-bonded to the 1-phosphate. To generate this figure, diacylglycerol has been attached to the 1-phosphate of the Ins (1,4,5) P_3 observed in the crystal structure of the PLC-δ_1/Ins (1,4,5) P_3 complex (FERGUSON et al. 1995) to give an impression of how the PH domain might bind PtdIns (4,5) P_2. The modeled PtdIns (4,5) P_2 is placed in a highly schematized bilayer. The three variable loops of the PH domain abut the membrane surface, and the face of the domain with positive electrostatic potential (see Fig. 6) associates with the membrane surface. Membrane association in this case may arise from a combination of specific recognition of the PtdIns (4,5) P_2 headgroup and a delocalized electrostatic attraction for the membrane surface

the additional six amino acids that participate in hydrogen bonding to the 4- and 5-phosphates are R40, which is analogous to R28 in the Btk PH domain at which XLA-associated mutations have been found (Fig. 5). R40 of the PLC-δ_1 PH domain forms two hydrogen bonds to the 5-phosphate of Ins (1,4,5) P_3. The backbone carbonyl of E54 in the PLC-δ_1 PH domain forms a water-mediated hydrogen bond with the 5-phosphate of Ins (1,4,5) P_3, and aligns with the position in the Btk PH domain (E41) at which mutation to lysine leads to a constitutively active form of Btk (LI et al. 1995). The 1-phosphate of Ins (1,4,5) P_3 is hydrogen bonded to the indole nitrogen of W36 in the PLC-δ_1 PH domain, and the inositol chair is in van der Waals contact with the W36 side-chain (FERGUSON et al. 1995). W36 of PLC-δ_1 aligns with F25 in the Btk PH domain (Fig. 5), another position at which XLA-associated mutations have been found (MATTSSON et al. 1996). The 1-phosphate of Ins (1,4,5) P_3 is solvent exposed, and can readily accommodate a glycerol moiety, as is evidenced by the fact that GPtdIns (4,5) P_2 binds with high affinity to the PLC-δ_1 PH domain (LEMMON et al. 1995). This in turn suggests a relationship between the PH domain and the membrane when bound to PtdIns (4,5) P_2, as is depicted schematically in Fig. 7. The positively charged surface of the PH domain abuts the membrane surface, specific interactions with the inositol phosphate headgroup and occur, in the center of the interface.

As mentioned above, the PH domain of PLC-δ_1 represents a specific noncatalytic binding site for the substrate of the enzyme. By tethering the enzyme to membranes that contain its substrate, the PH domain of PLC-δ_1 would permit processive hydrolysis without the requirement to dissociate from the membrane surface between catalytic cycles. A comparison of the effect of increasing the mole fraction of PtdIns (4,5) P_2 in vesicles upon its hydrolysis by intact and N-terminally proteolyzed PLC-δ_1 showed that removal of the PH domain abolishes processive catalysis by the enzyme (CIFUENTES et al. 1993). The ability of PLC-δ_1 to hydrolyze PtdIns (4) P is also markedly increased by the presence of PtdIns (4,5) P_2 in the same vesicles (LOMASNEY et al. 1996). When mutations are made in the PH domain of intact PLC-δ_1 that reduce its affinity for PtdIns (4,5) P_2, this effect is lost. Furthermore, Ins (1,4,5) P_3 inhibits this stimulatory effect of PtdIns (4,5) P_2 with an IC_{50} of less than 1 μM (LOMASNEY et al. 1996). These observations together argue that the PLC-δ_1 PH domain acts as a membrane tether that allows a processive or "scooting" mode of catalysis (RAMIREZ and JAIN 1991). The recent determination of the X-ray crystal structure of the catalytic domain of PLC-δ_1 and studies of crystals of the intact protein (ESSEN et al. 1996) provide strong evidence for a flexible connection between the PH domain and the remainder of the enzyme, as would be predicted for such a tether. A hypothetical view of how the PH domain may enhance PLC-δ_1 activity and how Ins (1,4,5) P_3 can prevent this is depicted schematically in Fig. 8.

In addition to providing an explanation for the ability of Ins (1,4,5) P_3 to inhibit PLC-δ_1 activity in vitro, these studies also provide a clear mechanism via which regulated PH domain-mediated membrane association is possible – through recognition of both a phosphoinositide and its soluble headgroup. When this soluble headgroup is a second messenger, as for Ins (1,4,5) P_3 and Ins (1,3,4,5) P_4, it might be speculated that inositol phosphate accumulation will lead to dissociation of the PH domain from the membrane surface when PtdIns (4,5) P_2 or PtdIns (3,4,5) P_3 is the membrane ligand. While this idea is attractive, the in vitro studies with PLC-δ_1 provide the only current evidence for this mechanism. In other cases,

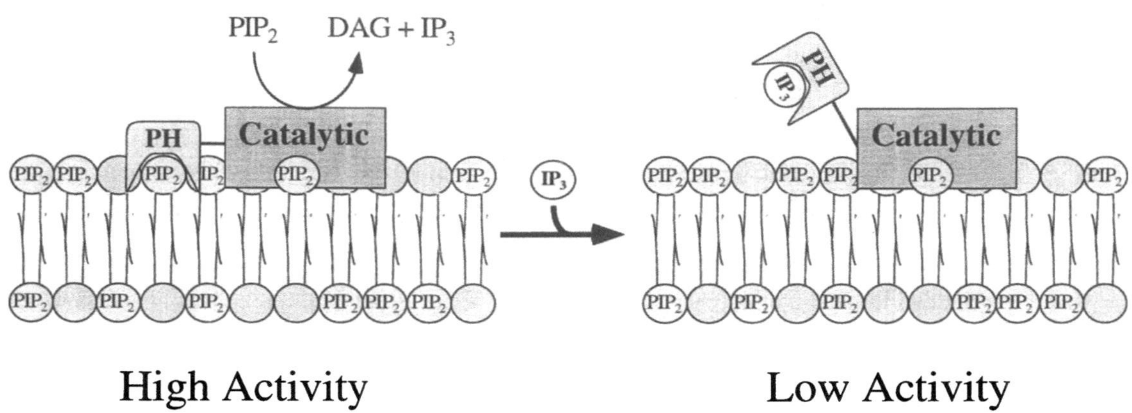

Fig. 8. A schematic view of how PtdIns (4,5) P_2 and Ins (1,4,5) P_3 binding by the PLC-δ_1 PH domain can affect enzyme activity. The PH domain tethers PLC-δ_1 to its substrate (PIP_2) in the left-hand part of the figure (high activity), allowing processive PIP_2 hydrolysis by the enzyme without dissociation from the membrane between catalytic cycles. If sufficient IP_3 (a product of the enzyme) accumulates, it will compete for binding to the PH domain, and thus remove the membrane tether. With IP_3 bound to its PH domain, the processive nature of PIP_2 hydrolysis will be lost, as is also observed when the PH domain is removed (CIFUENTES et al. 1993). Inhibition of PLC-δ_1 activity by IP_3 has been observed in vitro (KANEMATSU et al. 1992)

notably for PtdIns (4,5) P_2 binding by the N-terminal pleckstrin PH domain (HARLAN et al. 1994) and the Sos PH domain (KUBISESKI et al. 1997), rather high, nonphysiological, concentrations of Ins (1,4,5) P_3 are required to dissociate the PH domains from vesicles.

5.8 Recruitment of PH Domain-Containing Proteins to the Membrane by Interaction with 3-Phosphorylated Phosphoinositides

In addition to the PLC-δ_1 PH domain, two other examples have recently been identified in which PH domains bind to phosphoinositides with high affinity and specificity. One of these is the Ser/Thr kinase Akt (also known as PKB and RAC) (HEMMINGS 1997). The other is a molecule called Grp1 that also contains a region of homology with the yeast SEC7 gene product, and may be involved in regulation of cell adhesion (KOLANUS et al. 1996; KLARLUND et al. 1997). In both cases the PH domain specifically recognizes 3-phosphorylated phosphoinositides, significant levels of which are seen only in the membranes of cells activated with growth factors and other ligands (CARPENTER and CANTLEY 1996). We will discuss the findings for each of these proteins separately.

5.8.1 Akt Activation by PH Domain Binding to PtdIns (3,4) P_2

Akt is a Ser/Thr kinase that has a PH domain at its amino-terminus, and is activated upon stimulation of cells with PDGF (FRANKE et al. 1995), insulin (KOHN et al. 1995; CROSS et al. 1995) and other growth factors (BURGERING and COFFER 1995). A variety of studies show that Akt is a downstream target of PI 3-kinase, and its growth factor-dependent stimulation can be inhibited by a dominant-negative PI 3-kinase mutant or PI 3-kinase inhibitors (FRANKE et al. 1995; BURGERING and COFFER 1995; KOHN et al. 1995; CROSS et al. 1995; ANDJELKOVIC et al. 1996). Furthermore, expression of a membrane targeted, constitutively active, form of PI 3-kinase can activate Akt in the absence of growth factor stimulation (KLIPPEL et al. 1996; DIDICHENKO et al. 1996). Downstream targets of Akt include glycogen synthase kinase-3 (GSK-3), which is inactivated upon phosphorylation (CROSS et al. 1995), and possibly the p70 S6-kinase (BURGERING and COFFER 1995). Recent studies have shown that Akt plays a key role in the regulation of neuronal survival (DUDEK et al. 1997).

The PH domain of Akt appears to be responsible, at least in part, for recruitment to the plasma membrane where Akt is activated. Akt activation in response to growth factor stimulation or by expression of a constitutively active PI 3-kinase is significantly reduced when deletions or certain mutations are made in its PH domain (FRANKE et al. 1995, 1997; KLIPPEL et al. 1996; ANDJELKOVIC et al. 1996). It has further been shown that both intact Akt (JAMES et al. 1996) and its isolated PH domain (FRECH et al. 1997; FRANKE et al. 1997) bind with high affinity to PtdIns (3,4,5) P_3 and PtdIns (3,4) P_2, the major products of PI 3-kinase. The relative affinities of PtdIns (3,4) P_2 and PtdIns (3,4,5) P_3 differ between the reports, but it is clear that the 3-phosphorylated phosphoinositides are bound more tightly than PtdIns (4,5) P_2. FRECH et al. (1997) have also shown that the Ins (1,3,4,5) P_4 headgroup is recognized with high affinity, although it is not clearly preferred over Ins (1,4,5) P_3. As well as binding to the PH domain, PtdIns (3,4) P_2 directly activates Akt in vitro (KLIPPEL et al. 1997; FRANKE et al. 1997; FRECH et al. 1997), while PtdIns (4,5) P_2 has no such effect, and PtdIns (3,4,5) P_3 was actually found to be inhibitory (FRECH et al. 1997; FRANKE et al. 1997). Clear evidence for the specificity of this effect was provided by KLIPPEL et al. (1997), who showed that vesicles containing PtdIns (3,4,5) P_3 could activate Akt only after their treatment with an inositol 5′ phosphatase to produce PtdIns (3,4) P_2. Akt with its PH domain deleted or mutated could not be activated by PtdIns (3,4) P_2 or inhibited by PtdIns (3,4,5) P_3 in these studies (KLIPPEL et al. 1997; FRECH et al. 1997; FRANKE et al. 1997). As a caution, it should be noted that in one report Akt stimulation by PtdIns (3,4) P_2 was not detected, although high affinity binding of this phosphoinositide to Akt was observed (JAMES et al. 1996). In addition, one group has presented data indicating that the PH domain of Akt is not required for its activation upon treatment of cells with insulin (KOHN et al. 1995) or PDGF (KOHN et al. 1996). The origin of this disagreement is not clear.

Although not evidenced by every study, the majority of these data suggest that Akt can be recruited to the plasma membrane of growth factor-stimulated cell by interaction of its PH domain with PtdIns (3,4) P_2 or PtdIns (3,4,5) P_3. The finding that synthetic PtdIns (3,4) P_2 can activate the kinase led to the argument that this event is sufficient for Akt activation, possibly through binding-induced conformational changes (KLIPPEL et al. 1997) or by dimerization at the membrane surface (FRANKE et al. 1997). However, an additional PI 3-kinase-independent pathway for Akt activation has been reported (KONISHI et al. 1996), and it has been shown that full activation of Akt requires its Ser/Thr phosphorylation (apparently by a second kinase) at two specific sites (ALESSI et al. 1996). A mutated form of Akt that has its PH domain deleted and is targeted to the plasma membrane by myristoylation was found to be constitutively active (KOHN et al. 1996), and to be phosphorylated on Ser/Thr without growth factor-treatment. A model has been proposed for growth factor-induced Akt activation (HEMMINGS 1997) in which PtdIns (3,4) P_2 levels are increased upon stimulation of PI 3-kinase or phosphoinositide phosphatases and recruit Akt to the plasma membrane through interactions with its PH domain (Fig. 9). This interaction may activate Akt to some extent (as seen in vitro), but a primary effect of membrane recruitment is to bring Akt in close proximity with one or more kinases, normally resident at the plasma membrane, for which it is a substrate. These kinases may respond to other signals, such as cellular stress (KONISHI et al. 1996), and their phosphorylation of Akt is required for its full activation.

5.8.2 PH Domain-Mediated Recruitment of Proteins Containing Sec7 Homology Regions

Another class of PH domain-containing proteins implicated as targets of PI 3-kinase are those that contain a C-terminal PH domain preceded by a region

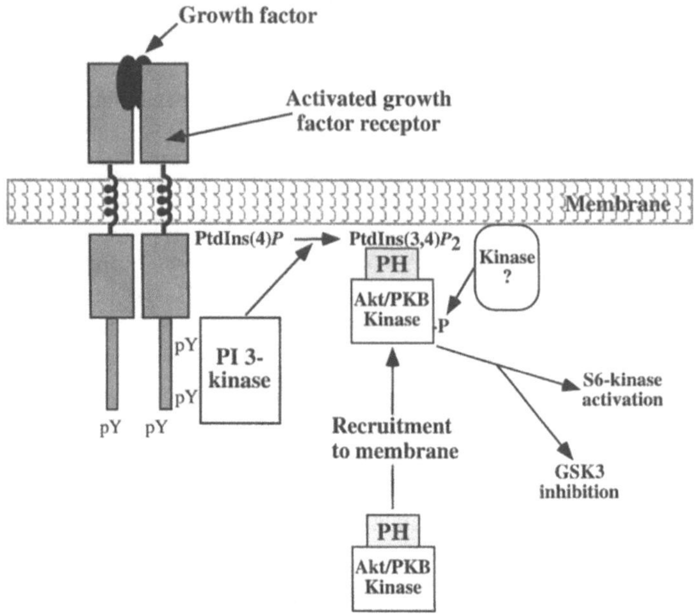

Fig. 9. A schematic view of the likely role of the PH domain in activation of Akt in response to growth factors (see text). Upon activation of a receptor tyrosine kinase such as the PDGF receptor, PI 3-kinase is recruited to the membrane surface and catalyzes the 3-phosphorylation of PtdIns (4) P and PtdIns (4,5) P_2. PtdIns (3,4) P_2 recruits Akt to the membrane surface by binding to its PH domain. At the membrane surface, Akt becomes activated. This activation may result in part from conformational changes induced by ligand binding to its PH domain. However, Akt also becomes phosphorylated, and this phosphorylation is required for full activation. It is thought that the kinase responsible is constitutively associated with the plasma membrane, and that PH domain-mediated recruitment of Akt to the vicinity of this kinase enhances its ability to phosphorylate Akt

homologous to the yeast SEC7 gene product. The first example of these proteins was named cytohesin-1, and was cloned in a yeast two-hybrid screen for proteins that interact with the cytoplasmic tail of the β2 integrin CD18 (Kolanus et al. 1996). Overexpression of cytohesin-1 or its isolated Sec7 homology domain in Jurkat cells resulted in their constitutive adhesion to dishes coated with the ICAM-1 extracellular domain, while stimulation of the T-cell receptor was required for this binding in control cells. By contrast, overexpression of the isolated PH domain inhibited the ability of T-cell receptor stimulation to enhance Jurkat cell binding to ICAM-1 (Kolanus et al. 1996). Several other PH domains tested had no such inhibitory effect, indicating that specific interactions of the cytohesin-1 PH domain are involved. It was shown that the Sec7 homology region of cytohesin-1 is sufficient for interaction of the molecule with the CD18 cytoplasmic domain, and this interaction is clearly implicated by the above data in 'inside-out' signaling of β2 integrins – modifying the ability of LFA-1 to bind to ICAM-1. The PH domain of cytohesin-1 may play a role in signal-dependent recruitment of the whole protein to the plasma membrane by binding to a specific phospholipid. This would enhance binding of the Sec7 homology region to the CD18 cytoplasmic tail through local concentration effects.

In strong support of this suggestion, KLARLUND et al. (1997) recently cloned a very similar protein by probing bacterial expression libraries with ^{32}P-labeled PI 3-kinase products. The protein, named Grp1 (for general receptor for phosphoinositides-1), is 84% identical in amino acid sequence to cytohesin-1, and binds specifically to PtdIns (3,4,5) P_3 and Ins (1,3,4,5) P_4 (with a micromolar-range K_D), but not PtdIns (3,4) P_2 or PtdIns (3) P through its C-terminal PH domain (which is 91% identical to the cytohesin-1 PH domain). The same binding characteristics were also found for the cytohesin-1 PH domain, while those from IRS-1 and Sos showed no significant PtdIns (3,4,5) P_3 binding (KLARLUND et al. 1997). This finding argues that stimulation of PI 3-kinase leads to recruitment of cytohesin-1 and Grp1 to the plasma membrane, where their Sec7 homology regions can exert their function. In the case of cytohesin-1, this function appears to be interaction with, and activation of, the β2 integrin (KOLANUS et al. 1996). Evidence has been presented that wortmannin and other PI 3-kinase inhibitors interfere with 'inside-out' β2 integrin signaling (NIELSEN et al. 1996).

A third example of this class of proteins, named ARNO, was cloned by CHARDIN et al. (1996) as an exchanger for the small G-protein ARF1 that is involved in vesicle budding from the Golgi. ARNO, cytohesin-1 and Grp-1 share 82% amino acid identity. Bacterially expressed ARNO was shown to stimulate nucleotide exchange on ARF1, for which only the Sec7 homology region is required. ARNO-mediated nucleotide exchange on vesicle-associated myristoylated ARF1 was found to be stimulated by the presence of PtdIns (4,5) P_2 in these vesicles, but only in the context of other negatively charged lipids (CHARDIN et al. 1996). The effect of PtdIns (3,4,5) P_3 has not yet been reported. The ability of ARNO to interact with integrin cytoplasmic tails has not been investigated, and nor has the ability of cytohesin-1 to act as a nucleotide exchange factor. Given the high level of identity, it is possible that both proteins function similarly. In both cases, PH domain-mediated recruitment to the plasma membrane will increase activity.

6 Conclusions: General Statements About PH Domains and their Ligands

From the examples discussed above, it is clear that there is still much to be learned about PH domains and their ligands. Only a small subset of the domains identified by sequence analysis has been studied to date. From these studies, a few general statements can be made regarding structure and function. Structurally, it is clear from the four cases studied that there is remarkable similarity considering the limited sequence identity. Since the domain is largely recognized on the basis of alternating hydrophobic and hydrophilic residues in six blocks of homology that correspond to β-strands, it is likely that the basic structure seen for the PH domains from dynamin, spectrin, pleckstrin and PLC-δ_1 will also be seen in other cases. It will be very interesting, as further structures are determined, to see whether the electrostatic polarization seen in these four cases is also common to other PH domains.

It is not yet clear whether a specific ligand should be expected for all PH domains. As discussed above, there are clearly some cases where a specific high-affinity ligand does exist (PLC-δ_1, Akt, Grp1/cytohesin-1 and Sos), and in each case the ligand is a phosphoinositide. There are fewer examples (PLC-δ_1, Akt, Grp1/cytohesin-1) where a strong case can be made for the physiological relevance of PH domain binding to a single ligand. In the few cases where specific PH domain interaction with a single protein ligand has been suggested, the physiological relevance is not yet clear. It is striking, however, that a number of PH domains have been reported to bind weakly to $G_{\beta\gamma}$-subunits, and many (apart from those with high-affinity phosphoinositide ligands) have been reported to bind weakly to phosphoinositides. Included in these is the βARK PH domain (plus C-terminal extension), which binds weakly to both $G_{\beta\gamma}$-subunits (Touhara et al. 1994) and to PtdIns (4,5) P_2 (Harlan et al. 1994). Neither one of these potential ligands is alone sufficient for association of βARK with defined vesicles in vitro. βARK binds strongly only to vesicles that contain both $G_{\beta\gamma}$-subunits and PtdIns (4,5) P_2 (Pitcher et al. 1995). This finding suggests that the two weak ligands are coordinately recognized by the βARK C-terminus, and this may turn out to be a feature of several PH domains. βARK binds to $G_{\beta\gamma}$ predominantly through the very C-terminus (and beyond) of the PH domain (Koch et al. 1993), while binding of the βARK PH domain to PtdIns (4,5) P_2 is likely to resemble that seen for the PH domains from pleckstrin (Harlan et al. 1994), spectrin (Hyvönen et al. 1995) and PLC-δ_1 (Ferguson et al. 1995), with the positively charged face adjacent to the negatively charged membrane surface (Figs. 6, 7). Similar coordinate binding to two distinct ligands (that individually bind only weakly) by other PH domains could explain the difficulties that have been experienced in identifying convincing specific binding partners for most PH domains.

Direct membrane binding by PH domains through their characteristic positively charged surface may involve specific recognition of lipid headgroups in some cases, as clearly seen with PLC-δ_1 (Fig. 7), Akt and Grp1. In other cases, delocalized nonspecific electrostatic attraction between the PH domain and the membrane surface may be of primary importance, and there are likely to be many examples that lie between the two extremes and show some specificity. Where phosphoinositide headgroups are specifically recognized, this binding alone appears to be sufficient for membrane association. It is likely that several more PH domains will be found to bind with high affinity to specific phosphoinositides, and there may also be examples that recognize other membrane components with similar specificity and affinity. However, where direct membrane binding by the PH domain is weak and nonspecific, which appears to be true in many cases, it alone will not be sufficient for membrane association of the host protein. In these cases, the PH domain can only contribute to membrane recruitment by cooperating with other interactions of the host protein with the membrane surface. These additional interactions may be mediated by other domains in the host protein, such as SH2, SH3 and PTB domains, alongside which PH domains are often found. In addition, the findings with βARK suggest that some PH domains are likely to coordinately recognize more than one ligand, and may associate with certain membrane-associated proteins through regions distinct from those that interact directly with the membrane. The region of the βARK PH domain implicated in $G_{\beta\gamma}$ binding is on the opposite face of the domain from that predicted to bind phosphoinositides. In the PTB domains from Shc (ZHOU et al. 1995) and IRS-1 (ECK et al. 1996), which have essentially the same structure as seen in PH domains (LEMMON et al. 1996), this face is of primary importance in phosphopeptide binding. The peptide binds primarily in a groove formed between the C-terminal α-helix (α1 in Fig. 4) and strand β5 (see Fig. 4) that is present in both PH and PTB domains. Given this fact, and the data for $G_{\beta\gamma}$ binding to βARK, it is tempting to speculate that other protein ligands might bind to this region of other PH domains, cooperating with membrane binding through the PH domain variable loops.

To conclude, it is clear that a number of PH domains participate in specific interactions that are important in signaling, and we have discussed several examples of these. For example, there is clear specificity in the ability of the isolated PH domains from cytohesin-1, Dbl, Sos and dynamin-1 to inhibit the function of their intact host proteins (Y. ZHENG et al. 1996; CHEN et al. 1997; KARLOVICH et al. 1995; ARTALEJO et al. 1997). The nature of the inferred specific ligands for the PH domains is not clear, except in a few cases where phosphoinositide binding is strong and specific. There may be other specific ligands that simply have not been identified. Alternatively, PH domains may be multifunctional in their ligand binding – requiring coordinate recognition of two (or more) different weakly binding ligands.

References

Adams JM, Houston H, Allen J, Lints T, Harvey R (1992) The hematopoietically expressed vav proto-oncogene shares homology with the dbl GDP-GTP exchange factor, the bcr gene and a yeast gene(CDC24) involved in cytoskeletal organization. Oncogene 7:611–618

Alessi DR, Andjelkovic M, Caudwell Cron P, Morrice N, Cohen P, Hemmings BA (1996) Mechanism of activation of protein kinase B by insulin and IGF-1. EMBO J 15:6541–6551

Andjelkovic M, Jakubowicz T, Cron P, Ming X-F, Han J-W, Hemmings BA (1996) Activation and phosphorylation of a pleckstrin homology domain containing protein kinase (RAC-PK/PKB) promoted by serum and protein phosphatase inhibitors. Proc Natl Acad Sci USA 93:5699–5704

Aronheim AD, Engelberg D, Li N, Al-Alawi N, Schlessinger J, Karin M (1994) Membrane targeting of the nucleotide exchange factor Sos is sufficient for activating the Ras signaling pathway. Cell 78: 949–961

Artalejo CR, Henley J, NcNiven M, Palfrey HC (1995) Rapid endocytosis coupled to exocytosis in adrenal chromaffin cells involves Ca^{2+}, GTP and dynamin, but not clathrin. Proc Natl Acad Sci USA 92:8328–8332

Artalejo CR, Lemmon MA, Schlessinger J, Palfrey HC (1997) Specific role for the PH domain of dynamin-1 in the regulation of rapid endocytosis in adrenal chromaffin cells. EMBO J 16:1565–1574

Buchsbaum R, Telliez J-B, Goonesekerra S, Feig LA (1996) The N-terminal pleckstrin, coiled-coil, and IQ domains of the exchange factor Ras-GRF act cooperatively to facilitate activation by calcium. Mol Cell Biol 16:4888–4896

Burgering BM, Coffer P (1995) Protein kinase B (c-Akt) in phosphatidylinositol-3-OH kinase signal transduction. Nature 376:599–602

Burgess WH, Dionne CA, Kaplow JM, Mudd R, Friesel R, Zilberstein A, Schlessinger J, Jaye M (1990) Characterization and cDNA cloning of phospholipase C-γ, a major substrate for heparin-binding growth factor 1 (acidic fibroblast growth factor)-activated tyrosine kinase. Mol Cell Biol 10:4770–4777

Carpenter LC, Cantley CL (1996) Phosphoinositide kinases. Curr Opin Cell Biol 8:153–158

Cerione RA, Zheng Y (1996) The Dbl family of oncoproteins. Curr Opin Cell Biol 8:216–222

Chardin P, Camonis JH, Gale NW, Van Aelst L, Wigler MH, Bar-Sagi D (1993) Human Sos 1: a guanine nucleotide exchange factor for Ras that binds to Grb2. Science 260:1338–1343

Chardin P, Paris S, Antonny B, Robineau S, Beraud-Dufour S, Jackson CL, Chabre M (1996) A human exchange factor for ARF contains Sec7- and pleckstrin-homology domains. Nature 384:481–484

Chen R-H, Cobalan-Garcia S, Bar-Sagi D (1997) The role of the PH domain in the signal-dependent membrane targeting of Sos. EMBO J 16:1351–1359

Cheng JQ, Godwin AK, Bellacosa A, Taguchi T, Franke TF, Hamilton TC, Tsichlis PN, Testa JR (1992) AKT2, a putative oncogene encoding a member of a subfamily of protein-serine/threonine kinases, is amplified in human ovarian carcinomas. Proc Natl Acad Sci USA 89:9267–9271

Chothia C (1984) Principles that determine the structure of proteins. Annu Rev Biochem 53:537–572

Chuang TT, Sallese, M, Ambrosini, G, Parruti, G, De Blasi A (1992) High expression of β-adrenergic receptor kinase in human peripheral blood leukocytes: isoproterenol and platelet activating factor can induce kinase translocation. J Biol Chem 267:6886–6892

Cifuentes ME, Honkanen L, Rebecchi MJ (1993) Proteolytic fragments of phosphoinositide-specific phospholipase C-δ_1: catalytic and membrane binding properties. J Biol Chem 268:11586–11593

Cifuentes ME, Delaney T, Rebecchi MJ (1994) D-Myo-inositol 1,4,5-trisphosphate inhibits binding of phospholipase C-δ_1 to bilayer membranes. J Biol Chem 269:1945–1994

Cook TA, Urrutia R, McNiven MA (1994) Identification of dynamin-2, an isoform ubiquitously expressed in rat tissues. Proc Natl Acad Sci USA 91:644–648

Cross DAE, Alessi DR, Cohen P, Andjelkovich M, Hemmings BA (1995) Inhibition of glycogen synthase kinase-3 by insulin mediated by protein kinase B. Nature 378:785–789

Davis LH, Bennett V (1994) Identification of two regions of β_G spectrin that bind to distinct sites in brain membranes. J Biol Chem 269:4409–4416

Didichenko SA, Tilton B, Hemmings BA, Ballmer-Hofer K, Thelen M (1996) Constitutive activation of protein kinase B and phosphorylation of $p47^{phox}$ by a membrane-targeted phosphoinositide 3-kinase. Curr Biol 6:1271–1278

Downing AK, Driscoll PC, Gout I, Salim K, Zvelebil MJ, Waterfield MD (1994) Three-dimensional solution structure of the pleckstrin homology domain from dynamin. Curr Biol 4:884–891

Dudek H, Datta SR, Franke TF, Birnbaum MJ, Yao R, Cooper GM, Segal RA, Kaplan DR, Greenberg ME (1997) Regulation of neuronal survival by the serine-threonine kinase Akt Science 275:661–665

Eck MJ, Dhe-Paganon S, Trüb T, Nolte RT, Shoelson SE (1996) Structure of the IRS-1 PTB domain bound to the juxtamembrane region of the insulin receptor. Cell 85:695–705

Essen L-O, Perisic O, Cheung R, Katan M, Williams R (1996) Crystal structure of a mammalian phosphoinositide-specific phospholipase Cd. Nature 380:595–602

Ferguson KM, Lemmon MA, Schlessinger J, Sigler PB (1994) Crystal structure at 2.2 Å resolution of the pleckstrin homology domain from human dynamin. Cell 79:199–209

Ferguson KM, Lemmon MA, Schlessinger J, Sigler PB (1995) Structure of a high affinity complex between inositol-1,4,5-trisphosphate and a phospholipase C pleckstrin homology domain. Cell 83:1037–1046

Font de Mora J, Guerrero C, Mahadevan D, Coque JJR, Rojas JM, Esteban LM, Rebecchi M, Santos E (1996) Isolated Sos1 PH domain exhibits germinal vesicle breakdown-inducing activity in Xenopus oocytes. J Biol Chem 271:18272–18276

Franke TF, Yang S-I, Chan TO, Datta K, Kazlauskas A, Morrison DK, Kaplan DR, Tsichlis PN (1995) The protein kinase encoded by the Akt proto-oncogene is a target of the PDGF-activated phosphatidylinositol 3-kinase. Cell 81:727–736

Franke TF, Kaplan DR, Cantley LC, Toker A (1997) Direct regulation of the Akt proto-oncogene product by phosphatidylinositol-3,4-bisphosphate. Science 275:665–668

Frech M, Andjelkovic M, Reddy KK, Falck JR, Hemmings BA (1997) High affinity binding of inositol phosphates and phosphoinositides to the pleckstrin homology domain of Rac/protein kinase B and their influence on the kinase activity. J Biol Chem 272: 8474–8481

Fukuda M, Mikoshiba K (1996) Structure-function relationships of the mouse Gap1^{m}: determination of the inositol 1,3,4,5 tetrakisphosphate-binding domain. J Biol Chem 271:18838–18842

Fukuda M, Kojima T, Kabayama H, Mikoshiba K (1996) Mutation of the pleckstrin homology domain of Bruton's tyrosine kinase in immunodeficiency impaired inisitol 1,3,4,5-tetrakisphosphate binding capacity. J Biol Chem 271:30303–30306

Fushman D, Cahill S, Lemmon MA, Schlessinger J, Cowburn D (1995) Solution structure of pleckstrin homology domain of dynamin by heteronuclear NMR spectroscopy. Proc Natl Acad Sci USA 92:816–820

Garcia P, Gupta R, Shah S, Morris AJ, Rudge SA, Scarlata S, Petrova V, McLaughlin S, Rebecchi MJ (1995) The pleckstrin homology domain of phospholipase C-δ_1 binds with high affinity to phosphatidylinositol 4,5-bisphosphate in bilayer membranes. Biochemistry 34:16228–16234

Gibson TJ, Hyvönen M, Musacchio A, Saraste M, Birney E (1994) PH domain: the first anniversary. Trends Biochem Sci 19:349–353

Habets GGM, Scholtes EHM, Zuydgeest D, van der Kammen RA, Stam JC, Collard JG (1994) Identification of an invasion inducing gene, termed Tiam-1, that encodes a protein with homology to GDP-GTP exchangers for Rho-like proteins. Cell 77:537–549

Harlan JE, Hajduk PJ, Yoon HS, Fesik SW (1994) Pleckstrin homology domains bind to phosphatidylinositol-4,5-bisphosphate. Nature 371:168–170

Haslam RJ, Koide HB, Hemmings BA (1993) Pleckstrin domain homology. Nature 363:309–310

Hemmings BA (1997) Akt signaling: linking membrane events to life and death decisions. Science 275:628–630

Horii Y, Beeler JF, Sakaguchi K, Tachibana M, Miki T (1994) A novel oncogene, ost, encodes a guanine nucleotide exchange factor that potentially links Rho and Rac signaling pathways. EMBO J 13:4776–4786

Hu R-J, Watanabe M, Bennett V (1992) Characterization of human brain cDNA encoding the general isoform of β-spectrin. J Biol Chem 267:18715–18722

Hyvönen M, Macias MJ, Nilges M, Oschkinat H, Saraste M, Wilmanns M (1995) Structure of the binding site for inositol phosphates in a PH domain. EMBO J 14:4676–4685

Inglese J, Koch WJ, Caron MG, Lefkowitz RJ (1992) Isoprenylation in regulation of signal transduction by G-protein-coupled receptor kinases. Nature 359:147–150

Inglese J, Koch WJ, Touhara K, Lefkowitz RJ (1995) $G_{\beta\gamma}$ interactions with PH domains and Ras-MAPK signaling pathways. Trends Biochem Sci 20:151–156

James SR, Downes CP, Gigg R, Grove SJA, Holmes AB, Alessi DR (1996) Specific binding of the Akt-1 protein kinase to phosphatidylinositol 3,4,5-trisphosphate without subsequent activation. Biochem J 315:709–713

Kanematsu T, Takeya H, Watanabe Y, Ozaki S, Yoshida M, Koga T, Iwanaga S, Hirata M (1992) Putative inositol 1,4,5-trisphosphate binding proteins in rat brain cytosol. J Biol Chem 267:6518–6525

Karlovich CA, Bonfini L, McCollam L, Rogge RD, Daga A, Czech MP, Banerjee U (1995) In vivo functional analysis of the Ras exchange factor Son of Sevenless. Science 268:576–579

Klarlund JK, Guilherme A, Holik JJ, Virbasius A, Czech MP (1997) Signaling by 3,4,5-phosphoinositide through proteins containing pleckstrin and Sec7 homology domains. Science (in press)

Klippel A, Reinhard C, Kavanaugh WM, Apell G, Escobedo M-A, Williams LT (1996) Membrane localization of phosphatidylinositol 3-kinase is sufficient to activate multiple signal-transducing pathways. Mol Cell Biol 16:4117–4127

Klippel A, Kavanaugh WM, Pot D, Williams LT (1997) A specific product of phosphatidylinositol 3-kinase directly activates the protein kinase Akt through its pleckstrin homology domain. Mol Cell Biol 17:338–344

Koch WJ, Inglese J, Stone WC, Lefkowitz RJ (1993) The binding site for the βγ subunits of heterodtrimeric G protein on the β-adrenergic receptor kinase. J Biol Chem 268:8256–8260

Koch WJ, Hawes BE, Inglese J, Luttrell LM, Lefkowitz RJ (1994) Cellular expression of the carboxy terminus of a G protein-coupled receptor kinase attenuates $G_{\beta\gamma}$-mediated signaling. J Biol Chem 269:6193–6197

Kohn AD, Kovacina KS, Roth RA (1995) Insulin stimulates the kinase activity of RAC-PK, a pleckstrin homology domain containing Ser/Thr kinase. EMBO J 14:4288–4295

Kohn AD, Takeuchi F, Roth RA (1996) Akt, a pleckstrin homology domain containing kinase, is activated primarily by phosphorylation. J Biol Chem 271:21920–21926

Kolanus W, Nagel W, Schiller B, Zeitlmann L, Godar S, Stockinger H, Seed B (1996) aLb2 integrin/ LFA-1 binding to ICAM-1 induced by cytohesin-1, a cytoplasmic regulatory molecule. Cell 86: 233–242

Konishi H, Kuroda S, Kikkawa U (1994) The pleckstrin homology domain of Rac protein kinase associates with the regulatory domain of protein kinase-C ζ. Biochem Biophys Res Commun 205:1770–1775

Konishi H, Kuroda S, Tanaka M, Matsuzaki H, Ono Y, Kameyama K, Haga T, Kikkawa U (1995) Molecular cloning and characterization of a new member of the RAC protein kinase family: association of the pleckstrin homology domain of three types of RAC protein kinase with protein kinase C subspecies and beta gamma subunits of G proteins. Biochem Biophys Res Commun 216:526–534

Konishi H, Matsuzaki H, Tanaka M, Ono Y, Tokunaga C, Kuroda S, Kikkawa U (1996) Activation of RAC-protein kinase by heat shock and hyperosmolarity stress through a pathway independent of phosphatidylinositol 3-kinase. Proc Natl Acad Sci USA 93:7639–7643

Kraulis P (1991) MOLSCRIPT: A program to produce both detailed and schematic plots of protein structures. J Appl Crystallogr 24:946–950

Kubiseski TJ, Chook YM, Parris WE, Rozakis-Adcock M, Pawson T (1997) High affinity binding of the pleckstrin homology domain of mSos1 to phosphatidylinositol (4,5)-bisphosphate. J Biol Chem 272:1799–1804

Langhans-Rajasekaran SA, Wan Y, Huang X-Y (1995) Activation of Tsk and Btk tyrosine kinase by G protein βγ subunits. Proc Natl Acad Sci USA 92:8601–8605

Lemmon MA, Ferguson KM, O'Brien R, Sigler PB, Schlessinger J (1995) Specific and high-affinity binding of inositol phosphates to an isolated pleckstrin homology domain. Proc Natl Acad Sci USA 92:10472–10476

Lemmon MA, Ferguson KM, Schlessinger J (1996) PH domains: diverse sequences with a common fold recruit signaling molecules to the cell surface. Cell 85:621–624

Li T, Tsukuda S, Satterthwiate A, Havlik MH, Park H, Takatsu K, Witte ON (1995) Activation of Bruton's tyrosine kinase (BTK) by a point mutation in its pleckstrin homology (PH) domain. Immunity 2:451–460

Lifshitz B, Fainstein E, Marcelle C, Shtivelman E, Amson R, Gale RP, Canaani E (1988) bcr genes and transcripts. Oncogene 2:113–117

Lomasney JW, Cheng H-F, Wang L-P, Kuan Y-S, Liu S-M, Fesik SW, King K (1996) Phosphatidylinositol 4,5-bisphosphate binding to the pleckstrin homology domain of phospholipase C-δ_1 enhances enzyme activity. J Biol Chem 271:25316–25326

Lombardo CR, Weed SA, Kennedy SP, Forget BG, Morrow JS (1994) βII-spectrin (fodrin) and βIS2-spectrin (muscle) contain NH_2- and COOH-terminal membrane association domains (MAD1 and MAD2). J Biol Chem 269:29212–29219

Luttrell LM, Hawes BE, Touhara K, van Biesen T, Koch WJ, Lefkowitz RJ (1995) Effect of cellular expression of pleckstrin homology domains on G_i-coupled receptor signaling. J Biol Chem 270:12984–12989

Macias MJ, Musacchio A, Ponstingl H, Nilges M, Saraste M, Oschkinat H (1994) Structure of the pleckstrin homology domain from β-spectrin. Nature 369:675–677

Mahadevan D, Thanki N, Singh J, McPhie P, Zangrilli D, Wang L-M, Guerrero C, LeVine H, Humblet C, Saldanha J, Gutkind JS, Najmabadi-Haske T (1995) Structural studies on the PH domains of Dbl, Sos1, IRS-1, and βARK1 and their differential binding to Gβγ subunits. Biochemistry 34:9111–9117

Mano H, Ishikawa F, Nishida J, Hirai H, Takaku F (1990) A novel protein-tyrosine kinase, tec, is preferentially expressed in liver. Oncogene 5:1781–1786

Margolis B, Silvennoinen O, Comoglio F, Roonprapunt C, Skolnik EY, Ullrich A, Schlessinger J (1992) High-efficiency expression cloning of epidermal growth factor receptor-binding proteins with Src homology domains. Proc Natl Acad Sci USA 89:8894–8898

Mattsson PT, Vihinen M, Smith CIE (1996) X-linked agammaglobulinemia (XLA): a genetic tyrosine kinase (Btk) disease. Bioessays 18:825–834

Mayer BJ, Ren R, Clark KL, Baltimore D (1993) A putative modular domain present in diverse signaling molecules. Cell 73:629–630

McCollam L, Bonfini L, Karlovich CA, Conway BR, Kozma LM, Banerjee U, Czech MP (1995) Functional roles for the pleckstrin and Dbl homology regions in the Ras exchange factor Son-of-sevenless. J Biol Chem 270:15954–15957

Musacchio A, Gibson T, Rice P, Thompson J, Saraste M (1993) The PH domain: a common piece in a pathwork of signalling proteins. Trends Biochem Sci 18:343–348

Myers MG Jr, Grammer TC, Brooks J, Glasheen EM, Wang L-M, Sun XJ, Blenis J, Pierce JH, White MF (1995) The pleckstrin homology domain in insulin receptor substrate-1 sensitizes insulin signaling. J Biol Chem 270:11715–11718

Nicholls A, Shap KA, Honig B (1991) Protein folding and association: insights from the interfacial and thermodynamic properties of hydrocarbons. Proteins Struct Funct Genet 11:281–296

Nielsen M, Svejgaard A, Skov S, Dobson P, Bendtzen K, Geisler C, Odum N (1996) IL-2 induces β2 integrin adhesion via a wortmannin/LY294002-sensitive, rapamycin-resistant pathway Phosphorylation of a 125-kilodalton protein correlates with induction of adhesion, but not mitogenesis. J Immunol 157:5350–5358

Parker PJ, Hemmings BA, Gierschik P (1994) PH domains and phospholipases – a meaningful relationship? Trends Biochem Sci 19:54–55

Paterson HF, Savopoulos JW, Perisic O, Cheung R, Ellis MV, Williams RL, Katan M (1995) Phospholipase C-δ_1 requires a pleckstrin homology domain for interaction with the plasma membrane. Biochem J 312:661–666

Pitcher JA, Inglese J, Higgins JB, Arriza JL, Casey PJ, Kim C, Benovic JL, Kwatra MM, Caron MG, Lefkowitz RJ (1992) Role of βγ subunits of heterotrimeric G proteins in targeting the β-adrenergic receptor kinase to membrane-bound receptors. Science 257:1264–1267

Pitcher JA, Touhara K, Payne ES, Lefkowitz RJ (1995) Pleckstrin homology domain-mediated membrane association and activation of the β-adrenergic receptor kinase requires coordinate interaction with $G_{\beta\gamma}$ subunits and lipid. J Biol Chem 270:11707–11710

Ramirez F, Jain MK (1991) Phospholipase A_2 at the bilayer interface. Proteins Struct Funct Genet 9: 229–239

Rebecchi M, Peterson A, McLaughlin S (1992) Phosphoinositide-specific phospholipase C-δ_1 binds with high affinity to phospholipid vesicles containing phosphatidylinositol 4,5-bisphosphate. Biochemistry 31:12742–12747

Richardson J (1977) β-Sheet topology and the relatedness of proteins. Nature 268:495–500

Ron D, Tronick SR, Aaronson SA, Eva A (1988) Molecular cloning and characterization of the human dbl proto-oncogene: evidence that its overexpression is sufficient to transform NIH/3T3 cells. EMBO J 7:2465–2473

Salim K, Bottomley MJ, Querfurth E, Zvelebil MJ, Gout I, Scaife R, Margolis RL, Gigg R, Smith CIE, Driscoll PC, Waterfield MD, Panayotou G (1996) Distinct specificity in the recognition of phosphoinositides by the pleckstrin homology domains of dynamin and the Bruton's tyrosine kinase. EMBO J 15:6241–6250

Schlessinger J (1994) SH2/SH3 signaling proteins. Curr Opin Genet Dev 4:25–30

Scott DL, Mandel AM, Sigler PB, Honig B (1994) The electrostatic basis for the interfacial binding of secretory phospholipases A_2. Biophys J 67:493–504

Shaw G (1993) Identification of novel pleckstrin homology (PH) domain provides a hypothesis for PH domain function. Biochem Biophys Res Commun 195:1145–1151

Shaw G (1996) The pleckstrin homology domain: an intriguing multifunctional protein module. Bioessays 18:35–46

Shou C, Farnsworth CL, Neel BG, Feig LA (1992) Molecular cloning of cDNAs encoding a guanine-nucleotide-releasing factor for Ras p21. Nature 358:351–354

Siliciano JD, Morrow TA, Desiderio SV (1992) Itk, a T-cell specific tyrosine kinase gene inducible by interleukin 2. Proc Natl Acad Sci USA 89:11194–11198

Sontag J-M, Fyske EM, Ushkaryov Y, Liu J-P, Robinson PJ, Südhof TC (1994) Differential expression and regulation of multiple dynamins. J Biol Chem 269:4747–4754

Suh P, Ryu S, Moon K, Suh H, Rhee S-G (1988) Cloning and sequence of multiple forms of phospholipase C. Cell 54:161–169

Sun XJ, Rothenberg P, Kahn CR, Backer JM, Araki E, Wilden PA, Cahill DA, Goldstein BJ, White MF (1991) Structure of the insulin receptor substrate IRS-1 defines a unique signal transduction protein. Nature 352:73–77

Thomas JD, Sideras P, Smith CIE, Vorechovsky I, Chapman V, Paul WE (1993) Colocalization of X-linked agammaglobulinemia and X-linked immunodeficiency genes. Science 261:355–358

Timm D, Salim K, Gout I, Guruprasad L, Waterfield M, Blundell T (1994) Crystal structure of the pleckstrin homology domain from dynamin. Nature Struct Biol 1:782–788

Touhara K, Inglese J, Pitcher JA, Shaw G, Lefkowitz RJ (1994) Binding of G protein βγ-subunits to pleckstrin homology domains. J Biol Chem 269:10217–10220

Trahey M, Wong G, Halenbeck R, Rubinfeld B, Martin GA, Ladner M, Long CM, Crosier WJ, Watt K, Koths K, McCormick F (1988) Molecular cloning of two types of GAP complementary DNA from human placenta. Science 242:1697–1700

Tsukuda S, Simon MI, Witte ON, Katz A (1994) Binding of βγ subunits of heterotrimeric G-proteins to the PH domain of Bruton's tyrosine kinase. Proc Natl Acad Sci USA 91:11256–11260

Tyers M, Rachubinski RA, Stewart MI, Varrichio AM, Shorr RGL, Haslam RJ, Harley CB (1988) Molecular cloning and expression of the major protein kinase C substrate of platelets. Nature 333:470–473

van der Bliek AM, Redelmeister TE, Damke H, Tisdale EJ Meyerowitz EM, Schmid SJ (1993) Mutations in human dynamin block, an intermediate stage in coated vesicle formation. J Cell Biol 122:553–563

Vetrie D, Vorechovsky I, Sideras P, Holland J, Davies A, Flinter F, Hammarstrom L, Kinnon C, Levinsky R, Bobrow M, Smith CIE, Bentley DR (1993) The gene involved in X-linked agammaglobulinaemia is a member of the src family of protein-tyrosine kinases. Nature 361:226–233

Vihinen M, Zvelebil MJJM, Zhu Q, Brooimans RA, Ochs HD, Zegers BJM, Nilsson L, Waterfield MD, Smith CIE (1995) Structural basis for pleckstrin homology domain mutations in X-linked agammaglobulinemia. Biochemistry 34:1475–1481

Voliovitch H, Schindler DG, Hadari YR, Taylor SI, Accili D, Zick Y (1995) Tyrosine phosphorylation of insulin receptor substrate-1 in vivo depends upon the presence of its pleckstrin homology region. J Biol Chem 270:18083–18087

Wang D-S, Shaw G (1995) The association of the C-terminal region of βIΣII spectrin to brain membrane is mediated by a PH domain, does not require membrane proteins, and coincides with a inositol-145-trisphosphate binding site. Biochem Biophys Res Commun 217:608–615

Wang D-S, Shaw R, Winkelmann JC, Shaw G (1994) Binding of PH domains of β-adrenergic receptor kinase and β-spectrin to WD40/b-transducin repeat containing regions of the β-subunit of trimeric G-proteins. Biochem Biophys Res Commun 203:29–35

Wang W, Fisher EMC, Jia Q, Dum JM, Porfiri E, Downward J, Egan SE (1995) The Grb2 binding domain of mSos1 is not required for downstream signal transduction. Nature Genetics 10:294–300

Wang D-S, Miller R, Shaw R, Shaw G (1996) The pleckstrin homology domain of human βIΣII spectrin is targeted to the plasma membrane in vivo. Biochem Biophys Res Commun 225:420–426

Whitehead I, Kirk H, Tognon C, Trigo-Gonzalez G, Kay R (1995) Expression cloning of lfc, a novel oncogene with structural similarities to guanine nucleotide exchange factors and to the regulatory region of protein kinase C. J Biol Chem 270:18388–18395

Woolfson DN, Evans PA, Hutchinson EG, Thornton JM (1993) Topological and stereochemical restrictions in β-sandwich protein structures. Protein Eng 5:461–470

Yagisawa H, Hirata M, Kanematsu T, Watanabe Y, Ozaki S, Sakuma K, Tanaka H, Yabuta N, Kamata H, Hirata H, Nojima H (1994) Expression and characterization of an inositol 1,4,5-trisphosphate binding domain of phosphatidylinositol-specific phospholipase C-δ_1. J Biol Chem 269:20179–20188

Yang W, Desiderio S (1997) BAP-135, a target for Bruton's tyrosine kinase in response to B cell receptor engagement. Proc Natl Acad Sci USA 94:604–609

Yao L, Kawakami Y, Kawakami T (1994) The pleckstrin homology domain of Bruton tyrosine kinase interacts with protein kinase C. Proc Natl Acad Sci USA 91:9175–9179

Yenush L, Makati KJ, Smith-Hall J, Ishibashi O, Myers MG, White MF (1996) The pleckstrin homology domain is the principal link between the insulin receptor and IRS-1. J Biol Chem 271:24300–24306

Yoon HS, Hajduk PJ, Petros AM, Olejniczak ET, Meadows RP, Fesik SW (1994) Solution structure of a pleckstrin-homology domain. Nature 369:672–675

Zhang P, Talluri S, Deng H, Branton D, Wagner G (1995) Solution structure of the pleckstrin homology domain of Drosophila beta-spectrin. Structure 3:1185–1195

Zheng J, Cahill SM, Lemmon MA, Fushman D, Schlessinger J, Cowburn D (1996) Identification of the binding site for acidic phospholipids on the PH domain of dynamin: implications for stimulation of GTPase activity. J Mol Biol 255:14–21

Zheng Y, Zangrilli D, Cerione RA, Eva A (1996) The pleckstrin homology domain mediates transformation by oncogenic Dbl through specific intracellular targeting. J Biol Chem 271:19017–19020

Zhou M-M, Ravichandran KS, Olejniczak ET, Petros AM, Meadows RP, Sattler M, Harlan JE, Wade WS, Burakoff SJ, Fesik SW (1995) Structure and ligand recognition of the phosphotyrosine binding domain of Shc. Nature 378:584–592

Structure and Function of LIM Domains

L.W. JURATA and G.N. GILL

1 Introduction

Conserved domains found in many proteins provide surfaces that interact with other proteins. Such domains fold independently of the remainder of the protein and are structurally similar within a class of domains. Individual members of a class have sufficient sequence variation to provide high specificity for particular target sequences. Thus, in evolution, conserved structures were retained but variations were selected that provided specific function. Molecular recognition is a fundamental property in biology, long recognized in immunology (antibodies, T and B cell receptors) and in endocrinology (hormone receptors) and more recently appreciated in intracellular signal transduction and transcription processes. Proteins often contain more than one domain either of the same or of a different class that provide for assembly of molecular complexes. The domains discussed in this monograph are in general 40–100 amino acids in length and recognize relatively short but specific sequences in target proteins. In combination, domains and

Department of Medicine, University of California San Diego, 9500 Gilman Drive, La Jolla, CA 92093-0650, USA

Table 1. Expression and mutant phenotypes of LMO and LIM Homeodomain proteins

Gene	Organism	Wild-type expression	Mutant phenotypes
LMO1 (*Ttg-1*, *Rbtn1*)	Human, mouse	Embryonic and adult CNS; ventral spinal cord and brain, retina, trigeminal and facial neurons (GREENBERG et al. 1990; BOEHM et al. 1991b)	Overexpression in T cells results in leukemia (MCGUIRE et al. 1992; FISCH et al. 1992)
LMO2 (*Ttg-2*, *Rbtn2*)	Human, mouse	Widespread expression, highest in embryonic liver and spleen; adult brain, kidney, liver and spleen (ROYER-POKORA et al. 1991; BOEHM et al. 1991b; FORONI et al. 1992)	Overexpression in T cells results in leukemia (FISCH et al. 1992; LARSON et al. 1995; NEALE et al. 1995b). Targeted deletion results in lethality at E10.5 due to failure of erythropoesis (WARREN et al. 1994)
lin-11	*C. elegans*		Mutation results in defects in vulval formation from failure of asymmetric cell division of secondary blast cells (FERGUSON et al. 1987)
mec-3	*C. elegans*	Mechanosensory neurons, FLP, PVD neurons (WAY and CHALFIE 1989)	Mutation results in insensitivity to gentle touch due to lack of touch receptor neurons (WAY and CHALFIE 1988)
apterous	*Drosophila*	Embryonic abdominal muscles, embryonic CNS; larval wing imaginal discs (COHEN et al. 1992; DIAZ-BENJUMEA and COHEN 1993; BOURGOUIN et al. 1992; LUNDGREN et al. 1995)	Mutation results in wingless, haltereless phenotype, lack of embryonic abdominal muscles, incorrect pathway selection of neurons which normally express *ap*. Genetic mosaic analysis results in ectopic wing margins (BOURGOUIN et al. 1992; LUNDGREN et al. 1995; DIAZ-BENJUMEA and COHEN 1993)
Xlim-1 *Lhx1* (*mLim-1*) *Lim1*	*Xenopus* Mouse Chick	Gastrula organizer, dorsal mesoderm and endoderm; embryonic prechordal plate, notochord, spinal cord and brain; adult brain and kidney (TAIRA et al. 1992; FUJII et al. 1994; BARNES et al. 1994)	Targeted deletion results in embryonic lethality at E10, absence of head structures anterior to rhombomere 3 (SHAWLOT and BEHRINGER 1995)

Lhx2 (*LH-2*)	Human, mouse	Embryonic nervous and immune systems, retina, pituitary, early B cell development (Xu et al. 1993; Roberson et al. 1994)	
Xlim-3 *Lhx3* (*mLim-3*, *P-lim*) *Lim3*	*Xenopus* Mouse Chick	Anterior and intermediate lobes of pituitary of the embryo, ventral motor neurons of the hindbrain and spinal cord, retina, pineal gland (Taira et al. 1992; Seidah et al. 1994; Bach et al. 1995; Zhadanov et al. 1995)	Targeted deletion results in lethality by P0, absence of anterior and intermediate lobes of the pituitary (Sheng et al. 1996)
Lhx4 (*Gsh-4*)	Mouse	Ventral hindbrain and spinal cord (Li et al. 1994)	Targeted deletion results in lethality at P0 due to defects in respiration (Li et al. 1994)
Xlim-2 *Lim5* *Lhx5* *Lim2*	*Xenopus* Zebrafish Mouse Chick	Blastula; gastrula ectoderm; anterior neural plate, ventral spinal cord (Toyama et al. 1995b; Tsuchida et al. 1994)	
Lmx1	Hamster, chick	Pancreatic islet cells, dorsal mesoderm of limb bud, notochord, spinal cord roof plate, floor plate, and dorsal neurons (German et al. 1992a; Riddle et al. 1995)	
Isl1	Zebrafish, mouse, chick	Pancreatic islet cells, anterior and intermediate lobes of the pituitary, thyroid, adrenomedullary chromaffin cells, dorsal root ganglia, ventral spinal cord motor neurons, hypothalamus, pineal gland, retina (Thor et al. 1991; Karlsson et al. 1990; Inoue et al. 1994)	Targeted deletion results in lethality by E11.5, absence of motor neurons and interneurons, endocrine and exocrine pancreatic defects (Pfaff et al. 1996; Ahlgren et al. 1997)

targets constitute a versatile and complex biological language for transfer of information.

LIM domains are composed of approximately 55 residues with the sequence $CX_2CX_{16-23}HX_2CX_2CX_{16-23}CX_2C$ (where X is any amino acid). Cysteine-rich LIM domains bind two atoms of Zn^{2+} with the most common tetrahedral coordination being S_3N and S_4 (Michelsen et al. 1993). The Zn^{2+} atoms bound at each end of the domain are essential to the structure of the domain (Perez-Alvarado et al. 1994, 1996). LIM domains were first recognized in three homeodomain proteins: *l*in-11 which functions in asymmetric division of *Caenorhabditis elegans* secondary vulval blast cells (Freyd et al. 1990); *I*sl1 which bound the rat insulin I gene enhancer (Karlsson et al. 1990) but has a major function in motor neuron development (Pfaff et al. 1996); and *m*ec-3, which is essential for differentiation of touch receptor neurons in *C. elegans* (Way and Chalfie 1988). Proteins containing LIM domains at their amino terminus and homeodomains at their carboxyl terminus constitute an important subset of homeodomain proteins. The first LIM domain sequence reported was contained in the cysteine-rich intestinal protein, CRIP (Birkenmeier and Gordon 1986). LIM domains were subsequently found in a variety of cytoplasmic proteins and are widely distributed in nature, being found in plants, yeast and a variety of metazoans.

2 Classes of LIM Domains

Based upon sequence comparisons, two classifications of LIM domains have been proposed. Dawid et al. (1995) classified LIM domains into five groups whereas Steinmetz and Evrard (1996) divided them into three groups with subdivisions in two of those. In both classifications, the N terminal LIM domains of the nuclear LIM homeodomain and LIM-only proteins are closely related as are the second LIM domains of these proteins. Sequence alignments indicate that LIM domains found in a particular position, i.e., LIM1 or LIM 2 are more similar than LIM domains found at different positions in a single protein (Dawid et al. 1995). Moreover, LIM domains located in a particular position in a particular protein from different species are the most highly conserved of LIM sequences. The convention of numbering LIM domains by their location in proteins from N to C termini has been adopted, i.e., the most N terminal LIM domain equals LIM1. Nomenclature is still being developed but by convention the nuclear LIM-only proteins are now designated LMO replacing the original names of Rhombotins (Greenberg et al. 1990) and T cell translocation genes (McGuire et al. 1989; Royer-Pokora et al. 1991). No convention has yet been adopted for the extranuclear LIM-only proteins such as cysteine-rich protein (CRP), CRIP and PINCH (Rearden 1994). Genome conventions are being adopted for many of the LIM-homeodomain proteins (Dawid et al. 1995). Thus, Lhx1 (*LIM h*omeobo*x*) of the mouse gene nomenclature replaces Lim1, Lhx2 replaces LH-2, Lhx3 replaces

Lim3/P-lim, Lhx4 replaces Gsh-4 and Lhx5 replaces Lim5 (see Table 1). Because some names such as Isl1 and apterous are well established, these have been retained. Cytoplasmic LIM-domain-containing proteins retain the quixotic names originally assigned them.

Sequence alignments for representative members of Steinmetz's classes A, B, and C are shown in Fig. 1. Class A contains the cytoplasmic LIM proteins Paxillin, Zyxin, Enigma and LIM kinase. Class B, which contains LIM homeodomain and nuclear LIM only proteins, contains subclasses of LIM1 and LIM2 domains. Class C contains proteins whose LIM domain structures have been solved, LIM2 of cCRP and rCRIP (Perez-Alvarado et al. 1994, 1996) and the plant LIM domains. Class C is also divided into subclasses for LIM1 and LIM2 domains. Sequences are anchored by the eight conserved residues that are ligands for the two bound Zn^{2+} atoms. In the alignment shown the first metal ligand Cys is designated residue 1 and the spacing between ligands 1 and 2, 3 and 4, and 7 and 8 is set at 2 although this spacing varies from two to four residues. Residues at an additional eight positions are conserved in LIM domains from all three classes. Of these residues, seven are hydrophobic. Some of these form the hydrophobic core located between the two Zn^{2+} modules (residues 27, 32, 41) and three contribute to a hydrophobic surface on one face of the domain (residues 50, 62, 67). Many of these are located at the ends of the four β sheets present in LIM domains.

Additional residues are conserved within the B and C classes but these conservations do not extend to the A class of LIM domains. For example, positively charged residues (Lys/Arg) are conserved at position 64 in the B and C classes but not in the A class. Lys/Arg are also conserved at position 29 in the C class whereas a hydrophobic residue (Leu/Ile/Val/Met) is conserved at this position in the LIM2 B subclass. A hydrophobic residue is conserved at position 17 in the B class and a Gly is conserved at position 55 in the C and LIM1 B subclasses. These sequence conservations indicated in Fig. 1 are the likely determinants of classes of target recognition specificities with nonconserved residues refining recognition.

3 Structure of LIM Domains

The Zn^{2+} coordinations of LIM domains are shown in the 2 Zn^{2+}-finger representation of LIM 3 of Enigma (Fig. 2) and in the three-dimensional structure of LIM2 of CRP (Fig. 3A). The NMR structures of the C terminal LIM domain of avian cysteine-rich protein (LIM2 of CRP) (Perez-Alvarado et al. 1994) and of the single LIM domain protein CRIP indicated that the two Zn^{2+} atoms are bound independently in the N terminal and C terminal liganding modules (Perez-Alvarado et al. 1996). In the first Zn^{2+} binding module, Cys residues occupy the first two Zn^{2+} ligands in all LIM domains and changing the second Cys ligand to His abolished Zn^{2+} binding of LIM2 of CRP (Michelsen et al. 1994). The third ligand position is usually His but Cys and Asp occur at this position in some LIM

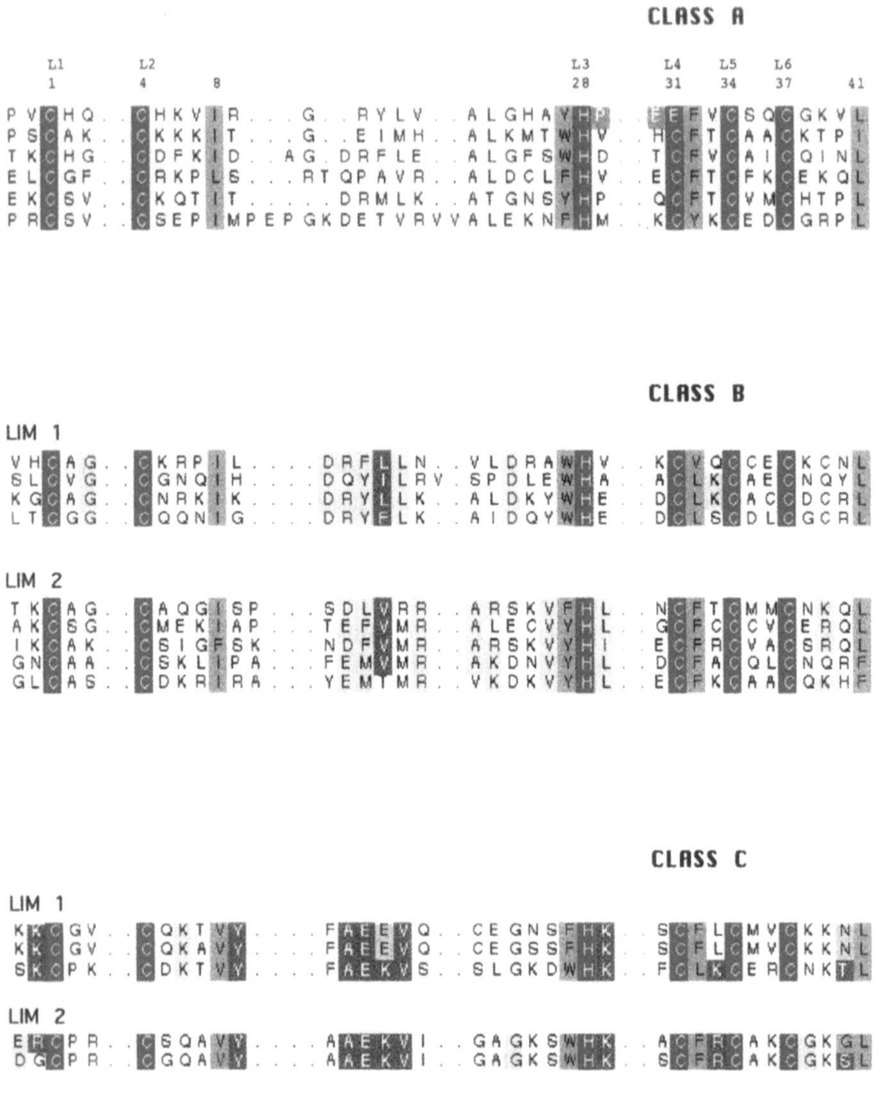

Fig. 1. Sequences of representative LIM domains. Alignments are from STEINMETZ and EVRARD (1996). For general convenience the first Cys that serves as a ligand to Zn^{2+} is numbered as *1* and the spacing between adjacent ligands is set at *2*. Reprinted with permission of the authors

domains. The fourth ligand is usually Cys but can be Glu or His, i.e., LIM domain 1 of the protein Enigma contains Glu at this position and binds two atoms of Zn^{2+} (GILL 1995). In the second Zn^{2+} binding module Cys residues occupy ligand positions 5, 6, and 7. A possible exception is the plant pollen protein PLIM where the spacing between ligands 5 and 6 may, in fact, be 4 rather than 2 which would retain this pattern (PEREZ-ALVARADO et al. 1996). Changing

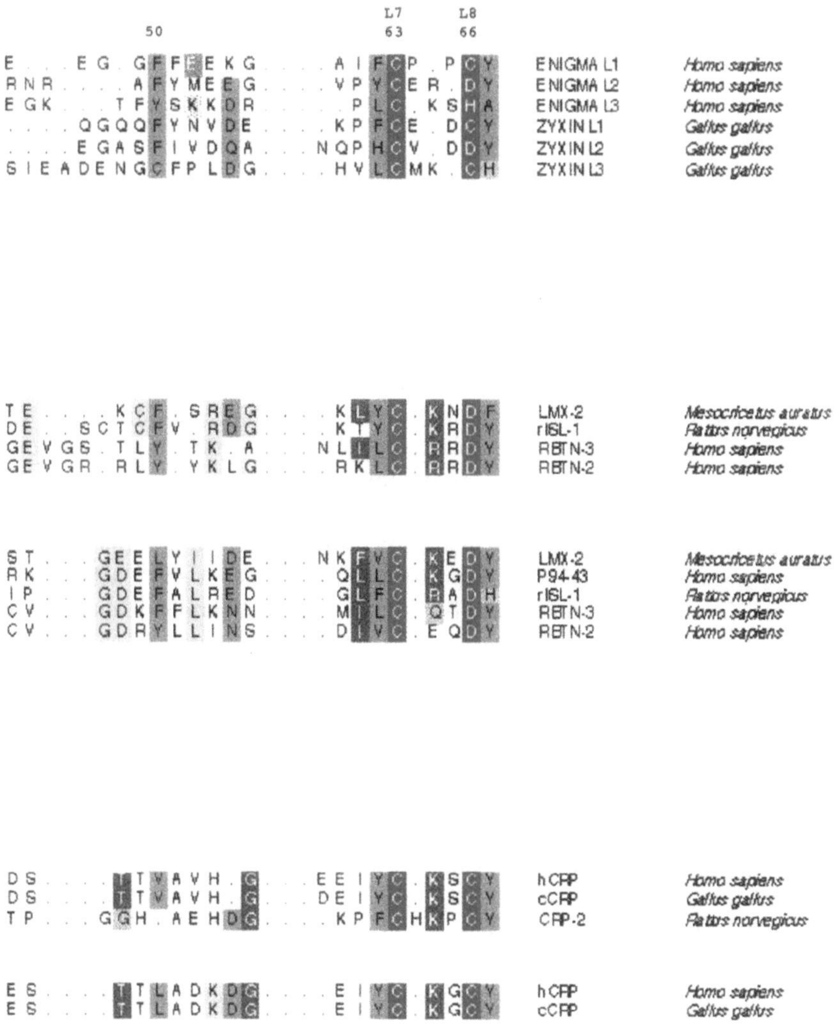

Fig. 1. (*Contd.*)

Cys at ligand position 5 to His abolished Zn^{2+} binding to LIM2 of CRP (MICHELSEN et al. 1994). Ligand position 8 can be Cys, His or Asp. The N terminal Zn^{2+} module thus consists of Cys-Cys-His/Cys/Asp-Cys/His/Glu and the C terminal Zn^{2+} module of Cys-Cys-Cys-Cys/His/Asp. Metal binding is essential for protein structure and renaturation studies indicate that binding is sequential with the C terminal module first occupied by Zn^{2+}, followed by the N terminal module (KOSA et al. 1994).

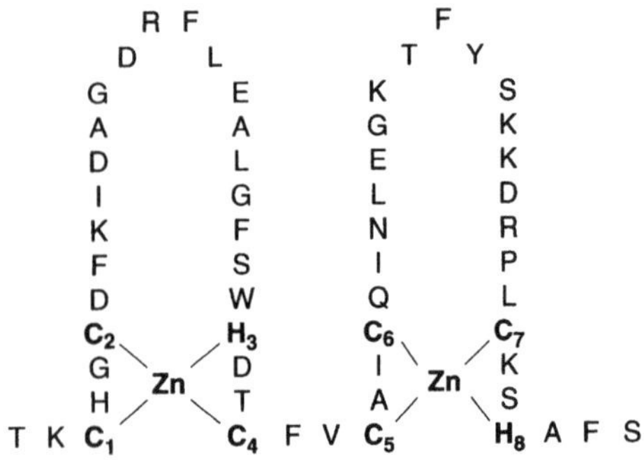

Fig. 2. Two Zn^{2+} finger structure of LIM domains. The sequence of LIM3 of Enigma (Wu and Gill 1994) is presented in the traditional double Zn^{2+} finger arrangement

The NMR structures of LIM2 of CRP and CRIP are similar and show that the N terminal and C terminal Zn^{2+} binding modules are packed together via a hydrophobic interface (Perez-Alvarado et al. 1996). Both structures contain four antiparallel β sheets (Fig. 3A). Two of the four β sheets (β1 and β3) contain a rubredoxin-type turn characteristic of the metal chelating domains of iron-binding rubredoxins (Blake et al. 1994). The first β sheet is followed by a turn that leads to the second β sheet oriented nearly perpendicular to the first. A short helical turn leads to the C terminal module that also contains two antiparallel β sheets. The third β sheet is followed by a turn that leads to the fourth antiparallel β sheet. Both LIM2 of CRP and CRIP contain C terminal α-helices but these have different orientations. The N and C terminal Zn^{2+} binding modules of the two proteins are similar but differ in orientation relative to the N terminal Zn^{2+} module (Perez-Alvarado et al. 1996).

The hydrophobic residues which constitute the core of LIM domains are conserved. Figure 3B shows how the side chains of some of these residues that are conservatively substituted in the extended LIM domain family pack against each other to form the hydrophobic core. Surface potential calculations for CRIP and LIM2 of CRP show hydrophobic surfaces surrounded by well-defined clusters of positive and negative charge. One hydrophobic surface is formed by the residues that also make up the core (Fig. 4A). A second hydrophobic surface is made up of conserved residues in the C terminal module (Tyr 62 and Tyr 67, Fig. 4A). Another hydrophobic surface formed by the C terminus of CRIP is not present in LIM2 of CRP (Perez-Alvarado et al. 1996). An interesting feature of both LIM domain structures that belong to the C class is the positively charged surface located in the upper module that follows an α-helical path (Fig. 4B). As discussed by Perez-Alvarado et al. (1994, 1996) this structure is very similar to the N terminal Cys 4 modules of the glucocorticoid hormone receptor and GATA-1 DNA binding

Fig. 3A, B. Structure of LIM domains. The structure of LIM2 of CRP was determined by NMR and is reprinted by permission of the authors (Perez-Alvarado et al. 1994). **A** emphasizes the two Zn^{2+} coordination modules at the two ends of the domain and **B** shows conserved residues that make up the hydrophobic core. *Residue numbers* correspond to those in CRP

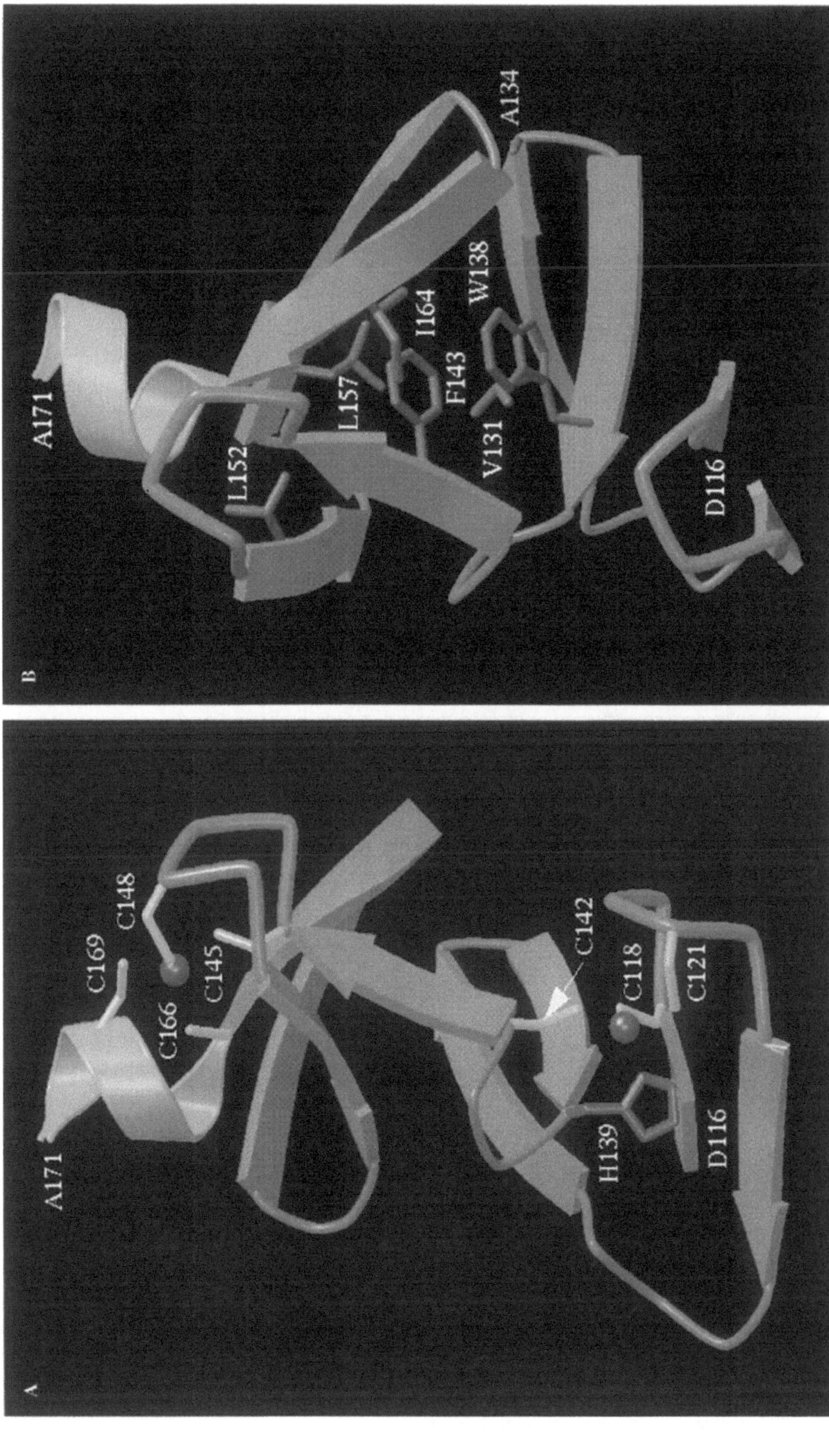
A
A171
C166
C169
C148
C145
C142
C118
C121
H139
D116
B
A171
L152
L157
F143
I164
W138
V131
A134
D116

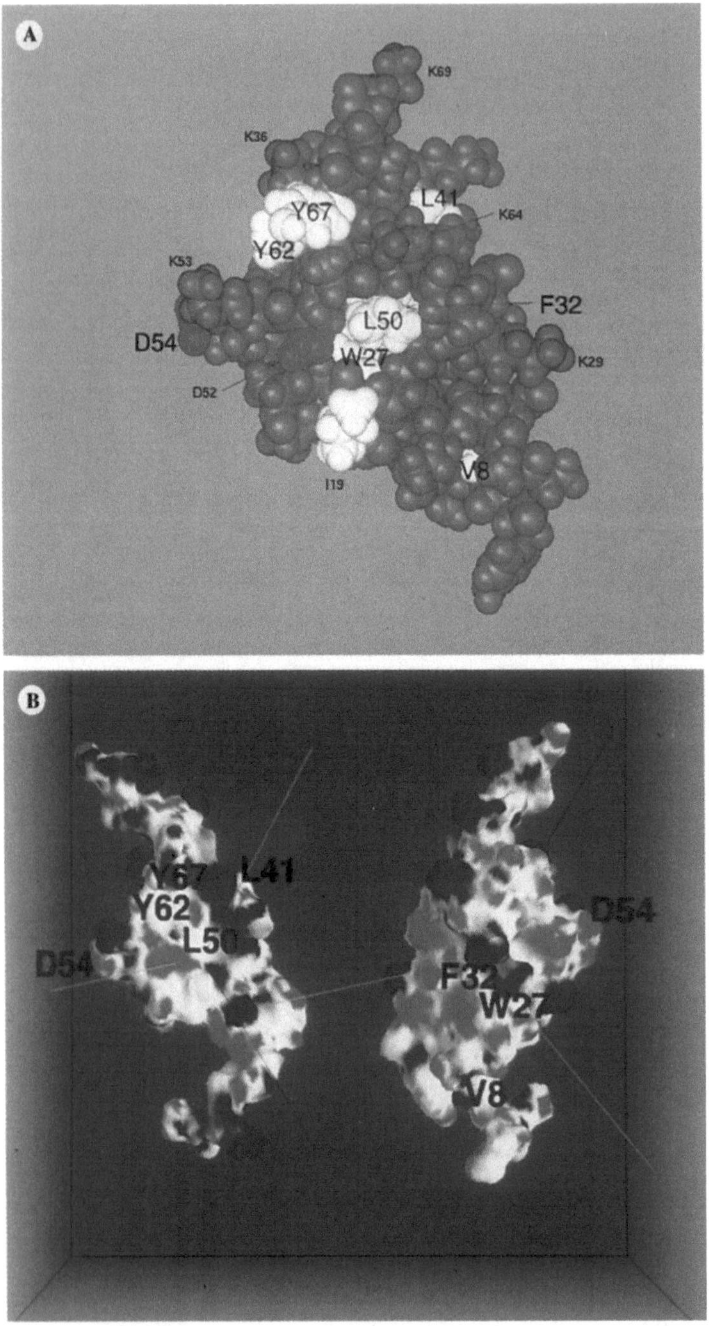

Fig. 4A, B. Space filling (**A**) and electrostatic (**B**) models of LIM2 of CRP. The data of PEREZ-ALVARADO et al. (1994) (PDB accession: 1CTL) were used to prepare these representations using the programs UHBD and GRASP. Analysis was kindly provided by Tom Diller, Department of Chemistry and Biochemistry, University of California, San Diego. Colors used: blue = basic, red = acidic, white = hydrophobic, green = unspecified amino acid residues

domains. This is not, however, conserved in the A class of LIM domains. Although most available data suggests that LIM domains direct protein-protein interactions, these authors raise the possibility that LIM domains could be multifunctional, interacting with both proteins and nucleic acids. One could also consider different protein interactions with the hydrophobic and positively charged faces.

4 Function of LIM Domains

Although three-dimensional structures of SH2 (Overduin et al. 1992; Waksman et al. 1992), SH3 (Koyama et al. 1993; Kohda et al. 1993; Lee et al. 1996), WW (Macias et al. 1996) and PDZ (Doyle et al. 1996) domains have been solved with bound target sequences, the structures of LIM2 of CRP and of CRIP do not contain target peptides. To date, neither a specific target for LIM2 of CRP and CRIP nor a general target peptide sequence for LIM domains has been identified. However, three types of target motifs have been described: dimerization with other LIM domains, tyrosine-containing tight turns and unspecified sequences in other proteins. By analogy with the structures of GATA-1 and the glucocorticoid receptor, some LIM domains have the potential to bind DNA (Perez-Alvarado et al. 1994, 1996).

Zyxin, a phosphoprotein present in sites of cell adhesion to the extracellular matrix, contains a proline-rich α-actinin binding domain at its N terminus and three LIM domains at its C terminus (Crawford and Beckerle 1992; Sadler et al. 1992). Zyxin interacts with a second cytoskeleton associated protein, CRP, via the first LIM domain of Zyxin (Schmeichel and Beckerle 1994). The interaction is specific because neither LIM2 nor LIM3 of Zyxin interacted with CRP. Zyxin is thus proposed to be an adaptor molecule binding different partners at its N and C termini. Although CRP contains little sequence beyond that encompassed in the two LIM domains, a specific sequence within CRP, i.e., a single LIM domain that interacts with LIM1 of Zyxin was not demonstrated. CRP was demonstrated to homodimerize, a reaction mediated by either of its two LIM domains, but not by an inter-LIM domain sequence fragment (Feuerstein et al. 1994).

Paxillin is another LIM domain-containing protein found in focal adhesions. At its N terminus, Paxillin contains binding sites for Vinculin and for the focal adhesion tyrosine kinase, FAK (Brown et al. 1996). Four LIM domains are located at the C terminus of Paxillin (Turner and Miller 1994). Although Paxillin was originally thought to localize at focal adhesions through binding to Vinculin, mutational analyses indicate that LIM3 is the domain that is necessary for Paxillin to localize at focal adhesions (Brown et al. 1996). Although the target for LIM3 of Paxillin has not been identified, these findings support the idea of recognition of specific targets by individual LIM domains.

A subtractive hybridization screen, designed to detect proteins enriched in denervated rat skeletal muscle, identified muscle LIM protein (MLP) which contains two LIM domains separated by a Gly-rich spacer region which is longer than that found between LIM domains in many proteins (ARBER et al. 1994). Overexpression of MLP enhanced and anti-sense expression decreased myogenic differentiation of C2 myoblasts. MLP and the structurally related proteins hCRP (WANG et al. 1992) and bCRP (WEISKIRCHEN and BISTER 1993) are located in the nucleus and at the actin cytoskeleton (ARBER and CARONI 1996). Using single LIM domains and various chimeras, ARBER and CARONI (1996) concluded that LIM2 of MLP was the major determinant required for association with actin filaments but that this association required a second LIM domain for cooperativity. Interestingly, a LIM domain from LMO1, a nuclear LIM-only protein, could provide the cooperative function exhibited by LIM1 of MLP but could not provide the actin targeting function of LIM2 of MLP. These results provide further support for the concept of specific targets for LIM domains. Analogous to the findings of FEUERSTEIN et al. (1994), ARBER and CARONI (1996) reported direct and indirect evidence for LIM-LIM interactions.

Enigma is a predominantly cytoplasmic protein that contains one PDZ domain at its N terminus and three LIM domains at its C terminus (WU and GILL 1994). LIM3 of Enigma specifically recognized the major endocytic code of the insulin receptor (InsR). Mutations in this target sequence that abolished ligand-induced endocytosis of InsR caused loss of recognition by LIM3 of Enigma. LIM3 of Enigma exhibited target specificity for InsR because it failed to interact with the tyrosine-containing endocytic codes of a variety of other receptors including those for EGF, LDL and tranferrin. Moreover, interaction of random peptide libraries with GST-LIM3 of Enigma revealed specific binding of the sequence Gly-Pro-Hyd-Gly-Pro-Hyd-Tyr-Ala that corresponds to the endocytic code of InsR (WU et al. 1996). NMR studies indicate that peptides containing the Gly-Pro-Leu-Tyr endocytic code of InsR adopt a tight turn conformation, a structure disrupted by mutations that prevent endocytosis of InsR (BACKER et al. 1992). Of the more than 12 LIM domains examined, only LIM3 of Enigma recognized the InsR endocytic code sequence. The hypothesis that LIM domains recognize specific tyrosine-containing tight turns was supported by finding that LIM2 of Enigma specifically recognized the C terminus of the tyrosine kinase receptor Ret (DURICK et al. 1996). Peptide competition indicated that the sequence Asn-Lys-Leu-Tyr was essential for this interaction (WU et al. 1996). Interestingly, this sequence is essential to mitogenic signaling by an oncogenic form of Ret, Ret/ptc2, that contains a fusion of the RI regulatory domain of A-kinase to the cytoplasmic domain of Ret (DURICK et al. 1996). LIM1 of Enigma and a variety of LIM domains from the A, B and C sequence classes failed to recognize these targets. Additionally, LIM2 of Enigma did not recognize InsR nor did LIM3 of Enigma recognize Ret. There is thus evidence for specific LIM domain binding to specific tyrosine-containing tight turn structures.

In attempts to design a "generic" endocytic code, the sequence Asn-Asn-Ala-Tyr-Phe was synthesized and shown to function in endocytosis when placed in mutant EGF receptors (CHANG et al. 1993). Two copies of this code separated by an α-helix were recognized by both LIM2 and LIM3 of Enigma (WU and GILL

1994). These results support the concept that LIM domains recognize tyrosine-containing tight turns but specificity depends on the amino acid sequence surrounding the Tyr residue. Like Zyxin, Enigma appears to serve an adaptor function binding to specific proteins via the individual LIM domains located at the C terminus and to other proteins via the N terminal PDZ domain. A subgroup of proteins that contain PDZ domains at their N termini and LIM domains at their C termini include Enigma, ENH (KURODA et al. 1996), ril (KIESS et al. 1995) and CLP 36 (WANG et al. 1995).

Using protein kinase C (PKC) in a two-hybrid screen, KURODA et al. (1996) isolated a rat protein termed ENH (*En*igma *H*omology). This protein contains a PDZ-like domain at its N terminus and three LIM domains at its C terminus, both exhibiting high homology to Enigma. The central region varies between the two proteins. The LIM domains of ENH, Enigma and LIM kinase specified interaction with the N terminal variable region of PKC. Thus, while there was little specificity for LIM domains there was high specificity for particular PKC isoforms. Interestingly, addition of TPA which activates PKC resulted in translocation of ENH from the membrane to cytoplasmic compartment suggesting that phosphorylation regulates cellular distribution of ENH. Moreover, PKC phosphorylated ENH in vitro. A protein kinase with specificity for serine and threonine residues that contains two LIM domains at its N terminus was independently cloned from kinase homology by two groups (MIZUNO et al. 1994; BERNARD et al. 1994). A second family member was subsequently identified (OKANO et al. 1995). LIM kinase is preferentially expressed in the nervous system (BERNARD et al. 1994; CHENG and ROBERTSON 1995). Chromosomal deletions involving LIM1 kinase occur in Williams syndrome which is characterized by poor visual-spatial constructive cognition (FRANGISKAKIS et al. 1996) suggesting that LIM kinase is necessary for development of specialized neuronal connections. The presence of LIM domains at the N terminus and a PDZ domain in the middle of the molecule suggests that protein interactions will specify subcellular localization and targets for LIM kinases.

The nuclear LIM-only proteins LMO1 and 2 (Rbtn 1 and Rbtn 2) are reported to bind to the transcription factors Tal1 and GATA-1. The observation that LMOs and Tal1 are activated by chromosomal translocation in human T cell leukemia (MCGUIRE et al. 1989; BOEHM et al. 1990a, 1991a; ROYER-POKORA et al. 1991; BAER 1993) and the observation that null mutations in LMO2 (WARREN et al. 1994), Tal1/Scl1 (SHIVDASANI et al. 1995; ROBB et al. 1995a; PORCHER et al. 1996), GATA-1 (PEVNY et al. 1991) and GATA-2 (TSAI et al. 1994) each block erythroid development suggested that interactions among these proteins were likely. Co-immunoprecipitation of Tal1 and LMO2 from MEL and HEL cells provided support for this idea (WADMAN et al. 1994; VALGE-ARCHER et al. 1994). Functional interactions were deduced from mammalian two hybrid analyses which showed transcriptional enhancement dependent on fusion proteins containing Tal1 and LMO2. The interaction was specific to the subset of basic helix-loop-helix (bHLH) proteins of the Tal1 family (Tal1, Tal2, LYL1) and was not observed with other bHLH proteins such as E47, Id1, Max, Myc and MyoD. The two hybrid interaction required the bHLH region of Tal1. Moreover, there was specificity for the LIM domains of LMO1 and 2 as CRP, CRIP and the LIM domains of Zyxin did not interact with Tal1. LMO2 was also shown to co-immunoprecipitate with GATA-1 (OSADA et al. 1995). Higher order complexes were deduced from mammalian cell two hybrid analysis where maximal increases in reporter gene activity were observed when GAL4-E47 (a heterodimeric partner of Tal1) and VP16-GATA-1 (which do not directly interact) were augmented by coexpression of Tal1 and LMO2. A tetrameric complex was proposed in which LMO2, which does not bind DNA, provided a bridge between the HLH heterodimeric E47/Tal1 complex and the Zn^{2+} finger containing GATA-1 (OSADA et al. 1995). Higher order complexes were not directly demonstrated however, nor were synergistic interactions between LMO2 and Tal1 shown using endogenous promoters.

Nuclear LIM interactor (NLI, also designated LIM-domain-binding protein, Ldb1) specifically binds to LMOs and to LIM-homeodomain proteins but not to cytoplasmic LIM domain containing proteins (JURATA et al. 1996; AGULNICK et al. 1996). The structure of NLI is highly conserved between *Xenopus* and mouse, and a second family member has been identified (AGULNICK et al. 1996). NLI binds to only one of the two LIM domains of Isl1, Lmx1, and LMO2, supporting the concept of target specificity for individual LIM domains; however, both LIM domains of Lhx1 appear to be required for the interaction (JURATA et al. 1996; AGULNICK et al. 1996). The site of interaction with LIM domains lies within a 38 amino acid fragment of NLI that lacks Tyr residues, indicating that target recognition differs from that exhibited by the LIM domains of the cytoplasmic protein Enigma (JURATA and GILL, 1997). NLI is coexpressed in ventral motor neuron nuclei with the LIM homeodomain protein Isl1 (JURATA et al. 1996). AGULNICK et al. (1996) observed that coinjection of NLI and Xlim-1 mRNA into *Xenopus* embryos caused formation of partial secondary axes especially notable as dorsalization and ectopic muscle formation. Neither Xlim-1 nor NLI alone were effective, demonstrating a functional interaction between the two proteins. The broad range of interactions between NLI and LIM-domain containing transcription factors suggests utilization of a common mechanism to impart unique cell fate instructions.

5 Nuclear LIM-Only Genes

The *LMO* genes are represented by three family members, *LMO1*, *LMO2*, and *LMO3*. They are distinct in function from the extranuclear proteins CRP, CRIP, and PINCH, whose structures also consist largely of LIM domains. Like the LIM homeodomain proteins, the LMO proteins appear to play a role in transcription as they localize in the nucleus and associate with other known transcription factors. How they participate in transcriptional processes at a molecular level is, however, unknown and target genes have yet to be described. LMO proteins function in development and differentiation and like their homeodomain-containing counterparts, are highly expressed in the embryonic and adult nervous systems.

LMO1 (formerly *Ttg-1*, *Rbtn1*) was first described as a gene disrupted by a t(11;14)(p15;q11) T-cell translocation event involving the TCRδ locus in RPMI8402 cells derived from a patient with T cell acute lymphoblastic leukemia (T-ALL) (BOEHM et al. 1988). The cDNA was isolated using the breakpoint region and was found to encode a nuclear, 18-kDa, cysteine-rich protein that was not expressed in normal thymus tissue or in T cell lines other than RPMI8402 (McGUIRE et al. 1989, 1991; BOEHM et al. 1990a). The *LMO1* gene utilizes two unrelated, alternative promoters (1a and 1) to generate transcripts of different length, 1.4 and 1.2 kb, and proteins which differ by only a single, conservative amino acid substitution (BOEHM et al. 1990a). The cysteine-rich regions of LMO1 were subsequently identified as LIM domains (BOEHM et al. 1990b; RABBITTS and BOEHM 1990), and murine and Drosophila *LMO1* homologues were isolated and found to be highly conserved (BOEHM et al. 1990a, b; ZHU et al. 1995).

More detailed analysis of *LMO1* expression revealed high levels in the embryonic and adult nervous systems, specifically in the branchiomotor trigeminal and facial neurons within rhombomeres two and four of embryonic day 9 (E9) mouse embryos, and the ventral spinal cord, ventral forebrain, and retina of E10 embryos (GREENBERG et al. 1990). In later stages of development (E15-16), *LMO1* mRNA was seen in the neocortex, diencephalon, mesencephalon, cerebellum, myelencephalon and the spinal cord (GREENBERG et al. 1990; BOEHM et al. 1991b). Very low levels were observed in non-nervous tissues, including the developing thymus. Interestingly, alternative promoter usage appeared to be regulated throughout development, with promoter 1a generating most of the expression from E12–E15, and promoter 1 primarily directing expression after E15 (BOEHM et al. 1991b).

The second *LMO* family member, *LMO2* (*Ttg-2*, *Rbtn2*), was isolated by two different groups using low stringency screening with *LMO1* (BOEHM et al. 1991a) and by the cloning of the separate, but linked, 11p13 breakpoint region in T-ALL, which is also involved in translocations with the TCRδ gene (ROYER-POKORA et al. 1991). LMO2 is 50% identical to LMO1, with the highest homology in the LIM domains and intervening linker region (BOEHM et al. 1991a; ROYER-POKORA et al. 1991). In contrast to *LMO1*, whose expression is primarily confined to the nervous system, *LMO2* expression is widespread although similarly low in the thymus and T cells (BOEHM et al. 1991a; ROYER-POKORA et al. 1991). *LMO2* mRNA is highest in embryonic liver and spleen, the sites of fetal hematopoiesis, as well as in adult brain, kidney, liver, and spleen (ROYER-POKORA et al. 1991; FORONI et al. 1992). Low levels of *LMO2* detected in the thymus were attributed to the medullary epithelial cells, not to T cells (NEALE et al. 1995a).

Like *LMO1*, the *LMO2* gene was found to utilize two alternate promoters to generate messages of different length. The most distal promoter was shown to contain two GATA-1 sites, suggesting erythroid-specific expression from the upstream promoter (ROYER-POKORA et al. 1995). The 11p13 breakpoint, which disrupts the chromosome 5′ to *LMO2* coding regions, was proposed to deregulate promoter 1, thereby activating promoter 2 in T-ALL. Translocations in this region are found more frequently in T-ALL patients than the rare 11p15 breakpoint involving *LMO1*, suggesting that *LMO2* may play a more common role in the pathogenesis of this childhood leukemia (Boehm et al. 1991a).

The importance of the erythroid specific expression of *LMO2* was illustrated by homozygous deletion of the gene in mice (Warren et al. 1994). Although resulting in embryonic lethality, a detailed study of developing embryos revealed that –/– mice contained no erythroid cells in the yolk sac or the fetal circulation and that the embryos died at approximately E10.5 presumably due to the failure of red blood cell development. In support of a requirement for *LMO2* in erythropoiesis, *LMO2* was found to be expressed normally in erythroid lineages, including adult bone marrow, and was shown to be required for erythroid differentiation of embryonic stem cells in vitro.

The final known member of the *LMO* gene family, *LMO3* (formerly *Rbtn3*), was also identified by homology to *LMO1* (Boehm et al. 1991a; Foroni et al. 1992). LMO3 is more closely related to LMO1 than to LMO2, with 89% identity between LMO1 and LMO3 (Foroni et al. 1992). Expression of LMO3 also resembled that of LMO1 in its restriction to tissues of the central nervous system and, like its relatives, showed undetectable levels in the thymus. In contrast to the other family members, however, *LMO3* does not map to chromosome 11 and thus far has not been identified in leukemia related translocation events.

Transgenic mice targeting expression of *LMO1* and *LMO2* to T cells and other tissues were used to address the question of whether translocation-mediated deregulation of these genes was a circumstantial or causative event in the pathogenesis of T cell acute lymphoblastic leukemia. *LMO1* expression from the lymphoid specific *lck* promoter or TCRβ enhancer resulted in clonal lymphoblastic T cell lymphomas similar to those observed in T-ALL (McGuire et al. 1992; Fisch et al. 1992). Overt malignancy was preceded by pre-leukemic changes in which the thymus and spleen became enlarged with an increased number of lymphocytes representing all stages of maturity (McGuire et al. 1992) but the immunophenotype of most tumor cells was $CD4^+CD8^+$, suggesting that these were derived from immature lymphocytes (McGuire et al. 1992; Fisch et al. 1992). Interestingly, tumor incidence was relatively low, but the frequency was proportional to the amount of transgene expression observed in T cells (McGuire et al. 1992). The long latency period preceding malignancy (approximately 10.5 months after birth) suggested that mutations in addition to overexpression of *LMO1* are required to accumulate prior to leukemogenesis (Fisch et al. 1992; McGuire et al. 1992). Directed expression of *LMO1* to the endocrine pancreas using an insulin promoter construct resulted in no insulinomas, indicating that misexpression in the thymus represents a specific case of *LMO1*-induced oncogenesis (Fisch et al. 1992).

Targeted expression of *LMO2* to T cells using either the CD2 promoter and enhancer (Fisch et al. 1992), or a thy1.1 expression cassette (Larson et al. 1995), also resulted in leukemogenesis resembling T-ALL; however, the incidence of tumor formation was much higher (72%) than for *LMO1* (averaging 19% for four transgenic lines) (Larson et al. 1994; McGuire et al. 1992; Fisch et al. 1992). Clonal tumors were preceded by changes in which populations of immature $CD4^-CD8^-$ lymphocytes were expanded at the expense of $CD4^+CD8^+$ cells to maintain homeostasis in the developing thymus (LARSON et al. 1995), but variable CD4/CD8 immunophenotypes were represented in tumors (LARSON et al. 1994). The latency period of *LMO2* induced leukemia was approximately 10 months, similar to that for *LMO1*, again suggesting the requirement for additional mutational events (LARSON et al. 1994). As with *LMO1*, the incidence of tumorigenesis but not the latency period was correlated to the amount of transgene expression in T cells (LARSON et al. 1994). Widespread overexpression of *LMO2* using the inducible metallothionein-1 promoter resulted in tumors that were only of T cell origin and whose development was similar to those induced by CD2 or thy1.1 driven *LMO2* misexpression (NEALE et al. 1995b). Similar to the case with *LMO1*, this suggests that only T cells provide a microenvironment which can be perturbed to oncogenesis by *LMO2* overexpression. Thus, although similar in latency period and specificity for induction of T cell leukemia, the pre-malignant changes and the tumors resulting from *LMO1* and *LMO2* misexpression are qualitatively different.

One of the additional mutational requirements for *LMO2* mediated leukemogenesis may be the upregulation of *Tal1* expression in the same cell. *Tal1*, which encodes a basic helix-loop-helix protein, is activated in over 25% of T-ALL cases, either by deletion of its promoter, or by translocation-mediated deregulation (BAER 1993) and some tumors show activation of both *LMO2* and *Tal1* (WADMAN et al. 1994). Like LMO2, Tal1 is normally expressed in erythroid cells and is required for erythropoiesis (SHIVDASANI et al. 1995; PORCHER et al. 1996), and LMO2 and Tal1 can be immunoprecipitated as a complex from erythroid cells, suggesting that they function together in directing erythroid differentiation (WADMAN et al. 1994; VALGE-ARCHER et al. 1994). Although no tumors resulted from misexpression of *Tal1* alone in the thymus (ROBB et al. 1995b; LARSON et al. 1996), coordinate misexpression of *Tal1* and *LMO2* using the CD2 promoter decreased the latency period of leukemogenesis from 10 months for *LMO2* alone, to 7 months for *LMO2*/*Tal1* (LARSON et al. 1996). The tumor phenotypes generated by *LMO2*/*Tal1* were generally $CD4^-CD8^-$, rather than the mixed phenotypes observed for *LMO2* alone, or $CD4^+CD8^+$ for *LMO1* alone, indicating that the presence of *Tal1* affects the developmental fate or origin of tumors. LMO2 and Tal1 could be co-immunoprecipitated from double transgenic thymuses, implying a coordinate involvement of these proteins in directing aberrant T cell development, as well as normal erythroid development. Although the co-misexpression of *Tal1* may answer some questions of how *LMO2* directs leukemic development, the relatively long 7-month latency period still implies the involvement of other, as yet unknown steps in the pathogenesis of T-ALL.

6 LIM Homeobox Genes

The initial isolation of the homeobox genes encoding the LIM domain-containing proteins, lin-11, Isl1, and mec-3, from the vastly divergent species of *C. elegans* (*mec-3* and *lin-11*) and rat (*Isl1*), illustrated the widespread use of this gene family across evolution. Additional genes were subsequently isolated from *Drosophila*, zebrafish, chinook salmon, *Xenopus*, chick, mouse, and human, suggesting that every animal species, from the lowest invertebrates to humans, utilizes LIM homeodomain proteins in their development and homeostasis.

LIM homeobox genes have restricted patterns of expression during embryogenesis and function in the specification of cell lineages. Most are expressed within regions of the developing nervous system but many have additional domains of expression in non-neuronal tissues, such as muscle, endocrine tissues, and sites of lymphocyte differentiation. Continued expression of most of these genes in adult tissues implies a role in the maintenance of differentiated cell phenotypes. The LIM homeodomain proteins share the same general structure, with two N terminal LIM domains, separated from a homeodomain by a linker region. Putative transcriptional activation domains have been found either within the linker region (apterous) or C terminal to the homeodomain.

6.1 Invertebrate LIM Homeodomain Proteins

The *lin-11* gene was isolated by the mapping of a transposon insertion in a *lin-11* mutant strain of *C. elegans* (Freyd et al. 1990). Such mutants have defects in vulval formation due to an inability to generate necessary vulval cells through the asymmetric cell division of secondary blast cells (Table 1). In mutant animals, this division generates two daughter cells which take on the same fate, rather than their normal disparate fates (Ferguson et al. 1987). The transposon insertion was found to interrupt a homeobox in the *lin-11* gene and conceptual translation of *lin-11* also identified a proline rich C terminus indicative of a transcriptional activation domain and two N terminal LIM domains (Freyd et al. 1990). Of interest, the LIM domains of lin-11 are more closely related to the LIM domains of the rat Isl1 than to the *C. elegans* LIM protein mec-3, although the homeodomains of lin-11 and mec-3 are much more similar to each other (43/60 residues identical) than they are to the homeodomain of Isl1 (27/60 from Isl1 to lin-11). Comparison of the homeodomain sequences of the LIM proteins lin-11, Isl1, and mec-3 with those of other homeodomain classes indicated that the LIM proteins were more similar to each other within this region than to the antennapedia and paired classes of homeodomain proteins. Notably, a conserved tyrosine found at position 25 of the homeodomains of antennapedia and paired class proteins was substituted with a basic residue in each of the LIM homeodomain proteins, distinguishing the LIM domain genes as a new homeobox class.

As with *lin-11*, the *mec-3* gene was identified by transposon tagging in a mutant *C. elegans* strain (Way and Chalfie 1988). *mec-3* mutants are insensitive to gentle touch as a result of improper development of six mechanosensory "touch" neurons. Analogous to the vulval defect in *lin-11* mutants, daughter cells which normally result from an asymmetric cell division become the same type of cell in *mec-3* mutants, precluding the normal development of one of the daughter cells into a touch cell. From their studies, the authors concluded that *mec-3* specifies a touch cell fate while repressing an alternative fate, that of the sister cell.

The first *Drosophila* LIM homeobox gene, *apterous* (*ap*), was simultaneously isolated by two different groups, one looking at its effects in wing development, and the other in muscle and neuronal development. *Apterous* was cloned in a search for the gene resulting in the wingless and haltereless apterous mutant strain of *Drosophila melanogaster* (Cohen et al. 1992). Mapping of a P element insertion in an *ap* mutant identified the LIM homeobox gene encoding a 459 amino acid protein with a proline-rich linker region which may serve as a transcriptional activation domain. High levels of *ap* expression were seen in the larval wing imaginal disks, in regions destined to become the dorsal surface of the wing, haltere blade, and wing hinge in the adult. In addition, *ap* was expressed in leg, eye, and antennal imaginal disks.

The onset of *ap* expression in the second instar larval wing imaginal disc coincided temporally and spatially with the restriction of dorsal and ventral lineages in the wing (Diaz-Benjumea and Cohen 1993). Genetic mosaic analysis resulted in the formation of *ap*$^-$ clones surrounded by *ap*$^+$ wild-type cells in the dorsal surface of the wing, resulting in an ectopic wing margin. Oddly, the ectopic margin resulted in the additional outgrowth of wild-type cells, suggesting that formation of the dorsal-ventral boundary between *ap*$^-$ and *ap*$^+$ cells normally produces a signal that induces proliferation. In agreement, the *ap* mutant flies in which the wing margin is never formed, are wingless, suggesting a requirement for dorsal-ventral boundary formation in wing development (also Williams et al. 1993). Thus the removal of *ap* from presumptive dorsal wing cells makes them ventral in character and the continued expression of *ap* is required to maintain a dorsal phenotype, suggesting a role for *ap* in the specification and maintenance of dorsal wing cell types (Blair et al. 1994).

Bourgouin et al. (1992) independently identified *apterous* using an enhancer trap strategy to identify genes expressed specifically in subsets of neurons and muscles. *ap* expression was seen in six muscles of each abdominal hemisegment in the embryo with overall expression peaking in mid-embryonic development. In *ap* mutants, these abdominal muscles are not formed suggesting a functional role for *ap* in muscle differentiation. In support of this hypothesis, ectopic expression of *ap* under control of a heat shock promoter resulted in the formation of extra muscles in this region. In addition, *ap* expression in five neurons near the *apterous* expressing abdominal muscles, suggested that these may represent motor neurons innervating the *ap*$^+$ abdominal muscles.

Ap expression in the CNS was also seen in the embryonic brain and ventral nerve cord (Cohen et al. 1992; Bourgouin et al. 1992; Lundgren et al. 1995). Neurons of the ventral nerve cord expressing *ap* were identified as interneurons, which began elongating axons soon after the onset of *ap* expression (Lundgren et al. 1995). *Ap*$^+$ interneuron axons all followed the same pathway within the ventral nerve cord, eventually fasciculating with one another. In *ap* mutants, however, the neurons that normally express *ap* projected axons along incorrect pathways and did not fasciculate with one another, perhaps explaining the uncoordinated nature of *ap* mutant flies (Altaratz et al. 1991).

6.2 Vertebrate LIM Homeodomain Proteins

Aside from *Isl1*, the first vertebrate LIM homeobox genes were cloned by TAIRA et al. (1992) from *Xenopus laevis* using a degenerate PCR approach based on the LIM class of homeobox sequences. *Xlim-1*, the first to be characterized, encoded a protein with similar structure to those described previously, and contained a C terminal proline rich region (TAIRA et al. 1992). Expression of *Xlim-1* began at the gastrula stage, with highest levels in the dorsal lip of the blastopore (Spemann's organizer) and the dorsal mesoderm and endoderm with only low levels seen in the ectoderm. mRNA levels peaked in the late gastrula/early neurula stage and declined, although a secondary activation of *Xlim-1* was observed at the tadpole stage. Activin A, a dorsal mesoderm inducer, as well as retinoic acid, activated *Xlim-1* expression, and these agents synergized with one another in the induction of *Xlim-1*. High levels of *Xlim-1* expression were localized in the prechordal plate and notochord, disappearing from the notochord during neural tube formation, at which time *Xlim-1* is expressed in the lateral regions of the neural tube itself (TAIRA et al. 1994a). At later stages, *Xlim-1* was seen in the forebrain, midbrain, and hindbrain. In addition, high levels of expression were seen in the dorsolateral region destined to become the pronephros and pronephric duct, and continued to be expressed in the adult kidney.

Injection of RNA encoding Xlim-1 with deleted or mutant inactive LIM domains into two cell embryos activated expression of the neuronal markers NCAM and *engrailed-2* as well as cement gland markers (TAIRA et al. 1994b). Interestingly, explant experiments showed that mutant *Xlim-1* injection could induce neuronal markers in adjacent, uninjected cells, suggesting a non-cell autonomous function for *Xlim-1*. Presumably, the direct or indirect activation of genes encoding secreted or cell surface neuralizing molecules results in neuronal differentiation of nearby cells. Injection of mutant RNA also induced expression of *goosecoid*, another homeobox gene expressed in the organizer region. Finally, ectopic expression of mutant *Xlim-1* in ventral regions of the embryo resulted in secondary axis formation similar to that seen with injection of *goosecoid* RNA. Coinjection of mutant *Xlim-1* and *goosecoid* resulted in the synergistic formation of secondary axes, often resulting in two ectopic notochords. In these studies, wild-type *Xlim-1* RNA only slightly activated muscle specific genes in the ventral mesoderm, and had no neuronal differentiation activity. The TGFβ family member *nodal*, whose expression is restricted to the node of gastrulating mouse embryos (ZHOU et al. 1993), may be upstream of *Lim 1* activation. In zebrafish, *nodal* coordinately upregulates *Lim1* and *goosecoid* and similarly induces ectopic secondary axes when misexpressed (TOYAMA et al. 1995a).

Lhx1, the murine homologue of *Xlim-1*, was found to be virtually identical in the LIM domains and homeodomain to its amphibian counterpart (BARNES et al. 1994; FUJII et al. 1994). As in *Xenopus*, *Lhx1* is expressed in gastrula staged embryos (E7-7.5) in the periphery of the node, the mammalian equivalent of the organizing blastopore lip (BARNES et al. 1994). Expression in the surrounding lateral and intermediate mesoderm was also evident and in later stages in the nephragenic cord, mesonephros, mesonephric duct, hindbrain, and the postmitotic commissural neurons of the dorsolateral neural tube (BARNES et al. 1994; FUJII et al. 1994). In the adult, *Lhx1* expression is maintained in the cerebellum, medulla, and the kidney. Thus, early gastrula expression of *Xlim-1* and *Lhx1* in the organizer and the mesoderm suggests a regionalizing role for this gene, while the specificity of later embryonic expression in the kidney, neural tube, and brain implies a cell type specification function similar to that of *lin-11*, *mec-3*, and *apterous*. Finally, expression in adult tissues indicates an additional role for the maintenance of differentiated cell phenotypes.

Compelling evidence for the role of *Lhx1* in embryonic anteriorization and cell type specification was illustrated by deletion of the gene by homologous recombination in mice (SHAWLOT and BEHRINGER 1995). This knockout resulted in embryonic lethality around E10, and recovered embryos were devoid of head structures anterior to rhombomere 3 of the hindbrain. Forebrain and midbrain markers were not expressed in the mutant embryos. In gastrula staged embryos, no organized node was formed in the head region and the node markers *Brachyury* and HNF3β were absent. Posterior to the hindbrain, development appeared to proceed normally, except for the absence of kidneys and gonads noted in the rare –/– mice that were carried full term.

Lhx2 (*LH-2*) was cloned by a subtractive hybridization approach to identify genes preferentially expressed in early B cell development (XU et al. 1993). In the homeodomain region, Lhx2 is most closely related to apterous, but divergence of the LIM domains prevents positive identification of *Lhx2* as the mammalian homologue of *apterous*. *Lhx2* is also expressed in T cell lines and shows highest levels of expression in the embryo, especially in the developing nervous and immune systems and the pituitary (XU et al. 1993; ROBERSON et al. 1994). In the embryonic nervous system, expression is concentrated in the differentiating neurons of the fore-, mid-, hindbrain, and spinal cord as well as in the developing retina and optic stalk (XU et al. 1993). Localization of *Lhx2* to the embryonic liver probably reflects pre-B cell development. In the adult, *Lhx2* is expressed in the cerebral cortex. Overexpression of *Lhx2* in chronic myelogenous leukemia (CML), a stem cell leukemia of myeloid lineage, implicates a role for *Lhx2* in the pathogenesis of this disease (WU et al. 1994). Of interest, *Lhx2* maps to human chromosome 9q34.1 with c-*Abl*, whose translocation with the B cell receptor gene to generate the Philadelphia chromosome results in the BCR-Abl fusion oncoprotein, the hallmark of CML (WU 1995). *Lhx2* is not translocated in the process, but disruption of surrounding regulatory regions results in upregulated gene expression.

Identification of *Xlim-3* using the same PCR based screen that isolated *Xlim-1* (TAIRA et al. 1992) revealed yet another LIM homeobox gene with high levels of specific expression in the embryonic nervous system. Highest levels of *Xlim-3* expression were found in the anterior and intermediate lobes of the pituitary, but postmitotic ventral motor neurons of the hindbrain and spinal cord, the retina, and the pineal gland also expressed *Xlim-3*. As with the other LIM homeobox genes, residual expression was evident in the adult, in the pituitary, eyes, and brain, suggesting a role for *Xlim-3* in the differentiation and maintenance of neuroendocrine and neuronal tissues.

The murine homologue of *Xlim-3* was shown to have an embryonic expression pattern very similar to that described for *Xlim-3* (SEIDAH et al. 1994; BACH et al. 1995; ZHADANOV et al. 1995). Expression of *Lhx3* in Rathke's pouch (E9.5), the primordium of the pituitary, preceded pituitary morphogenesis and the early marker α-GSU. However, in contrast to the expression in *Xenopus*, *Lhx3* expression in the adult mouse was limited to the anterior and intermediate lobes of the pituitary, but absent in the remainder of the nervous system. The role of *Lhx3* in development of the pituitary was demonstrated by targeted disruption of the gene which resulted in mice lacking both anterior and intermediate lobes of the pituitary (SHENG et al. 1996). Embryos were carried for the full gestation period but died either late in development or shortly after birth. In mutant embryos, Rathke's pouch formed but failed to grow and none of the hallmark pituitary genes (*TSH*β, *GH*, *LH*, or *Pit-1*) were expressed, suggesting that *Lhx3* is required for the generation or differentiation of a precursor cell common to the thyrotrope, gonadotrope, somatotroph, and lactotroph lineages.

A general search for homeobox genes identified *Lhx4* (*Gsh-4*) which is very similar to *Lhx3* within the homeodomain as well as in the region C terminal to the homeodomain (LI et al. 1994). However, in contrast to *Lhx3*, no expression was evident in the retina or Rathke's pouch, but like many other *Lhx* genes, *Lhx4* is expressed in ventral regions of the hindbrain and neural tube. Homozygous deletion of the gene resulted in lethality shortly after birth, but the cause of death was inconclusive. Newborn pups had apparently normal neural tubes and moved normally, yet had difficulty inflating their lungs and presumably died as a result. While lung development appeared to be relatively normal, expression of *Lhx4* in the ventrolateral medulla of wild-type animals, where the respiratory control centers are located, suggests a potential defect in the respiratory center of mutant mice.

Lhx5 is the homologue of *Xlim-2* isolated in the screen that identified *Xlim-1* and *Xlim-3* (TAIRA et al. 1992), but was first described in zebrafish as *Lim5* (TOYAMA et al. 1995b). Lim5 was found to be most similar to Xlim-1/Lhx1 with only one amino acid difference in the homeodomain. In both Xenopus and zebrafish, expression first appeared in the late blastula stage and peaked in the mid-gastrula. In contrast to *Xlim-1/Lhx1*, *Lim5* expression is localized in the ectoderm rather than the endoderm and mesoderm of the gastrula, but like *Xlim-1/Lhx1* shows restriction of expression within the central nervous system later in development. During neurulation, *Lim5* expression marked an anterior region of the neural plate, resulting in a stripe of expression possibly demarcating the future boundary between the diencephalon and midbrain. Unlike *Xlim-1*, which is upregulated in the presence of high levels of activin, *Xlim-5* is repressed by low concentrations of activin in dissociated cells, suggesting that differential regulation by external factors results in the differential expression patterns of these very similar genes.

Although initially isolated by virtue of its binding to the rat insulin I minienhancer, *Lmx1* appears to also play a role in dorsal-ventral patterning of the limb bud (RIDDLE et al. 1995). *Lmx1* expression is induced in the dorsal mesoderm of the limb bud by Wnt7a, which is secreted from the dorsal ectoderm immediately prior to expression of *Lmx1* in the underlying mesoderm. The causality of the *Lmx1* induction was determined by infecting the ventral ectoderm of a chick limb bud with a retrovirus containing *Wnt7a*, which induced ectopic *Lmx1* expression in the ventral mesoderm. Retroviral infection with *Lmx1* in the ventral mesoderm produced a dorsalization of the tendons, muscles, and digits of the limb bud, and in some cases, extra digits. Such dorsalization of ventral structures suggests a role for *Lmx1* in limb development analogous to that of *apterous* in the dorsalization of the fly wing. In each case, it appears that the default state is ventral, with the LIM homeobox genes providing dorsalizing signals to create dorsal-ventral polarity in the appendage. When extra digits were formed as a result of *Lmx1* infection, ectopic expression of *sonic hedgehog* (*shh*) was seen just proximal to the border of tissue expansion. The authors also noted *Lmx1* expression in the notochord and mesonephros, and uniform expression in the neural tube in early stages. Expression in the neural tube later became restricted to the roof plate, floor plate, and a subset of dorsal neurons. Notochordal expression of *Lmx1* is interesting, in that the notochord has been shown to be a source of *shh* secretion, which directs the differentiation of floor plate and motor neurons of the overlying neural tube (YAMADA et al. 1993; ERICSON et al. 1995).

Also isolated in a screen for proteins capable of binding to the rat insulin I minienhancer (KARLSSON et al. 1990), *Isl1* shows the most conservation across species of the known LIM homeobox genes with 98% amino acid identity from zebrafish to rat (INOUE et al. 1994). In addition to its expression in the pancreatic islets cells for which it was named, *Isl1* is found in many different endocrine and neuronal tissues. High levels of *Isl1* are expressed in the anterior and intermediate lobes of the pituitary, in calcitonin producing thyroid cells, and in adrenomedullary chromaffin cells (THOR et al. 1991). Widespread expression in the nervous system includes neurons of the autonomic peripheral nervous system (in the adrenal medulla), somatic sensory neurons of the dorsal root ganglia, motor neurons, cranial motor nuclei, neurons of the hypothalamus, pineal gland, and the retina and visual processing centers of the brain. As with the other LIM homeobox genes, *Isl1* expression continues through adult stages, implying a role in maintenance of differentiated phenotypes.

The involvement of *Isl1* in motor neuron differentiation has been the most widely studied aspect of this gene in development. In chick embryos, Isl1 precedes expression of SC1, an immunoglobulin-like surface molecule which was the earliest known marker of motor differentiation prior to the characterization of Isl1 (ERICSON et al. 1992). Isl1 expression in the ventral neural tube also precedes that of the other LIM homeodomain proteins expressed in this region (PFAFF et al. 1996). BrdU labelling, however, revealed that Isl1 is only expressed in post-mitotic cells of the neuroepithelia. Experiments in which notochord or floor plate cells were grafted to the dorsal region of the neural tube resulted in ectopic expression of Isl1 and SC1 in this region. In agreement with the notion of a notochord derived inductive signal, removal of the notochord from its usual position underlying the ventral midline of the neural tube prevented development of the floor plate and $Isl1^{+}/SC1^{+}$ motor neurons. Upregulation of dorsal markers in the ventral neural tube also occurred following removal of the notochord, suggesting that a notochord derived signal represses dorsal markers in addition to inducing ventral markers. It was subsequently shown that media conditioned by the notochord or floor plate was sufficient to induce *Isl1* in neural tube explants (YAMADA et al. 1993).

The identification of *sonic hedgehog* (*shh*), the vertebrate homologue of the *Drosophila* gene *hedgehog*, and its expression in the notochord and floor plate led to more elaborate studies of ventral neuron differentiation in the developing brain. ERICSON et al. (1995) showed that chick explants from the telencephalon, diencephalon, and rhombencephalon began expressing Isl1 when grown on COS cells expressing *shh*. In the rhombencephalon explants, induced cells were also $SC1^{+}$ indicating a normal motor neuron phenotype for these cells. In this region, $Isl1^{-}/SC1^{-}/FP1^{+}$ floor plate cells were also induced. In the diencephalon and telencephalon explants, SC1 was not induced (these cells are normally not motor neurons), but $Isl1^{+}$ cells were also Nkx-2.1^{+}, identifying them as forebrain neurons in agreement with their origin. Coexpression of Lhx1 in the ventral diencephalic neurons suggested that Lhx1 may serve as a marker for the diencephalon, although its expression does not appear to be mediated by *shh*. These experiments indicated that *shh* is sufficient to induce Isl1 expression and that early rostrocaudal distinctions are maintained in vitro.

Finally, targeted disruption of the *Isl1* gene in mice illustrated its requirement for motor neuron differentiation and surprisingly, for the differentiation of ventral interneurons which do not normally express *Isl1* (PFAFF et al. 1996). *Isl1* deficient embryos arrested in development around E9.5 and died by E11.5, perhaps due to defects in the dorsal aorta. In E9.5 embryos, typical motor neuron markers were undetectable indicating that motor neuron differentiation had not occurred, although *shh* expression and the gross development of the neural tube appeared normal. Thinning of the ventral neuroepithelia in *Isl1* –/– mice suggested that presumptive motor neurons underwent apoptosis rather than assuming the fate of another neuronal cell type. *Engrailed*$^{+}$ interneurons which typically arise just dorsal to the *Isl1*$^{+}$ motor neurons, also did not form in *Isl1* mutant embryos, indicating a motor neuron dependent step in interneuron development. Embryos deficient in *Isl1* expression also lacked pancreatic endocrine islet cells as well as dorsal exocrine tissues of the pancreas, which apparently resulted from the failure of early *Isl1* function in the pancreatic mesenchyme (AHLGREN et al. 1997).

While displaying impressive functions when studied in isolation, the LIM homeobox genes have even more interesting developmental roles when viewed in combination with one another. Studies of the combinatorial expression patterns of LIM homeobox genes in the chick neural tube suggested that these genes are not functionally redundant and that combinatorial expression may direct axonal migration and thereby specify target muscle innervation (TSUCHIDA et al. 1994). Individual motor columns organized in the ventral neural tube were shown to express specific subsets of Isl1, Isl2, Lhx1, and Lhx3, correlating to sites of future target muscle innervation. Thus the medial subdivision of the median motor column neurons which innervate axial muscles express Isl1, Isl2, and Lhx3, while the lateral subdivision of the median motor column which projects axons to the body wall muscles are defined by the expression of only Isl1 and Isl2. The medial subdivision of the lateral motor column neurons which project to ventrally derived limb muscles express Isl1 and Isl2, while the lateral subdivision of the lateral motor column which innervate the dorsally derived limb muscles selectively express Isl2 and Lhx1. Finally, the column of Terni sympathetic neurons express only Isl1, suggesting that expression of Isl1 in the absence of Isl2 specifies a visceral rather than somatic neuronal fate (for further review, see LUMSDEN 1995; TOSNEY et al. 1995; TANABE and JESSELL 1996). As virtually every cell in the differentiated neural tube has been shown to express at least one known LIM homeobox gene, it is likely that combinatorial expression of these genes in more dorsally located neurons will also specify their identity as well.

Considered as a whole, LIM class homeobox genes play a general role in dorsal-ventral patterning of appendages and the central nervous system and are confined to primarily post-mitotic cells of the nervous system and to cells of endocrine lineages. This is in contrast to the *HOX* genes which appear earlier in development and specify rostral-caudal patterning of the central nervous system (MCGINNIS and KRUMLAUF 1992; KRUMLAUF 1994). Notable exceptions are Lhx1 and Lhx5, whose early expression in the gastrula suggests more uncharacteristic regionalizing functions. Finally, the overlapping expression but apparently distinct functions of LIM homeobox genes in the same tissues suggests a non-redundant role for each of these genes in development. Every known vertebrate LIM homeobox gene has been shown to be expressed in the neural tube, and the anterior and intermediate lobes of the pituitary simultaneously express *Lhx2*, *Lhx3* and *Isl1*. Similarly, pancreatic islet cells express *Isl1* and *Lmx1*, and the thymus and retina coexpress *Lhx2* and *Isl1*. An additional level of complexity may involve a balance between LIM homeodomain proteins and non-LIM homeodomain proteins within the same cells, as some promoters have been shown to be regulated by both LIM and non-LIM homeodomains.

7 LIM Domain Proteins in Transcription

The identification of a homeobox in the genes of each of the three founding LIM members, *lin-11*, *Isl1* and *mec-3*, immediately indicated a functional role for LIM domain-containing proteins in transcription. Furthermore, the sequence of each of the original LIM proteins identified a putative C terminal transcription activation domain (reviewed by MITCHELL and TJIAN 1989): Isl1 was found to contain a glutamine-rich region (KARLSSON et al. 1990), lin-11 a proline-rich region (FREYD et al. 1990), and mec-3 an acidic domain (WAY and CHALFIE 1988). Indeed the acidic domain of mec-3 functioned as an activation domain when fused to the Gal4-DNA binding domain (LICHTSTEINER and TJIAN 1995). Although lacking an obvious DNA binding domain, the nuclear localization and interaction of the LMO family members with the DNA binding transcription factors Tal1 (Valge-Archer et al. 1994; Wadman et al. 1994) and GATA-1 (Osada et al. 1995), suggests a similar role for the LIM-only proteins in transcription. In contrast to the LIM homeodomain proteins, however, LMO1 and 2 contain an N terminal transcription activation domain with only slight homology to the classic VP16 activation domain. These studies suggested that, while incapable of binding DNA alone, LMO1 and 2 may activate transcription via interaction with other DNA bound factors such as Tal1 (Sanchez-Garcia et al. 1995).

7.1 Gene Targets

Surprisingly few gene targets for the LMO and LIM homeodomain proteins have been identified. To date, the only known promoters through which the LIM homeodomain proteins directly function regulate genes encoding secreted hormones in endocrine cell lineages. The sole exceptions are the regulation of *mec-3* expression by its own gene product, and regulation of the POU homeobox *Pit-1* gene by Lhx3. Notably absent are promoters of genes specifically expressed in neuronal and erythroid cell types, where the LIM homeodomain and LMO proteins are highly expressed and are required for proper development.

Although Isl1 was initially isolated from a pancreatic islet cell line by virtue of its homeodomain binding to the rat insulin I minienhancer element (Karlsson et al. 1990), the subsequent cloning and characterization of another LIM homeodomain protein, Lmx1, suggested that this factor plays a more physiological role in insulin gene regulation than Isl1 (German et al. 1992a) (Table 2). The E2 minienhancer extends from −247 to −198 of the rat insulin I gene and contains two elements known as Far and FLAT which bind distinct complexes from insulin producing pancreatic islet β cells that are interdependent for transcription (German et al. 1992). E47 and E12, the ubiquitously expressed basic helix-loop-helix proteins, were shown to occupy the E box containing Far element (German et al. 1991) while the FLAT element, which contains several overlapping homeodomain TAAT binding sites, was found to bind Lmx1 with much higher affinity than Isl1 (German et al. 1992a). Reporter gene assays showed only modest activation of expression from the minienhancer by transfected Lmx1 in a non-insulin producing hamster fibroblast cell line. However, co-transfection of E12 or E47 with Lmx1 resulted in a 100-fold synergistic response which was dependent on the presence of LIM1 of Lmx1 and was not seen when the LIM domains of Isl1 were substituted for those of Lmx1.

While it is unlikely that Isl1 plays a major role in insulin I gene regulation (German et al. 1992a; Dandoy-Dron et al. 1993), it may be involved in transcription of two other pancreatic endocrine hormone genes, somatostatin and proglucagon. Two groups studying somatostatin gene expression in pancreatic islet D cells identified a binding site for Isl1 or a closely related protein within the upstream element of the promoter which extends from −114 to −78 of the so-

Table 2. LIM Homeodomain gene targets

LIM protein	Target promoter		Binding site
Lmx1/Isl1	Rat insulin I		CTTCATCAGGCCATCTGGCCCCTTG*TTAATAATCTAATTA*CCCTAGGTCTA (KARLSSON et al. 1990; GERMAN et al. 1992a)
Isl1	Somatostatin		CTTCTTTGATTGATTTTGCGAGG*CTAATGG*TGCGTAAAAGCACTGGT (LEONARD et al. 1992; VALLEJO et al. 1992a)
Isl1	Proglucagon	Ga	GCG*TAATAT*CTGCAAGGCTAAACAG
		Gb	CCCATTATTTACAGATGAGAAATTTA
		Gc	ATTTATATTGTCAGCGTAATATCTG (WANG et al. 1995)
Lhx2/Lhx3	α-GSU		ATATCAGGTACTT*AGCTAATT*AAATGTG (ROBERSON et al. 1994; BACH et al. 1995)
Lhx3	TSH-β		GAATTTTCAATAGATGCTTTTCAGATAAGAAAGCAGCAATTCGAATGCAATTATATAAACA
	Prolactin		CCTGATTATATATATATTCATGAA (BACH et al. 1994)
mec-3	mec-3	CS1	TCATTCGAAATGCATTGCCCA*TAAT*GAATCGACCGAAAAACA
		CS2	TTCATAAGAAATGCATCT*ATTAT*CGTCAC
		CS3	CCAGTTTTAGCGCACA*TTAATAA*TCGATCAGGG (XUE et al. 1992)

matostatin gene (LEONARD et al. 1992; VALLEJO et al. 1992a). Full length recombinant Isl1 footprinted this region in DNAse I protection experiments, and deletion of the putative Isl1 binding site reduced expression in reporter gene assays performed in the somatostatin and Isl1 expressing Tu-6 pancreatic islet tumor cell line. Supporting the involvement of Isl1 in somatostatin gene expression, transfection of Isl1 increased transcription from a minimal somatostatin promoter to which four Isl1 binding sites were added in RIN-5AH cells, which normally do not express somatostatin or Isl1 (LEONARD et al. 1992). In addition to the upstream element Isl1 binding site, high levels of transcription in somatostatin expressing cell lines required a more proximal CREB binding site (−56 to −32), suggesting a synergistic relationship between these two factors in somatostatin gene expression (LEONARD et al. 1992; VALLEJO et al. 1992b). A second CREB binding site that overlapped the Isl1 site in the upstream element was required for maximal transcription, suggesting that Isl1 and CREB may either compete for binding to this site or heterodimerize on the DNA (VALLEJO et al. 1992a). Despite the evidence for Isl1 involvement in somatostatin gene expression, it is likely that homeodomain proteins other than Isl1 regulate the somatostatin gene, and it has been reported that the major homeodomain protein binding to the upstream element may be the antennapedia class IDX-1/STF-1 rather than a LIM homeodomain protein (LEONARD et al. 1993; MILLER et al. 1994). Interestingly, IDX1/STF-1 also binds to the rat insulin I FLAT element (MILLER et al. 1994).

More direct evidence for Isl1 mediated transcription comes from studies involving the proglucagon proximal promoter (Table 2). This gene encodes the precursor polypeptide for glucagon (MOJSOV et al. 1986), which is secreted from pancreatic islet α cells (THORENS and WAEBER 1993). The 100 base pairs immediately upstream of the proglucagon gene transcription start site contain three Isl1 binding sites, designated Ga, Gb, and Gc, each of which bound Isl1 with higher affinity than the rat insulin I minienhancer in electrophoretic mobility shift assays (EMSAs) (WANG and DRUCKER 1995). Full length in vitro translated Isl1 was capable of binding to these sequences, and importantly, protein-DNA complexes generated from nuclear extracts of Isl1-expressing InR1-G9 cells were supershifted with anti-Isl1 antisera. Additionally, transcriptional activation of reporter genes mediated by the Gb and Gc elements in InR1-G9 cells could be blocked by co-transfection of antisense Isl1 constructs, which effectively diminished Isl1 protein levels in these cells.

In addition to the Isl1 and Lmx1 mediated transcription of pancreatic endocrine hormone genes, the endocrine cells of the pituitary have been shown to utilize the LIM homeodomain proteins Lhx2 and Lhx3 in regulating the expression of secreted pituitary hormones. The pituitary glycoprotein hormone basal element (PGBE), which extends from −344 to −300 of the α-glycoprotein subunit (*α-GSU*) gene was shown to bind Lhx2 (ROBERSON et al. 1994). The *α-GSU* gene is expressed early in pituitary development (~E9), and encodes the common α-subunit shared by the pituitary glycoprotein hormones (PIERCE and PARSONS 1981). The PGBE is sufficient for directing maximal expression in gonadotrope lineages which express luteinizing hormone and follicle-stimulating hormone, and thyrotrope lineages which express thyroid-stimulating hormone (SCHODERBEK et al. 1992). Additionally, the PGBE showed a shift in EMSA experiments specific to gonadotrope and thyrotrope cell types (ROBERSON et al. 1994). DNAse I protection experiments indicated a footprinted region of the PGBE by recombinant Lhx2 homeodomain, which included a CTAATTA motif also found in the rat insulin I promoter, and mutations in this site which blocked Lhx2 binding also affected transcription of reporter genes driven by the full α-GSU promoter in a gonadotrope cell line. Cotransfected Lhx2 activated transcription from the α-GSU promoter or three copies of the PGBE in heterologous COS cells (ROBERSON et al. 1994).

The LIM homeodomain protein, Lhx3, also binds to the PGBE in vitro and activates transcription from the α-GSU promoter in CV-1 cells (BACH et al. 1995). In these studies, Lhx3 acted synergistically with the pituitary specific POU homeodomain protein Pit-1 in activation of transcription from the Pit-1, thyroid stimulating hormone-β and prolactin promoters, but not the α-GSU promoter. The LIM domains of Lhx3 were required to achieve the synergistic effect. Interestingly, Pit-1 and Lhx3 were shown to interact in vitro, through either LIM domain of Lhx3 and the POU domain of Pit-1, suggesting that direct association between these proteins on a DNA element results in enhancement of transcription of certain pituitary specific genes. Regulation of the *Pit-1* gene, therefore, appears to involve a positive feedback loop in which initial low levels of the Pit-1 gene product may synergize with Lhx3 to direct much higher levels of expression from the Pit-1 promoter.

A similar example of a synergistic transcriptional relationship between a LIM and a POU homeodomain protein comes from the study of *C. elegans mec-3* gene expression and its product, a LIM homeodomain protein. In early studies of mec-3 mutants, it was found that the POU protein unc-86 and the mec-3 protein itself were required for maintaining expression from a *mec-3* promoter using lacZ reporter constructs in mutant animals (WAY and CHALFIE 1989). Unc-86 appeared responsible for the initiation of *mec-3* gene expression, while both unc-86 and mec-3 proteins were required for the subsequent maintenance of expression (XUE et al. 1992, 1993). This finding makes sense in terms of the development of the *mec-3* expressing touch receptor neurons, which arise through the asymmetric cell division of an unc-86 expressing parent cell (WAY and CHALFIE 1989). Thus, similar to the Pit-1 and Lhx-3 relationship with the *Pit-1* promoter, unc-86 may initiate low levels of *mec-3* expression in a new daughter cell, which would subsequently synergize with unc-86 in maintaining high levels of expression throughout the life of the animal (XUE et al. 1992). Although these results do not discriminate between direct versus indirect regulation of the promoter by unc-86 and mec-3, further studies illustrated that both proteins bound to overlapping sites in at least three elements within the *mec-3* promoter (XUE et al. 1992, 1993). Heterodimerization increased the specificity and affinity of the unc-86/mec-3 complex binding to DNA (XUE et al. 1993; LICHTSTEINER and TJIAN 1995). Like Lhx3 and Pit-1, mec-3 and unc-86 were shown to bind each other in vitro, although in this case, the LIM domains of mec-3 were not required for the interaction which occurred through the POU domain of unc-86 (XUE et al. 1993; LICHTSTEINER and TJIAN 1995).

A continually surfacing theme in the literature describing LIM domain-containing transcription factors is the synergistic relationship between LIM proteins and other ubiquitously expressed or tissue specific transcription factors. In most cases studied thus far, expression of a LIM homeodomain protein alone with a responsive element showed only modest (less than ten fold) activation of reporter genes (GERMAN et al. 1992a; BACH et al. 1995; ROBERSON et al. 1994; WANG and DRUCKER 1995; LICHTSTEINER and TJIAN 1995) while coexpression of other factors shown to bind nearby DNA elements increased transcription by at least an order of magnitude over the homeodomain alone (GERMAN et al. 1992; BACH et al. 1995; LICHTSTEINER and TJIAN 1995; VALLEJO et al. 1992b). The most notable examples are the Lmx1/E47 synergism which increases transcription from the rat insulin I promoter at least 100-fold over Lmx1 or E47 alone (GERMAN et al. 1992a) and the Lhx3/Pit-1 synergy on the *Pit-1* promoter in which transcription is activated greater than 100-fold over that with either factor alone (BACH et al. 1995). Thus, it is likely that despite the presence of DNA binding homeodomains and transcription activation domains, the LIM transcription factors do not act alone to generate high levels of transcription from target genes. Instead, they appear to require additional factors, whether ubiquitously expressed (E47, CREB, or unc-86) or tissue specific (Pit-1 or Tal1 as may be the case for LMO1 and 2) to obtain maximal levels of transcriptional activation. These cofactors may (BACH et al. 1995; XUE et al. 1993; LICHTSTEINER and TJIAN 1995) or may not directly interact with the LIM proteins and in theory, these synergistic relationships can affect transcription by any of several possible mechanisms. Heterodimerization, indirect interactions, or binding to adjacent sites on a promoter element could affect DNA binding affinity, as appears to be the case for the mec-3 and unc-86 interaction (XUE et al. 1992, 1993; LICHTSTEINER and TJIAN 1995), or could affect local chromatin structure, allowing additional activating factors to bind nearby DNA elements. Synergism may also result from the removal of inhibitory factors which prevent DNA binding or transcription activation, or the concentration of activation domains near the basal transcription machinery. Finally, it is possible that synergizing factors recruit and concentrate additional non-DNA binding factors in close proximity to the basal machinery, thereby influencing transcription levels in an indirect manner. Several of these strategies are likely utilized by different promoters with their respective synergizing transcription factors, rather than one common mechanism acting at each distinct promoter.

While evidence has accumulated describing how LIM-containing transcription factors regulate the expression of a small set of genes, there is a frustrating lack of gene targets that can explain the phenotypes observed in genetic mutants. In addition to its own gene, mec-3 may regulate other genes identified by *C. elegans* mechanosensory mutants. Candidate targets for mec-3 include the *mec-7* gene, which encodes a β-tubulin used specifically by touch receptor neuron microtubules, and the *mec-4* gene, encoding a membrane protein whose expression is limited to the same six neurons as mec-3 (WAY and CHALFIE 1988; XUE et al. 1992). In the nervous system, it is possible that the *Drosophila* apterous protein and Isl1, Lmx1, and Lhx proteins may upregulate expression of secreted ligands or cell surface proteins that direct axonal migration of specific subsets of neurons to their target tissues (LUNDGREN et al. 1995; TSUCHIDA et al. 1994; TOSNEY et al. 1995). Alternatively, these proteins may self-regulate their own genes or the genes of other transcription factors affecting neuron specific gene expression. Finally, LMO2 may regulate erythroid differentiation factors such as Tal1 and GATA1 or erythroid specific genes such as the globins (WARREN et al. 1994).

7.2 Role of LIM Domains in Transcription

While it is evident that LIM domain containing transcription factors regulate gene expression, the role of the LIM domain itself in this process is unclear. Due to their similarity to GATA-1 zinc fingers, it has been suggested that LIM domains might be capable of binding DNA (LI et al. 1994; SANCHEZ-GARCIA and RABBITTS 1994; PEREZ-ALVARADO et al. 1994, 1996). LIM domains have generally been shown to lack affinity for DNA, and are therefore unlikely to be involved in DNA binding by themselves (SANCHEZ-GARCIA et al. 1993; XUE et al. 1993). On the other hand, it has become apparent that LIM domains function in the assembly of protein complexes through protein-protein interactions. In the case of Lmx1 and Lhx3, the LIM domains are required for synergistic activation of transcription, presumably by mediating protein-protein interactions between partners (GERMAN et al. 1992a; BACH et al. 1995). Coupling of the LIM-only proteins through the LIM domains to other DNA binding bHLH proteins such as Tal1 may serve to tether the non-DNA binding LIM protein to a promoter element, where it would be in a position to affect transcription (WADMAN et al. 1994; VALGE-ARCHER et al. 1994).

There is controversy in the literature regarding LIM domain-mediated inhibition of DNA binding by LIM homeodomain proteins. In DNA site selection experiments, intact LIM domains of recombinant Isl1 inhibited selection of the homeodomain consensus site TAAT, as well as transcription from a multimerized TAAT reporter construct in vivo (SANCHEZ-GARCIA et al. 1993). In addition, deletion of the LIM domains of Lhx3 seemed to increase the affinity of this protein for the α-GSU, TSH-β, and prolactin promoters (BACH et al. 1995). Likewise, deletion of the mec-3 LIM domains appeared to enhance DNA binding (XUE et al. 1993). In contrast, it has been clearly shown that full length recombinant or in vitro translated Isl1 and other LIM homeodomain proteins are quite capable of binding DNA (GERMAN et al. 1992; LEONARD et al. 1992; BACH et al. 1995; WANG and DRUCKER 1995; GONG and HEW 1994; XUE et al. 1992).

Additional evidence for LIM domain mediated inhibition of transcription comes from the functional studies of Xlim-1 (the homologue of murine Lhx1) in *Xenopus laevis*. Injection of Xlim-1 RNA encoding a protein with deleted or mutated LIM domains into two cell embryos upregulated expression of the neuronal genes NCAM and *engrailed* as well as cement gland markers. The mutated protein also activated expression of muscle specific markers in *Xbra* expressing ventral mesoderm. Surprisingly, wild-type Xlim-1 RNA had no effect on the expression of these genes and no obvious phenotypic effects. The conclusions drawn from these experiments were that the LIM domains of Xlim-1 inhibit transcription, and that other LIM binding factors may be required to relieve the inhibition (Taira et al. 1994). NLI/Ldb1 is proposed to be such a factor (Agulnick et al. 1996).

One can imagine two possible explanations for these apparently contradictory results. The first is that LIM domains inhibit DNA binding and transcription at some promoters and not others, presumably by blocking access of the homeodomain to the DNA or by otherwise altering its conformation to prevent DNA binding. This could be due to intramolecular interactions between LIM and homeodomains or intermolecular interactions through proteins that recognize specific LIM domains. Such inhibition might be relieved by another LIM binding protein or a phosphorylation event which imparts a conformational change, freeing the homeodomain. An alternative explanation is that LIM domains confer DNA binding specificity to the homeodomain. In this context, the LIM domains could function by constraining the conformation of the homeodomain region to impart specificity of a specific LIM homeodomain protein for a subset of TAAT containing promoters. LIM domains could also impart specificity to the homeodomains by requiring the interaction with another promoter specific transcription factor for stable binding to that promoter. Finally, although there is little data to support LIM-DNA interactions, the LIM domains could make additional contacts with DNA (suggested by Li et al. 1991), thus providing promoter specific recognition analogous to the function of the POU specific (POU_S) domain in POU homeodomain proteins. The POU_S domain of Pit-1 was shown to lack affinity for DNA in isolation, but when adjacent to the DNA binding homeodomain, increased the affinity and specificity of Pit-1 for DNA dramatically (Ingraham et al. 1990). Like LIM domains, the POU_S domain has also been shown to be involved in protein-protein interactions, allowing dimerization of Pit-1 on the prolactin promoter.

In support of the second hypothesis that LIM domains provide DNA binding specificity, isolated homeodomains and full length homeodomain proteins are capable of binding to a wide variety of TAAT containing elements with low affinity, and relatively unrelated homeodomain proteins are capable of binding to the same site in vitro (Hoey and Levine 1988; Hayashi and Scott 1990; Rosenfeld 1991; Kalionis and O'Farrell 1993). For example, without the POU specific domain which provides DNA binding specificity to the POU proteins, the Pit-1 homeodomain can bind to an engrailed site (Ingraham et al. 1990). Thus, overexpression of an isolated homeodomain or full length homeodomain-containing protein would be likely to non-specifically activate a variety of homeodomain responsive genes. Indeed, the studies of the LIM homeodomain proteins indicate promiscuity of homeodomain binding in vitro and in transfection assays in which protein levels are elevated over endogenous levels. For example, Isl1, Lmx1, and the non-LIM homeodomain proteins Idx-1/STF-1 and HNF-1α, have all been shown to bind to the rat insulin I FLAT element (German et al. 1992a; Miller et al. 1994; Leonard et al. 1993; Emens et al. 1992). In addition, Lhx2 and Lhx3 are both capable of binding to and activating expression from the same site on the α-GSU promoter (Roberson et al. 1994; Bach et al. 1995). It is possible that each factor plays a role in regulation of the same gene at particular phases of development or under specific cellular conditions, but it seems more likely that cofactors, DNA binding affinities, and tight regulation of protein concentrations dictate the factor that plays a physiological role at each promoter. Finally, the high degree of conservation between the homeodomain elements of various LIM domain proteins which appear to have distinct functions (Lhx1 and Lhx5, Lhx3 and Lhx4, and Isl1 and Isl2) argues that factors in addition to homeodomain sequences are necessary to discriminate between target promoters.

8 Conclusions

LIM domains are widely distributed in both nuclear and cytoplasmic proteins that have diverse functions. Although LIM domains have only about 30%–35% amino acid identity, they likely have very similar three-dimensional core structures. Studies of protein-protein interactions indicate that individual LIM domains have high specificity for particular target sequences in other proteins. Although not demonstrated, the LIM domains of the LIM homeobox class have the potential to augment DNA binding via the homeodomain.

Specific biological functions of LIM domains are yet to be defined. Mutational and deletional analyses of the nuclear LMO and LIM homeodomain proteins reveal essential requirements for these proteins in various developmental pathways but do not reveal the molecular mechanisms through which LIM domains act. Much less is known about the biological function of cytoplasmic LIM domain-containing proteins but these are likely involved in receptor trafficking, signal transduction and organization of the cytoskeleton. The conserved core of LIM domains consisting of N and C terminal Zn^{2+} coordination modules provides a platform upon which sequence variations provide for target specificity. Molecular recognition is the fundamental role for this versatile module conserved from plants to man.

Acknowledgments. The authors thank Tom Diller for helpful discussions and for preparation of Fig. 4 and Chun-Leung Chan for preparation of Fig. 1. We thank Mrs. Anne Casey for manuscript preparation. The authors' work cited is supported by NIH grant DK13149. L.W.J. is supported by NIH training grant DK07541.

References

Agulnick AD, Taira M, Breen JJ, Tanaka T, Dawid IB, Westphal H (1996) Interactions of the LIM-domain-binding factor Ldb1 with LIM homeodomain proteins. Nature 384:270–272

Ahlgren U, Pfaff SL, Jessell TM, Edlund T, Edlund H (1997) Independent requirement for ISL1 in formation of pancreatic mesenchyme and islet cells. Nature 385:257–260

Altaratz M, Applebaum SW, Richard DS, Gilbert LI, Segal D (1991) Regulation of juvenile hormone synthesis in wild-type and apterous mutant Drosophila. Mol Cell Endocrinol 81:205–216
Arber S, Caroni P (1996) Specificity of single LIM motifs in targeting and LIM/LIM interactions in situ. Genes Dev 10:89–300
Arber S, Halder G, Caroni P (1994) Muscle LIM protein, a novel essential regulator of myogenesis, promotes myogenic differentiation. Cell 79:221–231
Bach I, Rhodes SJ, Pearse RV, Heinzel T, Gloss B, Scully KM, Sawchenko PE, Rosenfeld MG (1995) P-Lim, a LIM homeodomain factor, is expressed during pituitary organ and cell commitment and synergizes with Pit-1. Proc Natl Acad Sci USA 92:720–72
Backer JM, Shoelson SE, Weiss MA, Hua QX, Cheatham RB, Haring E, Cahill DC, White MF (1992) The insulin receptor juxtamembrane region contains two independent tyrosine/beta-turn internalization signals. J Cell Biol 118:831–839
Baer R (1993) TAL1, TAL2 and LYL1: a family of basic helix-loop-helix proteins implicated in T cell acute leukaemia. Semin Cancer Biol 4:341–347
Barnes JD, Crosby JL, Jones CM, Wright CV, Hogan BL (1994) Embryonic expression of Lim-1, the mouse homolog of Xenopus Xlim-1, suggests a role in lateral mesoderm differentiation and neurogenesis. Dev Biol 161:168–178
Bernard O, Ganiatsas S, Kannourakis G, Dringen R (1994) Kiz-1, a protein with LIM zinc finger and kinase domains, is expressed mainly in neurons. Cell Growth Differ 5:1159–1171
Birkenmeier EH, Gordon JI (1986) Developmental regulation of a gene that encodes a cysteine-rich intestinal protein and maps near the murine immunoglobulin heavy chain locus. Proc Natl Acad Sci USA 83:2516–2520
Blair SS, Brower DL, Thomas JB, Zavortink M (1994) The role of apterous in the control of dorsoventral compartmentalization and PS integrin gene expression in the developing wing of Drosophila. Development 120:1805–1815
Blake PB, Lee B, Park J-B, Zhou Z-H, Adams MWW, Summers MF (1994) Heteronuclear magnetic resonance studies of Zn, ^{113}Cd and ^{119}Hg substituted P furiosus rubredoxin: implications for biological electron transfer. New J Chem 18:387–395
Boehm T, Baer R, Lavenir I, Forster A, Waters JJ, Nacheva E, Rabbitts TH (1988) The mechanism of chromosomal translocation t(11;14) involving the T-cell receptor C delta locus on human chromosome 14q11 and a transcribed region of chromosome 11p15. EMBO J 7:385–394
Boehm T, Greenberg JM, Buluwela L, Lavenir I, Forster A, Rabbitts TH (1990a) An unusual structure of a putative T cell oncogene which allows production of similar proteins from distinct mRNAs. EMBO J 9:857–868
Boehm T, Foroni L, Kennedy M, Rabbitts TH (1990b) The rhombotin gene belongs to a class of transcriptional regulators with a potential novel protein dimerisation motif. Oncogene 5:1103–1105
Boehm T, Foroni L, Kaneko Y, Perutz MF, Rabbitts TH (1991a) The rhombotin family of cysteine-rich LIM-domain oncogenes: distinct members are involved in T-cell translocations to human chromosomes 11p15 and 11p13. Proc Natl Acad Sci USA 88:4367–4371
Boehm T, Spillantini MG, Sofroniew MV, Surani MA, Rabbitts TH (1991b) Developmentally regulated and tissue specific expression of mRNAs encoding the two alternative forms of the LIM domain oncogene rhombotin: evidence for thymus expression. Oncogene 6:695–703
Bourgouin C, Lundgren SE, Thomas JB (1992) Apterous is a Drosophila LIM domain gene required for the development of a subset of embryonic muscles. Neuron 9:549–561
Brown MC, Perotta JA, Turner CE (1996) Identification of LIM3 as the principal determinant of Paxillin focal adhesion localization and characterization of a novel motif on paxillin directing vinculin and focal adhesion kinase binding. J Cell Biol 135:1109–1123
Chang CP, Lazar CS, Walsh BJ, Komuro M, Collawn JF, Kuhn LA, Tainer JA, Trowbridge IS, Farquhar MG, Rosenfeld MG et al. (1993) Ligand-induced internalization of the epidermal growth factor receptor is mediated by multiple endocytic codes analogous to the tyrosine motif found in constitutively internalized receptors. J Biol Chem 268:19312–19320
Cheng AK, Robertson EJ (1995) The murine LIM-kinase gene (limk) encodes a novel serine threonine kinase expressed predominantly in trophoblast giant cells and the developing nervous system. Mech Dev 52:187–197
Cohen B, McGuffin ME, Pfeifle C, Segal D, Cohen SM (1992) apterous, a gene required for imaginal disc development in Drosophila encodes a member of the LIM family of developmental regulatory proteins. Genes Dev 6:715–729
Crawford AW, Beckerle MC (1992) Purification and characterization of Zyxin, an 82,000-Dalton component of adherens junctions. J Biol Chem 266:5847–5853

Dandoy-Dron F, Deltour L, Monthioux E, Bucchini D, Jami J (1993) Insulin gene can be expressed in the absence of Isl-1. Exp Cell Res 209:58–63

Dawid IB, Toyama R, Taira M (1995) LIM domain proteins. C R Acad Sci III 318:295–306

Diaz-Benjumea FJ, Cohen SM (1993) Interaction between dorsal and ventral cells in the imaginal disc directs wing development in Drosophila. Cell 75:741–752

Doyle DA, Lee A, Lewis J, Kim E, Sheng M, MacKinnon R (1996) Crystal structures of a complexed and peptide-free membrane protein-binding domain: molecular basis of peptide recognition by PDZ. Cell 85:1067–1076

Durick K, Wu RY, Gill GN, Taylor SS (1996) Mitogenic signaling by Ret/ptc2 requires association with enigma via a LIM domain. J Biol Chem 271:12691–12694

Emens LA, Landers DW, Moss LG (1992) Hepatocyte nuclear factor 1 alpha is expressed in a hamster insulinoma line and transactivates the rat insulin I gene. Proc Natl Acad Sci USA 89:7300–7304

Ericson J, Thor S, Edlund T, Jessell TM, Yamada T (1992) Early stages of motor neuron differentiation revealed by expression of homeobox gene Islet-1. Science 256:1555–1560

Ericson J, Muhr J, Placzek M, Lints T, Jessell TM, Edlund T (1995) Sonic hedgehog induces the differentiation of ventral forebrain neurons: a common signal for ventral patterning within the neural tube. Cell 81:747–756

Ferguson EL, Sternberg PW, Horvitz HR (1987) A genetic pathway for the specification of the vulval cell lineages of Caenorhabditis elegans. Nature 326:259–267

Feuerstein R, Wang X, Song D, Cooke NE, Liebhaber SA (1994) The LIM/double zinc-finger motif functions as a protein dimerization domain. Proc Natl Acad Sci USA 91:10655–10659

Fisch P, Boehm T, Lavenir I, Larson T, Arno J, Forster A, Rabbitts TH (1992) T-cell acute lymphoblastic lymphoma induced in transgenic mice by the RBTN1 and RBTN2 LIM-domain genes. Oncogene 7:2389–2397

Foroni L, Boehm T, White L, Forster A, Sherrington P, Liao XB, Brannan CI, Jenkins NA, Copeland NG, Rabbitts TH (1992) The rhombotin gene family encode related LIM-domain proteins whose differing expression suggests multiple roles in mouse development. J Mol Biol 226:747–761

Frangiskakis JM, Ewart AK, Morris CA, Mervis CB, Bertrand J, Robinson BF, Klein BP, Ensing GJ, Everett LA, Green ED, Proschel C, Gutowski NJ, Noble M, Atkinson DL, Odelberg SJ, Keating MT (1996) LIM-kinase1 hemizygosity implicated in imparied visuospatial constructive cognition. Cell 86:59–69

Freyd G, Kim SK, Horvitz HR (1990) Novel cysteine-rich motif and homeodomain in the product of the Caenorhabditis elegans cell lineage gene lin-11. Nature 344:876–879

Fujii T, Pichel JG, Taira M, Toyama R, Dawid IB, Westphal H (1994) Expression patterns of the murine LIM class homeobox gene lim1 in the developing brain and excretory system. Dev Dyn 199:73–83

German MS, Blanar MA, Nelson C, Moss LG, Rutter WJ (1991) Two related helix-loop-helix proteins participate in separate cell-specific complexes that bind the insulin enhancer. Mol Endocrinol 5:292–299

German MS, Wang J, Chadwick RB, Rutter WJ (1992a) Synergistic activation of the insulin gene by a LIM-homeo domain protein and a basic helix-loop-helix protein: building a functional insulin minienhancer complex. Genes Dev 6:2165–2176

German MS, Moss LG, Wang J, Rutter WJ (1992b) The insulin and islet amyloid polypeptide genes contain similar cell-specific promoter elements that bind identical beta-cell nuclear complexes. Mol Cell Biol 12:1777–1788

Gill GN (1995) The enigma of LIM domains. Structure 3:1285–1289

Gong Z, Hew CL (1994) Zinc and DNA binding properties of a novel LIM homeodomain protein Isl-2. Biochemistry 33:15149–15158

Greenberg JM, Boehm T, Sofroniew MV, Keynes RJ, Barton SC, Norris ML, Surani MA, Spillantini M-G, Rabbitts TH (1990) Segmental and developmental regulation of a presumptive T-cell oncogene in the central nervous system. Nature 344:158–160

Hayashi S, Scott MP (1990) What determines the specificity of action of Drosophila homeodomain proteins? Cell 63:883–894

Hoey T, Levine M (1988) Divergent homeo box proteins recognize similar DNA sequences in Drosophila. Nature 332:858–861

Ingraham HA, Flynn SE, Voss JW, Albert VR, Kapiloff MS, Wilson L, Rosenfeld MG (1990) The POU-specific domain of Pit-1 is essential for sequence-specific, high affinity DNA binding and DNA-dependent Pit-1-Pit-1 interactions. Cell 61:1021–1033

Inoue A, Takahashi M, Hatta K, Hotta Y, Okamoto H (1994) Developmental regulation of islet-1 mRNA expression during neuronal differentiation in embryonic zebrafish. Dev Dyn 199:1–11

Jurata LW, Kenny DA, Gill GN (1996) Nuclear LIM interactor, a rhombotin and LIM homeodomain interacting protein, is expressed early in neuronal development. Proc Natl Acad Sci USA 93:11693–11698

Jurata LW, Gill GN (1997) Functional analysis of the nuclear LIM domain interactor NLI. Mol Cell Biol 61

Kalionis B, O'Farrell PH (1993) A universal target sequence is bound in vitro by diverse homeodomains. Mech Dev 43:57–70

Karlsson O, Thor S, Norberg T, Ohlsson H, Edlund E (1990) Insulin gene enhancer binding protein Isl-1 is a member of a novel class of proteins containing both a homeo- and a Cys-His domain. Nature 344:879–882

Kiess M, Scharm B, Aguzzi A, Hajnal A, Klemenz R, Schwarte-Waldhoff I, Schafer R (1995) Expression of ril, a novel LIM domain gene, is down-regulated in Hras-transformed cells and restored in phenotypic revertants. Oncogene 10:61–68

Kohda D, Hatanaka H, Odaka M, Mandiyan V, Ullrich A, Schlessinger J, Inagaki F (1993) Solution structure of the SH3 domain of phospholipase C-gamma. Cell 72:953–960

Kosa JL, Michelsen JW, Louis HA, Olsen JI, Davis DR, Beckerle MC, Winge DR (1994) Common metal ion coordination in LIM domain proteins. Biochemistry 33:468–477

Koyama S, Yu H, Dalgarno DC, Shin TB, Zydowsky LD, Schreiber SL (1993) Structure of the PI3 K SH3 domain and analysis of the SH3 family. Cell 72:945–952

Krumlauf R (1994) Hox genes in vertebrate development. Cell 78:191–201

Kuroda S, Tokunaga C, Higuchi O, Konishi H, Mizuno K, Gill GN, Kikkawa U (1996) Protein-protein interactions of zinc finger LIM domains with protein kinase C. J Biol Chem 271:31029–31032

Larson RC, Fisch P, Larson TA, Lavenir I, Langford T, King G, Rabbitts TH (1994) T cell tumours of disparate phenotype in mice transgenic for Rbtn-2. Oncogene 9:3675–3681

Larson RC, Osada H, Larson TA, Lavenir I, Rabbitts TH (1995) The oncogenic LIM protein Rbtn2 causes thymic developmental aberrations that precede malignancy in transgenic mice. Oncogene 11:853–862

Larson RC, Lavenir I, Larson TA, Baer R, Warren AJ, Wadman I, Nottage K, Rabbitts TH (1996) Protein dimerization between Lmo2 (Rbtn2) and Tal1 alters thymocyte development and potentiates T cell tumorigenesis in transgenic mice. EMBO J 15:1021–1027

Lee CH, Saksela K, Mirza UA, Chait BT, Kuriyan J (1996) Crystal structure of the conserved core of HIV-1 Nef complexed with a Src family SH3 domain. Cell 85:931–942

Leonard J, Serup P, Gonzalez G, Edlund T, Montminy M (1992) The LIM family transcription factor Isl-1 requires cAMP response element binding protein to promote somatostatin expression in pancreatic islet cells. Proc Natl Acad Sci USA 89:6247–6251

Leonard J, Peers B, Johnson T, Ferreri K, Lee S, Montminy MR (1993) Characterization of somatostatin transactivating factor-1, a novel homeobox factor that stimulates somatostatin expression in pancreatic islet cells. Mol Endocrinol 7:1275–1283

Li H, Witte DP, Branford WW, Aronow BJ, Weinstein M, Kaur S, Wert S, Singh G, Schreiner CM, Whitsett JA et al (1994) Gsh-4 encodes a LIM-type homeodomain, is expressed in the developing central nervous system and is required for early postnatal survival. EMBO J 13:2876–2885

Li PM, Reichert J, Freyd G, Horvitz HR, Walsh CT (1991) The LIM region of a presumptive Caenorhabditis elegans transcription factor is an iron-sulfur- and zinc-containing metallodomain. Proc Natl Acad Sci USA 88:9210–9213

Lichtsteiner S, Tjian R (1995) Synergistic activation of transcription by UNC-86 and MEC-3 in Caenorhabditis elegans embryo extracts. EMBO J 14:3937–3945

Lumsden A (1995) Neural development A 'LIM code' for motor neurons? Curr Biol 5:491–495

Lundgren SE, Callahan CA, Thor S, Thomas JB (1995) Control of neuronal pathway selection by the Drosophila LIM homeodomain gene apterous. Development 121:1769–1773

Macias MJ, Hyvonen M, Baraldi E, Schultz J, Sudol M, Saraste M, Oschkinat H (1996) Structure of the WW domain of a kinase-associated protein complexed with a proline-rich peptide. Nature 382:646–649

McGinnis W, Krumlauf R (1992) Homeobox genes and axial patterning. Cell 68:283–302

McGuire EA, Hockett RD, Pollock KM, Bartholdi MF, O'Brien SJ, Korsmeyer SJ (1989) The t(11;14)(p15;q11) in a T-cell acute lymphoblastic leukemia cell line activates multiple transcripts, including Ttg-1, a gene encoding a potential zinc finger protein. Mol Cell Biol 9:2124–2132

McGuire EA, Davis AR, Korsmeyer SJ (1991) T-cell translocation gene 1 (Ttg-1) encodes a nuclear protein normally expressed in neural lineage cells. Blood 77:599–606

McGuire EA, Rintoul CE, Sclar GM, Korsmeyer SJ (1992) Thymic overexpression of Ttg-1 in transgenic mice results in T-cell acute lymphoblastic leukemia/lymphoma. Mol Cell Biol 12:4186–4196

Michelsen JW, Schmeichel KL, Beckerle MC, Winge DR (1993) The LIM motif defines a specific zinc-binding protein domain. Proc Natl Acad Sci USA 90:4404–4408
Michelsen JW, Sewell AK, Louis HA, Olsen JI, Davis DR, Winge DR, Beckerle MC (1994) Mutational analysis of the metal sites in an LIM domain. J Biol Chem 269:11108–11113
Miller CP, McGehee RE Jr, Habener JF (1994) IDX-1: a new homeodomain transcription factor expressed in rat pancreatic islets and duodenum that transactivates the somatostatin gene. EMBO J 13:1145–1156
Mitchell PJ, Tjian R (1989) Transcriptional regulation in mammalian cells by sequence-specific DNA binding proteins. Science 245:371–378
Mizuno K, Okano I, Ohashi K, Nunoue K, Kuma K, Miyata T, Nakamura T (1994) Identification of a human cDNA encoding a novel protein kinase with two repeats of the LIM/double zinc finger motif. Oncogene 9:1605–1612
Mojsov S, Heinrich G, Wilson IB, Ravazzola M, Orci L, Habener JF (1986) Preproglucagon gene expression in pancreas and intestine diversifies at the level of post-translational processing. J Biol Chem 261:11880–11889
Neale GA, Mao S, Parham DM, Murti KG, Goorha RM (1995a) Expression of the proto-oncogene rhombotin-2 is identical to the acute phase response protein metallothionein, suggesting multiple functions. Cell Growth Differ 6:587–596
Neale GA, Rehg JE, Goorha RM (1995b) Ectopic expression of rhombotin-2 causes selective expansion of CD4-CD8- lymphocytes in the thymus and T-cell tumors in transgenic mice. Blood 86:3060–3071
Okano I, Hiraoka J, Otera H, Nunoue K, Ohashi K, Iwashita S, Hirai M, Mizuno K (1995) Identification and characterization of a novel family of serine/threonine kinases containing two N-terminal LIM motifs. J Biol Chem 270:31321–31330
Osada H, Grutz G, Axelson H, Forster A, Rabbitts TH (1995) Association of erythroid transcription factors: complexes involving the LIM protein RBTN2 and the zinc-finger protein GATA1. Proc Natl Acad Sci USA 92:9585–9589
Overduin M, Rios CB, Mayer BJ, Baltimore D, Cowburn D (1992) Three-dimensional solution structure of the src homology 2 domain of c-abl. Cell 70:697–704
Perez-Alvarado GC, Miles C, Michelsen JW, Louis HA, Winge DR, Beckerle MC, Summers MF (1994) Structure of the carboxy-terminal LIM domain from the cysteine rich protein CRP. Nat Struct Biol 1:388–398
Perez-Alvarado GC, Kosa JL, Louis HA, Beckerle MC, Winge DR, Summers MF (1996) Structure of the cysteine-rich intestinal protein, CRIP. J Mol Biol 257:153–174
Pevny L, Simon MC, Robertson E, Klein WH, Tsai S-F, D'Agati V, Orkin SH, Costantini F (1991) Erythroid differentiation in chimaeric mice blocked by a targeted mutation in the gene for transcription factor GATA-1. Nature 349:257–260
Pfaff SL, Mendelsohn M, Stewart CL, Edlund T, Jessell TM (1996) Requirement for LIM homeobox gene Isl1 in motor neuron generation reveals a motor neuron-dependent step in interneuron differentiation. Cell 84:309–320
Pierce JG, Parsons TF (1981) Glycoprotein hormones: structure and function. Annu Rev Biochem 50: 465–495
Porcher C, Swat W, Rockwell K, Fujiwara Y, Alt FW, Orkin SH (1996) The T cell leukemia oncoprotein SCL/tal-1 is essential for development of all hematopoietic lineages. Cell 86:47–57
Rabbitts TH, Boehm T (1990) LIM domains. Nature 346:418
Rearden A (1994) A new LIM protein containing an autoepitope homologous to "senescent cell antigen". Biochem Biophys Res Commun 201:1124–1131
Riddle RD, Ensini M, Nelson C, Tsuchida T, Jessell TM, Tabin C (1995) Induction of the LIM homeobox gene Lmx1 by WNT7a establishes dorsoventral pattern in the vertebrate limb. Cell 83:631–640
Robb L, Lyons I, Li R, Hartley L, Kontgen F, Harvey RP, Metcalf D, Begley CG (1995b) Absence of yolk sac hematopoiesis from mice with a targeted disruption of the scl gene. Proc Natl Acad Sci USA 92:7075–7079
Robb L, Rasko JE, Bath ML, Strasser A, Begley CG (1995b) Scl, a gene frequently activated in human T cell leukaemia, does not induce lymphomas in transgenic mice. Oncogene 10:205–209
Roberson MS, Schoderbek WE, Tremml G, Maurer RA (1994) Activation of the glycoprotein hormone alpha-subunit promoter by a LIM-homeodomain transcription factor. Mol Cell Biol 14:2985–2993
Rosenfeld MG (1991) POU-domain transcription factors: pou-er-ful developmental regulators. Genes Dev 5:897–907

Royer-Pokora B, Loos U, Ludwig WD (1991) TTG-2, a new gene encoding a cysteine-rich protein with the LIM motif, is overexpressed in acute T-cell leukaemia with the t(11;14)(p13;q11). Oncogene 6:1887–1893
Royer-Pokora B, Rogers M, Zhu TH, Schneider S, Loos U, Bolitz U (1995) The TTG-2/RBTN2 T cell oncogene encodes two alternative transcripts from two promoters: the distal promoter is removed by most 11p13 translocations in acute T cell leukaemia's (T-ALL). Oncogene 10:1353–1360
Sadler I, Crawford AW, Michelsen JW, Beckerle MC (1992) Zyxin and cCRP: two interactive LIM domain proteins associated with the cytoskeleton. J Cell Biol 119:1573–1587
Sanchez-Garcia I, Rabbitts TH (1994) The LIM domain: a new structural motif found in zinc-finger-like proteins. Trends Genet 10:315–320
Sanchez-Garcia I, Osada H, Forster A, Rabbitts TH (1993) The cysteine-rich LIM domains inhibit DNA binding by the associated homeodomain in Isl-1. EMBO J 12:4243–4250
Sanchez-Garcia I, Axelson H, Rabbitts TH (1995) Functional diversity of LIM proteins: amino-terminal activation domains in the oncogenic proteins RBTN1 and RBTN2. Oncogene 10:1301–1306
Schmeichel KL, Beckerle MC (1994) The LIM domain is a modular protein-binding interface. Cell 79:211–219
Schoderbek WE, Kim KE, Ridgway EC, Mellon PL, Maurer RA (1992) Analysis of DNA sequences required for pituitary-specific expression of the glycoprotein hormone alpha-subunit gene. Mol Endocrinol 6:893–903
Seidah NG, Barale JC, Marcinkiewicz M, Mattei MG, Day R, Chretien M (1994) The mouse homeoprotein mLIM-3 is expressed early in cells derived from the neuroepithelium and persists in adult pituitary. DNA Cell Biol 13:1163–1180
Shawlot W, Behringer RR (1995) Requirement for Lim1 in head-organizer function. Nature 374:425–430
Sheng HZ, Zhadanov AB, Mosinger B Jr, Fujii T, Bertuzzi S, Grinberg A, Lee EJ, Huang SP, Mahon KA, Westphal H (1996) Specification of pituitary cell lineages by the LIM homeobox gene Lhx3. Science 272:1004–1007
Shivdasani RA, Mayer EL, Orkin SH (1995) Absence of blood formation in mice lacking the T-cell leukaemia oncoprotein tal-1/SCL. Nature 373:432–434
Steinmetz A, Evrard J-L (1996) WWW web site, http://scillau-strasbgfr/
Taira M, Jamrich M, Good PJ, Dawid IB (1992) The LIM domain-containing homeo box gene Xlim-1 is expressed specifically in the organizer region of Xenopus gastrula embryos. Genes Dev 6:356–366
Taira M, Otani H, Jamrich M, Dawid IB (1994a) Expression of the LIM class homeobox gene Xlim-1 in pronephros and CNS cell lineages of Xenopus embryos is affected by retinoic acid and exogastrulation. Development 120:1525–1536
Taira M, Otani H, Saint-Jeannet JP, Dawid IB (1994b) Role of the LIM class homeodomain protein Xlim-1 in neural and muscle induction by the Spemann organizer in Xenopus. Nature 372:677–679
Tanabe V, Jessell TM (1996) Diversity and pattern in the developing spinal cord. Science 274:1115–1123
Thor S, Ericson J, Brannstrom T, Edlund T (1991) The homeodomain LIM protein Isl-1 is expressed in subsets of neurons and endocrine cells in the adult rat. Neuron 7:881–889
Thorens B, Waeber G (1993) Glucagon-like peptide-I and the control of insulin secretion in the normal state and in NIDDM. Diabetes 42:1219–1225
Tosney KW, Hotary KB, Lance-Jones C (1995) Specifying the target identity of motoneurons. Bioessays 17:379–382
Toyama R, O'Connell ML, Wright CV, Kuehn MR, Dawid IB (1995a) Nodal induces ectopic goosecoid and lim1 expression and axis duplication in zebrafish. Development 121:383–391
Toyama R, Curtiss PE, Otani H, Kimura M, Dawid IB, Taira M (1995b) The LIM class homeobox gene lim5: implied role in CNS patterning in Xenopus and zebrafish. Dev Biol 170:583–593
Tsai FY, Keller G, Kuo FC, Weiss M, Chen J, Rosenblatt M, Alt FW, Orkin SH (1994) An early haematopoietic defect in mice lacking the transcription factor GATA-2. Nature 371:221–226
Tsuchida T, Ensini M, Morton SB, Baldassare M, Edlund T, Jessell TM, Pfaff SL (1994) Topographic organization of embryonic motor neurons defined by expression of LIM homeobox genes. Cell 79:957–970
Turner CE, Miller JT (1994) Primary sequence of paxillin contains putative SH2 and SH3 domain binding motifs and multiple LIM domains: identification of a vinculin and pp125Fak-binding region. J Cell Sci 107:1583–1591
Valge-Archer VE, Osada H, Warren AJ, Forster A, Li J, Baer R, Rabbitts TH (1994) The LIM protein RBTN2 and the basic helix-loop-helix protein TAL1 are present in a complex in erythroid cells. Proc Natl Acad Sci USA 91:8617–8621

Vallejo M, Penchuk L, Habener JF (1992a) Somatostatin gene upstream enhancer element activated by a protein complex consisting of CREB, Isl-1-like, and alpha-CBF-like transcription factors. J Biol Chem 267:12876–12884

Vallejo M, Miller CP, Habener JF (1992b) Somatostatin gene transcription regulated by a bipartite pancreatic islet D-cell-specific enhancer coupled synergetically to a cAMP response element. J Biol Chem 267:12868–12875

Wadman I, Li J, Bash RO, Forster A, Osada H, Rabbitts TH, Baer R (1994) Specific in vivo association between the bHLH and LIM proteins implicated in human T cell leukemia. EMBO J 13:4831–4839

Waksman G, Kominos D, Robertson SC, Pant N, Baltimore D, Birge RB, Cowburn D, Hanafusa H, Mayer BJ, Overduin M et al (1992) Crystal structure of the phosphotyrosine recognition domain SH2 of v-src complexed with tyrosine-phosphorylated peptides. Nature 358:646–653

Wang H, Harrison-Shostak DC, Lemasters JJ, Herman B (1995) Cloning of a rat cDNA encoding a novel LIM domain protein with high homology to rat RIL. Gene 165:267–271

Wang M, Drucker DJ (1995) The LIM domain homeobox gene isl-1 is a positive regulator of islet cell-specific proglucagon gene transcription. J Biol Chem 270:12646–12652

Wang X, Lee G, Liebhaber SA, Cooke NE (1992) Human cysteine-rich protein A member of the LIM/double-finger family displaying coordinate serum induction with c-myc. J Biol Chem 267:9176–9184

Warren AJ, Colledge WH, Carlton MB, Evans MJ, Smith AJ, Rabbitts TH (1994) The oncogenic cysteine-rich LIM domain protein rbtn2 is essential for erythroid development. Cell 78:45–57

Way JC, Chalfie M (1988) mec-3, a homeobox-containing gene that specifies differentiation of the touch receptor neurons in C elegans. Cell 54:5–16

Way JC, Chalfie M (1989) The mec-3 gene of Caenorhabditis elegans requires its own product for maintained expression and is expressed in three neuronal cell types. Genes Dev 3:1823–1833

Weiskirchen R, Bister K (1993) Suppression in transformed avian fibroblasts of a gene (crp) encoding a cysteine-rich protein containing LIM domains. Oncogene 8:2317–2324

Williams JA, Paddock SW, Carroll SB (1993) Pattern formation in a secondary field: a hierarchy of regulatory genes subdivides the developing Drosophila wing disc into discrete subregions. Development 117:571–584

Wu H-K (1995) A structurally abnormal breakpoint cluster region gene in a transcription factor, hLH-2-negative human leukemia cell line. Cancer Lett 92:215–222

Wu H-K, Heng HHQ, Siderovski DP, Dong WF, Okuno Y, Shi XM, Minden MD, Tsui LC (1994) Consistent high level expression of a human LIM/Hox gene, hLH-2 in chronic myelogenous leukemia and chromosomal localization to 9q33-341. Blood [Suppl] 84:295a

Wu R, Durick K, Songyang Z, Cantley LC, Taylor SS, Gill GN (1996) Specificity of LIM domain interactions with receptor tyrosine kinases. J Biol Chem 271:15934–15941

Wu RY, Gill GN (1994) LIM domain recognition of a tyrosine-containing tight turn. J Biol Chem 269:25085–25090

Xu Y, Baldassare M, Fisher P, Rathbun G, Oltz EM, Yancopoulos GD, Jessell TM, Alt FW (1993) LH-2: a LIM/homeodomain gene expressed in developing lymphocytes and neural cells. Proc Natl Acad Sci USA 90:227–231

Xue D, Finney M, Ruvkun G, Chalfie M (1992) Regulation of the mec-3 gene by the C elegans homeoproteins UNC-86 and MEC-3. EMBO J 11:4969–4979

Xue D, Tu Y, Chalfie M (1993) Cooperative interactions between the Caenorhabditis elegans homeoproteins UNC-86 and MEC-3. Science 261:1324–1328

Yamada T, Pfaff SL, Edlund T, Jessell TM (1993) Control of cell pattern in the neural tube: motor neuron induction by diffusible factors from notochord and floor plate. Cell 73:673–686

Zhadanov AB, Bertuzzi S, Taira M, Dawid IB, Westphal H (1995) Expression pattern of the murine LIM class homeobox gene Lhx3 in subsets of neural and neuroendocrine tissues. Dev Dyn 202:354–364

Zhou X, Sasaki H, Lowe L, Hogan BL, Kuehn MR (1993) Nodal is a novel TGF-beta-like gene expressed in the mouse node during gastrulation. Nature 361:543–547

Zhu TH, Bodem J, Keppel E, Paro R, Royer-Pokora B (1995) A single ancestral gene of the human LIM domain oncogene family LMO in Drosophila: characterization of the Drosophila Dlmo gene. Oncogene 11:1283–1290

WW (WWP) Domains: From Structure to Function

D. ROTIN

1 Introduction

The WW domain, also known as WWP, or rsp5 domain, is a ~40 amino acid module which was identified in late 1994 by three different groups (BORK and SUDOL 1994; ANDRÉ and SPRINGAEL 1994; HOFMANN and BUCHER 1995). The name WW or WWP is based on the primary sequence of the domain, which includes two highly conserved tryptophans and an invariant proline. Like several other protein:protein or protein:lipid interaction domains, WW domains have been detected in numerous unrelated proteins, often alongside other domains, and often in multiple copies (reviewed in STAUB and ROTIN 1996) (Fig. 1). The most noted examples of WW-containing proteins are Nedd4 (*n*euronal precursor cell *e*xpressed *d*evelopmentally *d*ownregulated) and its yeast homologues rsp5 and pub1, YAP (*y*es *a*ssociated *p*rotein), dystrophin, FE65, ess1/dodo/pin1, CD45AP (CD45 *a*ssociated *p*rotein), formin binding proteins (FBPs), and several other less well

The Hospital for Sick Children, and Biochemistry Department, University of Toronto, 555 University Ave, Toronto, Ontario, Canada, M5G 1X8

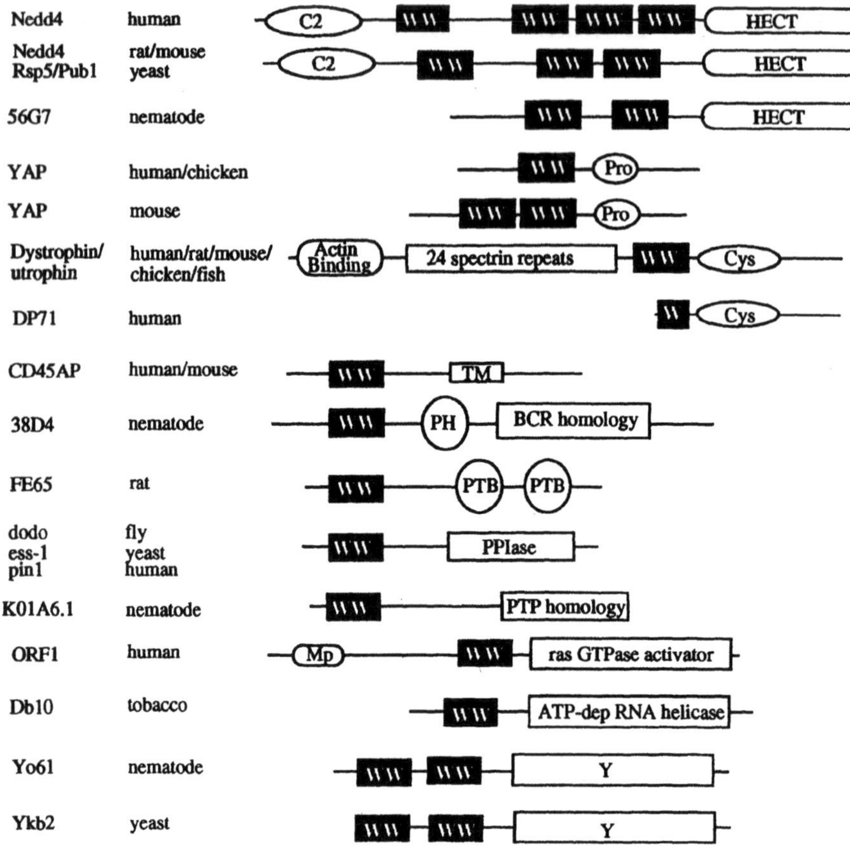

Fig. 1. Schematic representation of selected proteins containing WW domains. WW domains are in *black boxes*. *Single boxed W* (e.g., in DP71) represents a portion of the domain containing only the second conserved Trp. The C2 domain, known to mediate Ca^{2+}-dependent association with phospholipids/membranes, is found in Nedd4/rsp5/Pub1 and also in PKC, PLA_2, PLCγ, rasGAP, synaptotagmin I and other protein. The HECT domain in Nedd4/rsp5 and 56G7 is a ubiquitin protein ligase (E3) enzyme present also in E6-AP, the yeast ykl162, rat p100 and UreB1; *Pro*, proline-rich (SH3 binding) region; *Cys*, cystein rich region; Actin binding domain is homologous to actinin, calponin, vav and spectrin; spectrin repeats are 24 spectrin-like repeats; *TM*, transmembrane domain of CD45 associated protein (CD45AP); *PH*, pleckstrin homology domain also found in dynamin, SOS, PLCγ, IRS-1, rasGAP and Btk; BCR (breakpoint cluster region) homology domain, also shared by p85 of PI-3 kinase, rhoGAP and n-chimerin; PTB (phosphotyrosine binding) or PID (phosphotyrosine interacting) domains, recently suggested to be a subclass of PH domains, are also present in Shc, numb, X11 and IRS-1; PPIase, peptidylpropyl *cis-trans* isomerase (see text); *PTP*, protein tyrosine phosphatases; Mp domain, homologous to the fly muscle protein mp20; the Y domain, shared by Yo61 and Ykb2, has no known function. Sizes of domains and proteins are not to scale. (Reproduced and modified with permission from Staub and Rotin 1996)

characterized proteins (Figs. 1, 2). The presence of more than one WW domain in some of these proteins (e.g., Nedd4) suggests they interact with multiple targets. Phylogenetic analysis of the various WW domains reveals in some cases greater relatedness between WW domains from different proteins than those within the same protein (Sudol et al. 1995), implicating divergent origin. As is elucidated below, the WW domain is a protein:protein interaction module which likely functions in an analogous (yet distinct) fashion to SH3 domains.

2 WW Primary Sequence

Sequence alignment of various WW domains reveals it is a short module which generally includes the consensus sequence Lxx**GW**tx$_6$Gtx(Y/F)(Y/F)h(N/D)Hx(T/S)tT(T/S) t**W**xt**P**t (where x = any amino acid, t = turn like or polar residue, and h = hydrophobic amino acid. Bold letters indicate invariant residues) (Fig. 2). Subclassification of the various WW domains identified to date is based on a phylogenetic analysis using sequence alignment of the conserved residues within the domain (SUDOL et al. 1995). This analysis suggests a hierarchy of relatedness in which Nedd4/rsp5 and YAP WW domains are very closely related, and both more distantly related to dystrophin WW domain (SUDOL et al. 1995). Accordingly, the WW domains of both YAP and Nedd4, but not of dystrophin, can bind to the peptide sequences PGTPPPxY derived from WBP1 or ENaC (CHEN and SUDOL 1995; Rotin, unpublished) (see below); Specific amino acid substitutions within the first half of the Nedd4 WW(II) domain, generated to mimic the dystrophin WW domain sequence, indeed abrogate binding to this peptide (D. Rotin, unpublished). Further distant from the Nedd4/rsp5-YAP and dystrophin WW domains (in increasing evolutionary distance) are the ess1/dodo/pin1, msb1 and FE65 WW domains. A more comprehensive phylogenetic analysis of the various WW domains is detailed elsewhere (SUDOL et al. 1995).

3 WW Ligand Sequences: the PY Motif and the PPLP Motif

A screen of an expression library using the YAP-WW domain as a probe has identified two novel proteins termed WW binding protein (WBP) 1 and 2, each containing short proline-rich repeats which also include a tyrosine (PPPPY) (CHEN and SUDOL 1995). Based on subsequent in vitro binding studies, a preliminary consensus for WW domain binding was proposed, the PY motif (xPPxY, x = any amino acid). Similar proline-rich PY motifs, recently identified in the epithelial Na^+ channel (ENaC) (STAUB et al. 1996; SCHILD et al. 1996), were demonstrated to associate in vitro and in living cells with the WW domains of Nedd4 (STAUB et al. 1996). In the latter, each of the three subunits of the channel (αβγENaC) contains a single, C-terminal located PY motif (PPPAY, PPPNY, PPP(R/K)Y, for αβγENaC, respectively). The biological significance of these ENaC-Nedd4 interactions is discussed below. The PY motif differs from SH3 binding motifs (xPpxP, where p is usually also a Pro) (FENG et al. 1994; YU et al. 1994). Accordingly, work of CHEN

Protein/Species	Position	Sequences of WW domains	Accession #
Nedd4/rat-1	251	PSPLPPGWEERQDVL.GRTYYVNHE........SRTTQWKRPSPEDDLT	U50842
Nedd4/rat-2	407	SSGLPPGWEEKQDDR.GRSYYVDHN........SKTTTWSKPTMQDDPR	U50842
Nedd4/rat-3	464	LGPLPPGWEERTHTD.GRVFFINHN........IKKTQWEDPRMQNVAI	U50842
Nedd4/mouse-1	39	PSPLPPGWEERQDVL.GRTYYVNHE........SRRTQWKRPSPDDDLT	P46935
Nedd4/mouse-2	195	SSGLPPGWEEKQDDR.GRSYYVDHN........SKTTTWSKPTMQDDPR	P46935
Nedd4/mouse-3	250	LGPLPPGWEERTHTD.GRVFFINHN........IKKTQWEDPRLQNVAI	P46935
Nedd4/human-1	217	PSPLPPGWEERQDIL.GRTYYVNHE........SRRTQWKRPTPQDNLT	P46934
Nedd4/human-2	374	SSGLPPGWEEKQDER.GRSYYVDHN........SRTTTWTKPTVQATVE	P46934
Nedd4/human-3	447	QGFLPKGWEVRHAPN.GRPFFIDHN........TKTTTWEDPRLKIPAH	P46934
Nedd4/human-4	499	LGPLPPGWEERTHTD.GRIFYINHN........IKRTQWEDPRLENVAI	P46934
Rsp5/S.cerevisiae-1	228	YGRLPPGWERRTDNF.GRTYYVDHN........TRTTTWKRPTLDQTEA	P39940
Rsp5/S.cerevisiae-2	330	LGELPSGWEQRFTPE.GRAYFVDHN........TRTTTWVDPRRQQYIR	P39940
Rsp5/S.cerevisiae-3	386	LGPLPSGWEMRLTNT.ARVYFVDHN........TKTTTWDDPRLPSSLD	P39940
Pub1/S.pombe-1	205	YGRLPPGWERRTDNL.GRTYYVDHN........TRSTTWIRPNLSSVAG	U62795
Pub1/S.pombe-2	288	SGELPPGWEQRYTPE.GRPYFVDHN........TRTTTWVDPRRQQYIR	U62795
Pub1/S.pombe-3	344	LGPLPSGWEMRLTNT.ARVYFVDHN........TKTTTWDDPRLPSSLD	U62795
Yap/mouse-1	155	DVPLPAGWEMAKTSS.GQRYFLNHN........DQTTTWQDPRKAMLSQ	P46938
Yap/mouse-2	214	SGPLPDGWEQAMTQD.GEVYYINHK........NKTTSWLDPRLDPRFA	P46938
Yap/human	170	DVPLPAGWEMAKTSS.GQRYFLNHI........DQTTTWQDPRKAMLSQ	P46937
Yap/chicken	168	DVPLPPGWEMAKTPS.GQRYFLNHI........DQTTTWQDPRKAMLSQ	P46936
cl.3544mRNA/mouse	21	TVILPPGWHSYLSPQ.GRRYYVNTT........TNETTWERPSSSPGIS	U19860
F13E6.4/C.elegans	217	QLPMPQGWEMCYDSD.GVRYFKDHN........SKTTTWDDPRLKQQEQ	Z68105
Dodo/drosophila	4	AEQLPDGWEKRTSRSTGMSYYLNMY........TKESQWDQPTEPAKKA	U35140
Ess1/S.cerevisiae	29	STGLPTPWTARYSKSKKREYFFNPE........TKHSQWEEPEGTNKDQ	P22696
Pin1/human	4	EEKLPPGWEKRMSRS.GRVYYFNHI........TNASQWERPSGNSSSG	U49070
Msb1/human	249	IVLPPNWKTARDPE.GKIYYYHVI........TRQTQWDPPTWESPG	?

ORF1/human	679	VGDNNSKWVKHWVKG.GYYYYHNLE........TQEGGWDEPPNFVQN	D29640
K01A6.1/C.elegans	130	EGLLPPNWETAYTEN.GDKYFIDHN........TGTTTWDDPRELPPGW	Z68750
Fe65/rat	11	DSDLPAGWMRVQDTS.GT.YYWHIP........TGTTQWEPPGRASPSQ	X60468
Ykb2/S.cerevisiae-1	1	MSIWKEAKDAS.GRIYYYNTL........TKKSTWEKPKELISQE	P33203
Ykb2/S.cerevisiae-2	38	LLLRENGWKAAKTAD.GKVYYYNPT........TRETSWTIPAFEKKVE	P33203
Dystrophin/human	3954	STSVQGPWERAISPN.KVPYYINHE........TQTTCWDHPKMTELYQ	P11532
DP71/human	8	-------HE........TQTTCWDHPKMTELYQ	A45255
Utrophin/human	2811	STSVQLPWQRSISHN.KVPYYINHQ........TQTTCWDHPKMTELF	P46939
C38D4.5/C.elegans	95	RRDLLNGWFEYETDV.GRTFFFNKE........TGKSQWIPPRFIRTPA	P46941
P9659.21/S.cerevisiae	1	MRGEWQEFKTPA.GKKYYYNKN........TKQSRWEKPNLKKGSN	U40829
Yo61/C.elegans-1	77	SPSVESDWSVHTNEK.GTPYYHNRV........TKQTSWIKPDVLKTPL	P34600
Yo61/C.elegans-2	123	QPQQGQWKEFMSDD.GKPYYYNTL........TKKTQWEPPGRASPS	P34600
Yfx1/S.cerevisiae	8	PPQVPSGWKAVFDDEYQTWYYVDLS........TNSSQWEPPRGTTWPR	P43582
ZK1248.15/C.elegans	198	TENVSPPWKAWWHTEKKRKFYTNDK........TKESLWDHPNTRKNEE	U29244
K015c11/C.elegans	?	QNPDDAWNEFNAPD.GRKYYFNSI........TQENTWEKPKALIDQE	D34959
56G7/C.elegans	228	QTPPESHWKTYLDAK.KRKFYVNHV........TKETRWTKPDTLNNNH	Z46793
SPAC13C5.02/S.pombe	89	RIPNNDSWVVVFTKK.NRYFFHNLK........SHESYWEPPLEISKDLK	Q09685
CD45AP/human	48	TGLALAWRR.LSRD.SGGYYHPARLGAALWGRTRRLLWASPPGRWL	A55412
CD45AP/mouse	48	TALALAWCR.LSHA.SGGYYHPARLGAALWGHTRRLLWASPAGRWL	A49957
FBP11/mouse-1	?	WTEHKSPD.GRTYYYNTE........TKQSTWEKP	U40747
FBP11/mouse-2	?	WKEYKSDS.GKPYYYNSQ........TKESRWAKP	U40747
FBP21/mouse-1	?	WVEGVTAD.GHCYYYDLI........TGASQWEKP	U40746
FBP21/mouse-2	?	WVEGLSED.GYTYYYNTE........TGESKWEKP	U40746
FBP23/mouse	?	WVENKTPD.GKVYYYNAR........TRESAWTKP	U40748
FBP28/mouse	?	WTEYKTAD.GKTYYYNNR........TLESTWEKP	U40749
FBP30/mouse	?	WQEVWDEN.TGCYYYWNTQ.......TNEVTWELP	U40750
Consensus:		LPtGWE ttt Gt YYhNH TtTTtW tPt t	
		V FF D S S	

Fig. 2. Sequence alignment of WW domains from different proteins. Invariant residues are in *bold letters*; conserved residues in the consensus sequence are in *capital letters*; *h*, hydrophobic; *t*, polar or charged (turn like) residues. *Dots* represent spaces introduced to maximize homology; *dashes* (in DP71) represent an unrelated amino acid sequence. The list of WW domains shown is not exaustive. Several "WW-like" domains in which the second Trp is replaced by another hydrophobic residue (Y, F), such as those in Db10, YJL168c, 56G7 and SPAC13c5.02 (the latter two already have one WW domain), are not included in the alignment. (Modified and reproduced with permission from STAUB and ROTIN 1996)

and SUDOL (1995) and our own work (STAUB et al. 1996) has indicated that the PY motifs in WBP1, βENaC and γENaC were unable to bind in vitro to several SH3 domains, including those of c-abl, GAP, Fyn, Yes, PLCγ, c-src, p85 of PI3-kinase and α-spectrin. An exception to this observation has been reported: a proline-rich sequence devoid of tyrosines in the protein formin was recently found to associate with both SH3 and WW domains (see below). A search through the gene bank database performed over a year ago has already identified close to 5000 sequences in mainly unrelated proteins conforming to the xPPxY sequence motif (EINBOND and SUDOL 1996). Most of these are likely random sequences with so far no evidence for relevant WW binding. Some, however, appear to be conserved in several family members (Table 1): In addition to the PY motifs in the three ENaC subunits which we have identified (STAUB et al. 1996), several groups of proteins, including viral GAG proteins (e.g., RSV, FMLV, HTLV-1, end V, BLV, AEV), interleukin receptors (e.g., IL-2R, IL-7R) and several Ser/Thr kinases (e.g., MAPKAP2, CamKI) were recently found to contain conserved PY motifs (EINBOND and SUDOL 1996). A conserved sequence which includes a PY motif (Table 1) has been also identified in the MAD, MADR1 and MADR2 (but not DPC4 or sma2) proteins (EPPERT et al. 1996; LIU et al. 1996), which are downstream targets of TGFβ or BMP2 receptors. As these are newly characterized proteins, it is not yet known whether their PY motifs indeed interact with WW domains and, if so, of what proteins. Similarly, two PY-containing repeats (Table 1) have been identified recently in *inscuteable*, a *Drosophila* neuronal protein involved in asymmetric cell division, together with numb and prospero (KRAUT and ORTEGA 1996; KRAUT et al. 1996). *Inscuteable* also contains a putative PDZ binding sequence (C terminal SFV sequence), suggesting that both WW and PDZ domains may bind this protein. The identity of such putative binding proteins and their possible involvement in *inscuteable* function are not known.

Our understanding of the specificity of WW binding to the different PY motifs is rudimentary. Nevertheless, a recent study in which a series of amino acids were used to substitute the variable (x) amino acids of the *x*PP*x*Y sequence revealed inhibition of binding to the WW domains of YAP by substitution with acidic amino acids, whereas equivalent basic substitutions increased the affinity towards WW binding (M. Sudol, personal communications).

Recently, an expression library screen with a probe which includes the proline-rich sequence of formin (SPPAPPTPPPLPPPLIPPPPPLPPGLGPLPP) has indentified a set of formin binding proteins (FBPs) containing either SH3 or WW domains (Chan et al., 1996). Subsequent expression screens with the WW domain-containing FBP11 (see Fig.2) led to the isolation of 8 proline-rich ligands (WBP3-WBP10), each containing the sequence PPLP which was necessary to mediate binding (Bradford et al., 1997). FBP11 did not bind the PY motif of WBP1, and conversely, YAP-WW domain did not bind WBP4 or WBP5. Thus, at least two classes of WW-domain ligand sequences exist: the PY motif and the PPLP motif.

Table 1. Example of PY (xPPxY) motifs

Protein	Sequence
EnaC	
αr/hENaC	LTA**PPPAY**AT
βr/hENaC	PGT**PPPNY**DS
γrENaC	PGT**PPPRY**NT
γhENaC	PGT**PPPKY**NT
WBP	
WBP1	PGT**PPPPY**TV
WBP2	VQP**PPPPY**PG
	SQP**PPPPY**YP
Gag	
RSV	ASA**PPPPY**VG
HTLV1	DPQ**IPPPY**VE
AEV	MAR**DPPRY**LV
IL-R	
h/cIL-2Rγ	PYW**APPCY**TL
mIL-6R	TSP**PPPPY**SL
rIL-7R	DRN**RPPVY**QD
Kinases	
hMAPKAP2	LCG**YPPFY**SN
hCAMKI	LCG**YPPFY**DE
MAD	
dMAD	AGT**PPPAY**SP
hMADR1	ADT**PPPAY**LP
hMADR2	**TPPPGY**IS
Inscuteable	
Repeat1	APP**PPPPY**NN
Repeat2	LGE**APPPY**NN

c, canine; d, *Drosophila*; h, human; m, mouse; r, rat.
Citations are in the text.

4 WW Tertiary Structure

The three-dimensional structure of human YAP-WW domain in solution, complexed with the ligand peptide **GTPP$_4$P$_5$PY$_7$TV$_9$G** from WBP1, has been recently solved by NMR (Macias et al. 1996). The ligand sequence was modeled in, as the small number of intermolecular peptideprotein NOEs was only sufficient to orient the peptide and to provide limited binding information (see below). The WW domain is a compact module composed of a slightly curved three stranded antiparallel β sheet, with two turns flanking the β strands (Fig. 3A, B). The N and C termini meet on the convex side of the domain to form a hydrophobic buckle, in which the two conserved prolines P14 and P42 are opposing each other and are situated above the conserved W17 (Fig. 3B). A flat binding surface is formed by the

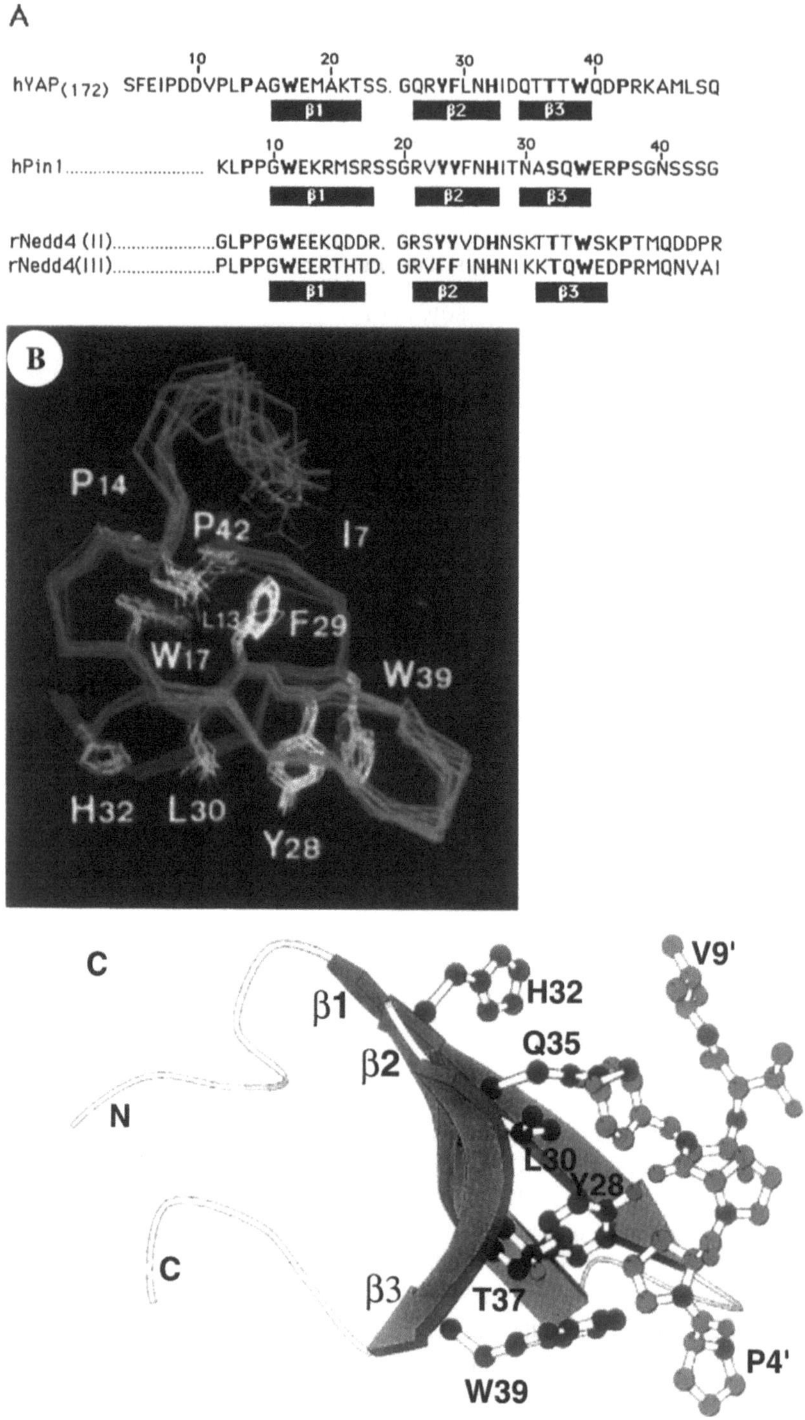
A
10 20 30 40
hYAP(172) SFEIPDDVPLPAGWEMAKTSS. GQRYFLNHIDQTTTWQDPRKAMLSQ
β1 β2 β3
10 20 30 40
hPin1 KLPPGWEKRMSRSSGRVYYFNHITNASQWERPSGNSSSG
β1 β2 β3
rNedd4 (II) GLPPGWEEKQDDR. GRSYYVDHNSKTTTWSKPTMQDDPR
rNedd4(III) PLPPGWEERTHTD. GRVFF INHNI KKTQWEDPRMQNVAI
β1 β2 β3
B
P14
P42
I7
L13
F29
W17
W39
H32
L30
Y28
C
β1
H32
V9'
Q35
β2
N
L30
Y28
C
β3
T37
P4'
W39

side chains of the hydrophobic residues Y28, W39 (which are highly conserved) and L30 on the concave surface of the domain (Fig. 3A, B). A hydrophobic residue at position 30 (usually L, I, V, Y or W) is found in most WW domains (Figs. 2, 3A). Chemical shift changes and peptide-protein NOEs indicate that residues Y28, L30, H32, D34-T38 and W39 on the concave side of the domain participate in peptide (ligand) binding. Specifically, interactions are suggested between $\mathbf{P_4}$ and $\mathbf{P_5}$ (peptide residues are bold) with W39, $\mathbf{Y_7}$ with L30 and H32 (and possibly Q35), and $\mathbf{V_9}$ with H32 (Fig. 3C). Based on these interactions, it has been predicted that replacing $\mathbf{Y_7}$ in the PY motif with a phosphotyrosine will abrogate binding to WW domains. This was indeed demonstrated recently, where placing a phosphotyrosine in the PY motif of WBP1 blocked binding to the WW domains of YAP (PIROZZI et al. 1996). Such an effect could allow for tyrosine phosphorylation-dependent inhibition of WW binding, functionally opposite to SH2 or PTB binding, which are enhanced by tyrosine phosphorylation of target proteins.

Recently, the crystal structure of the peptidyl-prolyl *cis-trans* isomerase (PPIase) Pin1 at 1.35A was determined by X-ray crystallography by J. Noel and colleagues (RANGANATHAN et al., 1997). Pin1 consists of two structurally independent domains organized around a largely hydrophobic, 23A deep cavity: an N terminal WW domain and a C terminal PPIase domain. The WW domain adopts a topology very similar to that of the YAP-WW domain, consisting of a triple stranded antiparallel β-sheet. The flat hydrophobic binding surface of the WW domain faces the above mentioned cavity and thus provides one wall of this cavity; the other wall is provided by the PPIase domain. The two domains "meet" at the tip of the cavity, where residues in the β2/β3 loop (I28, T29 and N30), β3 strand (S32) and P39 of the WW domain (see Fig. 3A), as well as residues in the α1 and α4 helices and β3 strand of the PPIase domain, form interdomain contacts and a well ordered interaction surface. The PPIase enzymatic activity of Pin1 is mediated by a region within the PPIase domain which is distinct from the PPIase/WW hydrophobic cavity. Similar to YAP (Fig. 3), the flat binding surface of the Pin1 WW domain includes the conserved amino acids Y23 (Y28 in YAP), S32 (T37 in YAP), W34 (W39 in YAP), and likely interact with a ligand peptide, as suggested by its sequestration of a hydrophobic PEG molecule. Ser16 in Pin1 also participates in

Fig. 3A–C. Backbone fold and three-dimensional structure of WW domains. **A** Sequence alignment of WW domains from YAP (MACIAS et al. 1996), Pin1 (RANGANATHAN et al. 1997) and Nedd4 WW II and WW III domains (Kanelis, Farrow, Rotin and Forman-Kay, unpublished) and their corresponding backbone β strands (*black boxes*). The preliminary assignment of the β strands in Nedd4 WW II and WW III domains is based on Cα and Cβ chemical shifts. The *numbers shown on top of the YAP and Pin1 sequences* represent numbers corresponding to the WW domain, as depicted also in **B** and **C**. **B** Tertiary structure of YAP-WW domain, showing backbone residues I7-P42, and the side chains of I7, P14, W17, Y28, F29, L30, H32, I33, W39 and P42. The corresponding primary sequence is depicted in **A** above. (Reproduced with permission from MACIAS et al. 1996). **C** A model depicting interactions of YAP-WW domain with the proline-rich peptide from WBP1. Residues showing contact to the protein ($\mathbf{P_4}$–$\mathbf{V_9}$) are shown. The WBP1 peptide was modeled by using ten intermolecular NOEs and restraining the conformation of the peptide sequence into a polyproline type II helix. The side chain of the WW domain (carbon atoms *in black*) and ligand (*green*) residues involved in the interaction are shown. (Reproduced with permission from MACIAS et al. 1996)

this interaction. An interacting protein or peptide to the WW domain of Pin1 has not been identified yet. As in the YAP-WW structure, the upper part (buckle) of the Pin1 WW domain is stabilized by the hydrophobic residues P8, W11, Y24 and P37 (equivalent to P14, W17, F29 and P42 in YAP, Fig. 3). The possible biological implications for this interesting structure of the Pin1 protein are discussed below.

Work in progress on the solution structure of the region encompassing the second plus third WW domains of rat Nedd4 (Fig. 3A), in the presence of a peptide corresponding to the PY motif of βENaC, reveals an overall topology of each domain similar to that of the YAP-WW domain (i.e., three β strands). However, titration experiments with the above peptide identified a strong preference for binding to the third WW over the second WW domain of Nedd4 (V. Kanelis, N. Farrow, D. Rotin and J. Forman-Kay, unpublished), suggesting specificity in the binding interface. This observation also supports the notion that multiple WW domains in a protein are likely targets of different substrates.

The structure of WW domains does not resemble that of SH3 domains, despite the presence of a hydrophobic binding surface which binds proline-rich sequences in both. Nevertheless, such a hydrophobic surface may explain the above-mentioned binding of both SH3 and WW domains to the proline-rich sequence of formin (Chan et al., 1996; Bradford et al., 1997). It is possible that hydrophobic residues replacing the Tyr in the PY motif could be accommodated on the surface of WW domains which contain hydrophobic residues other than Leu at position 30 (Fig. 3C) and therefore differ from YAP-WW domain (Macias et al. 1996). Indeed, all WW domains in the formin binding proteins (FBPs) isolated in an expression library screen with the above proline-rich formin probe contain a Tyr rather than Leu at the equivalent position (Chan et al. 1996). It is also possible that regions with overlapping sequences for both SH3 and WW domain binding may exist in some proteins. For example, the sequence PPLALTAPPPAYTL in αrENaC has been shown to bind both SH3 (Rotin et al. 1994; McDonald and Welsh 1995) and WW domains (Staub et al. 1996).

5 Biological Function of WW Domains

Because WW domains were first described only in late 1994, not much is known yet about their biological functions. However, some recent work has implicated this protein:protein interaction domain in several biological processes, some involving human genetic disorders. In this chapter, I will review these studies in some detail, as well as review other work which describes the isolation of WW containing proteins for which the biological function(s) of the domain is not yet known.

5.1 Nedd4/Rsp5/Pub1

5.1.1 Nedd4-ENaC Interactions: Role in a Hereditary Form of Hypertension (Liddle Syndrome)

Nedd4 (KUMAR et al. 1992), and its yeast homologues Rsp5 and Pub1, is a ubiquitin protein ligase composed of an N terminal C2/CaLB (*Ca*$^{2+}$ *l*ipid *b*inding) domain (OHNO et al. 1987), three or four WW domains, and a C terminal Hect (*h*omology to *E*6-AP *C* *t*erminus) domain (HUIBREGTSE et al. 1995) (Fig. 1). Ubiquitination of proteins serves to tag them for regulated degradation by proteasomes or possibly by lysosomes (CIECHANOVER 1994; HOCHSTRASSER 1996). We have recently demonstrated that Nedd4 binds to the (amiloride-sensitive) epithelial Na$^+$ channel (ENaC) (STAUB et al. 1996). ENaC is composed of three partially homologous subunits, αβγ, each containing a proline-rich region at its C terminus which also includes a PY motif (LTA*PPPAY*ATL, PGT*PPP*$_6$*NY*$_8$DSL$_{11}$, PGT*PPP(R/K)Y*NTL, for α, β, γENaC, respectively). Deletions or mutations within the proline rich regions of β or γENaC have been recently demonstrated by genetic linkage analysis to cause Liddle syndrome (SHIMKETS et al. 1994; HANSSON et al. 1995a, b; TAMURA et al. 1996). Liddle syndrome (pseudohyperaldosteronism) is a hereditary form of human hypertension, characterized by an early onset of severe hypertension caused by elevated Na$^+$ absorption in the distal nephron (LIDDLE et al. 1963) due to increased activity of the epithelial Na$^+$ channel (SCHILD et al. 1995). All mutations identified to date in the afflicted patients cause either a complete deletion of or point mutations within the PY motifs of β or γENaC. Using a yeast two hybrid screen we have recently identified Nedd4 as a binding partner for ENaC, and demonstrated that the binding occurs between the WW domains of Nedd4 and the PY motifs of αβγENaC (STAUB et al. 1996). Moreover, specific mutations within the PY motif of βENaC (in P_6 or Y_8 above), recently identified in Liddle syndrome patients (HANSSON et al. 1995b; TAMURA et al. 1996), led to both abrogation of Nedd4-WW domain binding (STAUB et al. 1996) and to elevation of ENaC activity (SNYDER et al. 1995; SCHILD et al. 1996). Mutating the C terminal Leu downstream of the βENaC-PY motifs (L_{11} above) to Ala also causes channel hyperactivation (SNYDER et al. 1995); By analogy to YAP-WBP1 interactions, we speculate such a mutation may abrogate association with the highly conserved His in the Nedd4-WW domains (equivalent to H32 in YAP, Figs. 2, 3A), which in YAP is interacting with V_9 of WBP1 (Fig. 3C). Since a sizeable component of ENaC hyperactivation associated with the Liddle phenotype is originating from an increase in channel numbers at the cell surface (FIRSOV et al. 1996), we proposed that by interacting with the ubiquitin ligase Nedd4, ENaC may be a short lived protein regulated by ubiquitination. In Liddle syndrome, where association with Nedd4 is impaired, channel stability likely increases, leading to the elevated Na$^+$ channel activity associated with the disease (STAUB et al. 1996). In support of this model, our recent data indeed demonstrate that ENaC is a very short lived protein which is regulated by ubiquitination (STAUB et al., submitted). An alternative explanation for the cell surface elevation of ENaC numbers in Liddle syndrome has been proposed, however (SNYDER et al. 1995); in that model, the sequences in β and γENaC which encompass also the PY motif may constitute an internalization signal which is mutated or lost in Liddle syndrome.

5.1.2 Rsp5 Requirement for Degradation of Amino Acid Permeases in Yeast

In *C. cerevisae* grown on a poor nitrogen source, the addition of NH_4^+ leads to a rapid degradation of the general amino acid permease Gapl (GRENSON 1992). Similarly, subjecting yeast cells to adverse conditions leads to stress-induced degradation of the uracil permease Fur4 (VOLLAND et al. 1994; GALAN et al. 1994). In both cases, the presence of intact Rsp5 (Npil) (Fig. 1) is required for this rapid vacuolar degradation (HEIN et al. 1995; GALAN et al. 1996). Rsp5, like its mammalian counterpart Nedd4, is a ubiquitin protein ligase of the hect family, which also contain a C2 domain and three copies of WW domains. Indeed, recent work has demonstrated that Fur4 is ubiquitinated, and that this ubiquitination is required for its subsequent degradation in the vacuoles, as *rsp5* mutants caused stabilization of the permeases which were no longer ubiquitinated (GALAN et al. 1996). Similarly, *rsp5* bearing a catalytic inactivating mutation (Cys to Ser substitution) at the hect domain failed to restore Gapl inactivation in yeast cells in which expression of the WT gene is limited (B. André, personal communications), suggesting that the ubiquitin ligase activity of Rsp5 is necessary for Gapl degradation. The nature of the interactions between Rsp5 and Gapl or Fur4 is not yet known, since unlike ENaC, these two permeases do not contain bona fide PY motifs (JAUNIAUX and GRENSON 1990; JUND et al. 1988). They do, however, contain short PxY sequences. Whether these are sufficient for WW binding (an unlikely possibility), or whether regions other than the WW domains are responsible for Rsp5-permease interactions, or whether these interactions are altogether indirect, remains to be demonstrated experimentally.

5.1.3 Pub1 Role in cdc25 Degradation and Cell Cycle Regulation

Recently, a gene similar to *RSP5* and likely its homologue, called *Pub1*, was identified in *S. pombe* (NEFSKY and BEACH 1996; SALEKI et al. 1997). *Pub1* has been found to be involved in the ubiquitination and degradation of cdc25; Disruption of the *pub1* gene leads to elevated levels of cdc25 concomitantly with its reduced ubiquitination (NEFSKY and BEACH 1996). cdc25 is a tyrosine phosphatase which dephosphorylates Tyr15 of cdc2, thereby activating cdc2 and initiating mitosis. Thus, *pub1*-mediated destruction of cdc25 leads to mitotic arrest, and accordingly, disruption of *pub1* results in early entry into mitosis due to accumulation of cdc25 (NEFSKY and BEACH 1996). Although the simplest explanation for the *pub1*-mediated cdc25 ubiquitination is that the two proteins interact directly allowing for direct ubiquitination of cdc25 by *pub1*, it is not obvious that these putative interactions are mediated by the WW domains, since the sequence of cdc25 does not contain PY motifs. It is also apparent that *pub1* has additional effects related to membrane transport of nutrients and protons (SALEKI et al. 1997), perhaps analogous to the effect of Rsp5 on amino acid permeases in *S. cerevisae*.

5.2 Dystrophin

Dystrophin is a large (427 kDa) rod like protein which is absent from muscles of patients afflicted with Duchenne muscular dystrophy (DMD) or altered in patients with Becker muscular dystrophy (BMD) (KOEING et al. 1987). Based on its primary structure, the protein can be divided into five regions: N-terminal actin binding domain, spectrin-like domain, a WW domain, Cys-rich domain and a C-terminal region (Fig. 1). Dystrophin is located at the plasma membrane of skeletal muscles, and it forms a complex with several extracellular (α-dystroglycan, laminin) and intracellular (syntrophins, actin) proteins, as well as with a group of transmembrane glycoproteins believed to provide a link between the extracellular matrix and the actin cytoskeleton. This dystrophin-glycoprotein complex (DGC) consists of several proteins, including the αβγ-sarcoglycans and the 43-kDa β-dystroglycan (reviewed in WORTON 1995). β-Dystroglycan contains several proline rich regions in its C terminus (IBRAGHIMOV-BESKROVNAYA et al. 1993), including PLP*PPEY*P, PPYQPPPPFTV and PYRS*PPPY*VPP. A recent in vitro study using peptide competition (JUNG et al. 1995) has demonstrated that binding between dystrophin and β-dystroglycan is mediated by the region encompassing amino acids 3054–3271 (which includes the WW domain and the Cys-rich domain) of dystrophin and the C-terminal final 15 amino acids of β-dystroglycan (KNMTPYRS*PPPY*VPP), which include the above PY motif. These results suggest that interaction between dystrophin and β-dystroglycan is mediated, at least in part, by the WW domain of dystrophin and the second PY motif of β-dystroglycan. An additional study from the same group has demonstrated that the N terminal SH3 domain of GRB2 also associates with the proline-rich C terminus of β-dystroglycan (YANG et al. 1995). Whether or not the interactions with the SH3 and WW domains occur on overlapping proline rich sequences (as seen in formin) remains to be determined.

Deletions/mutations within the 3′ region of dystrophin have been associated with severe forms of muscular dystrophy, although it is not known if any of them map to the WW domain. Nevertheless, the study by JUNG et al. (1995) underscores the importance of the WW domain in maintaining integral complex between dystrophin and its associated proteins. Interestingly, a smaller carboxy terminal isoform of the dystrophin protein, DP71 (apo-dystrophin), which is ubiquitously expressed unlike the muscle-specific dystrophin, lacks the first 19 amino acids of the WW domain (Figs. 1, 2). This causes a deletion of the first highly conserved tryptophan (in the putative β1 strand) and the double tyrosines (in the putative β2 strand) (Fig. 3A) and presumably renders the DP71-WW domain nonfunctional. It is anticipated that binding between DP71 and β-dystroglycan may be reduced or impaired.

5.3 FE65

FE65 is a neuron specific protein composed largely of an N terminal WW domain and two C terminal PTB (PI) domains (Duilio et al. 1991; Bork and Margolis 1995) (Fig. 1). It was recently demonstrated that FE65 interacts with the intracellular domain of the Alzheimer's amyloid precursor protein (βAPP) (Fiore et al. 1995). βAPP (Kang et al. 1987) is a transmembrane precursor of Aβ amyloid, the major constituents of the senile plaques characterizing Alzheimer's disease. In afflicted individuals, Aβ amyloid is cleaved off βAPP and is deposited extracellularly to form the senile plaques. The interaction between βAPP and FE65 have been narrowed down recently to the C terminal PTB domains of FE65, which binds to the sequence YENPTY in βAPP in a phosphotyrosine-independent manner (Fiore et al. 1995; Borg et al. 1996; Guénette et al. 1996). This sequence, which conforms to the internalization motif NPxY, may be crucial for βAPP processing, likely by regulating its endocytosis (De Strooper et al. 1993; Koo and Squazzo 1994). The role of the WW domain in FE65 is not yet known, but it suggests the existence of an additional protein(s), not yet identified, that interacts with the WW domain and may be involved in the regulation of FE65 and βAPP function.

5.4 YAP

YAP (YAP65) is a 65-kDa proline-rich phosphoprotein (Fig. 1) which binds in vitro to the SH3 domain of the tyrosine kinase yes, as well as to other SH3 containing proteins such as Nck, Crk and Src (Sudol 1994). It also contains either a single (human, chicken) or two copies (mouse) of WW domains. As described above, the human YAP-WW domain was recently shown to associate in vitro with WBP1 and WBP2 (Chen and Sudol 1995). In addition, it was also reported that this domain can bind in vitro to the L domain of the Gag protein of Rous sarcoma virus (RSV), which is responsible for retroviral budding. Similar to WBP1, the L domain of RSV contains the PPPPY sequence motif. L domains from Gags of other retroviruses (e.g., HIV, EIAV), which do not contain this PY motif, do not bind the YAP-WW domain (Garnier et al. 1996). Although the possibility that YAP and yes (and WBP1?) may be involved in retroviral budding is intriguing and potentially exciting, these interactions need to be demonstrated in vivo as well. It is therefore possible that the YAP-WW domains may have other, yet unidentified, biological targets and functions.

5.5 Pin1/ess1/dodo

Pin1 is a nuclear peptidyl-prolyl isomerase (PPIase) which also contains a WW domain (Lu et al. 1996) (Fig. 1) and is structurally and functionally related to the yeast Ess1 (Hanes et al. 1989; Hani et al. 1995) and the *Drosophila* dodo

(Maleszka et al. 1996). The *ESS1* gene in *S. cerevisae* (also known as *PTF1*) is an essential gene (Hanes et al. 1989) and mutant *ess1* can be rescued with the *Drosophila* dodo, suggesting evolutionary conservation (Maleszka et al. 1996). PPIases are enzymes which catalyze *cis*/*trans* isomerization of the peptidyl-prolyl bonds in oligopeptides, and are involved in protein folding and unfolding. Depletion/ablation of Pin1/Ess1 in HeLa or yeast cells results in mitotic arrest, while overexpression of Pin1 in HeLa cells leads to G2 arrest. The mitotic arrest can be reproduced by either truncation of the PPIase-containing C terminus of Pin1, or mutations of conserved residues within this domain, both of which abrogate PPIase activity (Lu et al. 1996). Likewise, mutation of both tryptophans within the Ess1-WW domain abrogates the cell cycle regulatory effect of this enzyme (reviewed in Ranganathan et al. 1997). Interestingly, the nuclear localization of Pin1 is abolished when each of its domains (the active PPIase and the WW domain) is expressed separately, even though the putative nuclear localization signal (present within the PPIase region) remains intact (reviewed in Ranganathan et al. 1997). This suggests that either the contact region between the two domains is necessary for the nuclear translocation, or that an intact hydrophobic cavity may be required for an interaction with as yet an unidentified protein which binds to both domains and participates in Pin1 mobilization. Unfortunately, the ligand(s) for the Pin1-WW domain has not been identified yet, nor a putative protein that fits into the PPIase/WW hydrophobic cavity.

Pin1 has been shown to interact with NIMA, a kinase essential for progression through mitosis (Osmani et al. 1988). NIMA and Pin1 co-immunoprecipitate and co-localize, and the interactions between them have been proposed to be direct, by binding of Pin1 to the newly identified protein interacting domain in NIMA called NID (Lu and Hunter 1995; Lu et al. 1996). The NID region contains several SP and TP motifs, and is phosphorylated by the S/T-P-specific CDK, CDC2 (Ye et al. 1995). Since the crystal structure of the Pin1 catalytic region suggests its Pro substrate would be best accommodated when preceded by a phosphoserine or phosphothreonine, Noel and colleagues have suggested that phosphorylated NIMA may be a biological substrate for Pin1 proline isomerization (Ranganathan et al. 1997). Taken together, these data suggest that Pin1 catalytic activity is involved in cell cycle regulation, and that the WW domain, together with the PPIase domain, is involved in proper targeting of the protein to the nucleus, where its target substrate(s) reside.

5.6 Other Examples

A WW domain-containing protein which binds to CD45 (called CD45AP or LPAP) was isolated (Takeda et al. 1994; Schraven et al. 1994). CD45 is a tyrosine phosphatase which dephosphorylates the src kinases lck and fyn, and is involved in T cell signaling and proliferation (Pingel and Thomas 1989; also reviewed in Weiss 1993). The interactions between CD45 and CD45AP/LPAP were recently demonstrated to be mediated by the transmembrane domains of both proteins and not by the WW domain (McFarland et al. 1995; Bruyns et al. 1995). It is therefore speculated that CD45AP/LPAP may bind via its WW domain to other protein(s), possibly important for CD45 signaling.

Work by K. Matsumoto's group has shown that Msb1 (mammalian suppressor of bck1) contains a single WW domain and encodes a suppressor of bck1, a yeast MAPKKK (MAP kinase kinase kinase). It also suppresses mpk1 (yeast MAP kinase) mutation. Since a PY motif (YPPFY) was recently identified in the MAP kinase-associated protein kinase 2 (MAPKAP2) (Einbond and Sudol 1996), it is possible that Msb1 may interact via its WW domain with MAPKAP2, and that this interaction may contribute to the suppressive effect of Msb1.

Acknowledgements. Work from the author's lab presented in this chapter was supported by the Medical Research Council of Canada, by the Canadian Cystic Fibrosis Foundation and by the international Human Frontier Science Program.

References

André B, Springael J-Y (1994) WWP, a new amino acid motif present in single or multiple copies in various proteins including dystrophin and the SH3-binding Yes-associated protein YAP65. Biochem Biophys Res Commun 205:1201–1205

Borg JP, Ooi J, Levy E, Margolis B (1996) The phosphotyrosine interaction domains of X11 and FE65 bind to distinct sites on the YENPTY motif of amyloid precursor protein. Mol Cell Biol 16:6229–6241

Bork P, Margolis B (1995) A phosphotyrosine interaction domain. Cell 80:693–694

Bork P, Sudol M (1994) The WW domain: a signalling site in dystrophin. Trends Biochem Sci 19:531–533

Bradford MT, Chan DC, Leder P (1997) FBP WW domains and the abl SH3 domain bind to a specific class of proline-rich ligands. EMBO J 16:2376–2383

Bruyns E, Hendricks-Taylor LR, Meuer S, Koretzky GA, Schraven B (1995) Identification of the sites of interaction between lymphocyte phosphatase-associated phosphoprotein (LPAP) and CD45. J Biol Chem 270:31372–31376

Chan DC, Bedford MT, Leder P (1996) Formin binding proteins bear WWP/WW domains that bind proline-rich peptides and functionally resemble SH3 domains. EMBO J 15:1045–1054

Chen HI, Sudol M (1995) The WW domain of Yes-associated protein binds a novel proline-rich ligand that differs from the consensus established for SH3-binding modules. Proc Natl Acad Sci USA 92:7819–7823

Ciechanover A (1994) The ubiquitin-proteasome proteolytic pathway. Cell 79:13–21

De Strooper B, Umans L, Van Lauven F, Van Den Berghe H (1993) Study of the synthesis and secretion of normal and artificial mutants of murine amyloid precursor protein (APP): cleavage of APP occurs in a late compartment of the default secretion pathway J Cell Biol 121:295–304

Duilio A, Zambrano N, Mogavero AR, Ammendola R, Cimino F, Russo T (1991) A rat brain mRNA encoding a transcriptional activator homologous to the DNA binding domain of retroviral integrases. Nucleic Acids Res 19:5269–5274

Einbond A, Sudol M (1996) Towards prediction of cognate complexes between the WW domain and proline-rich ligands. FEBS Lett 384:1–8

Eppert K, Scherer SW, Ozcelik H, Pirone R, Hoodless P, Kim H, Tsui L-C, Bapat B, Gallinger S, Andrulis IL, Thomsen GH, Wrana JL, Attisano L (1996) MADR2 maps to 18q21 and encodes a TGFβ-regulated MAD-related protein that is functionally mutated in colorectal carcinoma. Cell 86:543–552

Feng S, Chen JK, Yu H, Simon JA, Schreiber SL (1994) Two binding orientations for peptides to the Src SH3 domain: development of a general model for SH3-ligand interactions. Science 266:1241–1247

Fiore F, Zambrano N, Minopoli G, Donini V, Duilio A, Russo T (1995) The regions of the Fe65 protein homologous to the phosphotyrosine interaction/phosphotyrosine binding domain of Shc bind the intracellular domain of Alzheimer's amyloid precursor protein. J Biol Chem 270:30853–30856

Firsov D, Schild L, Gautschi I, Mérillat AM, Schneeberger E, Rossier B (1996) Cell surface expression of the epithelial Na channel and a mutant causing Liddle syndrome: a quantitative approach. Proc Natl Acad Sci USA 93:15370–15375
Galan JM, Volland C, Grimal DU, Haguenauer-Tsapis R (1994) The yeast plasma membrane uracil permease is stabilized against stress-induced degradation by a point mutation in a cyclin-like destruction box. Biochem Biophys Res Commun 201:769–775
Galan JM, Moreau V, Andre B, Volland C, Haguenauer-Tsapis R (1996) Ubiquitination mediated by the Npi1p/Rsp5p ubiquitin-protein ligase is required for endocytosis of the yeast uracil permease. J Biol Chem 271:10946–10952
Garnier L, Wills JW, Verderame MF, Sudol M (1996) WW domains and retroviral budding. Nature 381:744–745 (correspondence)
Grenson M (1992) Amino acid transporters in yeast: structure, function and regulation. In: De Pont JML (ed) Molecular aspects of transport proteins. Elsevier Science, Amsterdam, pp 219–245
Guénette SY, Chen J, Jondro PD, Tanzi RE (1996) Association of a novel human FEG5-like protein with the cytoplasmic domain of the β-amyloid precursor protein. Proc Natl Acad Sci USA 93:10832–10837
Hanes SD, Shank PR, Bostian KA (1989) Sequence and mutational analysis of ESS1, a gene essential for growth in Saccharomyces cerevisae. Yeast 5:55–72
Hani J, Stumf G, Domdey H (1995) PTF1 encodes an essential protein in Saccharomyces cerevisiae, which shows strong homology with a new putative family of ppI-ases. FEBS Lett 365:198–202
Hansson JH, Nelson-Williams C, Suzuki H, Schild L, Shimkets RA, Lu Y, Canessa C, Iwasaki T, Rossier BC, Lifton RP (1995a) Hypertension caused by a truncated epithelial sodium channel gamma subunit: genetic heterogeneity of Liddle syndrome. Nature Genet 11:76–82
Hansson JH, Schild L, Lu Y, Wilson TA, Gautschi I, Shimkets RA, Nelson-Williams C, Rossier BC, Lifton RP (1995b) A de novo missense mutation of the β subunit of the epithelial sodium channel causes hypertension and Liddle syndrome, identifying a proline-rich segment critical for regulation of channel activity. Proc Natl Acad Sci USA 25:11495–11499
Hein C, Springael JH, Volland C, Haguenauer-Tsapis R, Andre B (1995) NPI1, an essential yeast gene involved in induced degradation of Gap1 and Fur4 permeases, encodes the Rsp5 ubiquitin-protein ligase. Mol Microbiol 18:77–87
Hochstrasser M (1996) Protein degradation or regulation: Ub the judge. Cell 84:813–815
Hofmann K, Bucher P (1995) The rsp5-domain is shared by proteins of diverse functions. FEBS Lett 358:153–157
Huibregtse JM, Scheffner M, Beaudenon S, Howley PM (1995) A family of proteins structurally and functionally related to the E6-AP ubiquitin-protein ligase. Proc Natl Acad Sci USA 92:2563–2567
Ibraghimov-Beskrovnaya O, Milatovich A, Ozcelik T, Yang B, Koepnick K, Francke U, Campbell KP (1993) Human dystroglycan: skeletal muscle cDNA, genomic structure, origin of tissue specific isoforms and chromosomal localization. Hum Mol Genet 2:1651–1657
Jauniaux J C, Grenson M (1990) GAP1, the general amino acid permease gene of Saccharomyces cerevisiae. Nucleotide sequence, protein similarity with the other bakers yeast amino acid permeases, and nitrogen catabolite repression. Eur J Biochem 190:39–44
Jund R, Weber E, Chevallier MR (1988) Primary structure of the uracil transport protein of Saccharomyces cerevisiae. Eur J Biochem 171:417–424
Jung D, Yang B, Meyer J, Chamberlain JS, Campbell KP (1995) Identification and characterization of the dystrophin anchoring site on β-dystroglycan. J Biol Chem 270:27305–27310
Kang J, Lemaire HG, Unterbeck A, Salbaum JM, Masters CL, Grzeschik KH, Multhaup G, Beyreuther K, Muller-Hill B (1987) The precursor of Alzheimer's disease amyloid A4 protein resembles a cell surface receptor. Nature 325:733–736
Koenig M, Hoffman EP, Bertelson CJ, Monaco AP, Feener C, Kunkel LM (1987) Complete cloning of the Duchenne muscular dystrophy (DMD) cDNA and preliminary genomic organization of the DMD gene in normal and affected individuals. Cell 50:509–515
Kraut R, Ortega JA (1996) Inscuteable, a neural precursor gene of Drosophila, encodes a candidate for cytoskeleton adaptor protein. Dev Biol 174:65–81
Kraut R, Chia W, Jan LY, Jan YN, Knoblich JA (1996) Role of inscuteable in orienting asymmetric cell divisions in Drosophila. Nature 383:50–55
Koo EH, Squazzo SL (1994) Evidence that production and release of amyloid β-protein involves the endocytic pathway. J Biol Chem 269:17386–17389
Kumar S, Tomooka Y, Noda M (1992) Identification of a set of genes with developmentally down-regulated expression in the mouse brain. Biochem Biophys Res Commun 185:1155–1161

Liddle GW, Bledsoe T, Coppage WS Jr (1963) A familial renal disorder simulating primary aldosteronism but with negligible aldosterone secretion. Trans Assoc Am Physicians 76:199–213
Liu F, Hata A, Baker JC, Doody J, Cárcamo J, Harland RM, Massagué J (1996) A human Mad protein acting as a BMP-regulated transcriptional activator. Nature 381:620–623
Lu KP, Hunter T (1995) Evidence for a NIMA-like mitotic pathway in vertebrate cells. Cell 81:413–424
Lu KP, Hanes SD, Hunter T (1996) A human peptidyl-prolyl isomerase essential for regulation of mitosis. Nature 380:544–547
Macias MJ, Hyvonen M, Baraldi E, Schultz J, Sudol M, Saraste M, Oschkinat H (1996) Structure of the WW domain of a kinase-associated protein complexed with a proline-rich peptide. Nature 382:646–649
Maleszka R, Hanes SD, Hackett RL, De Conet HG, Gabor-Miklos GL (1996) The Drosophila melanogaster dodo (dod) gene, conserved in humans, is functionally interchangeable with the ESS1 cell division gene of Sacchromyces cerevisiae. Proc Natl Acad Sci USA 93:447–451
McDonald FJ, Welsh MJ (1995) Binding of the proline-rich region of the epithelial Na channel to SH3 domains and its association with specific cellular proteins. Biochem J 312:491–497
McFarland EDC, Thomas ML (1995) CD45 protein-tyrosine phosphatase associates with the WW domain-containing protein, CD45AP, through the transmembrane region. J Biol Chem 270:28103–28107
Nefsky B, Beach D (1996) Pub1 acts as an E6-AP like protein ubiquitin ligase in the degradation of cdc25. EMBO J 15:1301–1312
Ohno S, Kawasaki H, Imajoh S, Suzuki H (1987) Tissue-specific expression of three distinct types of rabbit protein kinase C. Nature 325:161–166
Osmani SA, Pu RT, Morris NR (1988) Mitotic induction and maintenance by overexpression of a G2-specific gene that encodes a potential protein kinase. Cell 53:237–244
Pawson T (1995) Protein modules and signalling networks. Nature 373:573–579
Pingel JT, Thomas ML (1989) Evidence that leukocyte-common antigen is required for antigen-induced T lymphocyte proliferation. Cell 58:1055–1065
Pirozzi G, McDonnell DM, Uveges AJ, Sparks AB, Carter JM, Kay BK, Fowlkes DM (1996) Identification of novel human WW domain – containing proteins by cloning of ligand targets. Mol Biol Cell 7 (abstract 1991)
Ranganathan R, Lu KP, Hunter T, Noel JP (1997) X-Ray crystal structure and functional properties of the mitotic peptidyl-prolyl isomerase Pin1 Cell, (in press)
Rotin D, BarSagi D, O'Brodovich H, Merilainen J, Lehto VP, Canessa C, Rossier BC, Downey GP (1994) An SH3 binding region in the epithelial Na channel (arENaC) mediates its localization at the apical membrane. EMBO J 13:4440–4450
Saleki R, Jia Z, Karagiannis J, Young P (1997) Low pH tolerance in Schizosaccharomyces pombe requires a functioning pub1 ubiquitin ligase. Mol Gen Genet 254:520–528
Schild L, Canessa CM, Shimkets RA, Warnock DG, Lifton RP, Rossier BC (1995) A mutation in the epithelial sodium channel causing Liddle's disease increases channel activity in the Xenopus laevis oocyte expression system. Proc Natl Acad Sci USA 92:5699–5703
Schild L, Lu Y, Gautschi I, Schneeberger E, Lifton RP, Rossier BC (1996) Identification of a PY motif in the epithelial Na channel subunits as a target sequence for mutations causing channel activation found in Liddle syndrome. EMBO J 15:2381–2387
Schraven B, Schoenhaut D, Bruyns E, Koretzky G, Eckerskorn C, Wallich R, Kirchgessner H, Sakorafas P, Labkovsky B, Ratnofsky S, Heuer S (1994) LPAP, a novel 32-kDa phosphoprotein that interacts with CD45 in human lymphocytes. J Biol Chem 269:29102–29111
Shimkets RA, Warnock DG, Bositis CM, Nelson-Williams C, Hansson JH, Schambelan M, Gill JR, Ulick S, Milora RV, Findling JW, Canessa CM, Rossier BC, Lifton RP (1994) Liddle's syndrome: heritable human hypertension caused by mutations in the β subunit of the epithelial sodium channel. Cell 79:407–414
Snyder PM, Price MP, McDonald FJ, Adams CM, Volk KA, Zeiher BG, Stokes JB, Welsh MJ (1995) Mechanism by which Liddle's syndrome mutations increase activity of a human epithelial Na^+ channel. Cell 83:969–978
Staub O, Rotin D (1996) WW domains. Structure 4:495–499
Staub O, Dho S, Henry P, Correa J, Ishikawa T, McGlade J, Rotin D (1996) WW domains of Nedd4 bind to the proline-rich PY motifs in epithelial Na channel deleted in Liddle's syndrome. EMBO J 15:2371–2380
Staub O, Firsov D, Gantschi I, Ishikawa T, Breitschopf K, Ciechanover A, Schild L, Rotin D (1997) Regulation of stability and function of the epithelial Na^+ channel (ENaC) by ubiquitination (submitted)

Sudol M (1994) Yes-associated protein (YAP65) is a proline-rich phosphoprotein that binds to the SH3 domain of the Yes-oncogene product. Oncogene 9:2145–2152

Sudol M, Chen HI, Bougeret C, Einbond A, Bork P (1995) Characterization of a novel protein-binding module – the WW domain. FEBS Lett 369:67–71

Tamura M, Schild L, Enomoto N, Matsui N, Marumo F, Rossier B, Sasaki S (1996) Liddle disease caused by a missense mutation of beta subunit of the epithelial sodium channel gene. J Clin Invest 97:1780–1784

Takeda A, Maizel AL, Kitamura K, Ohta T, Kimura S (1994) Molecular cloning of the CD45-associated 30-kDa protein. J Biol Chem 269:2357–2360

Volland C, Urban-Grimal D, Geraud G, Haguenauer-Tsapis R (1994) Endocytosis and degradation of the yeast uracil permease under adverse conditions. J Biol Chem 269:9833–9841

Weiss A (1993) A cell antigen receptor signal transduction: a tale of tails and cytoplasmic protein-tyrosine kinases. Cell 73:209–212

Worton R (1995) Muscular dystrophies: diseases of the dystrophin-glycoprotein complex. Science 270:755–756

Yang B, Jung D, Motto D, Meyer J, Koretzky G, Campbell KP (1995) SH3 domain-mediated interaction of Dystroglycan and GRB2. J Biol Chem 270:11711–11714

Ye XS, Xu G, Pu RT, Fincher RR, McGuire SL, Osmani AH, Osmani SA (1995) The NIMA protein kinase is hyperphosphorylated and activated downstream of p34cdc2/cyclinB: coordination of two mitosis promoting kinases. EMBO J 14:986–994

Yu H, Chen JK, Feng S, Dalgarno DC, Brauer AW, Schreiber SL (1994) Structural basis for the binding of proline-rich peptides to SH3 domains. Cell 76:933–945

Modular Domains of Focal Adhesion-Associated Proteins

J.M. TAYLOR, A. RICHARDSON, and J.T. PARSONS

1 Introduction

Attachment of cells to the extracellular matrix is mediated by structures called "focal adhesions." Focal adhesions can be readily visualized by reflection contrast microscopy in cells grown in culture either on plastic or a defined extracellular matrix (ECM; e.g., fibronectin, collagen), appearing as spear tip-like structures connecting the extracellular matrix with the ventral plasma membrane (JOCKUSCH et al. 1995). The cytoplasmic face of such structures is decorated with the termini of numerous actin stress filaments, indicating a direct linkage with the actin cytoskeletal network. Focal adhesions not only provide the cell with a point of adhesive contact with the extracellular matrix, but also function to mediate signals that regulate cell growth, migration and apoptosis (BURRIDGE and CHRZANOWSKA-WODNICKA 1996; HYNES 1992; JOCKUSCH et al. 1995). The regulated assembly of focal adhesions at the leading edge of the cell is a central step in cell migration and more specialized processes such as axonal growth (HYNES 1992; LAUFFENBERGER and HORWITZ 1996). Loss of adhesion to the ECM can lead to cell cycle arrest or in some cells the induction of programmed cell death (ROUSLAHTI and REED 1994).

Department of Microbiology, Box 441, Health Sciences Center, University of Virginia, Charlottesville, Virginia 22908, USA

Thus, adhesive interactions with the ECM play a critical role in regulating a variety of intracellular signaling pathways.

The integrin family of transmembrane, cell surface receptors are responsible for mediating cell contact with the ECM. Integrins are a family of heterodimeric transmembrane receptors containing extracellular ligand binding domains (with specificity for specific components of the ECM) and a short cytoplasmic domain that serves to couple integrins with the actin cytoskeleton (BURRIDGE and CHRZANOWSKA-WODNICKA 1996; HYNES 1992). The regulated assembly of focal adhesions has been the source of intense investigation for number of years. Recent information indicates that focal adhesion assembly is regulated by the activation of a number of familiar signaling pathways, among which are pathways controlling activation of protein tyrosine kinases (PTKs), small GTP binding proteins, serine/threonine and lipid kinases (SCHALLER and PARSONS 1994; SCHWARTZ et al. 1995; BURRIDGE and CHRZANOWSKA-WODNICKA 1996). In this review we summarize recent information regarding the structural properties of the major components of focal adhesions. In addition, we speculate how modular domains identified in individual focal adhesion proteins may serve to organize the assembly of focal adhesion components in response to specific cellular signals. A number of excellent reviews will provide the reader with additional information on integrin structure and signaling (SCHWARTZ et al. 1995; BURRIDGE and CHRZANOWSKA-WODNICKA 1996).

2 Proteins Localized to Focal Adhesions

Attachment of cells to the ECM results in the clustering of integrin receptors and initiates the recruitment of numerous cytoplasmic proteins to the focal adhesion complex, including both structural and regulatory molecules. Although proteins are generally identified as "focal adhesion" proteins, by virtue of their localization to cell-ECM contacts, little information is available about the mechanisms or structural properties that target proteins to focal adhesion complexes. Proteins such as α-actinin, talin and focal adhesion kinase (FAK) appear to interact with the short cytoplasmic tail of integrin receptors (OTEY et al. 1993; KNEZEVIC et al. 1996; SCHALLER and PARSONS 1995). Talin and α-actinin as well as other structural focal adhesion proteins such as filamin, radixin, tensin, vinculin, gelsolin, profilin, VASP and Mena bind to actin (BURRIDGE and CHRZANOWSKA-WODNICKA 1996; JOCKUSCH et al. 1995). Additional focal adhesion proteins including the LIM domain containing proteins, paxillin, zyxin, and cysteine rich protein (CRP) do not bind to actin or integrins, but may localize to focal adhesions indirectly, by binding to actin- or integrin-binding proteins (BROWN et al. 1996; ARBER and CARONI 1996; BEKERLE 1986; SADLER et al. 1992).

Recently, a variety of studies have shown that focal adhesions are richly populated with catalytically active signaling proteins, for example, protein tyrosine kinases such as FAK, Src (and Src family members), Csk and Abl; the serine/threonine kinases PKCα and δ and; the tyrosine phosphatases, LAR,

PTP1B; and the catalytic subunit (Cδ) for the ubiquitous serine/threonine phosphatase, PP1 (Burridge and Chrzanowska-Wodnicka 1996; Jockusch et al. 1995). Other regulatory molecules contained within these complexes include polyphosphatidyl inositide 4,5-bis phosphate (PIP2), phosphatidyl inositide 3-kinase (PI3 kinase) and the protease calpain II. Adapter proteins such as paxillin, p130CAS and Grb2 are also found in focal adhesions and contain a number of docking sites for additional regulatory proteins (Burridge and Chrzanowska-Wodnicka 1996; Jockusch et al. 1995). The large repertoire of signaling proteins present in focal adhesions underscores the importance of these structures in propagating "downstream" signals (Burridge and Chrzanowska-Wodnicka 1996).

Because focal adhesions are dynamic structures, their formation and breakdown is regulated by many different extracellular stimuli. Therefore, it is not surprising that many of the focal adhesion proteins exhibit relative low affinity for their individual binding partners. However, the low affinity interactions exhibited by these proteins are often offset by multiple interactions with several different binding partners, permitting the formation of stable, although potentially dynamic structures. For example, multiple binding partners have been described for vinculin, which binds actin, paxillin, α-actinin, talin, and tensin; talin, which binds, integrins, actin and vinculin; and zyxin, which binds to CRP, α-actinin and VASP (Burridge and Chrzanowska-Wodnicka 1996; Jockusch et al. 1995). A number of these proteins also bind phospholipids, which may help to anchor focal adhesion complexes to the plasma membrane.

3 Modular Domains of Focal Adhesion Proteins

3.1 SH2 and SH3 Domains

Several focal adhesion proteins contain Src homology-2 (SH2) domains (Src, tensin, p85PI3 K and Grb2) or Src homology-3 (SH3) domains (p130CAS, Src, Grb2 and p85PI3 K). Upon adhesion to the ECM or growth factor stimulation, a number of focal adhesion proteins (including FAK, p130CAS, paxillin, tensin and Src) become tyrosine phosphorylated and interact with SH2 containing proteins. Other focal adhesion proteins contain proline rich sequences (paxillin, vinculin, p130CAS, FAK, p85PI3 K) which serve as docking motifs for SH3-containing proteins (Table 1). Since the SH2 and SH3 domains are discussed in detail in other chapters of this book, we will confine our discussion of these domains to their possible role in the regulation of focal adhesion complex assembly (see below).

3.2 Actin Binding Domains

Focal adhesions are rich in actin and serve to anchor stress filaments at the plasma membrane. Because of this architecture, it is not surprising that many of the pro-

Table 1. Structural domains in focal adhesion proteins

Domain	Focal adhesion protein
SH2	Src, P13K, Grb 2, tensin
PTyr	Src, P13K, Grb 2, P130, paxillin
SH3	Src, P13K, Grb 2, P130, FAK
PRO	P13K, P130, FAK, paxillin, zyxin
KINASE	Src, FAK, PKC α, δ
ABD-1	α-Actinin, filamin, tensin
ABD-2	Tensin, radixin, talin, gelsolin
ABD-3	VASP
LIM	Paxillin, zyxin, CRP
FP-4	Zyxin, vinculin
EVH1	VASP, Mena
GP-5	VASP, Mena
PIP2	Gelsolin, profilin, vinculin
EF-HAND	α-Actinin
ERM	Radixin, tensin

Pro, proline-rich binding site for SH3 domains; Ptyr, binding site for SH2 domains; PIP2, binding site for PIP2.

teins within focal adhesions bind actin and exert a wide variety of effects on actin organization. Several focal adhesion proteins including α-actinin, filamin and tensin bind laterally to polymerized filamentous actin (F-actin) and serve to cross-link filaments. Vinculin also binds laterally to F-actin and connects the actin filaments to other focal adhesion proteins. Capping proteins, such as radixin, talin, and tensin, bind to the ends of the actin filaments and regulate filament growth. The cross-linking and capping proteins both have the ability to anchor filament ends to the plasma membrane. Monomeric actin binding proteins, profilin and gelsolin, also affect actin polymerization. The former controls the pool of unpolymerized actin and the later severs polymerized actin and promotes formation of new filaments (for review see Schafer and Cooper 1995). Below we discuss the properties of each class of actin binding proteins found in focal adhesions and discuss the role of modular domains in regulating the dynamics of actin remodeling within focal adhesions.

3.2.1 Actin Cross-linking Proteins: α-Actinin, Filamin and Tensin

The length and concentration of actin filaments and the geometry of their cross-links govern changes in cell structure and motility. Structures defined by bundles of cross-linked actin include filopodial extensions, lamellipodia, stress fibers, focal adhesions and actin networks (Zigmond 1996; Lauffenburger and Horwitz 1996). The morphology of these actin structures is determined in large part by the size and structural organization of the cross-linking proteins.

Filopodia are finger-like projections characterized by long parallel arrays of closely packed actin filaments (Fig. 1). The actin filaments filopodia are cross-linked by small monomeric proteins such as fimbrin and villin. These globular proteins link the actin filaments relatively close together (approximately 14 nm).

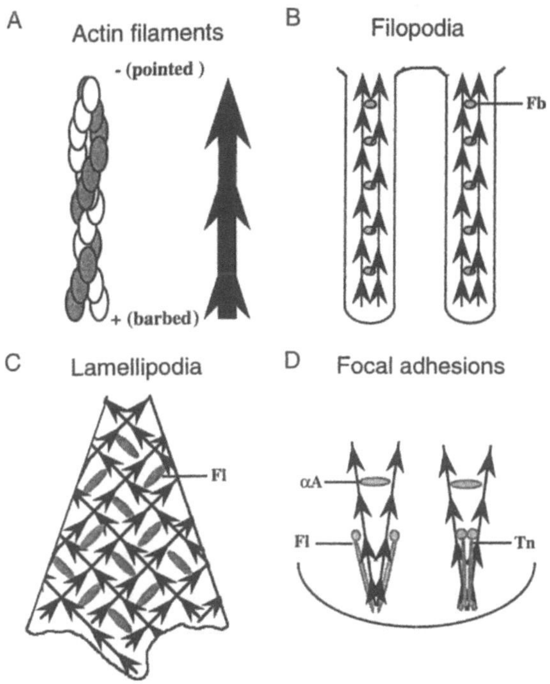

Fig. 1A–D. Schematic view of filamentous actin and actin-containing structures. **A** Actin subunits oriented in a filament structure are shown schematically. The – (*pointed end*) is at the top and the + (*barbed end*) is at the bottom. **B** The organization of actin filaments into filopodial extensions is shown. The barbed end (+) is proximal to the plasma membrane. The cross-linking protein fimbrin (*Fb*) is indicated by *shaded circles*. **C** The organization of actin filaments into lamellipodia is shown. The cross-linking protein filamin (*Fn*) is indicated by the *shaded ovals*. **D** The organization of actin filaments into focal adhesion complexes is shown. The cross-linking proteins filamin (*Fl*), tensin (*Tn*) and α-actinin (αA) are indicated by *the shaded structures*

Actin in stress fibers is cross-linked by dimeric α-actinin, which separates these fibers by about 40 nm (the width of α-actinin). The actin networks in lamellipodia (broad sheet-like projections) are even less tightly packed (approximately 80 nm) and contain shorter filaments which are orthogonally cross-linked by the large, flexible protein filamin (Fig. 1; MATSUDAIRA 1991). Although the actin in focal adhesions is tightly bundled, α-actinin and even larger cross-linking proteins including filamin and tensin are localized to these structures. Perhaps the flexible nature of proteins such as filamin and tensin allows more efficient bundling of filaments at the tip of the focal adhesion (closest to the membrane), whereas the rigid α-actinin bundles at the tail (at the connection to the stress-fibers; Fig 1). In addition, tensin has filament capping activity and filamin binds to membrane glycoproteins, suggesting a potential role for these proteins in binding and cross-linking the membrane proximal actin filaments (CHUANG et al. 1995; EZZELL et al. 1988).

Structural and functional analysis of the actin cross-linking proteins reveals a common 250 amino acid domain that exhibits actin binding activity (designated here as ABD-1 or actin binding domain-1). Within this domain, a conserved 26 amino acid sequence appears to be essential for in vitro actin binding (Matsudaira 1991) (Table 2). For example, a peptide representing this region in α-actinin binds efficiently to actin with a K_d of 4 μ*M* (Hemmings et al. 1992; Kuhlman et al. 1992). This sequence appears to be shared amongst a variety of actin binding proteins. Other cross-linking proteins which contain this consensus sequence include spectrin (fodrin), dystrophin, fimbrin, and the focal adhesion proteins filamin and tensin (Matsudaira 1991).

Table 2. Consensus sequences for actin binding domains

ABD-1

1 125 250

α-Actinin	113	LVSIGAEEIVDGNVKMTLGMIWTIILR	138
Filamin	127	LVSIDSKAIVDGNLKLILGLIWTIILH	152
Tensin	420	LENGPSVSV-DYNTSDPLIR-WDSYE	443

ABD-2

h Capz	243	LRSVQTFADKSKQEALKNDLVEALKRKQ	270
m gelsolin	89	LGNECSQDESGAAAIFTVQLDDYLNGRA	116
A profilin	97	VGVYNEKIQPGTAANVVEKLADYLIGQG	125
c tensin	1109	SPSSLTSGGVRSPPGLAKPTLSALGLKP	1137
c Capz	243	LRSIDAIPDNQKYKQLQRELSQVLTQRQ	272
m radixin	550	LHANEVKAGRDKYKTLRQIRQGNTKHAI	577
m talin	2874	ARKKLAQIRQQQYKFLPSELRDEH-COOH	2901

ABD-3

Thymosin b4	2	KPDMAEIEKFDKSKLKKT	19
Villin	805	PAAFSALPRWKQQNLKKE	822
VASP	221	AGAPGLAAAIAGAKLRKV	238

h, human; m, mouse; A, *Acanthamoeba*; C, chicken

By definition, cross-linking proteins must contain two actin-binding sites, thus promoting interactions with two separate actin filaments. Cross-linking proteins such as fimbrin contain two such motifs within their primary sequence, while other cross-linking proteins such as α-actinin form homodimers as a means to generate two binding domains (MATSUDAIRA 1991). Since bundled actin exhibits the same polarity (Fig. 1), many of the cross-linking proteins form antiparallel dimers. In addition, F-actin is largely α-helical; thus only certain actin subunits will be properly oriented to interact with such an antiparallel dimer (MCLAUGHLIN and WEEDS 1995). Therefore, the stoichiometry of cross-linking protein to an actin monomeric unit is generally 1:4–1:6. To date, the regions on F-actin which bind cross-linking proteins with conserved ABD-1 are unidentified. However, the ability of a number of cross-linking proteins to compete for F-actin binding suggests that the binding sites are overlapping.

F-actin binding proteins have been divided into two groups based on their ability to compete for sites on actin (POPE et al. 1994). The first group includes F-actin binding domains of villin, gelsolin, α-actinin, dystrophin, filamin, fibrin and other spectrin homologues. The second group includes ADF/cofilin, the head domain of villin and dematin (POPE et al. 1994). The competitive binding exhibited by these two groups of proteins suggests that there may be a limited number of potential binding sites for cross-linking proteins on actin filaments.

α-Actinin is an abundant cross-linking protein which localizes to focal adhesions in fibroblasts, in membrane associated dense plaques and dense bodies in smooth muscle, the fascia adherens of the intercalated disks and Z-disks in cardiac muscle, and adherens cell-cell junctions (ZIGMOND et al. 1979; SANGER et al. 1987). α-Actinin is a 90–100 kDa protein which forms head to tail antiparallel dimers. It contains two major structural domains, an N-terminal ABD-1 and C-terminal calcium binding region or EF-hand motif (Fig. 2; SCHLEICHER et al. 1988). The central region of α-actin is composed of a series of rigid α-helical repeat sequences, similar to those observed in spectrin and related proteins. These repeat domains are predicted to stabilize the dimeric rod-like structures of α-actinin, spectrin and dystrophin, the number of α-helical segments determining the distance between the actin- and calcium-binding domains (MATSUDAIRA 1991).

The EF-hand domain is a calmodulin-like calcium binding motif (HEIZMANN and HUNZIKER 1991). Calcium inhibits nonmuscle α-actinin binding to actin, while the muscle form is calcium insensitive (BURRIDGE and FERAMISCO 1981). Since the calcium binding domain of one α-actinin molecule interacts with the ABD of its dimerization partner (Fig. 2), it has been suggested that calcium may disrupt dimerization, and, therefore, block cross-linking (MATSUDAIRA 1991). An amino acid sequence comparison between the muscle and nonmuscle forms of α-actinin reveals that the muscle form is alternatively spliced within the first EF-hand region, yielding a form of α-actinin that is insensitive to calcium (PARR et al. 1992). Fibroblasts contain both muscle and nonmuscle α-actinin and both forms localize to focal adhesions (WAITES et al. 1992). The significance of calcium sensitive and insensitive forms of α-actinin in focal complexes is unclear; however, it has been proposed that more mature focal adhesions contain higher levels of the calcium

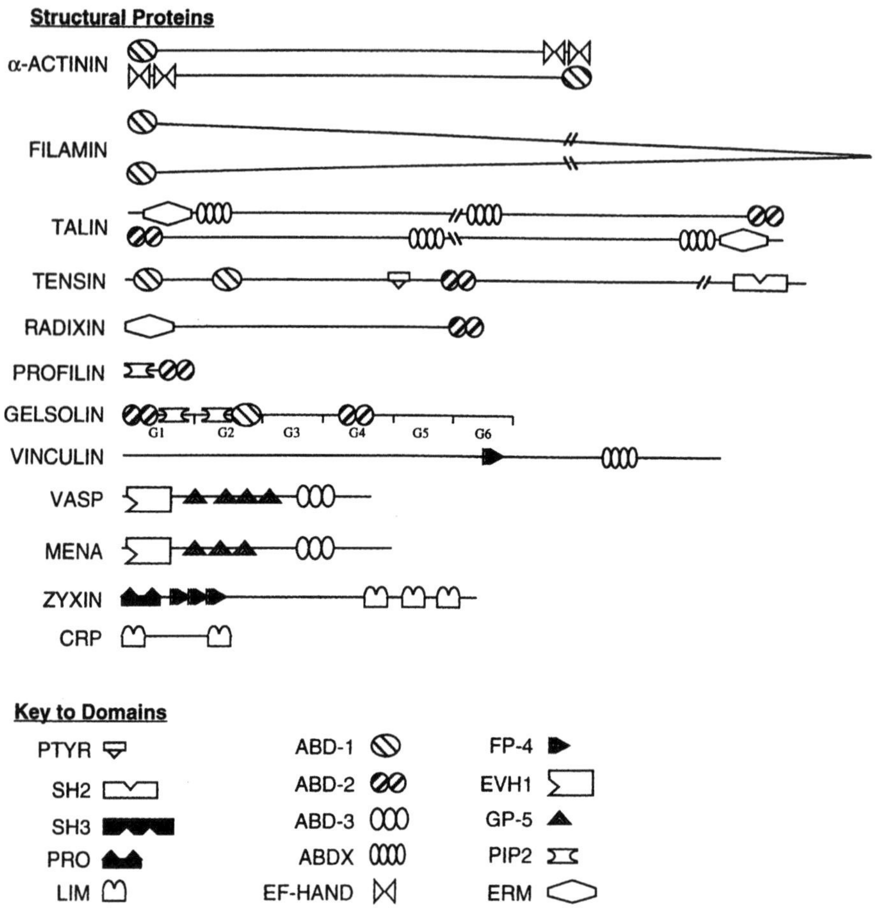

Fig. 2. Topology of focal adhesion structural proteins. Focal adhesion proteins which provide structural support to the complex are shown. Approximate locations of the individual structural domains in each protein are *indicated by the symbols*. *Pro*, proline-rich binding site for SH3 domains; *PTyr*, binding site for SH2 domain; *ABDX*, a region shown to bind actin which does not have homology to other actin binding domains. Each of the other domains is defined in the text

insensitive form (Waites et al. 1992). Interestingly, calcium is required to observe normal filopodia formation (as opposed to loops of actin bundles) and for the transition from filopodia to lamellipodia. Thus calcium may play a role in modulating the extent of actin cross-linking under certain circumstances (Janmey 1994).

PIP2 is another potential regulator of α-actinin. PIP2 co-purifies with α-actinin and has been shown to enhance the ability of α-actinin to cross-link actin in vitro (Fukami et al. 1992). Other known focal adhesion binding partners for α-actinin include the β1 integrin cytoplasmic domain, zyxin (see LIM domains below) and vinculin (Otey et al. 1993; Crawford et al. 1992; Belkin and Koteliansky 1987).

Filamin is a ubiquitous actin cross-linking protein that interconnects actin filaments within cortical actin and links actin filaments at sites of cell-matrix adhesion (PAVALKO et al. 1989). Filamin is a 280-kDa dimer which contains an N-terminal ABD-1 domain (Fig. 2). The central region of filamin is composed of tandem repeats which form a β-sheet motif. The β-sheet motif repeats are also found in ABP-120, another actin binding protein (GORLIN et al. 1990). Like the α-helical repeats in α-actinin, the β-sheet motifs are proposed to function as structural spacers. However, the β-sheets do not appear to mediate dimerization. Rather, filamin is a V-shaped dimer in which two molecules self-associate via a C-terminal domain. Thus filamin is a flexible hinge-like molecule with the potential to tightly pack actin bundles into focal adhesion contacts (Fig. 1).

The C-terminal domain of filamin also functions to anchor actin filaments to the plasma membrane. In platelets, filamin has been shown to bind to the transmembrane glycoprotein 1β/IX (EZZEL et al. 1988). In myeloid cells, it has been shown to bind the CD64 Fc receptor (OHTA et al. 1991). However, in cells that form focal adhesions (e.g., fibroblasts) the protein(s) responsible for linking filamin to the membrane have not been identified. In vitro evidence suggests that in addition to binding to transmembrane proteins, filamin may also directly interact with membrane phospholipids (TEMPEL et al. 1994). The actin cross-linking activity of filamin may be regulated by phosphorylation. Filamin is phosphorylated on serine residues by calcium calmodulin-dependent protein kinase II at sites proximal to the dimerization and/or membrane localization domain. Phosphorylation of these sites in vitro inhibits actin cross-linking (OHTA et al. 1995), indicating that like α-actinin, calcium may play a role in regulating filamin cross-linking activity.

Tensin is a large (200 kDa) actin cross-linking and capping protein that localizes in the dense plaques of smooth muscle, the Z-lines of cardiac muscle and focal adhesion plaques of fibroblasts (BOCKHOLT et al. 1992). Tensin forms parallel dimers by intermolecular interactions involving C-terminal sequences. Three distinct actin binding domains have been identified within tensin, two amino-terminal proximal ABD-1 domains and a central ABD-2 domain (see below). In addition, tensin contains a C-terminal SH2 domain (Fig. 2). Unlike α-actinin and filamin, tensin lacks large repeated spacer domains. In vitro binding experiments indicate that tensin binds F-actin via its N-terminal ABD-1 domains (LO et al. 1994). Capping of the membrane proximal barbed ends of actin, however, is thought to occur by actin binding to the central ABD-2 domain (LO et al. 1994). Hence, dimerization via the C-terminal domains could generate a U-shaped structure allowing for cross-linking and capping of two F-actin filaments (Fig. 1).

Tensin is reported to associate with vinculin; however, sites of interaction within tensin have not been determined (WILKINS et al. 1987). Tensin is tyrosine phosphorylated in fibroblasts growing in culture (BOCKHOLT et al. 1992). While several putative phosphotyrosine binding sites for SH2 containing proteins including Src, Abl, Crk, and p85 PI3 K (LO et al. 1994) have been noted, the binding of SH2 containing proteins to tyrosine phosphorylated tensin remains to be clarified. In addition, the binding specificity of the tensin SH2 domains remains unresolved. Tensin also contains nine PEST sequences (signals for rapid proteolysis), three of which surround the ABD-2 domain (Lo et al. 1994). The role of these sequences in regulating tensin activity or degradation is unclear.

3.2.2 Capping Proteins: Radixin, Tensin, and Talin

While the cross-linking proteins affect the organization and stiffness of the actin fibers, capping proteins regulate the rate of growth or polymerization of existing fibers. Addition of actin monomers to the growing stress filament takes place at the membrane proximal "+," or barbed end as opposed to the "−," or pointed end (Fig. 1; McLaughlin and Weeds 1995). Since polymerization at the barbed end is driven by the concentration of unpolymerized actin in the cell as well as the availability of free ends, capping proteins, by binding with high affinity to the barbed end, can regulate actin polymerization and help maintain the cellular pools of G-actin required for rapid remodeling (Schafer and Cooper 1995; Sun et al. 1995).

Focal adhesion complexes contain several actin capping proteins including talin, radixin and tensin. The structural similarity among capping proteins is not as high as that observed between actin cross-linking proteins. However, a leucine-rich sequence which is present within the actin binding domains of gelsolin, profilin, tensin, and radixin is also found within the capping domains of nonfocal adhesion proteins CapZ, villin, severin and fragmin (Table 2; Barron-Casella et al. 1995). In addition, radixin, talin and the nonfocal adhesion proteins ezrin, moesin, the chicken CapZ β isoform and several myosin heavy chains contain a basic amino acid sequence (KYKXL; K = K/R) present in the center of the leucine-rich capping domain (Turunen et al. 1994). Table 2 illustrates the extent of the conservation of this basic amino acid region in several capping proteins. We refer to the combination of these conserved sequences as ABD-2.

Radixin is a 68-kDa protein with an N-terminal ERM motif (see below) and a C-terminal ABD-2 domain (Funayama et al. 1991). The actin binding region in radixin (and the family member ezrin) resides within the C-terminal 25 amino acids, a region that contains the potential actin binding site KYKXL (Turunen et al. 1994). Recent evidence indicates that the N- and C-terminal regions of radixin self-associate in vitro. Thus radixin binding to actin may be regulated by modulating homodimerization (Magendantz et al. 1995).

As discussed above, tensin has both actin cross-linking and capping activities. The central domain of tensin (Fig. 2) contains the ABD-2 domain and exhibits capping activity in vitro (Chuang et al. 1995). Within this region the leucine residues are positionally conserved when compared to similar regions in other actin capping proteins (Table 2). However, the significance of the conservation of these leucine residues for actin capping has not been tested. The central domain of tensin also shares significant sequence homology to a group of previously purified peptides called "insertins." Insertins exhibit barbed-end binding activity which slows but does not completely inhibit actin polymerization (Ruhnau et al. 1989; Gaertner and Wegner 1991). This has led to speculation that insertins are proteolytic fragments of tensin (Lo et al. 1994). However, recent data demonstrate that the actin capping domain in tensin causes complete inhibition of actin polymerization, suggesting a mechanistic difference between tensin and insertins (Chuang et al. 1995).

Talin is localized in the myotendenous junctions and intercalated disks in muscle tissues and in focal adhesions in cultured cells (Burridge and Connell 1983). Although talin was initially defined as a capping protein, more recent reports describe talin as an actin nucleating protein with weak cross-linking abilities (Kaufmann et al. 1992). The nucleating activity of talin results from its capacity to bind three G-actin monomers to form a tetramer. Therefore, talin can promote the association of actin monomers into de novo filaments. Talin has also been shown to promote a marked increase in α-actinin mediated cross-linking activity (Muguruma et al. 1995). In the absence of α-actinin, talin cross-links actin into compact bundles (13 nm spacing), although this effect is observed in vitro at relatively low pH (pH 6.4) and low ionic strength, and thus may not be physiologically relevant (Zhang et al. 1996).

Talin (a 270-kDa protein) forms anti-parallel dimers (Fig. 2). Talin contains an amino-terminal domain that shows sequence similarity to the membrane binding region of ezrin (see below) (Rees et al. 1990). The secondary structure of talin predicts an extended C-terminal α-helical tail, resembling that of the elongated cross-linking proteins α-actinin and fimbrin; however, the primary sequence does not reveal a conserved ABD-1 domain found in other cross-linking proteins (McLachlan et al. 1994). Three actin binding domains within talin have been identified (Fig. 2; Hemmings et al. 1996). One of the regions, located in the extreme C-terminus of talin, displays sequence similarity to the ABD-2 domain of other actin binding proteins (Table 2). In addition to binding to actin, talin also binds to acidic phospholipids, to the focal adhesion proteins vinculin and FAK, and to the cytoplasmic tail of the β integrin (Goldman et al. 1992; Gilmore et al. 1993; Chen et al. 1995; Knezevic et al. 1996).

3.2.3 G-Actin Binding Proteins: Gelsolin and Profilin

The focal adhesion associated proteins profilin and gelsolin bind monomeric G-actin and play a regulatory role in actin polymerization (Schafer and Cooper 1995). Gelsolin contains a leucine rich ABD-2 domain which mediates the binding to G-actin. Whereas profilin contains a similar region (Table 2), the significance of this domain is unclear, due to lack of conservation of this domain among different species (Barron-Casella et al. 1995). The crystal structure of the G1 subunit of gelsolin in complex with actin has been solved. The structure predicts that the ABD-2 domain forms a long (four turn) α-helix that contains most of the actin-contacting residues (Schutt et al. 1993). Within this-helix ten amino acids make direct contact with actin, including two of the four conserved leucines of the core sequence (Table 2). These observations suggest that the conserved leucines may be important in stabilizing the interactions of this-helix domain with actin (McLaughlin and Weeds 1995). The crystal structure of profilin reveals a structure virtually identical to that of the G1 domain of gelsolin, although these two proteins differ significantly in amino acid sequence (Schutt et al. 1993).

Gelsolin belongs to the "severin" family of actin severing proteins. Included in this family are villin, severin, fragmin, and Cap G. Gelsolin (an 81-kDa protein) comprises six repeated segments (G1-6) of approximately 150 amino acids (MATSUDAIRA et al. 1988). A comparison of the crystal structure of the G-actin binding domains in severin and villin with gelsolin reveals a conserved tertiary structure, consistent with the finding that these domains mediate similar activities in vivo (SCHUTT et al. 1993).

Gelsolin has two repeat segments that bind monomeric G actin (G1 and G4) and one domain that binds F-actin (G2) (CHAPONNIER et al. 1986; BRYAN 1988). Both G1 and G4 contain ABD-2 consensus sites for actin binding. The F-actin binding domain contained in G2 does not fit the ABD-1 consensus; however, it does compete with the F-actin binding domain of α-actinin (and related proteins) for actin binding (POPE et al. 1994).

Gelsolin can sever, cap and nucleate actin filaments in a calcium-regulated manner (POPE et al. 1995). Maximal severing activity requires the coordinated side-binding of filaments (via G2) and capping (via G1), whereas nucleation requires the cooperative binding of segments G1 and G4 to actin monomers. The side-binding segment, G2, appears to target the capping domain, G1, to actin and may cause a conformational change that primes the actin filament for G1 binding (ORLOVA et al. 1995). Interaction of G1 with residues that are important for actin-actin interactions destabilizes F-actin and produces a break in the filament. Severing of F-actin produces a newly formed barbed end which remains tightly capped by gelsolin (SCHAFER and COOPER 1995).

Both the actin monomer binding segments require calcium for binding and severing actin. Two calcium binding sites have been mapped to the regions within the C-terminal subdomains of G4-G6 (POPE et al. 1995). However, these regions do not contain consensus calcium binding motif, such as the EF hand present in α-actinin. In contrast to calcium, PIP2 inhibits gelsolin-actin interactions. Two PIP2 binding sites are localized within the C-terminal region of segment G1 and N-terminal portion of G2 (YU et al. 1992). PIP2 binding appears to block both G1- and G2-actin interactions and therefore, inhibits the steps required for the actin severing process. Since the PIP2 and actin binding sites on gelsolin do not overlap, it is possible that PIP2 binding may induce a conformational change in the structure of the actin binding site. Evidence for the PIP2 induced conformational alteration in gelsolin comes from circular dichroism studies (XIAN et al. 1995). In addition to blocking actin binding, PIP2 dissociates actin-gelsolin complexes and, thus, can release gelsolin from the newly formed barbed end (BARKALOW et al. 1996).

Src is a potential focal adhesion targeting protein for gelsolin. Src and gelsolin transiently interact in osteoclasts and Src (but not the related tyrosine kinase, Lck) phosphorylates gelsolin in vitro (CHELLAIAH and HRUSKA 1996; DE CORTE et al. 1997). Src-dependent phosphorylation was dramatically enhanced by addition of PIP2 but not other phosphatidylinositides (DE CORTE et al. 1997). The physiological significance of tyrosine phosphorylation of gelsolin is not known.

The monomer-binding protein, profilin, was originally characterized as an actin sequestering protein, but since has been recognized as a multifunctional regulator of actin polymerization (SOHN and GOLDSCHMIDT-CLERMONT 1994). Sequestering proteins bind to actin monomers and prevent them from associating with other actin molecules at filament ends, either by directly masking the actin-actin binding sites or by inducing a conformational change in the bound actin. Therefore, sequestering proteins effectively reduce levels of monomeric actin concentrations and act in concert with capping proteins to inhibit spontaneous actin polymerization (SUN et al. 1995). Microinjection of profilin does reduce F-actin content, consistent with a role for profilin as an actin sequestering protein. However, microinjection of profilin-actin complexes increases cellular F-actin (CAO et al. 1992). Thus, depending on the relative amounts of profilin and actin in the cell, profilin can either promote or inhibit actin polymerization (SOHN and GOLDSCHMIDT-CLERMONT 1994).

The enhancement of actin polymerization has been suggested to be due to the ability of profilin to regulate the amount of ATP-G-actin available for polymerization. G-actin monomers are released from polymerized actin in the ADP bound state, but must be converted to the ATP bound form to be incorporated into an elongating filament. The rate-limiting step of this cycle is the exchange of bound ADP for ATP. Profilin has been shown to increase the off-rate of ADP from G-actin 1000-fold, thus enhancing exchange. Profilin, therefore, can increase the rate of actin polymerization by increasing the effective concentration of ATP bound G-actin (GOLDSCHMIDT-CLERMONT et al. 1991a). In addition, the profilin-G-actin complex can associate with F-actin, thereby shuttling ATP-actin monomers to the elongating filament (SOHN and GOLDSCHMIDT-CLERMONT 1994).

Profilin is a single polypeptide chain of 120–140 amino acids. The primary sequence of profilin is poorly conserved among species; however, the predicted secondary and tertiary structure is highly conserved and is very similar to that of the gelsolin family (as described above) (MCLAUGHLIN and WEEDS 1995). Profilin binds efficiently to small clusters of four to five molecules of PIP and PIP2 (LASSING et al. 1985). This interaction dissociates the profilin-actin complex and inhibits the hydrolysis of PIP2 by phospholipase C (PLC; GOLDSCHMIDT-CLERMONT et al. 1991b).

3.2.4 Other F-Actin Binding Proteins: VASP and Vinculin

VASP (vasodilator-stimulated phosphoprotein, see also GP-5, below) is a 46–50 kDa protein that localizes to focal adhesions and binds F-actin (REINHARD et al. 1992). Although the site of interaction with actin has not been mapped, VASP does contain a KLKR motif (ABD-3; see Table 2). This sequence is identical to the actin binding regions in thymosin β4, villin and dementin, suggesting that the actin binding domain in VASP may be localized to these residues (VAN TROYS et al. 1996).

Vinculin also binds F-actin and has been proposed to attach actin microfilaments to the plasma membrane and other focal adhesion proteins (BURRIDGE and

Feramisco 1980). The sequence of vinculin (116-kDa polypeptide) predicts a 90-kDa acidic N-terminal α-helical domain and a 25-kDa basic C-terminus separated by a 50 residue Pro-rich region (FP-4 domain, see below). Although the binding of vinculin to actin was originally controversial, co-sedimentation and fluorescence energy transfer experiments clearly demonstrate direct binding. The region responsible for actin binding has been located within residues at the extreme C-terminus (Menkel et al. 1994). However, this domain shows no sequence homology to either ABD-1, ABD-2, or ABD-3. Vinculin has no effect on actin polymerization, viscoelasticity or bending stiffness of actin filaments (Gotter et al. 1995).

Sequences in the N-terminal domain of vinculin direct binding to the focal adhesion proteins, talin and α-actinin, whereas sequences within the C-terminal domain of vinculin mediate the binding to tensin and paxillin (Gilmore et al. 1992; McGregor et al. 1994; Wilkins et al. 1987; Wood et al. 1994). Residues contained within the paxillin binding site have been shown to be required for vinculin localization to focal adhesions, suggesting that paxillin may direct vinculin to the focal adhesion complex (Wood et al. 1994). Vinculin has been shown to self-associate through head to tail interactions, a process that blocks talin and actin binding (Johnson and Craig 1994). In vitro, PIP2 binding to vinculin (to residues within the proline-rich FP-4 region, see below) enhances the binding to talin and actin, suggesting that PIP2 binding may be responsible for inducing conformational alterations in vinculin structure and modulate interactions with actin and talin (Gilmore and Burridge 1996; Weekes et al. 1996).

Despite the ability of vinculin to bind actin and a number of other focal adhesion proteins, vinculin deficient F9 cells exhibit focal adhesions that are morphologically indistinguishable from those in control cells (Volberg et al. 1995). However, these cells display reduced spreading, have longer filopodia and show increased cell motility. Similarly, PC12 cells expressing antisense vinculin RNA extend unstable filopodia and lamellipodia (Varnum-Finney and Reichardt 1994). These data suggest a role for vinculin in regulating or stabilizing the assembly of actin networks and/or filaments (Varnum-Finney and Reichardt 1994; Goldman et al. 1995; Coll et al. 1995).

3.3 ERM Motifs in Radixin and Talin

The ERM motif was initially identified as a region of high homology shared among the actin-binding, membrane associated proteins ezrin, radixin and moesin (hence the term ERM; Rees et al. 1990) (Table 1). The ERM motif is also present in the band 4.1 protein, an abundant 80-kDa polypeptide in red cell membranes. The 4.1 protein mediates the association between spectrin and actin in vitro and is one of the proteins that links the spectrin-actin lattice to the lipid bilayer (Takeuchi et al. 1994). ERM-domain containing proteins also include the focal adhesion protein talin and the nonfocal adhesion-associated tyrosine phosphatases PTP-H1 and

PTP-MEG (Takeuchi et al. 1994). Many of the ERM-domain containing proteins are localized to the interface between the plasma membrane and cytoskeleton. The targeting of these proteins to membranes appears to involve ERM-mediated protein-protein interactions with the integral membrane glycoproteins such as glycophorin C (which binds to protein 4.1) and CD44 (which binds to radixin, ezrin, and moesin; Bennett 1989, Tsukita et al. 1994).

The actin capping protein radixin contains an N-terminal ERM motif (Funayama et al. 1991). Radixin is reported to bind PIP2 and PIP2 binding appears to markedly enhance in vitro association of CD44 with ERM motif (Hirao et al. 1996). The interactions of radixin with CD-44 appear also to be regulated by the activation of the small GTP binding protein Rho. The association of radixin with cell membranes is enhanced by treatment of cell extracts with GTPγS and blocked by pretreatment of cells of the Rho-inhibitor, C3 exotransferase (Hirao et al. 1996). A possible explanation of this Rho effect is the reported ability of Rho to elevate PIP2 levels by activating a PIP 5-kinase (Ren et al. 1996). Alternatively, the association of radixin and CD44 could be regulated by Rho-dependent activation of a protein kinase. In the case of ezrin, another member of the family, serine and tyrosine phosphorylation of ezrin in response to EGF treatment of cells, leads to its redistribution from the cytosol to membrane microvilli (Krieg and Hunter 1992).

The actin nucleating and cross-linking protein talin contains an ERM motif proximal to the N-terminus (Fig. 2; Rees et al. 1990). The ERM domain is required for association of talin with the plasma membrane; however, the membrane targets of talin binding have not yet been identified. Like ezrin, talin is phosphorylated upon growth factor stimulation or plating cells on the ECM protein fibronectin (Bretscher 1989). It is tempting to speculate that phosphorylation of talin (as well as radixin and ezrin) may regulate either the availability or binding properties of the ERM domain.

Table 3. Consensus sequence for the ERM motif

N-term			
Radixin	63	WLKNKKVTQQDVKKENPLQFKFRAKFFPED	93
Talin	179	WLDHGRTILREQGVEEHETLLLRRKFFYSD	207
C-term			
Radixin	181	HEEHRGMLREDSMMEYLKIAQDLEMYGVNY	210
Talin	288	HKNCGQMSEIEAKVRYVKLARSLKTYGVSF	317

3.4 LIM Domains of Zyxin, CRP and Paxillin

The LIM domain is a cysteine rich, zink containing, protein binding module (SCHMEICHEL and BECKERLE 1994). LIM domains were originally identified as conserved modular domains in the transcription factors *C. elegans* L-11, rat Isl-1 and *C. elegans* Mec-3 from which the acronym was derived. However, LIM domains also appear to be important interaction domains for several focal adhesion proteins including zyxin, CRP (cysteine rich protein) and paxillin (SCHMEICHEL and BECKERLE 1994; CRAWFORD et al. 1994; TURNER and MILLER 1994).

Zyxin is a 69-kDa protein with an N-terminal proline-rich region containing three FP-4 motifs (see below) followed by three C-terminal LIM domains (Fig. 2; BECKERLE 1986). Although LIM domains all exhibit a characteristic consensus sequence, the sequence differences within each LIM domain may be important in providing specificity to individual LIM domain interactions. For example, analysis of GST fusion proteins containing individual LIM domains of zyxin reveals that only the first LIM domain is necessary and sufficient for the binding of zyxin to CRP (SCHMEICHEL and BECKERLE 1994). Whether the second and third LIM domains direct the specific binding to other proteins remains to be determined. The proline-rich N-terminus of zyxin contains proline directed binding sites for the SH3 domain of the nucleotide exchange protein vav (HOBERT et al. 1996) and also binds the focal adhesion proteins α-actinin and Mena (see below). The major binding partner of zyxin is CRP, a 192 amino acid "LIM domain only" protein. CRP comprises an amino-terminal LIM domain followed by a short spacer and a C-terminal LIM domain (Fig. 2; CRAWFORD et al. 1994). Binding partners for CRP other than zyxin have yet to be identified.

Paxillin is a 68-kDa protein containing an amino-terminal proline domain and four C-terminal LIM domains (Fig. 3; TURNER and MILLER 1994). Deletion mutagenesis experiments provide evidence that the third LIM domain is required to direct paxillin to focal adhesions (BROWN et al. 1996). Presently, no binding partners for any of the paxillin LIM domains have been identified. Paxillin does not appear to interact with either zyxin or CRP, indicating that perhaps other LIM-domain binding proteins remain to be discovered.

Paxillin has been demonstrated to interact with the focal adhesion proteins FAK and vinculin as well as with numerous SH2-containing proteins (TURNER and MILLER 1994; HILDEBRAND et al. 1995; TURNER et al. 1990). The binding to FAK and vinculin requires sequences in the middle domain of paxillin and appears to be constitutive, i.e., vinculin and FAK binding is not regulated by growth factor stimulation or adhesion of cells to the ECM (HILDEBRAND et al. 1995). On the other hand, the binding of other SH2-containing proteins is dependent upon the tyrosine phosphorylation of paxillin (SABE et al. 1994; SCHALLER and PARSONS 1995; BIRGE et al. 1993). Tyrosine phosphorylation of paxillin is observed following adhesion of cells to ECM proteins, stimulation of cells with growth factors, or following transformation by oncogenic tyrosine kinases, such as v-Src and v-Abl (SALGIA et al. 1995). Tyrosine phosphorylation of paxillin occurs predominantly on a YXXP motif, and results in the formation of a high affinity binding site for the

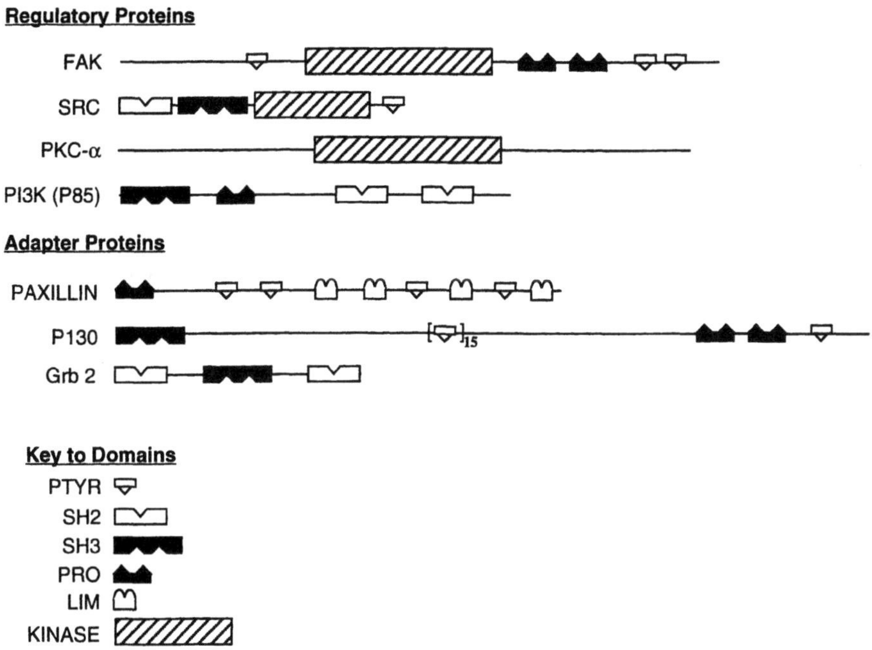

Fig. 3. Topology of focal adhesion signaling proteins. Focal adhesion proteins which are catalytically active or are adapter proteins (and recruit regulatory proteins) are shown. Approximate locations of the individual structural domains in each protein are *indicated by the symbols*

SH2-SH3 containing linker proteins, such as Crk or Crkl (Birge et al. 1993; Salgia et al. 1995). Transformation of cells with the oncogenic forms of v-Crk increases the tyrosine phosphorylation of paxillin and formation of stable complexes of paxillin with v-Crk (Birge et al. 1993).

Paxillin also contains an N-terminal proline motif that has been shown to bind to the SH3 domains of the tyrosine kinases, Src and Csk (Weng et al. 1993; Sabe et al. 1994; Schaller and Parsons 1995). Thus paxillin appears to serve as a multifunctional linker protein, mediating interactions with focal adhesion proteins (e.g., FAK and vinculin) as well as other SH2/SH3 domain-containing linker and signaling molecules.

3.5 Other Conserved Domains in Focal Adhesion Proteins: GP-5, FP-4 and EVH1 Motifs

The GP-5 domain contains repeats of a G-P-P-P-P-P sequence. This motif has been identified in a family of structurally similar cytoskeletal proteins including VASP, Mena (mammalian Ena, *enabled*), and Evl (Ena-VASP-like; Gertler et al. 1996). GP-5 motifs have also been identified in WASP, the protein implicated in Wiskott-Aldrich syndrome, which bears some homology to VASP (Symons et al. 1996).

Both VASP and Mena have been localized to focal adhesions by immunofluorescence.

Mena and VASP both contain an N-terminal EVH1 domain (see below), and a central proline rich region containing GP-5 repeats. In the case of human VASP, the first GP-5 repeat is separated by 46 amino acid from 3 downstream tandem GP-5 repeats (Fig. 2; HAFFNER et al. 1995). In Mena, a motif which differs from the GP-5 consensus by a single amino acid (an Ala to Pro substitution) is separated by 19 amino acids from two downstream GP-5 repeats each separated by five amino acids (GERTLER et al. 1996). There are likely to be additional examples of conservative substitutions within the GP-5 motif and variations in the spacing of these motifs. However, the significance of the spacing between these motifs and the importance of the maintaining the precise sequence of this motif has not been tested.

Since the GP-5 motifs have been identified in relatively few proteins, one can only speculate about their likely binding partners. VASP binds to profilin and this interaction can be blocked with a GP-5-containing peptide. The binding site for VASP on profilin is likely to be identical to the site that binds poly-L-proline. This region consists of an aromatic, hydrophobic interface between two helices and a β sheet (METZLER et al. 1994; ARBER and CARONI 1994). Although this binding site is structurally distinct from other binding sites that interact with proline-rich peptides such as SH3 domains and WW domains, each is characterized by the presence of a number of aromatic acid residues (YU et al. 1992; MACIAS et al. 1996). Over 200 proteins have been identified which contain at least 6 tandem proline residues, raising the possibility that the binding of GP-5 sequences to profilin-like domains represent a widespread protein-protein interaction domain.

The localization of VASP and profilin to focal adhesions, the association of VASP and profilin in vitro and the observations that VASP can bind to other focal adhesion proteins including vinculin and zyxin suggests that the GP-5-profilin interactions likely serve to recruit profilin to regions of actin assembly (discussed below; REINHARD et al. 1995a, b; BRINDLE et al. 1996).

The FP-4 motif, F-P-P-P-P, has been identified in the focal adhesion proteins vinculin and zyxin as well as in the surface protein ActA from *Listeria monocytogenes* (GERTLER et al. 1996; BRINDLE et al. 1996). ActA is required for *Listeria* to induce cytoplasmic F-actin formation, a process that leads to increased bacterial motility within the cell (SMITH et al. 1996). Zyxin has three FP-4 repeats, while vinculin contains a single FP-4 motif (SADLER et al. 1992; PRICE et al. 1989). Like the GP-5 motifs, the FP-4 motifs are distinct from the proline rich sequences recognized by SH3 domains and WW domains, but the extent to which FP-4 and GP-5 motifs share binding sites has yet to be fully explored.

A putative target for FP-4 binding has been identified and is referred to as the Ena-VASP homology domain, or EVH1 (GERTLER et al. 1996). The EVH1 domain is a region of homology contained in the extreme amino terminus of VASP, Mena, WASP and the *S. cerevisiae* protein verpolin (which is also implicated in cytoskeletal reorganization; SYMONS et al. 1996). This domain is predicted to form an extended α-helix (DERRY et al. 1994). Biochemical experiments have shown that zyxin, vinculin and ActA can bind directly to fusion proteins containing the EVH1 domain from Mena (GERTLER et al. 1996). In addition, a short peptide containing the FP-4 motif from vinculin binds to VASP (BRINDLE et al. 1996). Also, microinjection of a peptide containing the FP-4 motif displaces Mena and VASP from focal adhesions (GERTLER et al. 1996), suggesting that the EVH1 domain may function to direct Mena and related proteins (VASP, WASP) to focal adhesions via the interaction with other focal adhesion proteins.

4 Regulation of Focal Adhesion Assembly

Because focal adhesions are multiprotein complexes, their assembly requires that individual focal adhesion proteins interact with multiple binding partners (see Table 4; Fig. 4). The assembly of focal adhesions is undoubtedly complex. How are proteins recruited to the complex, how are actin stress filaments anchored within the complex and how are the assembly and disassembly of such structures regulated? Integrin binding to extracellular matrix (or the cross-linking of integrin subunits with antibodies that mimic ligand binding) is sufficient to initiate many of the early steps in the formation of focal adhesions (such as the recruitment of FAK, tensin, talin, vinculin and α-actinin) (MIYAMOTO et al. 1995a, b). Thus it is likely that the clustering of integrin subunits at the cytoplasmic face is an early and necessary event in focal adhesion formation. However, it is becoming increasingly clear that assembly of both focal adhesions and stress fibers is regulated by the activity of small GTP binding proteins of the Rho family and protein tyrosine kinases (RIDLEY 1996).

Table 4. Interactions between focal adhesion proteins

	Int	Act	aAct	Tal	Ten	Pro	Vinc	Vasp	Zyx	Crp	Fak	Src	p85	Pax	Cas	grb2
α-Actinin	+	+	–	–	–	–	–	–	+	–	–	–	–	+	–	–
Talin	+	+	–	–	–	–	+	–	–	–	+	–	–	–	–	–
Tensin	–	+	–	–	–	–	+	–	–	–	–	–	–	–	–	–
Profilin	–	+	–	–	–	–	–	+	–	–	–	–	–	–	–	–
Vinculin	–	+	+	+	+	–	–	+	–	–	–	–	–	+	–	–
Vasp	–	+	–	–	–	+	+	–	+	–	–	–	–	–	–	–
Zyxin	–	–	+	–	–	–	–	+	–	+	–	–	–	–	–	–
Fak	+	–	–	+	–	–	–	–	–	–	–	+	+	+	+	+
Src	–	–	–	–	–	–	–	–	–	–	+	–	+	+	+	–
P13 kinase	–	–	+	–	–	–	–	–	–	–	+	+	–	–	–	–
Paxillin	–	–	–	–	–	–	+	–	–	–	+	+	–	–	–	–
Cas	–	–	–	–	–	–	–	–	–	–	+	+	–	–	–	–

Int, integrin; Act, actin; aAct, α-actinin; Tal, talin; Ten, tensin; Pro, profilin; Vinc, vinculin; Zyx, zyxin; p85, p85 PI-3 kinase; Pax, paxillin; Cas, p130 Cas.

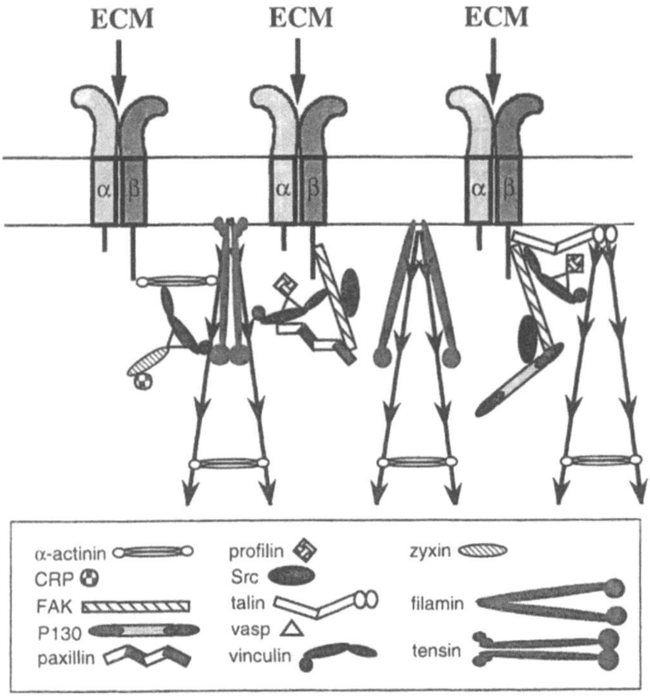

Fig. 4. Schematic view of a focal adhesion. This model shows the predicted organization of both structural and signaling proteins in focal adhesion complexes. Many of the known protein-protein interactions are shown

Swiss 3T3 cells deprived of serum and/or growth factors lose stress fibers and focal adhesions. Microinjection of activated Rho into starved cells rapidly induces the formation of stress fibers and focal adhesions, whereas treatment of cells with the Rho-specific inhibitor, C3 exotransferase, which ADP-ribosylates and inactivates Rho, inhibits formation of both stress fibers and focal adhesions (NOBES and HALL 1995). Similarly, treatment of cells with activators of G-protein coupled receptors such as lysophosphatidic acid (LPA), endothelin, thrombin and bombesin also stimulate the Rho dependent formation of stress fibers and focal adhesions (MACHESKY and HALL 1996). What are the downstream effectors of Rho which lead to assembly of stress fibers and focal adhesion? Two candidate effector pathways show intriguing activities. Rho activates at least two serine/threonine kinases, PKN and ROCK (WATANABE et al. 1996; AMANO et al. 1996a, b; LEUNG et al. 1996). Although the complete spectrum of substrates for PKN and ROCK is not known, one substrate is the myosin-binding, regulatory subunit of the myosin phosphatase (PP1-M). Phosphorylation of PPI-M by ROCK inactivates the phosphatase and in turn enhances myosin-light chain (MLC) phosphorylation, an effect that stimulates actin-myosin mediated contraction (KIMURA et al. 1996; AMANO et al. 1996a). Considerable evidence links the formation of focal adhesions and stress fibers with actin-myosin contractility (BURRIDGE and CHRZANOWSKA-WODNICKA 1996). Thus the regulation of contractility by Rho and the effects of MLC kinase inhibitors on focal adhesion formation are consistent with a role for actin-myosin in focal adhesion assembly (Burridge and Chrzanowska-Wodnicka 1996).

Rho is also reported to affect the synthesis and accumulation of PIP2. Cells held in suspension lack focal adhesions and stress fibers and contain low levels of PIP2 (McNamee et al. 1993). Integrin clustering or microinjection of activated Rho leads to the elevation of intracellular PIP2 (Chong et al. 1994). Rho binds to and stimulates a lipid kinase, phosphatidylinositol 4-phosphate-5-kinase, PIP5-K (Ren et al. 1996). PIP5-K converts phosphatidylinositol-4-phosphate (PIP) to PIP2. As discussed above, PIP2 regulates the binding interactions of a number of focal adhesion proteins and enhances actin polymerization.

Activation of tyrosine kinases is also a prerequisite for focal adhesion assembly since inhibitors of tyrosine kinase activity block cell adhesion and formation of focal adhesions (Ridley and Hall 1994; Barry and Critchley 1994). Clustering of integrins appears sufficient to recruit FAK and tensin, and results in autophosphorylation of FAK (Schaller and Parsons 1994; Cobb et al. 1994). FAK may be recruited directly to clustered integrin by a direct interaction with the cytoplasmic domain of the integrin subunit (Schaller and Parsons 1995). Tyrosine phosphorylation of FAK occurs at a major site, Tyr397, and phosphorylation of Tyr397 creates a binding site for the SH2 domain of another nonreceptor tyrosine kinase, Src, resulting in formation of a bipartite complex of two tyrosine kinases (Schaller and Parsons 1994; Cobb et al. 1994). The formation of the FAK/Src complex results in the activation of Src and the activation of downstream signals (Schlaepfer et al. 1994). One candidate substrate for the FAK/Src complex is the LIM domain containing protein, paxillin. Signals that stimulate the tyrosine phosphorylation of FAK, such as adhesion to ECM proteins and growth factor stimulation (LPA, endothelin, bombesin) also induce the tyrosine phosphorylation of paxillin (Turner et al. 1993; Zachary et al. 1993; Steufferlein and Rozengurt 1994; Rankin and Rozengurt 1994). Inhibition of FAK-dependent signaling by co-expression of the C-terminal domain of FAK results in the inhibition of the rate of cell spreading and the inhibition of paxillin tyrosine phosphorylation (Richardson and Parsons 1996). The co-localization and co-activation of paxillin and FAK, their stable interaction, the interaction of paxillin with vinculin and the potential for multiple protein-protein interactions via LIM domains argues that paxillin may be a key regulator of focal adhesion complex assembly.

Integrin clustering may lead to formation of focal adhesions by a mechanism independent of FAK activation (Wilson et al. 1995). For example, both α-actinin and talin have been reported to bind directly to the cytoplasmic domain of integrins (Otey et al. 1993; Knezevic et al. 1996). Thus, recruitment of α-actinin (or talin) may initiate the recruitment of other focal adhesion proteins and the formation of stable focal adhesions (Miyamoto et al. 1995a, b). One might speculate that these focal adhesions would lack the capacity to "signal" to downstream kinase cascades and thus might function in cells where adhesive interactions are more important.

Table 5. Focal adhesion proteins regulated by PIP2

Protein	Function	PIP2 effect
α-Actinin	Cross-links actin filaments	Enhances cross-linking activity
Radixin	Caps actin filaments and attaches filament to membrane	Enhances binding to membrane protein CD44
Vinculin	Connects FAPs to actin filaments	Enhances actin and talin binding
Profilin	Caps actin filaments and nucleates polymerization	Dissociates profilin from barbed end (favors actin polymerization)
Gelsolin	Actin severing protein	Dissociates gelsolin from barbed end (favors actin polymerization)

Table 6. Focal adhesion proteins regulated by calcium

Protein	Function	$CaCl_2$ effect
α-Actinin	Cross-links actin filaments	Inhibits cross-linking activity (favors depolymerization)
Gelsolin	Actin severing protein	Enhances severing activity (favors depolymerization)
Calpain II	Protease	Enhances proteolytic activity

Cell motility requires the formation of new focal adhesions at the leading edge of the cell and the breakdown of the same structures at the rear of the cell. Two major regulators of this process are likely to be PIP2 and calcium. As discussed above, PIP2 appears to regulate a number of actin binding proteins and thus promotes actin polymerization (Table 5). Calcium also is a putative regulator of actin binding proteins and may contribute to the process of actin depolymerization (Table 6; Janmey 1994). PIP2 itself has been localized to focal adhesions and has been shown to bind and regulate actin binding proteins including vinculin, the cross-linking protein α-actinin, the capping protein radixin, the sequestering protein profilin, and the severing protein gelsolin (Gilmore and Burridge 1996; Fukami et al. 1992; Hirao et al. 1996; Lassing et al. 1985). As noted above, PIP2 binding to vinculin prevents self-association, promotes actin and talin binding and may regulate the recruitment of proteins to focal adhesions (Gilmore and Burridge 1996). PIP2 binding to α-actinin enhances the ability of α-actinin to cross-link actin in vitro (Fukami et al. 1992). In addition, cell fractionation experiments indicate that the pool of α-actinin associated with the cytoskeleton is bound to PIP2 whereas the α-actinin in the cytosol is not, suggesting a role for PIP2 regulating α-actinin in vivo (Fukami et al. 1992). PIP2 binding to the capping (and ERM-containing) protein radixin enhances the interaction of radixin with the plasma membrane anchor protein CD44 (Hirao et al. 1996). Although the effect of PIP2 on the capping activity of radixin has not been tested, PIP2 binding to other capping proteins including gCap39, Cap 100 and MCP causes dissociation from the barbed end which exposes polymerization sites and enhances filament elongation (Janmey 1994). PIP2 binding to profilin and gelsolin also causes dissociation of these proteins from the barbed ends of actin, again resulting in increased actin polymerization (Goldschmidt-Clermont et al. 1991b; Barkalow et al. 1996). Therefore it is very likely that the synthesis of PIP2 at the site of integrin clustering may be a critical regulatory step in both the assembly and dynamic control of focal adhesions.

The regulation of local PIP2 levels within the focal adhesion complex may allow for selective polymerization and depolymerization of actin that is in contact with the plasma membrane. Interestingly, PIP2 when bound to profilin and gelsolin is a poor substrate for PLCs (Goldschmidt-Clermont et al. 1990). However, when PLCγ1 becomes tyrosine phosphorylated in response to growth factor stimulation, it can cleave profilin- and gelsolin-bound PIP2, suggesting a possible point of negative regulation of PIP2 levels (Goldschmidt-Clermont et al. 1991b).

Both the noncontractile form of α-actinin and gelsolin bind to and are regulated by calcium (Parr et al. 1992; Pope et al. 1995). Calcium-binding to α-actinin inhibits its association with actin, whereas calcium binding to gelsolin enhances its association with actin. Thus increased calcium would generate smaller filaments with less actin cross-links and favor depolymerization of the actin network. In addition, the focal adhesion localized protease calpain II is activated by calcium and may be involved in degradation of associated actin binding proteins (Beckerle et al. 1987). For example, cleavage of both talin and filamin by calpain II could result in dissociation of the actin binding domains from the membrane attachment site in each protein, destabilizing the actin network.

Although considerable progress has been made in elucidating the events leading to the assembly of focal adhesions, less is known about their disassembly. However, as the assembly of focal adhesions appears to involve tyrosine phosphorylation, tyrosine phosphatases are likely to play an important role in disassembly. PKA has been shown to promote the disassembly of stress fiber focal adhesions, an event that may be catalyzed by the dephosphorylation of paxillin (Han and Rubin 1996). cAMP causes a rapid tyrosine dephosphorylation of paxillin and a concomitant translocation of paxillin from focal adhesions to the cytoplasm. These events appear to precede disassembly of focal adhesion complexes (Han and Rubin 1996). The PKA effect appears to be coupled to protein-tyrosine phosphatase activity, since pervanadate inhibits PKA-mediated focal adhesion disassembly (Han and Rubin 1996). Interestingly, a recently identified tyrosine phosphatase, LAR, has been shown to be localized in focal adhesions (Serra-Pages et al. 1995). It remains to be determined whether this protein contributes to the breakdown of focal adhesions.

Future experiments must explore many difficult issues, including the structural interactions of proteins needed for focal adhesion complex formation, the role of tyrosine kinases and small GTP binding proteins in the regulation of actin network and stress fiber formation and how the generation of phospholipids regulates dynamic properties of focal adhesion complexes. As discussed in this review, multiple protein-protein interactions through distinct modular binding domains play an important role in the assembly of both the focal adhesion complex and actin stress fibers. The regulation of assembly and the regulation of the ordered breakdown of these structures will continue to be of importance for many cellular functions, and the understanding of these mechanisms will be central to our understanding of normal and abnormal cell growth.

Acknowledgments. We wish to thank our colleagues for many useful discussions and Kelly Mays for help in the preparation of the manuscript. J.M.T. is supported by a National Research Service Award, H1-F32-GM18297-01. Research in the laboratory of J.T.P. is supported by grants from the DHHS-National Cancer Institute, CA29243, CA40042, and the Council for Tobacco Research USA, Inc. #4491.

References

Amano M, Ito M, Kimura K, Fukata Y, Chihara K, Nakano T, Matsuura Y, Kaibuchi K (1996a) Phosphorylation and activation of myosin by Rho-associated kinase (Rho-kinase). J Biol Chem 271:20246–20249

Amano M, Mukai H, Ono Y, Chihara K, Matsui T, Hamajima Y, Okawa K, Iwamatsu A, Kaibuchi K (1996b) Identification of a putative target for Rho as the serine-threonine kinase protein kinase N. Science 271:648–650

Arber S, Caroni P (1996) Specificity of single LIM motifs in targeting and LIM/LIM interaction in situ. Genes Dev 10:289–300

Barkalow K, Witke W, Kwiatkowski DJ, Hartwig JH (1996) Coordinated regulation of platelet actin filament barbed ends by gelsolin and capping protein. J Cell Biol 134:389–399

Barron-Casella EA, Torres MA, Scherer SW, Heng HH, Tsui LC, Casella JF (1995) Sequence analysis and chromosomal localization of human Cap Z. Conserved residues within the actin-binding domain may link Cap Z to gelsolin/severin and profilin protein families. J Biol Chem 270:21472–21479

Barry ST, Critchley DR (1994) The RhoA-dependent assembly of focal adhesions in Swiss 3T3 cells is associated with increased tyrosine phosphorylation and the recruitment of both pp125FAK and protein kinase C- to focal adhesions. J Cell Sci 107:2033–2045

Beckerle MC (1986) Identification of a new protein focalized at sites of cell-substrate adhesion. J Cell Biol 103:1679–1687

Beckerle MC, Burridge K, DeMartino GN, Corall DE (1987) Colocalization of calcium-dependent protease II and one of its substrates at sites of cell adhesion. Cell 51:569–577

Belkin AM, Koteliansky VE (1987) Interaction of iodinated vinculin, metavinculin and -actinin with cytoskeletal proteins. Fed Eur Biochem Soc 220:291–294

Bennett V (1989) The spectrin-actin junction of erythrocyte membrane skeletons. Biochim Biophys Acta 988:107–121

Birge RB, Fajardo FJ, Reichman C, Shoelson SE, Songyang Z, Cantley LC, Hanafusa H (1993) Identification and characterization of a high-affinity interaction between v-Crk and tyrosine-phosphorylated paxillin in CT10-transformed fibroblasts. Mol Cell Biol 13:4648–4656

Bockholt SM, Otey CA, Glenney JRJ, Burridge K (1992) Localization of a 215-kDa tyrosine-phosphorylated protein that cross-reacts with tensin antibodies. Exp Cell Res 203:39–46

Bretscher A (1989) Rapid phosphorylation and reorganization of ezrin and spectrin accompany morphological changes induced in A-431 cells by epidermal growth factor. J Cell Biol 108:921–930

Brindle NP, Holt MR, Davies JE, Price CJ, Critchley DR (1996) The focal-adhesion vasodilator-stimulated phsophoprotein (VASP) binds to the proline-rich domain in vinculin. Biochem J 318:753–757

Brown MC, Perrotta JA, Turner CE (1996) Identification of LIM3 as the principal determinant of paxillin focal adhesion localization and characterization of a novel motif on paxillin directing vinculin and focal adhesion kinase binding. J Cell Biol 135:1109–1123

Bryan J (1988) Gelsolin has three actin-binding sites. J Cell Biol 106:1553–1562

Burridge K, Connell L (1983) A new protein of adhesion plaques and ruffing membranes. J Cell Biol 97:359–367

Burridge K, Feramisco JR (1980) Microinjection and localization of a 130 K protein in living fibroblasts: a relationship to actin and fibronectin. Cell 19:587–595

Burridge K, Feramisco JR (1981) Non-muscule α-actinins are calcium-sensitive actin-binding proteins. Nature 294:565–567
Burridge K, Chrzanowska-Wodnicka M (1996) Focal adhesions, contractility, and signaling. Annu Rev Cell Dev Biol 12:463–519
Cao LG, Babcock GG, Rubenstein PA, Wang YL (1992) Effects of profilin and profilactin on actin structure and function in living cells. J Cell Biol 117:1023–1029
Chaponnier C, Janmey PA, Yin HL (1986) The actin filament-severing domain of plasma gelsolin. J Cell Biol 103:1473–1481
Chellaiah M, Hruska K (1996) Osteopontin stimulates gelsolin-associated phosphoinositide levels and phosphatidylinositol trisphosphate-hydroxyl kinase. Mol Biol Cell 7:743–753
Chen H-C, Appeddu PA, Parsons JT, Hildebrand JD, Schaller MD, Guan J-L (1995) Interaction of focal adhesion kinase with cytoskeletal protein talin. J Biol Chem 270:16995–16999
Chong LD, Traynor-Kaplan A, Bokoch GM, Schwartz MA (1994) The small GTP-binding protein Rho regulates a phosphatidylinositol 4-phosphate 5-kinase in mammalian cells. Cell 79:507–513
Chuang J-Z, Lin DC, Lin S (1995) Molecular cloning, expression, and mapping of the high affinity actin-capping domain of chicken cardiac tensin. J Cell Biol 128:1095–1109
Cobb BS, Schaller MD, Leu TH, Parsons JT (1994) Stable association of pp60src and pp59fyn with the focal adhesion-associated protein tyrosine kinase, pp125FAK. Mol Cell Biol 14:147–155
Coll JL, Ben-Ze'ev A, Ezzell RM, Fernandez JLR, Baribault H, Oshima RG (1995) Targeted disruption of vinculin genes in F9 and embryonic stem cells changes cell morphology, adhesion, and locomotion. Proc Natl Acad Sci USA 92:9161–9165
Crawford AW, Michelsen JW, Beckerle MC (1992) An interaction between zyxin and α-actinin. J Cell Biol 116:1381–1393
Crawford AW, Pino JD, Beckerle MC (1994) Biochemical and molecular characterization of the chicken cysteine-rich protein, a developmentally regulated LIM-domain protein that is associated with the actin cytoskeleton. J Cell Biol 124:117–127
De Corte V, Gettemans J, Vanderkerckhove J (1997) Phosphatidylinositol 4,5-bisphosphate specifically stimulates pp60c-src catalyzed phosphorylation of gelsolin and related actin-binding proteins. FEBS Lett 401:191–196
Derry JMJ, Ochs HD, Francke U (1994) Isolation of a novel gene mutated in W-Aldrich Syndrome. Cell 78:635–644
Ezzell RM, Kenney DM, Egan S, Stossel TP, Hartwig JH (1988) Localization of the domain of actin-binding protein that binds to membrane glycoprotein Iβ and actin in human platelets. J Biol Chem 263:13303–13309
Fukami K, Furuhashi K, Inagaki M, Endo T, Hatano S, Takenawa T (1992) Requirement of phosphatidylinositol 4,5-bisphosphate for α-actinin function. Nature 359:150–152
Funayama N, Nagafuchi A, Sato N, Tsukita S, Tsukita S (1991) Radixin is a novel member of the band 4.1 family. J Cell Biol 115:1039–1048
Gaertner A, Wegner A (1991) Mechanism of the insertion of actin monomers between the barbed ends of actin filaments and barbed end-bound insertin. J Muscle Res Cell Motil 12:27–36
Gertler FB, Niebuhr K, Reinhard M, Wehland J, Soriano P (1996) Mena, a relative of VASP and drosophila enabled, is implicated in the control of microfilament dynamics. Cell 87:227–239
Gilmore AP, Burridge K (1996) Regulation of vinculin binding to talin and actin by phosphatidylinositol-4-5-bisphosphate. Nature 381:531–535
Gilmore AP, Jackson P, Waites GT, Critchley DR (1992) Further characterisation of the talin-binding site in the cytoskeletal protein vinculin. J Cell Sci 103:719–731
Gilmore AP, Wood C, Ohanian V, Jackson P, Patel B, Rees DJ, Hynes RO, Critchley DR (1993) The cytoskeletal protein talin contains at least two distinct vinculin binding domains. J Cell Biol 122:337–347
Goldmann WH, Niggli V, Kaufmann S, Isenberg G (1992) Probing actin and liposome interaction of talin and talin-vinculin complexes: a kinetic, thermodynamic and lipid labeling study. Biochemistry 31:7665–7671
Goldmann WH, Schindl M, Cardozo TJ, Ezzell RM (1995) Motility of vinculin-deficient F9 embryonic carcinoma cells analyzed by video, laser confocal, and reflection interference contrast microscopy. Exp Cell Res 221:311–319
Goldschmidt-Clermont PJ, Machesky LM, Baldassara JJ, Pollard TD (1990) The actin-binding protein profilin binds to PIP2 and inhibits its hydrolysis by phospholipase C. Science 247:1578–1578
Goldschmidt-Clermont PF, Machesky LM, Doberstein SK, Pollard TD (1991a) Mechanism of the interaction of human platelet profilin with actin. J Cell Biol 113:1081–1089

Goldschmidt-Clermont PJ, Kim JW, Machesky LM, Rhee SG, Pollard TD (1991b) Regulation of phospholipase C-gamma 1 by profilin and tyrosine phosphorylation. Science 251:1231–1233
Gorlin JB, Yamin R, Egan S, Steward M, Stossel TP, Kwiatkowski DJ, Hartwig JH (1990) Human endothelial actin-binding protein (ABP-280, nonmuscle filamin): a molecular leaf spring. J Cell Biol 111:1089–1105
Gotter R, Goldmann WH, Isenberg G (1995) Internal actin filament dynamics in the presence of vinculin: a dynamic light scattering study. FEBS Lett 359:220–222
Haffner C, Jarchau T, Reinhard M, Hoppe J, Lohmann SM, Walter U (1995) Molecular cloning, structural analysis and functional expression of the proline-rich focal adhesion and microfilament-associated protein VASP. EMBO J 14:19–27
Han J-D, Rubin CS (1996) Regulation of cytoskeleton organization and paxillin dephosphorylation by cAMP. Studies on murine Y1 adrenal cells. J Biol Chem 271:29211–29215
Heizmann CW, Hunziker W (1991) Intracellular calcium-binding proteins: more sites than insights. Trends Biochem Sci 16:98–103
Hemmings L, Kuhlman PA, Critchley DR (1992) Analysis of the actin-binding of α-actinin by mutagenesis and demonstration that dystrophin contains a functionally homologous domain. J Cell Biol 116:1369–1380
Hemmings L, Rees DJG, Ohanian V, Bolton SJ, Gilmore AP, Patel B, Priddle H, Trevithick JE, Hyans RO, Critchley DR (1996) Talin contains three actin-binding sites each of which is adjacent to a vinculin-binding site. J Cell Sci 109:2715–2726
Hildebrand JD, Schaller MD, Parsons JT (1995) Paxillin, a tryosine phosphorylated focal adhesion-associated protein binds to the carboxyl terminal domain of focal adhesion kinase. Mol Biol Cell 6:637–647
Hirao M, Sato N, Kondo T, Yonemura S, Monden M, Sasaki T, Takai Y, Tsukita S, Tsukita S (1996) Regulation mechanism of ERM (Ezrin/Radixin/Moesin) protein/plasma membrance association: possible involvement of phosphatidylinositol turnover and Rho-dependent signaling pathway. J Cell Biol 135:37–51
Hobert O, Schilling JW, Beckerle MC, Ullrich A, Jallal B (1996) SH3 domain-dependent interaction of the proto-oncogene product Vav with the focal contact protein zyxin. Oncogene 12:1577–1581
Hynes RO (1992) Integrins: versatility, modulation, and signaling in cell adhesion. Cell 69:11–25
Janmey PA (1994) Phosphoinositides and calcium as regulators of cellular actin assembly and disassembly. Annu Rev Physiol 56:169–191
Jockusch BM, Bubeck P, Giehl K, Kroemker M, Moschner J, Rothkegel M, Rudiger M, Schluter K, Stanke G, Winkler J (1995) The molecular architecture of focal adhesions. Annu Rev Cell Dev Biol 11:379–416
Johnson RP, Craig SW (1994) An intramolecular association between the head and tail domains of vinculin modulates talin binding. J Biol Chem 269:12611–12619
Kaufmann S, Kas J, Goldmann WH, Sackmann E, Isenberg G (1992) Talin anchors and nucleates actin filaments at lipid membranes a direct demonstration. Fed Eur Biochem Soc 314:203–205
Kimura K, Ito M, Amano M, Chihara K, Fukata Y, Nakafuku M, Yamamori B, Feng J, Nakano T, Okawa K, Iwamatsu A, Kaibuchi K (1996) Regulation of myosin phosphatase by Rho and Rho-associated kinase (Rho-kinase). Science 273:245–248
Knezevic I, Leisner TM, Lam SC-T (1996) Direct binding of the platelet integrin (GPIIb-IIIa) to talin. J Biol Chem 271:16416–16421
Krieg J, Hunter T (1992) Identification of the two major epidermal growth factor-induced tyrosine phosphorylation sites in the microvillar core protein ezrin. J Biol Chem 267:19258–19268
Kuhlman PA, Hemmings L, Critchley DR (1992) The identification of an actin-binding site in α-actinin by mutagenesis. FEBS Lett 304:201–206
Lassing I, Lindberg U (1985) Specific interaction between phosphatidylinositol 4,5-biphosphate and profilactin. Nature 314:472–474
Lauffenburger DA, Horwitz AF (1996) Cell migration: a physically integrated molecular process. Cell 84:359–369
Leung T, Chen X-Q, Manser E, Lim L (1996) The p160 RhoA-binding kinase ROKa is a member of a kinase family and is involved in the reorganization of the cytoskeleton. Mol Cell Biol 16:5313–5327
Lo SH, Janmey PA, Hartwig JH, Chen LB (1994) Interactions of tensin with actin and identification of its three distinct actin-binding domains. J Cell Biol 125:1067–1075
Machesky LM, Hall A (1996) Rho: a connection between membrane receptor signaling and the cytoskeleton. Trends Cell Biol 6:304–310
Macias MJ, Hyvonen M, Baraldi E, Schultz J, Sudol M, Saraste M, Oschkinat H (1996) Structure of the WW domain of a kinase-associated protein complexed with proline-rich peptide. Nature 382:646–649

Magendantz M, Henry MD, Lander A, Solomon F (1995) Interdomain interactions of radixin in vitro. J Biol Chem 270:25324–25327
Matsudaira P, Janmey P (1988) Pieces in the actin-severing protein puzzle. Cell 54:139–140
Matsudaira P (1991) Molecular organization of actin crosslinking proteins. Trends Biochem Sci 16:87–92
McGregor A, Blanchard AD, Rowe AJ, Critchley DR (1994) Identification of the vinculin-binding site in the cytoskeletal protein α-actinin. Biochem J 301:225–233
McLachlan AD, Stewart M, Hynes RO, Rees DJG (1994) Analysis of repeated motifs in the talin rod. J Mol Biol 235:1278–1290
McLaughlin PJ, Weeds AG (1995) Actin-binding protein complexes at atomic resolution. Annu Rev Biophys Biomol Struct 24:643–675
McNamee HP, Ingber DE, Schwartz MA (1993) Adhesion to fibronectin stimulates inositol lipid synthesis and enhances PDGF-induced inositol lipid breakdown. J Cell Biol 121:673–678
Menkel AR, Kroemker M, Bubeck P, Ronsiek M, Nikolai G, Jockusch BM (1994) Characterization of an F-actin-binding doamin in the cytoskeletal protein vinculin. J Cell Biol 126:1231–1240
Metzler WJ, Bell AJ, Ernst E, Lavoie TB, Mueller L (1994) Identification of the Poly-L-proline-binding site on human profilin. J Biol Chem 269:4620–4625
Miyamoto S, Akiyama SK, Yamada KM (1995a) Synergistic roles for receptor occupancy and aggregation in integrin transmembrane function. Science 267:883–885
Miyamoto S, Teramoto H, Coso OA, Gutkind JS, Burbelo PD, Akiyama SK, Yamada KM (1995b) Integrin function: molecular hierarchies of cytoskeletal and signaling molecules. J Cell Biol 131:791–805
Muguruma M, Nishimuta S, Tomisaka Y, Ito T, Matsumura S (1995) Organization of the functional domains in membrane cytoskeletal protein talin. J Biochem 117:1036–1042
Nobes CD, Hall A (1995) Rho, Rac, and Cdc42 GTPases regulate the assembly of multimolecular focal complexes associated with actin stress fibers, lamellipodia, and filopodia. Cell 81:53–62
Ohta Y, Stossel TP, Hartwin JH (1991) Ligand-sensitive binding of actin-binding protein to immunoglobulin G Fc receptor I (Fc gamma RI). Cell 67:275–282
Orlova A, Prochniewicz E, Egelman EH (1995) Structural dynamics of F-actin: cooperativity in structural transitions. J Mol Biol 245:598–607
Otey CA, Vasquez GB, Burridge K, Erickson BW (1993) Mapping of the α-actinin binding site within the beta 1 integrin cytoplasmic domain. J Biol Chem 268:21193–21197
Parr T, Waites GT, Patel B, Millake DB, Critchley DR (1992) A chick skeletal-muscle α-actinin gene gives rise to two alternatively spliced isoforms which differ in the EF-hand Ca(2+)-binding domain. Eur J Biochem 210:801–809
Pavalko FM, Otey CA, Burridge K (1989) Identification of a filamin isoform enriched at the ends of stress fibers in chicken embryo fibroblasts. J Cell Sci 94:109–118
Pope B, Way M, Matsudaira PT, Weeds A (1994) Characterisation of the F-actin binding domains of villin: classification of F-actin binding proteins into two groups according to their binding sites on actin. FEBS Lett 338:58–62
Pope B, Maciver S, Weeds A (1995) Localization of the calcium-sensitive actin monomer binding site in gelsolin to segment 4 and identification of calcium binding sites. Biochemistry 34:1583–1588
Price GJ, Jones P, Davison MD, Patel B, Bendori R, Geiger B, Critchley DR (1989) Primary sequence and domain structure of chicken vinculin. Biochem J 259:453–461
Rankin S, Rozengurt E (1994) Platelet-derived growth factor modulation of focal adhesion kinase (p125FAK) and paxillin tyrosine phosphorylation in Swiss 3T3 cells. Bell-shaped dose response and cross-talk with bombesin. J Biol Chem 269:704–710
Rees DJ, Ades SE, Singer SJ, Hynes RO (1990) Sequence and domain structure of talin. Nature 347:685–689
Reinhard M, Giehl K, Abel K, Haffner C, Jarchau T, Hoppe V, Jockusch BM, Walter U (1995a) The proline-rich focal adhesion and microfilament protein VASP is a ligand for profilins. EMBO J 14:1583–1589
Reinhard M, Halbrugge M, Scheer U, Wiegand C, Jockusch BM, Walter U (1992) The 46/50 kDa phosphoprotein VASP purified from human platelets is a novel protein associated with actin filaments and focal contacts. EMBO J 11:2063–2070
Reinhard M, Jouvenal K, Tripier D, Walter U (1995b) Identification, purification, and characterization of a zyxin-related protein that binds the focal adhesion and microfilament protein VASP (vasodilator-stimulated phosphoprotein). Proc Natl Acad Sci USA 92:7956–7960
Ren X-D, Bockoch GM, Traynor-Kaplan A, Jenkins GH, Anderson RA, Schwartz MA (1996) Physical association of the small GTPase rho with a 68-kDa phosphatidylinositol 4-phosphate 5-kinase in Swiss 3T3 cells. Mol Biol Cell 7:435–442

Richardson A, Parsons JT (1996) A mechanism for regulation of the adhesion-associated protein tyrosine kinase pp125^{FAK}. Nature 380:538–540
Ridley AJ (1996) Rho: theme and variations. Curr Biol 6:1256–1264
Ridley AJ, Hall A (1994) Signal transduction pathways regulating rho-mediated stress fibre formation: requirement for a tyrosine kinase. EMBO J 13:2600–2610
Ruhnau K, Gaertner A, Wegner A (1989) Kinetic evidence for insertion of actin monomers between the barbed ends of actin filaments and barbed end-bound insertin, a protein purified from smooth muscle. J Mol Biol 210:141–148
Ruoslahti E, Reed JC (1994) Anchorage dependence, integrins, and apoptosis. Cell 77:477–478
Sabe H, Hata A, Okada M, Nakagawa H, Hanafusa H (1994) Analysis of the binding of the Src homology 2 domain of Csk to tyrosine phosphorylated proteins in the suppression and mitotic activation of c-Src. Proc Natl Acad Sci USA 91:3984–3988
Sadler I, Crawford AW, Michelsen JW, Beckerle MC (1992) Zyxin and cCRP: two interactive LIM domain proteins associated with the cytoskeleton. J Cell Biol 119:1573–1587
Salgia R, Uemura N, Okuda K, Li JL, Pisick E, Sattler M, de Jong R, Druker B, Heisterkamp N, Chen LB (1995) CRKL links p210 BRCABL with paxillin in chronic myelogenous leukemia cells. J Biol Chem 270:29145–29150
Sanger JM, Mittal B, Pochapin MB, Sanger JW (1987) Stress fiber and cleavage furrow formation in living cells microinjected with fluorescently labeled α-actinin. Cell Motil Cytoskeleton 7:209–220
Schafer DA, Cooper JA (1995) Control of actin assembly at filament ends. Annu Rev Cell Dev Biol 11:497–518
Schaller MD, Parsons JT (1994) Focal adhesion kinase and associated proteins. Curr Opin Cell Biol 6:705–710
Schaller MD, Parsons JT (1995) pp125FAK-dependent tyrosine phosphorylation of paxillin creates a high-affinity binding site for Crk. Mol Cell Biol 15:2635–2645
Schaller MD, Hildebrand JD, Shannon JD, Fox JW, Vines RR, Parsons JT (1994) Autophosphorylation of the focal adhesion kinase, pp125FAK, directs SH2-dependent binding of pp60src. Mol Cell Biol 14:1680–1688
Schaller MD, Otey CA, Hildebrand JD, Parsons JT (1995) Focal adhesion kinase and paxillin bind to peptides mimicking beta integrin cytoplasmic domains. J Cell Biol 130:1181–1187
Schlaepfer DD, Hanks SK, Hunter T, van der Geer P (1994) Integrin-mediated signal transduction is linked to Ras pathway by GRB2 binding to focal adhesion kinase. Nature 372:786–791
Schleicher M, Andre E, Hartmann H, Noegel AA (1988) Actin-binding proteins are conserved from slime molds to man. Dev Genet 9:521–530
Schmeichel KL, Beckerle MC (1994) The LIM domain is a modular protein-binding interface. Cell 79:211–219
Schutt CE, Myslik JC, Rozychi MD, Goonesekere NC, Lindberg U (1993) Structure of crystalline profilin-beta-actin. Nature 365:810–816
Schwartz MA, Schaller MD, Ginsberg MH (1995) Integrins: emerging paradigms of signal transduction. Annu Rev Cell Dev Biol 11:549–600
Serra-Pages C, Kedersha NL, Fazikas L, Medley Q, Debant A, Streuli M (1995) The LAR transmembrane protein tyrosine phosphatase and a coiled-coil LAR-interacting protein co-localize at focal adhesions. EMBO J 14:2827–2838
Smith GA, Theriot JA, Portnoy DA (1996) The tandem repeat domain in the Listeria monocytogenes ActA protein controls the rate of actin-based motility, the percentage of moving bacteria, and the localization of vasodilator-stimulated phosphoprotein and profilin. J Cell Biol 135:647–660
Sohn RH, Goldschmidt-Clermont PJ (1994) Profilin: at the crossroads of signal transduction and the actin cytoskeleton. Bioessays 16:465–472
Steufferlein T, Rozengurt E (1994) Lysophosphatidic acid stimulates tyrosine phosphorylation of focal adhesion kinase, paxillin, and p130. J Biol Chem 269:9345–9351
Sun HQ, Kwiatkowska K, Yin HL (1995) Actin monomer binding proteins. Curr Opin Cell Biol 7:102–110
Symons M, Derry JMJ, Karlak B, Jiang S, Lemahieu V, McCormick F, Francke U, Abo A (1996) Wiskott-Aldrich Syndrome Protein, a novel effector for the GTPase CDC42Hs, is implicated in actin polymerization. Cell 84:723–734
Takeuchi K, Kawashima A, Nagafuchi A, Tsukita S (1994) Structural diversity of band 4.1 superfamily members. J Cell Sci 107:1921–1928
Tempel M, Goldmann WH, Dietrich C, Niggli V, Weber T, Sackmann E, Isenberg G (1994) Insertion of filamin into lipid membranes examined by calorimetry, the film balance technique, and lipid photolabeling. Biochemistry 33:12565–12572

Tsukita S, Oishi K, Sato N, Sagara J, Kawai A, Tsukita S (1994) ERM family members as molecular linkers between the cell surface glycoprotein CD44 and actin-based cytoskeletons. J Cell Biol 126:391–401

Turner CE, Miller JT (1994) Primary sequence of paxillin contains putative SH2 and SH3 domain binding motifs and multiple LIM domains: identification of a vinculin and pp125FAK-binding region. J Cell Sci 107:1583–1591

Turner CE, Glenney JRJ, Burridge K (1990) Paxillin: a new vinculin-binding protein present in focal adhesions. J Cell Biol 111:1059–1068

Turner CE, Schaller MD, Parsons JT (1993) Tyrosine phosphorylation of the focal adhesion kinase pp125-FAK during development: relation to paxillin. J Cell Sci 105:637–645

Turunen O, Wahlstrom T, Vaheri A (1994) Ezrin has a COOH-terminal actin-binding site that is conserved in the ezrin protein family. J Cell Biol 126:1445–1453

Van Troys M, Dewitte D, Goethals M, Carlier MF, Vandekerckhove J, Ampe C (1996) The actin binding site of thymosin beta 4 mapped by mutational analysis. EMBO J 15:201–210

Varnum-Finney B and Reichardt, LF (1994) Vinculin-deficient PC12 cell lines extend unstable lamellipodia and filopodia and have a reduced rate of neurite outgrowth. J Cell Biol 127:1071–1084

Volberg T, Geiger B, Kam Z, Pankov R, Simcha I, Sabanay H, Coll J-L, Adamson E, Ben-Ze'ev A (1995) Focal adhesion formation by F9 embryonal carcinoma cells after vinculin gene disruption. J Cell Sci 108:2253–2260

Waites GT, Graham IR, Jackson P, Millake DB, Patel B, Blanchard AD, Weller PA, Eperon IC, Critchley DR (1992) Mutually exclusive splicing of calcium-binding domain exons in chick α-actin. J Biol Chem 267:6263–6271

Watanabe G, Saito Y, Madaule P, Ishizaki T, Fujisawa K, Morii N, Mukai H, Ono Y, Kakizuka A, Narumiya S (1996) Protein kinase N (PKN) and PKN-related protein rhophilin as targets of small GTPase Rho. Science 271:645–647

Weekes J, Barry ST, Critchley DR (1996) Acidic phospholipids inhibit the intramolecular association between the N- and C-terminal regions of vinculin, exposing actin-binding and protein kinase C phosphorylation sites. Biochem J 314:827–832

Weng Z, Taylor JA, Turner CE, Brugge JS, Seidel-Dugan C (1993) Detection of Src homology 3-binding proteins, including paxillin, in normal and v-Src-transformed Balb/c 3T3 Cells. J Biol Chem 268:14956–14963

Wilkins JA, Risinger MA, Coffey E, Lin S (1987) Purification of a vinculin binding protein from smooth muscle. J Cell Biol 104:A130

Wilson L, Carrier MJ, Kellie S (1995) pp125FAK tyrosine kinase activity is not required for the assembly of F-actin stress fibres and focal adhesions in cultured mouse aortic smooth muscle cells. J Cell Sci 108:2381–2391

Wood CK, Turner CE, Jackson P, Critchley DR (1994) Characterisation of the paxillin-binding site and the C-terminal focal adhesion targeting sequence in vinculin. J Cell Sci 107:709–717

Xian W, Vegners R, Janmey PA, Braunlin WH (1995) Spectroscopic studies of a phosphoinositide-binding peptide from gelsolin: behavior in solutions of mixed solvent and anionic micelles. Biophys J 69:2695–2702

Yu F-X, Sun H-Q, Janmey PA, Yin HL (1992) Identification of a polyphosphoinositide-binding sequence in an actin monomer-binding domain of gelsolin. J Biol Chem 267:14616–14621

Zachary I, Sinnett-Smith J, Turner CE, Rozengurt (1993) Bombesin, vasopressin, and endothelin rapidly stimulate tyrosine phosphorylaiton of the focal adhesion-associated protein paxillin in Swiss 3T3 cells. J Biol Chem 268:22060–22065

Zhang J, Robson RM, Schmidt JM, Stromer MH (1996) Talin can crosslink actin filaments into both networks and bundles. Biochem Biophys Res Commun 218:530–537

Zigmond SH (1996) Signal transduction and actin filament organization. Curr Opin Cell Biol 8:63–73

Zigmond SH, Otto JT, Bryan J (1979) Organization of myosin in a submembranous sheath in well-spread human fibroblasts. Exp Cell Res 119:205–219

Physiological Function of Receptor-SH2 Interactions

C. Ponzetto

1 Introduction

Most tyrosine kinase receptors were first identified in the early 1980s, as the cellular counterpart of viral oncogenes, whose isolation was made possible by the transforming ability of their constitutively active products (Bishop 1983). During the following 10 years, together with the cloning and characterization of an increasing number of tyrosine kinase receptors (RTKs), came the realization that the prerequisite for stimulation of their kinase activity and consequent autophosphorylation is ligand-induced stabilization of receptor dimers (Ullrich and Schlessinger 1990). However, it was only in the early 1990s with the identification of SH2 (Src homology 2) domains as phosphotyrosine binding modules that it was finally appreciated how receptors work in transmitting signals to downstream effectors (Koch et al. 1991). It became clear that activated receptors interact via specific phosphotyrosine residues with SH2-containing cytoplasmic molecules, effectively relocating them to the cell membrane. This act of recruitment, sometimes followed by activation via tyrosine phosphorylation, brings effectors into the best position to generate second messengers from phospholipid precursors and to activate signaling cascades initiated by membrane-anchored proteins.

The presence of SH3 (Src homology 3) domains together with SH2s in many cytoplasmic molecules (SH3s were recognized as distinct protein-protein interaction

Department of Biomedical Sciences and Oncology, University of Turin, Corso Massimo d'Azeglio 52, 10126 Torino, Italy

modules soon after SH2s) indicated that additional partners could be indirectly recruited to the receptor, thus assembling multimolecular complexes capable of triggering intricate signaling circuits. As the key role of SH2 and SH3 domains in cell signaling was being unraveled, the search for new conserved domains began and, within a very short time, enough of them have emerged (reviewed by PAWSON 1995) to justify the call for the present volume. In this review I will focus on some aspects of the initial events in RTK mediated signal transduction, particularly on receptor-SH2 interactions, highlighting their relevance in the physiological response to activated receptors, both in vitro and in vivo.

2 Specificity of Receptor-SH2 Interactions is Determined by Both Members of the Pair

Work on PDGF (phabelet-derived growth factor) receptor, originally carried out using a combination of synthetic phosphopeptides and site-directed mutagenesis, revealed that phosphotyrosine-SH2 interactions are specific and that the residues in positions +1, +2 and +3 with respect to the phosphotyrosine are important for SH2 recognition (FANTL et al. 1992). This was soon confirmed by the development and use of degenerate phosphopeptide libraries, which allowed the identification of optimal binding motifs for each SH2 (SONGYANG et al. 1993). These studies, together with the solution of the crystal structures of SH2 domains completed with their phosphopeptide ligands, indicate that most SH2 domains fall into two main categories, named groups I and III. Group I SH2s, found in Src-like cytoplasmic protein tyrosine kinases (PTKs) prefer the motif pTyr-hydrophilic-hydrophilic-hydrophobic. Group III SH2s, found in phospholipase Cγ (PLCγ), p85, SH2 phosphatase 1 and 2 (SHP-1 and 2) and Shc, select for pTyr-hydrophobic-X-hydrophobic (SONGYANG and CANTLEY 1995). Substituting a conserved aliphatic residue found in position βD5 of group III SH2s with a Tyr switches their specificity to that of group I SH2s (SONGYANG et al. 1995a). Furthermore, the residue at the EF1 position of the Src SH2 binding pocket is a Thr, while that in the SH2 of the Sem5/Drk/Grb2 adaptor is a Trp. Changing this residue from Thr to Trp in the Src SH2 domain switches its selectivity to that of the Sem5/Drk/Grb2 SH2 (MARENGERE et al. 1994).

These and other studies show that single point mutations have the potential to disrupt or modify SH2 specificity. The adaptor Sem-5 provides an example of the consequences of such mutations in vivo. In *C. elegans*, the Sem-5 gene product participates in the induction of the hermaphrodite vulva by binding to the Let-23 receptor tyrosine kinase and linking it via its SH3 domains to SoS, thus activating the Ras pathway (CLARK et al. 1992). Mutations in the Sem-5 SH2 domain which interfere with phosphotyrosine binding produce a vulvaless phenotype, demonstrating that the SH2-receptor interaction is essential for Let-23-mediated activation of the Ras pathway in vivo. On the other hand, a chimeric Sem-5 protein containing a Src SH2 domain with the Thr to Trp mutation described above restores a normal vulva when expressed in a vulvaless *C. elegans* strain carrying a strong Sem-5 allele (MARENGERE et al. 1994). These results confirm in vivo that the biological activity of an SH2 domain depends on its binding specificity.

The use of degenerate phosphopeptide libraries was subsequently extended to the identification of optimal peptide substrates for PTKs (SONGYANG et al. 1995b). These studies indicate that the catalytic domain of RTKs is also selective. While both soluble and receptor-like tyrosine kinases prefer substrates with acidic residues at positions −3 and −2 to the Tyr and with a large hydrophobic amino acid at position +3, there is a major difference between the two groups in the preferred amino acid at the −1 and +1 positions. Cytoplasmic PTKs select for peptides with Ile or Val in −1, and for Glu, Asp or residues with small side chains (Gly or Ala) in +1. Receptor PTKs select for peptides with Glu in −1, and for amino acids with large hydrophobic side chains in +1. By considering the selectivities of cytoplasmic and receptor PTKs together with those of SH2 domains, it appears that receptor PTKs are predisposed to auto- or transphosphorylate residues which will act as docking sites for proteins with group III SH2 domains (phosphatidylinositol 3-kinase: PI3K, PLCγ, SHP-2 tyrosine phosphatase), while SH2-containg PTKs will phosphorylate tyrosines recognized by proteins with group I SH2 domains (other SH2 containing PTKs, Crk) (reviewed by SONGYANG and CANTLEY 1995). There are, however, exceptions. Some receptors, among which is the PDGFR, recruit Src-family PTKs (MORI et al. 1993). Furthermore, the Grb2 SH2, whose selectivity is determined mainly by the presence of an Asn in +2 (pYXNX), binds sites that are good substrates for both cytosolic or receptor PTKs (SONGYANG and CANTLEY 1995). Finally, it should be noted that although some natural sites conform surprisingly well to an optimal binding motif for a particular SH2 domain, the binding specificity of docking sites is not always absolute, and there may be more than one SH2-containing effector within a cell capable of binding with similar high affinity to the same site. For example, the EGFR (epidermal growth factor receptor) binding sites for PLCγ, RasGAP (GTPase activating protein) and p85 are not strictly specific (SOLER et al. 1994), and one of the two PI3K binding sites on the PDGFR can also potentially bind the Nck adaptor (NISHIMURA et al. 1993).

Furthermore, by mutating the residues in +3, +4 and +5 of the PLCγ binding site of the PDGF receptor, it is possible to convert a site that is specific for PLCγ and does not bind PI3K, into a site that complexes with both PLCγ and PI3K (LAROSE et al. 1995).

A particular example of such mixed binding specificity is given by two members of the Met family of receptors, the HGF (hepatocyte growth factor) and MSP (macrophage stimulating protein) receptors (Met and Ron). Both of these receptors signal through a bidentate docking site, which appears to be truly "multifunctional" (PONZETTO et al. 1994; IWAMA et al. 1996). The two phosphotyrosines of the bidentate site are part of a conserved motif permissive for several effectors which bind to both of them with affinities in the nanomolar range, with the exception of p85, which binds with a tenfold lower affinity (PONZETTO et al. 1993, 1994). High affinity phosphopeptide-SH2 interactions are characterized by very rapid dissociation rates (approx. 0.1/s), which allow for rapid exchange of competing SH2-containing proteins (FELDER et al. 1993; PANAYOTOU et al. 1993). Presumably the ultimate signaling output in this and other cases is influenced by the local concentration of effectors, which may vary from cell to cell. Interestingly, HGF has been shown to promote survival, growth, and motility (reviewed in BOROS and MILLER 1995). These effects are elicited alone or in combination in different target cells. Similarly, Ron recruits different effectors depending on cell type and mediates biological responses as diverse as growth and apoptosis (IWAMA et al. 1996).

3 Mutations Subverting Receptor Kinase Specificity are Oncogenic

The best evidence that correct recognition of potential phosphorylation sites is important for ensuring fidelity in intracellular signaling and control of cell growth is provided by the gain-of-function effect of mutations in the catalytic domain of Ret and Kit. Both receptors become oncogenic when they acquire single point mutations that subvert their substrate specificity. Structural and homology studies have identified a residue in subdomain VIII of tyrosine kinases important for substrate selection, which, as mentioned above, differs between receptor and non-receptor PTKs. In most RTKs this residue is a methionine, while in cytoplasmic PTKs there is a threonine in place of the methionine. A Met to Thr germline mutation in the Ret receptor gene, involving precisely this residue, is responsible for the autosomal dominantly inherited cancer syndrome multiple endocrine neoplasia 2B (MEN2B) (SANTORO et al. 1995). The mutation switches the substrate specificity of the Ret receptor to mimick that of cytoplasmic PTKs. This alters the phosphorylation pattern of the receptor itself, resulting in the appearance of a novel autophosphorylation site and in the loss of others (LIU et al. 1996). The pattern of phosphorylation of cytoplasmic proteins also varies in cells expressing the MEN2B Ret variant versus wild type Ret (SANTORO et al. 1995).

Another mutation which alters substrate specificity by converting a residue conserved in receptor PTKs into a residue typical of cytoplasmic PTKs has been identified in an oncogenic form of Kit, found in rodent and human mast cell leukemias (PIAO et al. 1996). The Kit mutation involves a conserved aspartic acid, located in subdomain VII (20 amino acids distant from the residue involved in MEN2B), which is mutated into a cytoplasmic PTK-like neutral residue. Similar to the effects of the Ret MEN2B mutation, this change results in aberrant autophosphorylation of the Kit receptor, and in preferential phosphorylation of a novel substrate which may contribute to the uncontrolled growth of mast cells (PIAO et al. 1996). Interestingly, the mutation also results in the selective ubiquitin-mediated degradation of the SHP-1 tyrosine phosphatase, which normally associates with wild type Kit receptor and is involved in negative regulation of its signal (PIAO et al. 1996). The identity of the variant autophosphorylation sites in the oncogenic forms of Ret and Kit has not yet been determined. It seems reasonable, however, to suppose that these newly acquired phosphotyrosines may contribute to altering the signaling capability of the receptors by working as additional docking sites. In both cases, aberrant signaling rather than constitutive activation of the receptor kinase may ultimately be responsible for the establishment of neoplasia.

4 Abrogation of Receptor Docking Sites Causes a Partial Loss-of-Function in the Biological Response to the Ligand

Proof that phosphotyrosine-SH2 interactions are required for the biological response to activated RTKs is provided by the loss-of-function phenotypes obtained in target cells by abrogating the docking sites via site-directed mutagenesis. Since many receptors have multiple docking sites relatively specific for a particular SH2-containing effector, single or combined Tyr-Phe mutations have been used to establish the role of distinct signaling pathways and their synergies in the biological response to receptors. Mutation of a single docking site may have either subtle or dramatic effects depending on the receptor, but generally does not completely abrogate the biological response. The most striking effects of all are observed with the oncogenic forms of receptors, which often lose their transforming ability after removal of a single tyrosine directly or indirectly involved in Ras activation. This is the case with activated ErbB-2, where mutation of the C-terminal autophosphorylation site (a "dual" binding site for the phosphotyrosine binding (PTB) domain of Shc, which connects the receptor to the Ras pathway, and for the SH2 domain of PLCγ) completely abolishes transformation (Ben-Levy et al. 1994). In Tpr-Met, the oncogenic form of the HGF receptor, a Tyr-Phe mutation of the multifunctional tyrosine responsible for binding Grb-2, among other effectors, drastically lowers its transforming ability and impairs the motility response to the receptor (Ponzetto et al. 1994, 1996). A more subtle mutation in this docking site (YVNV→YVHV), aimed at selectively disrupting binding with Grb2, greatly reduces transformation but is fully permissive for motility (Ponzetto et al. 1996; Fixman et al. 1996; Fournier et al. 1996). Similarly, mutation of the Grb2 binding site in Bcr-Abl (the product of the chimeric gene found in chronic myelogenous leukemia) results in the loss of its ability to transform primary bone marrow cultures, but does not interfere with its ability to protect lymphoid precursors from apoptosis (Cortez et al. 1995).

In more physiological systems, mutation of a single docking site has generally more subtle effects. Loss of the PI3K binding site in the Kit receptor interferes with Steel ligand-induced mast cell adhesion to fibronectin, but has only a partial effect on cell proliferation and survival (Serve et al. 1995). In PC12 cells, where stimulation of an ectopic PDGFR has a differentiative effect, a PDGFR mutant lacking one of the two PI3K binding sites can still promote differentiation, but loses the ability to sustain survival after serum deprivation (Yao and Cooper 1995), implicating activation of PI3K in protection from apoptosis. The endogenous mediator of survival and differentiation signals in PC12 cells is the Trk receptor. Removal of the binding site for the Shc PTB domain in the juxta-membrane of Trk causes a severe defect in neurite outgrowth, suggesting an essential role for the Ras/Map kinase pathway in the regulation of neuronal differentiation (OBERMEIER et al. 1994). Mutation of the high affinity binding site for PLCγ in the carboxy-terminal tail of the Trk receptor does not seem to have any effect. However, "re-introduction" of the PLCγ binding site in a Trk mutant lacking association sites for PI3K, Shc and PLCγ results in moderate neurite outgrowth, providing support for a complementary role for PLCγ with Shc in the Trk-mediated differentiative response (OBERMEIER et al. 1994). This "add back" approach is often more revealing than mutating single docking sites, especially when the latter does not appear to cause any obvious loss of function. The response to receptor mutants lacking docking sites is generally evaluated at saturating concentrations of ligand. This may lead to an underestimate of the contribution of effectors which may play an important role at more physiological ligand concentrations. Using the "add back" approach it has been shown that there are at least two discrete pathways by which the PDGF receptor can transduce a mitogenic signal. In fact, recruitment of either PLCγ or PI3K rescues the response to a PDGFR carrying five Tyr-Phe mutations and unable to trigger DNA synthesis (VALIUS and KAZLAUSKAS 1993). The general picture which emerges from the studies on docking site mutagenesis in the various receptors (only a few examples of which were reported above) is, as expected, one of cooperation among the signaling molecules recruited by the receptors. A conclusion which seems to be valid for every receptor is that docking sites are essential for receptor signaling, and that activation of the kinase domain is not sufficient by itself to elicit a response.

The simple model of "linear" signaling so far outlined, however, is not an adequate representation of signal transduction as it occurs in a real cell. There are, in fact, several ways for RTKs to effectively expand the range of their downstream effectors, or to amplify their signaling. One example is that of the ErbB family of receptors (reviewed in CARRAWAY and BURDEN 1995), for which the term "lateral" signaling has been coined, alluding to the unique modality of signal diversification made possible by heterodimerization among different members of the same family (PINKAS-KRAMARSKI et al. 1996; TZAHAR et al. 1996). ErbB-2, which has no known ligand and has the highest intrinsic kinase activity, represents a shared subunit found in heterodimers with ErbB-1 (cognate ligands EGF and transforming growth factor-α; TGF-α) and ErbB-3 or ErbB-4 (cognate ligands Neu differentiation factors). The various members of the family differ in their potential SH2 docking sites; thus heterodimerization enables variation of signaling outputs and results in graded proliferative and survival responses (CARRAWAY and CANTLEY 1994).

Remarkably, ErbB-3, which appears to confer the most potent mitogenic activity to an ErbB-2 heterodimer, is kinase defective but contains several consensus motifs capable, upon transphosphorylation, of binding the p85 regulatory subunit of PI3K. Thus ErbB-3 serves more as an adaptor than as a receptor kinase. "Lateral" signaling can explain the paradox of an EGFR mutant lacking all known autophosphorylation sites which is apparently still competent in mitogenic signaling (GOTOH et al. 1994). In this case endogenous ErbB-2 or other members of the EGFR family supply in trans the necessary docking sites for SH2-containing proteins (SASAOKA et al. 1996).

Another means of signal amplification, which may be more widespread than originally thought, is the transphosphorylation of multiadaptor-like proteins, whose prototypes are insulin receptor substrate-1 and 2 (IRS-1 and 2). Following activation of the insulin or IGF-1 receptors, the IRS molecule is phosphorylated on multiple tyrosine residues (MYERS et al. 1994). These include potential docking sites for Grb2, PI3K, PLCγ, and SHP-2, which are not found on the receptor itself. It has recently emerged that IRS-1 and IRS-2 are also phosphorylated by various cytokine receptors coupled to JAK family kinases (WATERS and PESSIN 1996); thus their role is broader than originally thought. The insulin receptor, however, seems to phosphorylate an additional multiadaptor protein, GAB-1, which is also a substrate of the EGF and HGF receptors (HOLGADO-MADRUGA et al. 1996). While GAB-1 associates with the insulin and EGF receptors through Grb2, GAB-1 binds Met through a novel phosphotyrosine binding domain, which seems to be specific for the bidentate docking site in the carboxy-terminal tail of the Met receptor (WEIDNER et al. 1996). It is thus likely that the receptors for insulin, EGF, HGF and other growth factors may rely on seemingly redundant ways to activate signaling pathways. The HGFR, for example, can activate Ras by recruiting Grb2-SoS directly, through the YVNV consensus, or indirectly, via Shc or GAB-1. It should be noted that a docking site seems to be necessary in all cases, since each of the three adaptors interacts with the receptor via a phosphotyrosine binding domain (an SH2 for Grb2, a PTB domain for Shc, and the novel Met-binding domain for GAB-1).

5 Phosphotyrosine-SH2 Interactions are Involved in Mediating Termination of Signaling

So far, most of the work on recruitment of SH2-containing molecules by RTKs suggests a role in mediating the "activation" of signaling pathways. However, it is probable that such interactions are also involved in termination of receptor signaling. The existence of cytosolic protein tyrosine phosphatases containing SH2 domains (reviewed by STREULI 1996) suggests that these molecules can be recruited by receptors as well, presumably to shut off their signal by dephosphorylation. However, the SHP-2 phosphatase and its *Drosophila* homolog Corkscrew (Csw, PERKINS et al. 1992) have been implicated in "positive" rather than "negative" signaling by RTKs. SHP-2 associates with several receptors and becomes phosphorylated on a residue which is an optimal binding site for Grb2. Thus at least one of its functions seems to be that of working as an adaptor molecule to recruit Grb2 to receptors (LI et al. 1994). Whether SHP-2 has any role as a phosphatase in signal transduction is not clear. On the other hand, there is now strong evidence that SHP-1, which is mainly confined to hematopoietic tissues, works as a negative regulator of signaling by PTKs (IMBODEN and KORETSKY 1995). Recruitment of SHP-1 to the erythropoietin (EPO) receptor, which is itself devoid of tyrosine kinase activity but is phosphorylated by the nonreceptor PTK JAK2, causes termination of EPO-induced proliferative signals by de-phosphorylation and inactivation of JAK2 (KLINGMÜLLER et al. 1995). SHP-1 also associates with the Kit receptor in response to ligand stimulation (YI and IHLE 1993). Kit is encoded by the *Dominant White Spotting (W)* locus in the mouse, and SHP-1 by the *motheaten (me)* gene. Both mutants are characterized by hematological defects. Homozygosity for mutations in both *W* and *me* ameliorates aspects of both phenotypes, demonstrating that Kit plays a role in the pathology of the *me* phenotype and, conversely, that SHP-1 negatively regulates Kit signaling in vivo (PAULSON et al. 1996). Furthermore, recent studies show that SHP-1 is involved in modulating B, T, and NK cell function by being recruited via phosphotyrosine-SH2 interactions by "inhibitory" immunoreceptors (reviewed in SCHARENBERG and KINET 1996). This has the effect of terminating in each of these cells signaling of PTK-associated "activating" immunoreceptors. Altogether these studies suggest that recruitment of SH2-containing protein tyrosine phosphatases by the RTKs themselves, or by associated molecules, may prove to be a common mechanism for terminating their signaling.

6 Loss of Docking Sites Impairs Development

Most of the data supporting the current model of signal transduction by RTKs have been gathered by work done in cultured cells. Evidence validating this model in vivo came initially from simple organisms such as *C. elegans* and *Drosophila*. RTKs are involved in controlling many developmental events, both in lower and higher organisms. As mentioned previously, mutations in the SH2 domain of Sem-5 that disrupt phosphotyrosine binding cause a vulvaless phenotype in *C. elegans* (CLARK et al. 1992). In the nematode the Let-23 RTK, besides being involved in vulval induction, also controls larval viability, hermaphrodite fertility, and the development of the male tail and posterior ectoderm. The carboxy-terminal tail of Let-23 contains several autophosphorylation sites, one of which is in the context of an optimal binding motif for the Sem-5 SH2 domain. The sy97 allele of Let-23 is due to a frame-shift mutation which causes the loss of the Sem-5 binding site (AROIAN et al. 1994). Sy97 homozygotes display severe phenotypes. They have very low viability, improper differentiation of the male tail, and are vulvaless. Interestingly, they retain fertility, showing that loss of the link with Sem-5 disrupts Let-23 function in some but not all cells. This indicates that the specific function served by the Let-23 receptor in different cell types may vary in terms of signaling requirements.

The importance of direct coupling with the Ras pathway has also been analyzed in vivo in *Drosophila* (RAABE et al. 1995, and this volume). In this case, a Tyr-Phe mutation of the Drk docking site was introduced in the cytoplasmic domain of the Sevenless (Sev) receptor, which controls the development of the R7 precursor cells in the *Drosophila* ommatidium. The mutation did not completely block Sev function in vivo. Its effects became apparent only using a genetically sensitized system whereby small changes in the efficiency of signal transduction could be detected (RAABE et al. 1995). This result indicates that in R7 cells the signaling output coming from the Sev receptor lacking the Drk docking site is adequate to specify neural development. Subsequent studies have shown that Drk is still critical for signaling from this mutant receptor, just as are Ras, Dos (Daughter of sevenless, a multiadaptor protein homologous to Gab-1), and Csw (RAABE et al. 1996). Both Dos and Csw interact with the mutant receptor and possess Drk binding sites. Thus Dos and Csw may, through Drk, provide an indirect link with the Ras pathway, explaining the redundancy of the Drk docking site in R7 cells.

Given the technical feasibility of introducing point mutations in the mouse genome, similar analyses are currently being extended to various receptors known to be active during mouse development. The first such study, done on the HGF receptor, has graphically confirmed the importance of docking sites for receptor function in vivo (MAINA et al. 1996). A double Tyr-Phe mutation was introduced by the *knock-in* approach into the carboxy-terminal tail of the Met receptor, to eliminate the bidentate docking site responsible for mediating receptor signaling (PONZETTO et al. 1994). Homozygous mice carrying this mutant HGF receptor, which is fully competent in kinase activity but is unable to recruit its effectors, display a phenotype, lethal in midgestation, which essentially reproduces that of the corresponding *knock-out* (SCHMIDT et al 1995; UEHARA et al. 1995; BLADT et al. 1995). In the absence of HGF/Met signaling the placenta does not develop properly and the embryos begin to die at E13.5. In the liver, parenchymal cells do differentiate, but soon undergo massive apoptosis (F. Maina, unpublished). In these embryos there is also a complete lack of muscles deriving from migratory precursors, such as those of limbs, body walls, diaphragm, and tip of the tongue, while deep axial muscles seem to be normal. This phenotype, together with the complementary pattern of expression of HGF mRNA (SONNENBERG et al. 1993), suggests that the function of HGF in the development of migratory muscles is that of mediating the detachment of Met-expressing myogenic cells from the somites and of guiding their migration to the appropriate final locations in the embryo.

The same study also describes the effects of a point mutation selective for the loss of Grb-2 binding (YVNV→YVHV), which, in cultured cells was permissive for motility but not for transformation (PONZETTO et al. 1996). This mutation should not interfere with binding of any other effector (including Shc and Gab-1); thus it is functionally similar to the abrogation of the Grb2 docking site in the Sev receptor which was described above. In line with results obtained for Sev in the R7 cells, this mutation does not completely block Met function in homozygous mouse embryos, which in fact develop to birth with normal placenta and liver. However, the mutation does interfere with muscle development. Appendicular and diaphragm muscles are strongly reduced in these mutants, which die perinatally of respiratory failure. Muscle cell progenitors seem not to be impaired in their ability to detach from the somite and to start their migration into the limb bud. They rather seem to manifest a partial survival or proliferative defect while on route, or upon arrival at their site of destination. In these mutants, deep axial muscles are normal up to mid-gestation, but later on, during secondary myogenesis, they too develop a defect. In fact the proliferation of fetal myoblasts, which are the precursors of secondary fibers, is reduced in the absence of Met signaling. This proliferative defect could not be seen in the severe mutants which, like the *knock-outs,* die at an earlier stage. The fact that uncoupling Met from Grb2 has different consequences depending on the cell type suggests that the HGF/Met pair fulfills distinct roles in placenta and liver (survival?) and in myogenic cells (survival, proliferation, and motility).

Based on the cell-type-specific behavior of mutations in the *Let-23* gene in *C. elegans*, it was hypothesized that cell-specific action may be a general feature of RTKs, and that by site-specific mutagenesis it might be possible to alter and/or target their function in a cell-specific fashion also in vertebrates (Aroian et al. 1994). The mice expressing the Met signaling mutants demonstrate that this is indeed feasible. It will be particularly appropriate to use this approach when the *knock-out* of a receptor gene yields a lethal phenotype. Introduction of partial loss-of-function mutations will result in a milder phenotype, more informative of the full spectrum of the receptor physiological roles. The opportunity of manipulating signaling pathways in vivo and thus of studying signal transduction in the whole organism (rather than at the cellular level) will most likely prompt the generation of a number of interesting murine models in the near future. These studies could well bring signal transduction to new heights in molecular medicine, both from a diagnostic and a therapeutic perspective.

Acknowledgements. The research in my laboratory has been supported by the Italian Association for Cancer Research (AIRC), Telethon Italy (Project 805), and the National Council for Research (CNR).

References

Aroian RV, Lesa GM, Sternberg PW (1994) Mutations in the Caenorhabditis elegans let23 EGFR-like gene define elements important for cell type specificity and function. EMBO J 13:360–366

Bishop JM (1983) Cellular oncogenes and retroviruses. Ann Rev Biochem 52:301–354

Ben-Levy R, Paterson HF, Marshall CJ, Yarden Y (1994) A single autophosphorylaion site confers oncogenicity to the Neu/ErbB-2 receptor and enables coupling to the MAP kinase pathway. EMBO J 13:3302–3311

Bladt F, Riethmacher D, Isenmann S, Aguzzi A, Birchmeler C (1995) Essential role for the c-met receptor in the migration of myogenic precursor cells into the limb bud. Nature 376:768–771

Boros P, Miller CM (1995) Hepatocyte growth factor: a multifunctional cytokine. Lancet 345:293–295

Carraway KL, Cantley LC (1994) A Neu acquaintance for ErbB3 and ErbB4: a role for receptor heterodimerization in growth signaling. Cell 78:5–8

Carraway KL, Burden SJ (1995) Neuregulins and their receptors. Curr Opin Neurobiol 5:606–612

Clark SG, Stern MJ, Horvitz HR (1992) *C. elegans* cell-signalling gene Sem-5 encodes a protein with SH2 and SH3 domains. Nature 356:340–344

Cortez D, Kadelc L, Pendergast AM (1995) Structural and signalling requirements for BCR-ABL-mediated transformation and inhibition of apoptosis. Mol Cell Biol 15: 5531–5541

Fantl WJ, Escobedo JA, Martin GA, Turck CW, Rosario M, McCormick F, Williams LT (1992) Distinct phosphotyrosines on a growth factor receptor bind to specific molecules that mediate different signaling pathways. Cell 69:413–423

Felder S, Zhou M, Hu P, Urefia J, Ullrich A, Chaudhuri M, White M, Shoelson SE, Schlessinger J (1993) SH2 domains exhibit high affinity binding to tyrosine phosphorylated peptides yet also exhibit rapid dissociation and exchange. Mol Cell Biol 13:1449–1455

Fixman ED, Fournier TM, Kamikura DM, Naujokas MA, Park M (1996) Pathways downstream of Shc and Grb2 are required for cell transformation by the Tpr-Met oncoprotein. J Biol Chem 271:13116–13122

Fournier TM, Kamikura D, Teng K, Park M (1996) Branching tubulogenesis but not scatter of Madin-Darby canine kidney cells requires a functional Grb2 binding site in the Met receptor tyrosine kinase. J Biol Chem 271:22211–22217

Gotoh N, Tojo A, Muroya K, Hashimoto Y, Hattori S, Nakamura S, Takenawa T, Yazaky Y, Shibuya M (1994) Epidermal growth factor-receptor mutant lacking the autophosphorylation sites induces phosphorylation of Shc protein and Shc-Grb2/ASH association and retains mitogenic activity. Proc Natl Acad Sci USA 91:167–171

Holdago-Madruga, M, Emlet DR, Moscatello DK, Godwin AK, Wong AJ (1996) A Grb-2 associated docking protein in EGF- and insulin-receptor signalling. Nature 379:560–564

Imboden JB, Koretsky GA (1995) Switching off signals. Curr Biol 5:727–729

Iwama A, Yamaguchi N, Suda T (1996) STK/RON receptor tyrosine kinase mediates both apoptotic and growth signals via the multifunctional docking site conserved among the HGF receptor family. EMBO J 15:5866–5875

Klingmüller U, Lorenz U, Cantley LC, Neel BJ, Lodish HF (1995) Specific recruitment of SH-PTP 1 to the erythropoietin receptor causes inactivation of JAK2 and termination of proliferative signals. Cell 80:729–738

Koch CA, Anderson D, Moran MF, Ellis C, Pawson T (1991) SH2 and SH3 domains: elements that control interactions of cytoplasmic signalling proteins. Science 252:668–674

Larose L, Gish G, Pawson T (1995) Formation and analysis of an SH2 domain binding site with mixed specificity. J Biol Chem 269:12320–12324

Li W, Nishimura R, Kashishian A, Bazter AG, Kim WJH, Cooper JA, Schlessinger J (1994) A new function for a phosphotyrosine phosphatase: linking Grb2-Sos to a receptor tyrosine kinase. Mol Cell Biol 14:509–517

Liu Y, Vega QC, Decker RA, Pandley A, Worby CA, Dixon JE (1996) Oncogenic RET receptors display different autophosphorylation sites and substrate binding specificities. J Biol Chem 271:5309–5312

Maina F, Casagranda F, Audero E, Simeone A, Comoglio PM, Klein R, Ponzetto C (1996) Uncoupling of Grb2 from the Met receptor in vivo reveals complex roles in muscle development. Cell 87:531–542

Marengere LEM, Songyang Z, Gish G, Schaller MD, Parsons JT, Stem MJ, Cantley LC, Pawson T (1994) SH2 domain specificity and activity modified by a single residue. Nature 369:502–505

Mori S, Ronnstrand L, Yokote K, Engstrom A, Courtneidge SA, Claesson-Welsh L, Heldin C-H (1993) Identification of two juxtamembrane autophosphorylation sites in the PDGF-P receptor involvement in the interaction with Src family tyrosine kinases. EMBO J 12:2257–2264

Myers MG, Jian Sun Y, White MF (1994) The IRS-1 signaling system. TIBS 19:289–293

Nishimura R, Li W, Kastushian A, Mondmo A, Zhou M, Cooper J, Schlessinger J (1993) Two signalling molecules share a phosphotyrosine-containing binding site in the platelet-derived growth factor receptor. Mol Cell Biol 13:6889–6896

Obermeier A, Bradshaw RA, Seedorf K, Choidas A, Schlessinger J, Ullrich A (1994) Neuronal differentiation signals are controlled by nerve growth factor receptor/Trk binding sites for SHC and PLCγ. EMBO J 13:1585–1590

Panayotou G, Gish G, End P, Truong O, Gout I, Dhand R, Fry MJ, Hiles I, Pawson T, Waterfield MD (1993) Interactions between SH2 domains and tyrosine-phosphorylated platelet-derived growth factor receptor sequences: analysis of kinetic parameters by a novel biosensor-based approach. Mol Cell Biol 13:3567–3576

Paulson RF,Vesely S, Siminovitch KA, Bernstein A (1996) Signalling by the W/Kit receptor tyrosine kinase is negatively regulated in vivo by the protein tyrosine phosphatase Shpl. Nature Gen 13:309–315

Pawson T (1995) Protein modules and signalling networks. Nature 373:573–579

Perkins LA, Larsen I, Perrimon N (1992) Corkscrew encodes a putative protein tyrosine phosphatase that functions to transduce the terminal signal from the receptor tyrosine kinase torso. Cell 70:225–236
Piao X, Paulson R, van der Geer P, Pawson T, Bernstein A (1996) Oncogenic mutation in the Kit receptor tyrosine kinase alters substrate specificity and induces degradation of the protein tyrosine phosphatase SHP-1. Proc Natl Acad Sci USA 93:14665–14669
Pinkas-Kramarski, Soussan L, Waterman H, Levkovitz G, Alroy I, Klapper L, Lavi S, Seger R, Ratzkin BJ, Sela M, Yarden Y (1996) Diversification of Neu differentiation factor and epidermal growth factor signaling by combinatorial receptor interactions. EMBO J 15:2452–2467
Ponzetto C, Bardelli A, Maina F, Longati P, Panayotou G, Dhand R, Waterfield MD, Comoglio P (1993) A novel recognition motif for phosphoinosytol 3-kinase binding mediates its association with hepatocyte growth factor/scatter factor receptor. Mol Cell Biol 13:4600–4608
Ponzetto C, Bardelli A, Zhen Z, Maina F, Dalla Zonca P, Giordano S, Graziani A, Panayotou G, Comoglio P (1994) A multifunctional docking site mediates signalling and transformation by the Hepatocyte Growth Factor/Scatter Factor Receptor family. Cell 77:261–271
Ponzetto C, Zhen Z, Audero E, Maina F, Bardelli A, Basile ML, Giordano S, Narsiman R, Comoglio P (1996) Specific uncoupling of GRB2 from the Met receptor: differential effect on transformation and motility. J Biol Chem 271:14119–14123
Raabe T, Olivier JP, Dickson B, Liu X, Gish GD, Pawson T, Hafen E (1995) Biochemical and genetic analysis of the Drk SH2/SH3 adaptor protein of Drosophila. EMBO J 14: 2509–2518
Raabe T, Riesgo-Escovar J, Liu X, Bausenwein BS, Deak P, Maröy P, Hafen E (1996) DOS, a novel Plackstrin homology domain-containing protein required for signal transduction between Sevenless and Ras1 in Drosophila. Cell 85:911–920
Santoro M, Carlomagno F, Romano A, Bottaro DP, Dathan NA, Grieco M, Fusco A, Vecchi G, Matodkova B, Kraus MH, Di Fiore PP (1995) Activation of RET as a dominant transforming gene by germline mutations of MEN2A and MEN2B. Science 267:381–383
Sasaoka T, Langlois WJ, Bai F, Rose DW, Leitner JW, Decker SJ, Saltiet AR, Gill GN, Kobayashi M, Draznin B, Olefsky JM (1996) Involvement of ErbB2 in the signalling pathway leading to cell cycle progression from a truncated epidermal growth factor receptor lacking the C-terminal autophosphorylaton sites. J Biol Chem 271:8338–8344
Scharenberg AM, Kinet J-P (1996) The emerging field of receptor-mediated inhibitory signaling: SHP or SHIP? Cell 87:961–964
Schmidt C, Bladt F, Goedecke S, Brinkmann VZ, Schlesche W, Sharpe M, Gherardi E, Birchmeier C (1995) Scatter Factor/Hepatocyte Growth Factor is essential for liver development. Nature 373:699–702
Serve H, Yee NS, Stella G, Sepp-Lorenzino L, Tan JC, Besmer P (1995) Differential roles of PI3-kinase and Kit tyrosine 821 in Kit receptor-mediated proliferation, survival and cell adhesion in mast cells. EMBO J 14:473–483
Soler C, Beguinot L, Carpenter G (1994) Individual epidermal growth factor receptor autophosphorylation sites do not stringently define association motifs for several SH2-containing proteins. J Biol Chem 269:12320–12324
Songyang Z, Cantley LC (1995) Recognition and specificity in protein tyrosine kinase mediated signaling. TIBS 20:470–475
Songyang Z, Shoelson SE, Chaudhuri M, Gish G, Pawson T, Haser WG, King F, Roberts T, Ratnofsky S, Lechleider RJ, Neel BG, Birge RB, Fajardo JE, Chou MM, Hanafusa H, Schaffhausen B, Cantley LC (1993) SH2 domains recognize specific phosphopeptide sequences. Cell 72:1–20
Songyang Z, Gish G, Mbamalu G, Pawson T, Cantley LC (1995a) A single point mutation switches specificity of Group III Src homology (SH)2 domains to that of Group I SH2 domains. J Biol Chem 270: 26029–26032
Songyang Z, Carraway KL, Eck MJ, Harrison SC, Feldman RA, Mohammadi M, Schlessinger J, Hubbard SR, Smith DP, Eng C, Lorenzo MJ, Ponder BAJ, Mayer BJ, Cantley LC (1995b) Catalytic specificity of protein-tyrosine kinases is critical for selective signalling. Nature 373:536–539
Sonnenberg E, Meyer D, Weidner KM, Birchmeier C (1993) Scatter factor/hepatocyte growth factor and its receptor, the c-*met* tyrosine kinase, can mediate a signal exchange between mesenchyme and epithelia during mouse development. J Cell Biol 123:223–235
Streuli M (1996) Protein tyrosine phosphatases in signalling. Curr Opin Cell Biol 8:182–188
Tzahar E, Waterman H, Chen X, Levkowitz G, Karunagaran D, Lavi S, Ratzkin BJ, Yarden Y (1996) A hierarchical network of interreceptor interactions determines signal transduction by Neu differentiation factor neuregulin and epidermal growth factor. Mol Cell Biol 16:5276–5287

Uehara Y, Minowa O, Mori C, Shiota K, Kuno J, Noda T, Kitamura N (1995) Placental defects and embryonic lethality in mice lacking Hepatocyte Growth Factor/Scatter Factor. Nature 373:702–705
Ulhich A, Schlessinger J (1990) Signal transduction by receptor with tyrosine kinase activity. Cell 61:203–212
Valius M, Kazlauskas A (1993) Phospholipase C-γ 1 and phosphatidylinositol 3 kinase are the downstream mediators of the PDGF receptor's mitogenic signal. Cell 73:321–334
Waters SB, Pessin JE (1996) Insulin Receptor Substrate 1 and 2 (IRS1 and IRS2): what a tangled web we weave. Trends Cell Biol 6:1–4
Weidner KM, Di Cesare S, Sachs M, Brinkmann V, Beherens J, Birchmeier W (1996) Interaction between Gab-1 and the c-Met receptor tyrosine kinase is responsible for epithelial morphogenesis. Nature 384:173–176
Yao R, Cooper GM (1995) Requirement for phosphatidylinositol-3 kinase in the prevention of apoptosis by nerve growth factor. Science 267:2003–2005
Yi T, Hile JN (1993) Association of hematopoietic cell phosphatase with c-kit after stimulation with c-kit ligand. Mol Cell Biol 13:3350

The IRS-Signaling System: A Network of Docking Proteins That Mediate Insulin and Cytokine Action

M.F. WHITE and L. YENUSH

1 Introduction

The integration of multiple transmembrane signals is especially important during development and maintenance of the nervous system, communication between cells of the immune system, evolution of transformed cells, and metabolic control (HUNTER 1997). Tyrosine phosphorylation plays a key role in many of these processes by directly controlling the activity of receptors or enzymes at early steps in

---

Research Division, Joslin Diabetes Center, 1 Joslin Place, and the Graduate Program in Biomedical and Biological Sciences, Harvard Medical School, Boston, MA 02215, USA

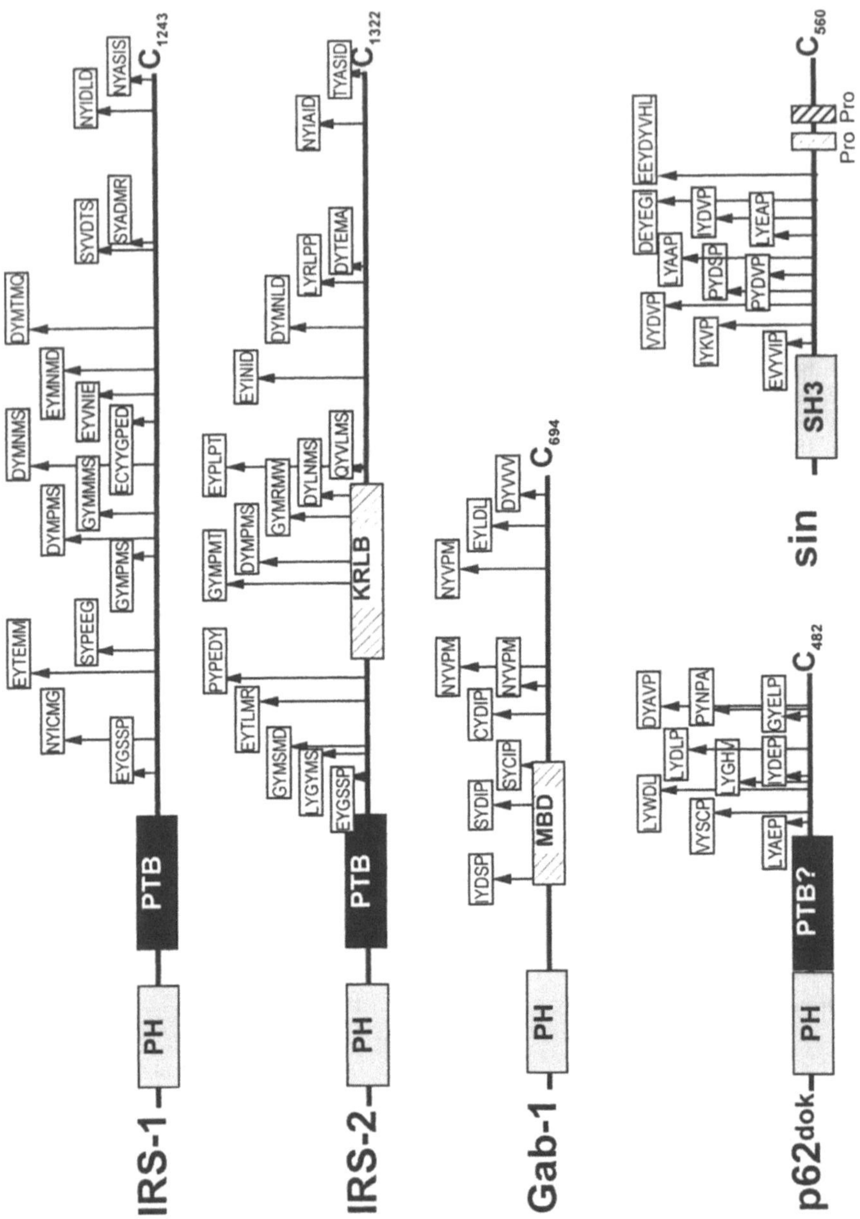

Fig. 1. IRS protein family members. Linear depiction of IRS-1, 2, Gab-1, p62[dok], and sin showing receptor interaction domains and tyrosine phosphorylation motifs. *PH*, pleckstrin homology domain; *PTB*, phosphotyrosine binding domain; *KRLB*, kinase regulatory loop binding domain; *MBD*, c-Met binding domain; *SH3*, src-homology-3 domain; *Pro*, proline-rich domain

signaling cascades, or by the assembly of multicomponent signaling complexes around activated receptors or their cellular substrates (PAWSON 1995). In most if not all cases, initialization of the signaling cascade controlled by growth factor and cytokine receptors originates with multisite tyrosine phosphorylation catalyzed directly by kinases activated during ligand-induced dimerization of specific membrane receptors (SCHLESSINGER 1988; HELDING 1995). In many cases, tyrosine autophosphorylation sites in activated receptors directly bind signaling proteins containing Src homology-2 domains (SH2 proteins). In other cases, tyrosine autophosphorylation increases the activity of the receptor kinase, which mediates tyrosine phosphorylation of cytosolic substrates or *docking proteins* that recruit SH2 proteins into multipotential signaling complexes (MYERS and WHITE 1995). The network is further elaborated through other modules which mediate protein-protein or protein-lipid interactions, including PTB, PDZ, SH3, WW, and PH domains.

This review focuses on a growing family of proteins that link receptor tyrosine kinases to downstream signaling molecules. These docking proteins provide a common interface between multiple receptor complexes and various signaling proteins with Src homology 2 domains. The insulin receptor substrate, IRS-1, was the first docking protein identified in mammalian systems, and serves as the prototype for this class of molecules (SUN et al. 1991). Three proteins related to IRS-1 include IRS-2, Gab-1 and $p62^{dok}$ (SUN et al. 1995; HOLGADO-MADRUGA et al. 1996; YAMANASHI and BALTIMORE 1997) (Fig. 1); a 60-kDa insulin receptor substrate in adipocytes which binds strongly to the PI-3 kinase has recently joined this family (LAVAN et al. 1997). These docking proteins are not related by extensive amino acid sequence identities, but are related functionally as insulin receptor substrates (IRS proteins). They contain several common structures, including an NH_2-terminal PH and/or PTB domain that mediates protein-lipid or protein-protein interactions; multiple COOH-terminal tyrosine residues that create SH2-protein binding sites; proline-rich regions to engage SH3 or WW domains; and serine/threonine-rich regions which may regulate overall function through other protein-protein interactions (Fig. 1).

Other classes of docking proteins include Shc, $p130^{cas}$ and sin (Fig. 1). Like IRS proteins, these are composed of protein-protein interaction domains and multiple tyrosine phosphorylation sites. Shc was one of the first docking proteins identified (PELLICI et al. 1992). It is composed of an NH_2-terminal PTB domain, a COOH-terminal SH2 domain, a proline-rich motif and a few tyrosine phosphorylation sites which engage Grb-2. Together, the PTB domain and the SH2 domain provide Shc with considerable promiscuity during its interactions with activated receptors; the PTB domain mediates the interaction with the insulin receptor, whereas the SH2 domain binds to the activated EGF receptor (VAN DER GEER et al. 1996; ISAKOFF et al. 1996). $p130^{cas}$ and sin were recognized recently as activators and substrates of $pp60^{src}$ (SAKAI et al. 1994; ALEXANDROPOULOS and BALTIMORE 1996). They are composed of an SH3 domain that binds to Src and other elements of the cytoskeleton and a tail of tyrosine phosphorylation sites that bind to various SH2 proteins (SAKAI et al. 1994; ALEXANDOPOULOS and BALTIMORE 1996). $p130^{cas}$

and sin have not been found to be substrates for the insulin receptor, probably owing to the absence of the required coupling modules; however, indirect coupling through adapter proteins may occur allowing access to the insulin receptor.

2 Identification of IRS Proteins

2.1 Identification of IRS-1 and IRS-2

Insulin stimulates tyrosine phosphorylation of a 185-kDa phosphoprotein in all insulin-sensitive tissues and cells. This protein, originally called pp185, provided the earliest support for the idea that insulin signals are mediated through tyrosine phosphorylation of a cytoplasmic docking protein (WHITE et al. 1985). Using immobilized antiphosphotyrosine antibodies, sufficient protein was purified from rat liver extracts to obtain partial amino acid sequence to design optimized oligonucleotides for cDNA cloning (SUN et al. 1991; ROTHENBERG et al. 1991). The first cloned insulin receptor substrate, IRS-1, is encoded by a single gene on human chromosome 2q36-37 (mouse chromosome 1) and has a calculated molecular mass of 132 kDa (SUN et al. 1991; ARAKI et al. 1994).

A related protein (IRS-2) was immediately predicted because specific antibodies against IRS-1 bound rather weakly to pp185 in hepatoma cells, and the unreactive protein was called $pp185^{HMW}$ because it is migrated at a slightly higher molecular weight during SDS-PAGE (MIRALPEIX et al. 1992). Subsequent findings revealed a similar size protein in myeloid progenitor cells (the IL-4 receptor substrate, 4PS), and in hepatocytes and skeletal muscle from the mice lacking IRS-1 (pp190) (SUN et al. 1995; TOBE et al. 1995). The purification and cloning of 4PS from FDC-P2 cells revealed a protein similar to IRS-1, and immunologically identical to $pp185^{HMW}$ and pp190; the cloned protein was called IRS-2 (SUN et al. 1995). The IRS-2 gene resides on the short arm of mouse chromosome 8 near the insulin receptor gene (SUN et al. 1997).

A comparison of the amino acid sequences of IRS-1 and IRS-2 reveals several common features, including a well conserved pleckstrin homology (PH) domain at the extreme NH_2-terminus, followed immediately by a phosphotyrosine binding (PTB) domain that binds to phosphorylated NPXY-motifs (Fig. 1). However, the COOH-terminal regions of IRS-1 and IRS-2 are rather poorly conserved, displaying only 35% identity overall. Several tyrosine phosphorylation sites in IRS-2 align with similar motifs in IRS-1, and in a few cases the motifs are nearly identical. By contrast, half of the motifs display unique acidic or hydrophobic residues which may alter their interaction with upstream kinases or downstream SH2 proteins. IRS-1 binds several SH2 proteins, including p85, Grb-2, SHP2, nck, crk, fyn and others (MYERS et al. 1994; SKOLNIK et al. 1993b; SUN et al. 1996; KUHNE et al. 1993; BEITNER-JOHNSON et al. 1996; LEE et al. 1993). The comparative association of IRS-1 and IRS-2 with various SH2-domains suggests overlapping and distinct signaling potential (SUN et al. 1997). Interestingly, immortalized fibroblasts lacking IRS-1 while overexpressing IRS-2 do not display normal insulin signaling, suggesting that these docking proteins are functionally distinctly identical (BRUNING et al. 1997).

2.2 Small IRS Proteins: Gab-1, DOS and $p62^{DOK}$

Recently, two additional proteins with limited amino acid sequence identity have been identified that are functionally similar to IRS-1 and IRS-2 (Fig. 1). One of them, Gab-1, occurs in mammalian tissues and was cloned by expression screening with a Grb-2 probe (HOLGADO-MADRUGA et al. 1996). In common with IRS-1 and IRS-2, Gab-1 contains a similar NH_2-terminal PH domain followed immediately by a short COOH-terminal tail with multiple tyrosine phosphorylation sites, including motifs that bind PI-3 kinase (Fig. 1). Unlike IRS-1 and IRS-2, there is no intervening PTB domain in Gab-1, which apparently diminishes the coupling affinity with the insulin receptor (HOLGADO-MADRUGA et al. 1996). Thus signals mediated by Gab-1 during insulin stimulation may occur only at relatively high insulin concentrations.

Drosophila contains a docking protein called DOS, which is similar to Gab-1 (RAABE et al. 1996). It was identified by screening for mutations that suppress retinal development by a constitutively activated sevenless receptor protein tyrosine kinase (SEV). The presence of an NH_2-terminal pleckstrin homology domain with considerable similarity to that in mammalian IRS protein suggests that it may mediate signals from the activated *Drosophila* insulin/IGF-1 receptor, but this has not been reported. Like Gab-1, DOS contains many potential tyrosine phosphorylation sites that can bind SH2 proteins. Genetic analysis demonstrates that DOS functions upstream of Ras1 and defines a signaling pathway that is independent of direct binding of the DRK (Dro homologue of Grb-2) to the SEV receptor tyrosine kinase (HERBST et al. 1996).

Two other small docking proteins phosphorylated by the insulin receptor substrates have been identified in various mammalian cells and tissues, including a 62-kDa protein that binds rasGAP (originally called $p62^{rasGAP}$) and a 60-kDa protein that binds PI-3 kinase (pp60) (HOSOMI et al. 1994; OGAWA et al. 1994; KAPLAN et al. 1990; ELLIS et al. 1990; ROTH et al. 1992). The $p62^{rasGAP}$ is a common target of several protein-tyrosine kinases, including v-Abl, v-Src, v-Fps, v-Fms, and activated receptors for IGF-1, EGF, csf-1 as well as the insulin receptor (YAMANASHI and BALTIMORE 1997). Recently, a substrate for bcr-abl that binds to ras-Gap was purified and cloned (YAMANASHI and BALTIMORE 1997; CARPINO et al. 1997). This protein, called $p62^{dok}$, reacts with monoclonal antibodies raised against $p62^{rasGAP}$, suggesting they are identical (YAMANASHI and BALTIMORE 1997). The $p62^{dok}$ contains a recognizable PH domain at its NH_2-terminus that is distantly similar to the PH domain in IRS-1, IRS-2 and Gab-1 (Fig. 1). $p62^{dok}$ may contain a PTB domain but it contains little amino acid sequence similarity to IRS-1; however, two conserved arginine residues appear to be correctly positioned to bind phosphotyrosine, by comparison to the PTB domain of IRS-1 (ECK et al. 1996a). The COOH-terminus of $p62^{dok}$ contains multiple tyrosine phosphorylation sites in motifs that recognize various SH2 proteins. Interestingly, none of the phosphorylation sites are predicted to bind PI-3 kinase (Fig. 1).

In rat adipocytes a 60-kDa insulin receptor substrate (pp60) is especially sensitive to insulin stimulation (MONOMURA et al. 1988; THIES et al. 1990). In adipocytes lacking IRS-1, pp60 is the predominant insulin receptor substrate as IRS-2 is not enhanced (SMITH-HALL et al. 1997). It binds strongly to SH2 domains in p85, suggesting that it contains phosphorylated YXXM-motifs; it also binds to immobilized peptides containing phosphorylated NPXY-motifs, suggesting that it contains a phosphotyrosine binding domain. We propose the name $pp60^{IRS3}$ to reflect these functional similarities to IRS-1 and IRS-2 (SMITH-HALL et al. 1997). The recent cloning of $pp60^{IRS3}$ reveals a new small member of the IRS protein family, which may play an important role during insulin-stimulated glucose uptake (LAVAN et al. 1997).

2.3 IRS Proteins Coordinate a Diverse and Flexible Signaling System

What signaling advantages could the IRS proteins provide? First, IRS proteins provide a means for signal amplification by eliminating the stoichiometric constraints encountered by receptors which directly recruit SH2 proteins to their autophosphorylation sites. For example, the activated PDGF receptor assembles SH2 proteins around its autophosphorylation sites, so the intensity of these signals is restricted by the number of receptors in the plasma membrane. By contrast, tyrosine phosphorylation of IRS proteins is catalytic, since transient but specific formation of the enzyme/substrate complex facilitates phosphorylation of multiple proteins.

Second, IRS proteins dissociate the intracellular signaling complex from the endocytic pathways of the activated receptor. During ligand binding, receptors typically undergo internalization. The exact mechanism and fate of each receptor is cell context and receptor dependent, but, in general, the activated receptors migrate into clatherin-coated vesicles which form endosomes that are acidified and either return the receptor to the plasma membranes or targeted to lysosomes for degradation (BACKER et al. 1992c). Signaling proteins associated with receptors are presumably obligated to follow the intracellular itinerary of the receptor. By contrast, IRS proteins associate transiently with activated receptors and then migrate to other cellular compartments or are targeted by associated partners. Itinerant docking proteins may be essential for various biological effects, such as insulin-stimulated translocation of GLUT-4 in adipocytes, or neurite formation in differentiating neurons.

Third, the ability of a single receptor to engage multiple IRS proteins expands the repertoire of regulated signaling pathways. During insulin stimulation, IRS-1, IRS-2, Gab-1 and $p62^{dok}$ are all potential substrates of the activated receptor. Each substrate will be engaged by the receptor with certain characteristics and possess common and unique tyrosine phosphorylation motifs. Moreover, the presence of unique phosphorylation sites on distinctly targeted substrates provides a broader capacity to regulate multiple cellular responses. Similarly, during IL4 signaling IRS-2 mediates signals controlling cellular growth, whereas IL-4-specific gene expression is largely regulated by a direct association between STAT6 and the IL-4 receptor (RYAN et al. 1996).

Finally, the shared use of docking proteins by multiple receptors integrates diverse signals into a coordinated cellular response. In addition to the insulin and IGF-1 receptors, IRS proteins are phosphorylated by the receptors for growth hormone and prolactin (UDDIN et al. 1995; PLATANIAS et al. 1996; ARGETSINGER et al. 1996; BERLANGA et al. 1997), several interleukins (IL-2, IL-4, IL-9, IL-13 and IL-15) (BEITNER-JOHNSON et al. 1996; CHEN et al. 1997; JOHNSON et al. 1995), interferons (IFNα/β and IFNγ) (UDDIN et al. 1995; PLATANIAS et al. 1996), members of the IL-6 receptor family (ARGETSINGER et al. 1996), and the angiotensin receptor (VELLOSO et al. 1996). Similarly, Gab-1 is a substrate for the insulin receptor, the hepatocyte growth factor receptor (c-Met) and the EGF receptor; $p62^{dok}$ is a target for v-Abl, v-Src, v-Fps, v-Fms and the receptors for insulin, IGF-1, EGF, PDGF, CSF-1 and VEGF (YAMANASHI and BALTIMORE 1997). Thus, docking proteins provide a common interface to integrate signals from a variety of activated tyrosine kinases involved in metabolic and growth control.

3 Coupling Mechanisms Between IRS Proteins and Activated Receptors

3.1 Introduction

Separation of the docking proteins (IRS proteins) from the activated receptors creates a necessity for kinase/substrate coupling. Specific interactions are essential, but these must be transient to allow the receptor to engage multiple molecules to amplify and diversify the signal. These requirements probably exclude SH2 domain interactions, as these tend to be too strong and stoichiometric, which could diminish the catalytic turnover. Several solutions appear to be employed by the IRS proteins. A pleckstrin homology (PH) domain provides one of the interaction domains found at the NH_2-terminus of all IRS proteins; however, the exact mechanism for coupling is unknown. A second means for receptor coupling is provided by the phosphotyrosine binding (PTB) domain. This domain binds specifically but weakly to the phosphorylated NPXY-motifs located in the receptors for insulin, IGF-1 and IL-4 (KEEGAN et al. 1994). However, all receptors which engage the IRS protein do not contain NPXY motifs, so the PTB domain may not be universally employed. In these latter cases, the PH domain must be adequate or other modules may contribute (SAWKA-VERHELLE et al. 1996; HE et al. 1996).

Together these interaction domains provide specific mechanisms for receptor substrate coupling.

3.2 PTB Domains Bind Phosphorylated NPXY Motifs

A role for the $NPXY_{960}$ motif in the juxtamembrane region of the insulin receptor for substrate recognition was proposed 10 years ago, because mutations at this motif diminish tyrosine phosphorylation of IRS-1 while simultaneously reducing the biological activity of insulin (SUN et al. 1991; WHITE et al. 1988). These results lead to the hypothesis that the NPXY motif in the insulin receptor mediates substrate recognition (WHITE et al. 1988). This idea was later extended to the IGF-1 and IL-4 receptors (KEEGAN et al. 1994; YAMASAKI et al. 1992). Alignment of the cytoplasmic regions of the IL-4 receptor with the juxtamembrane region of the insulin and IGF-1 receptors reveals a conserved motif (LxxxxNPxYxSxP), that mediates the interaction with IRS-1 (KEEGAN et al. 1994).

Phosphorylated NPXY motifs are now known to bind on the surface of phosphotyrosine binding (PTB) domains. PTB domains are regions of 100–150 residues that were first identified at the NH_2-terminus of Shc (VAN DER GEER et al. 1996; KAVANAUGH et al. 1955). Similar domains are found in IRS-1 and IRS-2, which provide one of the mechanisms for coupling to the activated insulin or IGF-1 receptors, and the IL-4 receptor (SUN et al. 1995; YEANUSH et al. 1996). A similar PTB domain may also exist in $p62^{dok}$, whereas Gab-1 does not appear to contain a PTB domain (HOLGADO-MADRUGA et al. 1996; YAMANASHI and BALTIMORE 1997). Although PTB domains bind phosphotyrosine they are not SH2 domains, but structurally resemble PH domains (LEMMON et al. 1996; HARRISON 1996). In vitro, recombinant PTB domains of IRS-1 or Shc bind directly to activated insulin receptors, as well as phosphopeptides containing NPXY motifs (YENUSH et al. 1996; WOLF et al. 1995). The binding selectivity of the PTB domain in IRS1/IRS2 and Shc is different, owing largely to the interaction of residues located on the NH_2-terminal side of the NPXY motif. A structural basis for this binding selectivity has been proposed (ECK et al. 1996a; HARRISON 1996; ZHOU et al. 1995, 1996).

3.3 The PH Domain

Pleckstrin homology (PH) domains were originally identified in pleckstrin, but are now known to occur in many signaling proteins and display a broad array of ligand binding selectivity (HASLAM et al. 1993; GIBSON et al. 1994). However, detailed structural determinations of the PH domains from β_G-spectrin, pleckstrin, dynamin, PLCγ and others reveal a highly conserved tertiary structure; each domain is composed of two antiparallel β-sheets forming a sandwich, with one corner covered by an amphipathic COOH-terminal α-helix (LEMMON et al. 1996). The discovery that PTB domains are structurally identical emphasized that amino acid sequence does not predict the existence of a PH domain (ECK et al. 1996a). However, PTB domains contain an L-shaped cleft that is absent in most PH domains to provide a point of contact with the phosphorylated NPXY motif (HARRISON 1996).

PH domains bind to a variety of ligands. For example, the PH domain in the β-adrenergic receptor kinase (βark) interacts with βγ-subunits of an activated trimeric G-protein, providing a point for recruitment to the β-adrenergic receptor complex (LUTTRELL et al. 1995). Furthermore, the PH domain in PKB, a serine kinase activated downstream of the PI-3 kinase, is important for its recruitment to the plasma membrane (HEMMINGS 1997). Although the specific ligand for each PH domain is difficult to predict, they clearly play important roles during assembly of signaling complexes at membrane surfaces.

The NH_2-terminal PH domain is essential for coupling the insulin receptor to IRS-1 and this is probably true for IRS-2 and Gab-1 since their PH domains are 62% similar; the PH domain in $p62^{dok}$ is relatively distinct and may interact with a different target. Without the PH domain, IRS-1 is poorly tyrosine phosphorylated during insulin stimulation, especially at low receptor expression encountered in ordinary cells (YENUSH et al. 1996; MYERS et al. 1995). By contrast, the PH domain is sufficient in IRS-1 without the PTB domain, although the efficiency of the interaction is diminished (YENUSH et al. 1996). At high levels of insulin receptor, either the PH domain or the PTB domain is adequate to couple the insulin receptor to IRS-1 (YENUSH et al. 1996; MYERS et al. 1995). Thus, PH and PTB domain appear to function independently and through distinct ligand interactions.

The ligand which binds to the PH domain in the IRS proteins is not known. Since the PH domains from IRS-2 or Gab-1 are fully functional when substituted into IRS-1, they probably have similar binding specificity (D. Burks et al., unpublished results). Moreover, PH domains from unrelated proteins do not substitute for the endogenous PH domain in IRS-1, further supporting the specificity of this module. Since the insulin receptor does not interact directly with the PH domain, other membrane associated elements may act as the interface (O'NEILL et al. 1994; GUSTAFSON et al. 1995; CRAPARO et al. 1995). Membrane proteins or specific phospholipids preferentially located near activated insulin receptors are ideal candidates, but their identities are unknown.

3.4 Novel Phosphotyrosine Binding Domains in IRS-2 and Gab-1

Recently, yeast two hybrid analysis revealed novel phosphotyrosine interaction domains in both IRS-2 and Gab-1 (SAWKA-VERHELLE et al. 1996; HE et al. 1996). Unlike PH and PTB domains, the binding modules reside in the tyrosine-rich tails. In IRS-2, a region between residues 591 and 786 binds to the phosphorylated regulatory loop of the β-subunit of the insulin receptor (SAWKA-VERHELLE et al. 1996; HE et al. 1996). This kinase regulatory loop binding (KRLB) domain only binds to activated insulin receptors and requires the presence of all three tyrosine phosphorylation sites in the regulatory loop (SAWKA-VERHELLE et al. 1996). The exact nature of the KRLB domain is not known, but the Tris-phosphorylated regulatory loop may constitute an important binding site. Although phosphorylation of the regulatory loop may expose another receptor region which mediates the interaction, recent experiments show that a synthetic phosphopeptide with the sequence of the insulin receptor regulatory loop binds directly to the KRLB domain (SAWKA-VERHELLE et al. 1997). A region with similar binding specificity has not been detected in IRS-1, highlighting an important functional difference between these related proteins. This distinction may be important during substrate discrimination. Since trisphosphorylation of the regulatory loop fully activates the kinase, the KRLB domain could serve to preferentially localize IRS-2 with the most active receptors. Alternatively, IRS-2 may be less reliant on the interaction between the NPXY motif and the PTB domain, or a functional PH domain. Finally, two points of attachment between IRS-1 and the insulin receptor may reduce the degrees of freedom in the complex and restrict access to the available phosphorylation sites.

Using a similar approach, Gab-1 was found to bind to the c-Met receptor through a novel phosphotyrosine binding domain called the c-Met binding domain (MBD). The MBD occurs between residues 450 and 532 of Gab-1 and binds to autophosphorylation sites in the COOH-terminus of c-Met (WEIDNER et al. 1996). This interaction is blocked by a 24 residue synthetic phosphopeptide based on the sequence around these phosphorylation sites. By contrast, the MBD did not bind to TrkA, c-Ros, c-Neu, the insulin receptor or several other tyrosine kinases (WEIDNER et al. 1996). These results suggest that different kinases may engage docking protein differently to assemble unique signaling complexes with specific signaling potential.

4 Downstream Elements Engaged by IRS Proteins

4.1 IRS Proteins Activate PI-3 Kinase

One of the major mechanisms used by IRS proteins to generate downstream signals is the direct binding to the SH2 domains of various signaling proteins. Several enzymes and adapter proteins have been identified which associate with IRS-1, including PI-3 kinase, SHP2, Fyn, Grb-2, nck, and Crk (MYERS et al. 1994; SKOLNIK et al. 1993b; SUN et al. 1996; KUHNE et al. 1993; BEITNER-JOHNSON et al. 1996; LEE et al. 1993); other partners associate through unknown mechanisms which do not depend on tyrosine phosphorylation, including SV40 large T antigen, 14-3-3 and $\alpha_{v\beta 3}$ (VUORI and RUOSLAHTI 1994; FEI et al. 1995; CRAPARO et al. 1997).

PI-3 kinase is the best studied signaling molecule activated by IRS-1. It plays an important role in the regulation of a broad array of biological responses by various hormones, growth factors and cytokines, including mitogenesis (VALIUS and KAZLAUSKAS 1993; YAO and COOPER 1995), differentiation (KIMURA et al. 1994), chemotaxis (KUNDRA et al. 1994; OKADA et al. 1994), membrane ruffling (WENNSTROM et al. 1994), and insulin-stimulated glucose transport (OKADA et al.

1994). Moreover, PI-3 kinase activity is required for neurite extension and inhibition of apoptosis in PC12 cells (YAO and COOPER 1995) and cerebellar neurons (DUDEK et al. 1997), suggesting that it plays an important role in neuronal survival.

PI-3 kinase was originally identified as a dimer composed of a 110-kDa catalytic subunit (p110α or p110β) associated with an 85-kDa regulatory subunit (p85α or p85β). During a search for new SH2 proteins that bind to IRS-1, we cloned a smaller regulatory subunit that occurs predominantly in brain and testis, called $p55^{PIK}$ (Fig. 2). The COOH-terminal portion of $p55^{PIK}$ is similar to p85, including a proline-rich motif, two SH2 domains, and a consensus binding motif for p110 (DHAND et al. 1994). However, $p55^{PIK}$ contains a unique 30-residue NH_2-terminus which replaces the Src homology-3 (SH3) domain and the Bcr-homology region found in p85 (PONS et al. 1995). Like p85, $p55^{PIK}$ associates with tyrosine phosphorylated IRS-1, and this association activates the PI-3 kinase (PONS et al. 1995). Two other small regulatory subunits (p55α and p50α) are encoded by alternative splicing of the p85α gene (ANTONETTI et al. 1996; INUKAI et al. 1996; FRUMAN et al. 1996). These various regulatory subunits confer considerable variety to PI-3 kinase.

PI-3 kinase plays an important role in many insulin-regulated metabolic processes, including glucose uptake, general and growth-specific protein synthesis, and

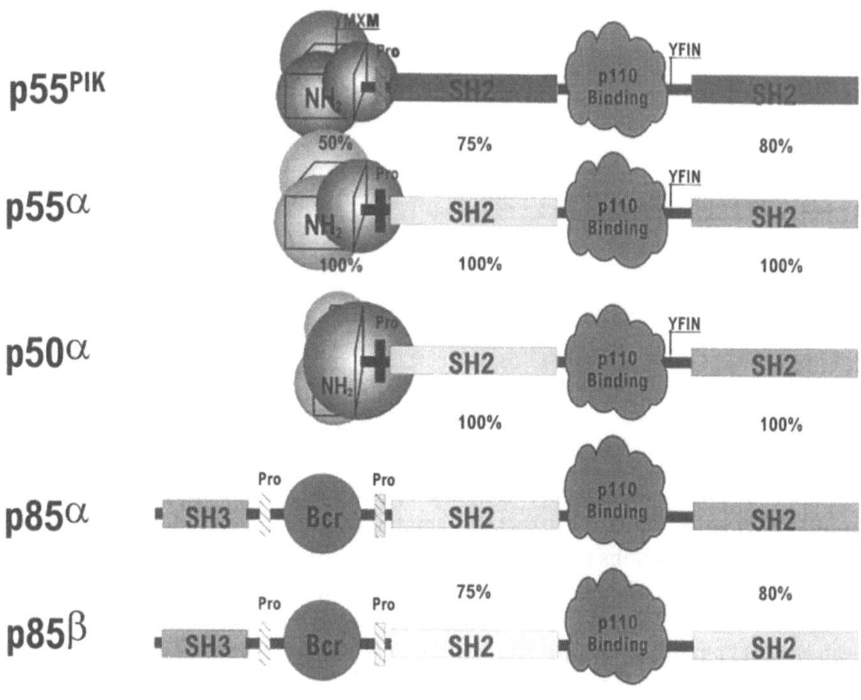

Fig. 2. PI-3 kinase regulatory subunits. The five regulatory subunits of PI-3 kinase are pictured, highlighting the common and unique structural features. *Pro*, proline-rich region; NH_2, amino-terminal regions; *Bcr*, Bcr-homology region. *Percentages* refer to the percentage identity relative to p85α

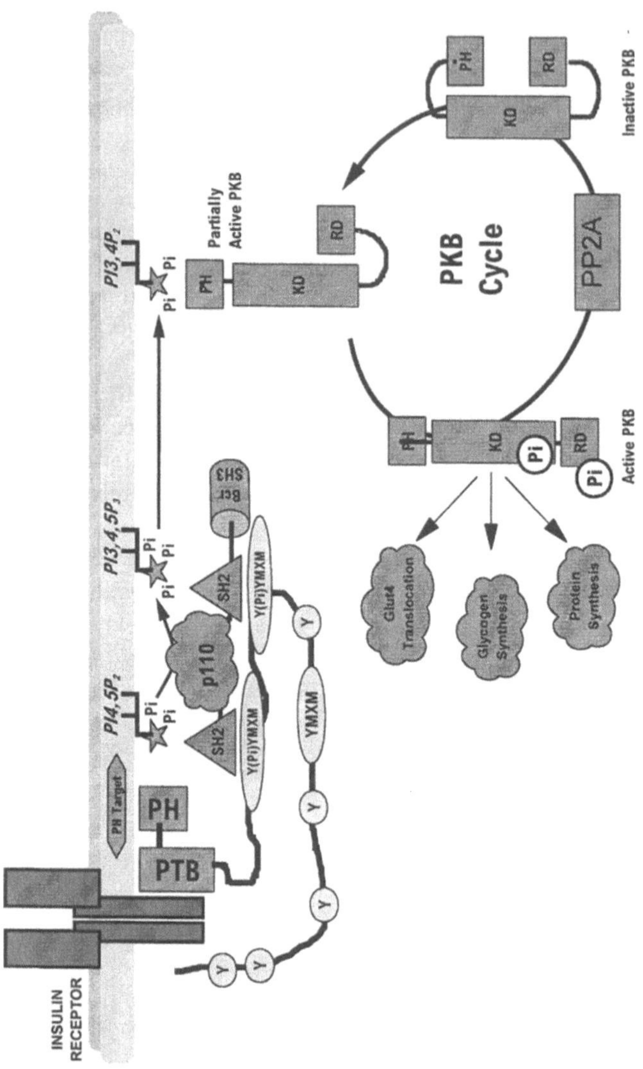

Fig. 3. Model of PKB activation. Schematic representation of a working model of the signal transduction pathways and molecular mechanism of the cycle of PKB activation. *KD*, kinase domain; *RD*, regulatory domain

glycogen synthesis. Binding of the SH2 domain in p85 to phosphorylated YMXM motifs in IRS-1 activates the associated catalytic domain, and this is maximal when both of the SH2 domains are occupied (BACKER et al. 1992b). Double occupancy is easily accomplished when both YMXM motifs are located within the same peptide, suggesting that the second binding event occurs more readily through an intramolecular reaction (RORDORF-NIKOLIC et al. 1995). Since IRS-1 and IRS-2 contain about nine YMXM-motifs, they are ideal docking proteins to activate PI-3 kinase. This is the major mechanism used by insulin to activate PI-3 kinase in 32D cells, where the experiment can be conducted in a background without IRS proteins.

Based on the results of various inhibitor studies, several enzymes appear to carry the signal initiated by PI-3 kinase activation to its final destinations (Fig. 3). The $p70^{s6k}$, PKB, PKCζ and others are thought to be downstream of PI-3 kinase (MYERS et al. 1996; FRANKE et al. 1995; DIAZ-MECO et al. 1994). The regulation of PKB is complex, involving occupancy of its PH domain and serine/threonine phosphorylation (ALESSI et al. 1996). PKB and PKCζ are implicated in various biological responses, including translocation of GLUT4 to the plasma membrane, general and growth-regulated protein synthesis, and glycogen synthesis (HEMMINGS 1997; ALESSI et al. 1996; CROSS et al. 1996; BANDYOPADHYAY et al. 1997). Moreover, activation of PI-3 kinase and its downstream partners is important for survival and differentiation of neurons (see below).

4.2 IRS-1 Activates SHP2

SHP2 is a phosphotyrosine phosphatase with two SH2 domains that is expressed in most mammalian cells (FREEMAN et al. 1992). Several growth factor receptors, including the EGFr, the PDGFr, and c-kit bind specifically to the SH2 domains in SHP2 (FENG et al. 1993; LECHLEIDER et al. 1993a, b; TAUCHI et al. 1994); a homologue in *Drosophila*, csw, mediates signals from the PDGF receptor homologue, *torso* (PERKINS et al. 1992). During insulin stimulation, SHP2 binds to two tyrosine residues at the extreme COOH-terminus of IRS-1 (KUHNE et al. 1993; ECK et al. 1996b); it also binds to phosphorylated Tyr_{1146} in the regulatory loop of the insulin receptor (KHARITONENKOV et al. 1995). In addition, SHP2 associates with a 115-kDa protein during insulin stimulation (ECK et al. 1996b). This protein is located in the plasma membrane, contains YXX(L/VI) motifs and may be a direct substrate for the insulin receptor and other tyrosine kinase receptors (YAMAO et al. 1997). The association of SHP2 with various docking protein may serve to localize this phosphatase in various subcellular regions where it modulates distinct signaling pathways.

Several reports suggest that SHP2 mediates downstream signals from the insulin receptor, since a catalytically inactive mutant inhibits insulin-stimulated MAP kinase and c-*fos* transcription in intact cells (YAMAUCHI et al. 1995; NOGUCHI et al. 1994; SASAOKA et al. 1994). This dominant negative effect is partially reversed by co-expression of v-ras or Grb2, indicating that SHP2 may act upstream of Ras, possibly as an adapter protein (NOGUCHI et al. 1994). Alternatively, SHP2 may diminish the tyrosine phosphorylation of IRS-1, providing a mechanism to attenuate certain signals. This hypothesis is supported by recent findings that reduced levels of SHP2 in mice increase IRS-1 tyrosine phosphorylation and its associated PI-3 kinase activity during insulin stimulation (ARRANDALE et al. 1996). Consistent with this, expression in 32D cells of mutant IRS-1 lacking the SHP2 binding motifs increases IRS-1 tyrosine phosphorylation and associated PI-3 kinase activity, resulting in twofold enhanced insulin-stimulated $[^{32}S]$methionine-incorporation (MENDEZ et al. 1996). Thus, SHP2 may be essential to balance the converging and opposing signals essential for insulin and cytokine action.

4.3 IRS Proteins Engage Multiple Adapter Molecules

In addition to PI-3 kinase and SHP2, tyrosine phosphorylated motifs in the IRS proteins bind to the SH2-domains in several small adapter proteins, including Grb-2, nck and crk (SKOLNIK et al. 1993b; BEITNER-JOHNSON et al. 1996; LEE et al. 1993). In addition to SH2 domains, these proteins contain multiple SH3 domains that bind various downstream signaling molecules that regulate metabolism, growth and differentiation (PAWSON 1995; SCHLESSINGER 1993) (Fig. 4). Flanking its SH2 domain, Grb-2 contains two SH3-domains that associate constitutively mSOS, a guanine nucleotide exchange protein that stimulates GDP/GTP exchange on $p21^{ras}$ (SKOLNIK et al. 1993a, b; GALE et al. 1993). The recruitment by growth factor receptors of Grb2/mSos to membranes containing $p21^{ras}$ is one of the mechanisms employed to activate the MAP kinase cascade (SCHLESSINGER 1993). In addition to mSOS, Grb-2 associates with other proteins such as dynamin (ANDO et al. 1994). Dynamin is a GPTase that plays a critical role in the earliest stages of endocytosis, which may contribute to the internalization of membrane proteins during insulin stimulation (SHPETNER et al. 1996).

During insulin stimulation, Grb-2 engages IRS-1, IRS-2, Shc or SHP2, although the preferred interactions depend on the cell background. For example, in skeletal muscle IRS-1 is the dominant GRB2 binding protein (YAMAUCHI et al.

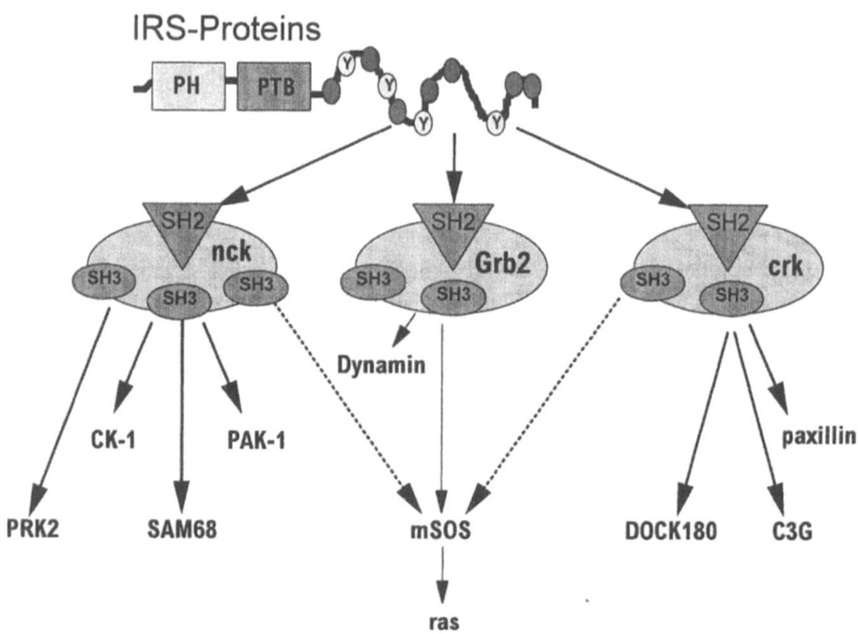

Fig. 4. Signaling pathways potentially mediated by IRS-binding adapter proteins. Nck, Crk, and Grb-2 and their interacting proteins are shown. The physiological relevance of some of these interactions is still being defined. IRS proteins regulate some, but probably not all, of the pathways shown here

1996). Thus, mice lacking IRS-1 display an 80% reduction in insulin-stimulated MAP kinase even though Shc phosphorylation is normal (YAMAUCHI et al. 1996). Apparently, Grb2 binds poorly to Shc in this background. By contrast, Shc plays a major role during insulin stimulation of MAP kinase in cultured cells, and in many systems Shc is believed to the major Grb2/Sos activator during insulin stimulation (MYERS et al. 1994; SKOLNIK et al. 1993b).

Although the MAP kinase cascade is a well documented insulin signaling pathway, it is not very sensitive to insulin. First, the Grb-2 binding site (Tyr_{895}) in IRS-1 is weakly phosphorylated, requiring relatively high insulin receptor expression; similarly, insulin-stimulated Shc tyrosine phosphorylation requires a strong insulin signal, usually achieved by overexpression of the insulin receptor (MYERS et al. 1994). In 32D myeloid cells, which express few insulin receptors and no murine IRS proteins, ectopic expression of IRS-1 alone is insufficient to stimulated Grb-2 binding, even though IRS-1 becomes tyrosine phosphorylated and bound to p85 (MYERS et al. 1994). By contrast, overexpression of the insulin receptor alone mediates insulin-stimulated MAP kinase activation without IRS proteins, apparently through tyrosine phosphorylation of endogenous Shc (MYERS et al. 1994). Thus low insulin receptor levels are sufficient to mediate IRS-1 phosphorylation, but only on a few sites that activate PI-3 kinase. By contrast, Grb-2 binding to IRS-1 occurs at high receptor levels. Since insulin-stimulated tyrosine phosphorylation of Shc also requires relatively high receptor levels, Grb2/Sos activation is generally insensitive during insulin stimulation.

Two other small SH2/SH3 adapter proteins, nck and crk, bind to tyrosine phosphorylated IRS-1. Nck is a 47-kDa adapter protein composed of three SH3 domains and a single SH2 domain; crk contains two SH3 domains and a single SH2 domain (PAWSON 1995). These adapter proteins bind through their SH3 domains to a variety of signaling proteins, and are targeted to specific subcellular locations by tyrosine phosphorylated membrane receptors or docking proteins (Fig. 4).

4.4 Phosphotyrosine-Independent Partners for IRS-1

The role of Ca^{2+} and calmodulin in modulating the insulin signal has been implicated by various studies, but the mechanisms remain obscure. Recently, it was observed that IRS-1 co-immunoprecipitates with calmodulin from lysates of Chinese hamster ovary cells expressing IRS-1 (MUNSHI et al. 1996). In vitro the association of purified calmodulin and recombinant IRS-1 increases with the Ca^{2+} concentration; and in vivo treatment of cells with A23187 to increase cytosolic Ca^{2+} increases the association. In contrast, trifluoperazine, a cell-permeable calmodulin antagonist, decreases binding of calmodulin to IRS-1. Insulin-stimulated tyrosine phosphorylation of IRS-1 does not significantly alter the interaction between calmodulin and IRS-1. Calmodulin appears to bind to IQ motifs between residues 106–126 and 839–859 of IRS-1. Complete IQ motifs consist of approximately 23 residues with the consensus sequence IQXXXRGXXXR (CHENEY and MOOSEKER 1992). Synthetic peptides based on these sequences inhibit the association between IRS-1 and calmodulin (MUNSHI et al. 1996). These data suggest that calmodulin links IRS proteins to calcium-sensitive pathways.

Another tyrosine-phosphate independent mechanism for protein-protein interaction may involve 14-3-3 proteins, which are a group of small ubiquitously and highly conserved proteins that bind to a variety of signaling proteins (MORRISON 1994). 14-3-3 exists as a dimer which may mediate activation of associated kinases, including Raf-1 (MORRISON and CUTLER 1997). Several 14-3-3 isoforms bind IRS-1[3] (White et al., unpublished results). 14-3-3 was proposed to binding phosphoserine in RSxpSxP motifs (MUSLIN et al. 1996). Similar motifs occur in IRS-1 (Ser_{265} and Ser_{637}) and IRS-2 (Ser_{303} and Ser_{429}), but our results do not indicate a role for serine phosphorylation of IRS-1 in the binding of 14-3-3. Whatever the mechanism, 14-3-3 proteins may cause dimerization of IRS proteins, or provide a bridge between IRS proteins and heterologous signaling molecules such as Raf-1.

5 Inhibition of IRS Protein Signaling by Serine Phosphorylation

IRS-1 and IRS-2 contain over 30 potential serine/threonine phosphorylation sites in motifs recognized by various kinases, including casein kinase-2, MAP kinases, PKC and cdc2 (SUN et al. 1991, 1995). IRS proteins are heavily serine/threonine phosphorylated in the basal state and insulin stimulates additional serine phosphorylation (SUN et al. 1993). Although the relevant phosphorylation sites have been difficult to determine for most of these kinases, casein kinase-2 phosphorylates rat IRS-1 at Ser_{99} and Thr_{502}. The physiological effects of these phosphorylation events are unknown, especially as Ser_{99} is absent from human IRS-1 (TRANISIJEVIK et al. 1993).

Serine and threonine phosphorylation of IRS-1 may downregulate IRS-dependent signaling by inhibiting tyrosine phosphorylation during insulin stimulation (JULLIEN et al. 1993; KANETY et al. 1995; TANTI et al. 1994). In 3T3-L1 adipocytes, okadaic acid treatment generally increases serine phosphorylation of IRS-1, while decreasing tyrosine phosphorylation, PI 3-kinase activation and deoxyglucose uptake (TANTI et al. 1994). Serine to alanine point mutations of consensus MAP kinase phosphorylation sites in IRS-1 which overlap p85-binding-YMXM motifs increase both IRS-1 tyrosine phosphorylation and associated PI-3 kinase activity (MOTHE et al. 1996). These data suggest that serine phosphorylation sites may inhibit the insulin-stimulated tyrosine phosphorylation of the proximal YMXM motifs, reducing p85/IRS-1 interactions.

Although the physiological relevance of the modulation of serine/threonine phosphorylation has been difficult to investigate, recent work suggests that TNFα stimulates IRS-1 serine phosphorylation, which correlates closely with decreased IRS-1 tyrosine phosphorylation and insulin receptor tyrosine kinase activity. Since adipocytes secrete TNFα, a molecular explanation for insulin resistance in obesity and diabetes may involve this pathway (HOTAMISLIGIL and SPIEGELMAN 1994;

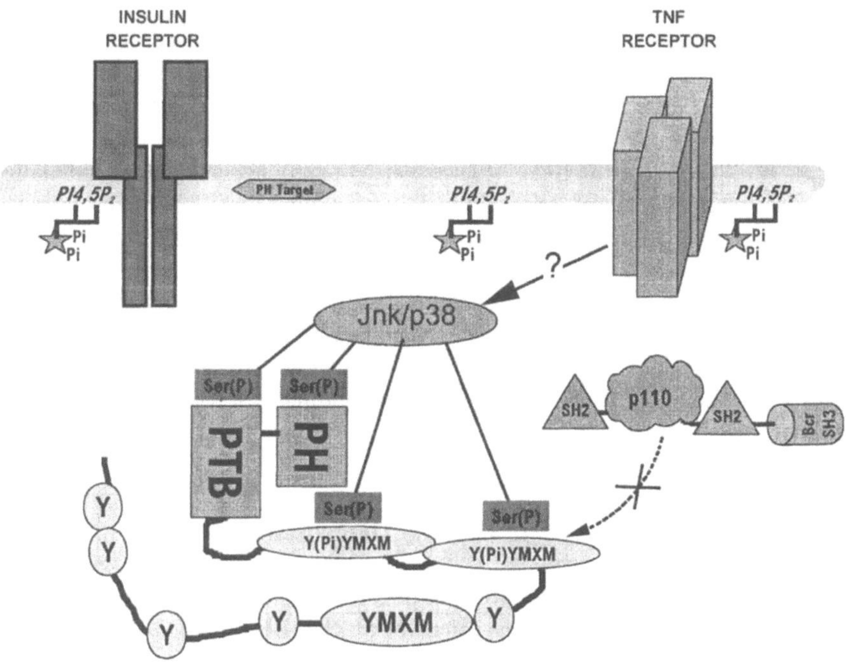

Fig. 5. Model of TNF-α-mediated downregulation of insulin action. TNF-α stimulates decreased insulin receptor kinase activity and increased serine and decreased tyrosine phosphorylation of IRS proteins. This downregulation may occur through p38 and/or Jnk-mediated serine phosphorylation of IRS proteins on motifs adjacent to p85-binding tyrosine phosphorylation sites, decreasing its efficiency of tyrosine phosphorylation and subsequent PI-3 kinase binding activity

HOTAMISLIGIL et al. 1995, 1996). Inhibition of the insulin receptor kinase by TNFα requires expression of IRS-1, but the mechanism for this effect is unclear (HOTAMISLIGIL et al. 1996). Identification of the serine phosphorylation sites that occur during TNFα stimulation will provide new information about the putative kinases involved. Jnk and p38 serine kinases are excellent candidates, as these are stimulated by TNFα (VERHEIJ et al. 1996; SAKLATVALA et al. 1996; HIRAI et al. 1996). Future experiments in physiological systems, including human samples, will be required to support these hypotheses (Figs. 4, 5).

6 Role of IRS Proteins in Mammalian Physiology

6.1 Introduction

IRS proteins regulate many biological processes, including the control of glucose metabolism, protein synthesis, and cell survival, growth and transformation. Although not all insulin signals are mediated by the IRS proteins, major physiological responses to insulin are probably absent without them. Mice lacking IRS-1 survive and display mild insulin resistance, suggesting that IRS-2 or other related proteins compensate adequately to maintain nearly normal carbohydrate metabolism. The importance of IRS-1 for carbohydrate metabolism is best demonstrated in mice double heterozygous for null alleles in the insulin receptor and IRS-1 (BRUNING et al. 1997). Whereas the IRS1$^{(-/-)}$ mice display nearly normal glucose metabolism, they are markedly smaller throughout life, suggesting that IRS-1 plays important roles during early development that are not rescued by other docking proteins.

Although best known for their role in insulin signaling, the IRS proteins may emerge as important points for treatment of cancer or other degenerative diseases. The IRS proteins play a central role in signaling by IGF-1, which strongly inhibits apoptosis in various cellular backgrounds (BASERGA 1996). IGF-1 is also essential for normal neuronal development, suggesting indirectly a role for IRS proteins in this process (DUDEK et al. 1997).

6.2 Role of IRS Proteins in Glucose Homeostasis

The movement of glucose across the plasma membrane is largely accomplished by a family of facilitated glucose transporters, of which GLUT1 and GLUT4 are largely responsible for removing glucose from the bloodstream (LIENHARD et al. 1992). GLUT1 accumulates constitutively at the plasma membrane of most cells and tissues, and is relatively insensitive to acute insulin treatment. Activation of the p21^{ras}/MAP kinase pathway increases expression of GLUT1, which stimulates glucose uptake (FINGAR and BIRNBAUM 1994). In skeletal muscle, IRS-1 mediates this process, whereas Shc plays a prominent role in cultured cells (MYERS et al. 1994; YAMAUCHI et al. 1996; OUWENS et al. 1994; SASAOKA et al. 1994).

The regulation of GLUT4 translocation to the plasma membrane is essential for the rapid effect of insulin on glucose uptake in adipocytes and muscle. It is sequestered in an intracellular vesicular compartment under basal conditions, and insulin stimulates its accumulation at the plasma membrane which stimulates glucose influx (LIENHARD et al. 1992). Considerable evidence now indicates that activation of PI-3 kinase during association with IRS proteins provides one of the essential signals for GLUT-4 translocation (SUN et al. 1991; OKADA et al. 1994; XU and SONNTAG 1996; RIDDERSTRALE and TORNQVIST 1994; OKADA et al. 1994; HARA et al. 1994; BACKER et al. 1992a). It may be the only signal required as expression of constitutively active p110 seems to circumvent the insulin requirement (TANTI et al. 1996). Moreover, expression of activate PKB also stimulates translocation of GLUT4 to the plasma membrane, suggesting that a serine phosphorylation step may be involved (KOHN et al. 1996). Cellular substrates of PKB appear to be good candidates to explore to extend the insulin signaling pathway to GLUT4 translocation. The exact role of IRS proteins in this pathway has not been formally demonstrated, and whether IRS-1/2 or pp60 will predominate is unknown.

The synthesis of glycogen in skeletal muscle is important for glucose disposal following a meal. Two kinase cascades are implicated as upstream mediators of glycogen synthase, including the MAP kinase cascade and the PI-3 kinase/PKB pathway (CROSS et al. 1995; DENT et al. 1990); in skeletal muscle, IRS-1 is the main regulator of both pathways during insulin stimulation (YAMAUCHI et al. 1996). The insensitivity of glycogen synthesis to the MEK inhibitor, and its inhibition by PI-3 kinase inhibitors (LY294002), suggests that PI-3 kinase/PKB plays a prominent role (CROSS et al. 1996).

6.3 Regulation of Protein Synthesis

General protein synthesis is one of the principal physiologic responses to insulin, particularly in the skeletal muscle. Insulin also mediates the synthesis of proteins specifically involved in growth and survival, including myc and fos (MENDEZ et al. 1996). The insulin receptor and IRS-1 are essential elements for both types of protein synthesis. Insulin stimulated protein synthesis in 32D cells requires ectopic expression of both the insulin receptor and IRS-1, suggesting that this insulin effect requires at least two pathways, one dependent on MAP kinase activation and the other on PI-3 kinase (MENDEZ et al. 1996).

General and growth-specific protein synthesis are controlled during insulin stimulation through two distinct branches. One branch, needed for myc translation for instance, is sensitive to rapamycin, suggesting that it involves mTOR and $p70^{s6k}$ (Fig. 6); insulin-stimulated phosphorylation of PHAS-I and eIF4E play an important role in this pathway, and PHAS-I may be a direct substrate of mTOR (J. Lawrence, University of Virginia, personal communication). General protein synthesis is unaffected by rapamycin, as it weakly inhibits this [^{35}S]methionine incorporation during insulin stimulation (MENDEZ et al. 1996). By contrast, insulin-stimulated general protein synthesis is sensitive to certain PKC inhibitors, and constitutively active PKCζ compensates for the absence of IRS-1 in 32D cells expressing the insulin receptor (MENDEZ et al. 1997). These results suggest that PKCζ, or a kinase with similar substrate range, is downstream of the PI-3 kinase during insulin stimulated protein synthesis (Fig. 6).

6.4 IRS Proteins and Diabetes

Mice lacking IRS-1 are mildly insulin-resistant, confirming that IRS proteins mediate insulin signals in the intact mammals. Without IRS-1, glucose levels are maintained at normal levels during fasting by elevated circulating insulin, and after a glucose challenge the serum glucose is reduced slowly even by an exaggerated insulin release (ARAKI et al. 1994; TAMEMOTO et al. 1994). However, the absence of IRS-1 does not cause NIDDM (BRUNING et al. 1997). Apparently, insulin resistance caused by the absence of IRS-1 is adequately compensated through IRS-2 and by elevated insulin production and secretion.

A few studies have investigated insulin responses in various tissues of the $IRS1^{(-/-)}$ mouse to determine the nature of the compensatory signaling. In liver,

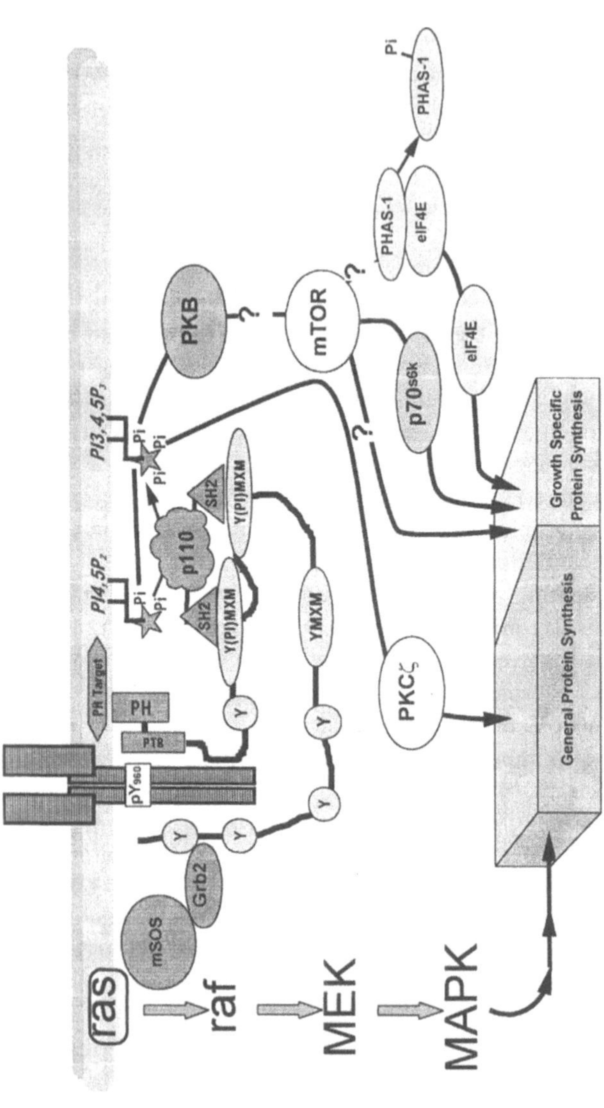

Fig. 6. Insulin-stimulated protein synthesis. Model of the molecular pathways leading to insulin-stimulated general and growth regulated protein synthesis is shown. This bioeffect of insulin involves at least two major pathways, one mediated by MAP kinase and one mediated by PI 3-kinase. Other possible downstream events are also depicted

insulin signaling is nearly normal even though IRS-2 expression is not elevated; however, IRS-2 tyrosine phosphorylation increases, which mediates a typical PI-3 kinase response (PATTI et al. 1995). Perhaps the increase in phosphorylation of IRS-2 occurs because competition from IRS-1 is absent; however, this compensation does not occur in skeletal muscle (YAMAUCHI et al. 1996). Muscle from the IRS1$^{(-/-)}$ mice retains only a 20% response to insulin, including PI-3 kinase, MAP kinase and p70^{s6k} glucose uptake, glycogen synthesis and protein synthesis, which reflects the low level of IRS-1 phosphorylation that occurs in this tissue (YAMAUCHI et al. 1996). The respectable control of glucose homeostasis must arise from a nearly normal inhibition of hepatic gluconeogenesis (although this has not been measured directly), and slow but reasonable glycogen production in muscle; although insulin weakly stimulates glucose uptake in skeletal muscle, once inside its conversion to glycogen is largely substrate driven.

Although in isolation, the heterozygous disruption of the insulin receptor is slightly more severe than a heterozygous disruption of IRS-1, these mice are generally euglycemic throughout their lives. At birth, the compound heterozygous mice ($IR^{\pm}/IRS^{\pm}$) are slightly more resistant to insulin than the individual heterozygotes as various parameters of insulin signaling, such as IRS-1 tyrosine phosphorylation and PI-3 kinase activity, are reduced as expected. Before the age of 4 months, insulin secretion adequately compensates to maintain normal glycemia. However, between 4 and 6 months of age about half of the mice develop diabetes owing mainly to severe insulin resistance, which is not overcome by high insulin levels (Bruning et al. 1997). At this point, the β cell mass and serum insulin levels increase in parallel to the insulin resistance. The molecular basis for this latter transition may eventually provide a better understanding of the pathophysiology of NIDDM. The compound heterozygous mouse model provides the best evidence that the enzyme:substrate relation between the insulin receptor and IRS-1 is important for carbohydrate metabolism.

Mutations in the insulin receptor and IRS-1 are rare in humans, and probably do not contribute significantly to the disease. In a few cases, polymorphisms in IRS-1 have been identified in human IRS-1, including Ser_{513}, Ala_{972}, and Arg_{1221} (Almind et al. 1996; Clausen et al. 1995). The Ala_{972} mutation occurs in 10.7% of NIDDM subjects from various ethnic backgrounds, but also at 5.8% in control subjects. Subsequent analysis of this mutation in 32D cells suggests that it partially reduces the ability of IRS-1 to activate PI-3 kinase (Almind et al. 1996). Although mutations may not be the main cause of NIDDM in humans, other mechanisms may result in a partial reduction of functional insulin receptor and IRS-1. Independent of the mechanism, NIDDM may be the outcome. If NIDDM is even partially due to reduced IRS protein expression, then it may be possible to identify drugs to enhance its expression or reduce its degradation to rescue a normal signaling capacity.

6.5 Cellular Survival, Growth and Cancer

Considerable evidence indicates that the IRS proteins provide a common intermediate for the regulation of metabolic and growth-related signals. Historically, IGF-1 has been thought to have an important effect on cell growth, whereas the effect of insulin is small. Consistent with this view, disruption of the IGF-1 receptor in mice causes severe developmental abnormalities (Baker et al. 1993; Liu et al. 1993). In rare cases when people survive without IGF-1 receptors, they are mentally impaired, demonstrating the importance of IGF-1 receptor function in neural development. However, insulin is also essential for embryonic development, since human offspring without functional insulin receptors are severely deformed and die shortly after birth (Accili et al. 1996). IGF-1 is usually studied for its growth promoting potential, as both a stimulator of mitogenesis and an inhibitor of apoptosis in various cellular backgrounds (PRISCO et al. 1997; BASERGA 1994). Since insulin also stimulates mitogenesis and inhibits apoptosis, the main distinction in vivo may lie with the tissue distribution of each receptor. However, the IGF-1 receptor may possess a greater potential for transformation than the insulin receptor, owing to unique interaction motifs in its COOH-terminus (SURMACZ et al. 1995). The exact nature of these IGF-1 receptor signals remains to be defined.

IGF-1 promotes cell survival and in some contexts this may be important for the development of cancer. The ability of insulin/IGF-1 to inhibit apoptosis may be essential. Induction of apoptosis in murine BALB/c3T3 cells with topoisomerase I was significantly inhibited by IGF-1 treatment and cell survival under these conditions is considerably enhanced by overexpression of the IGF-1 receptor (SELL et al. 1995). However, in 32D cells that undergo spontaneous apoptosis upon removal of IL-3, IGF-1 and insulin cannot restore cell viability without ectopic expression of an IRS protein, strongly implicating IRS proteins in this response (L. Yenush et al., unpublished observation; PRISCO et al. 1997).

IGF-1 mediates cellular transformation and IRS proteins appear to play an important role in this process. Murine 3T3-like fibroblasts lacking the IGF-1 receptor or expressing dominant negative IGF-1 receptors cannot be transformed by SV40 T antigen, which ordinarily transforms wild-type cells. However, the co-expression of IRS-1 and the SV40 T antigen induces transformation (D'AMBROSIO et al. 1995). In contrast, IRS-1 induces a mitogenic response to insulin in 3T3-cells lacking IGF-1 receptors, but it does not promote transformation, suggesting that additional IGF-1 receptor signals are required (SURMACZ et al. 1995). Thus, the transforming competence of the IGF-1 receptor requires an IRS-1-dependent signal and one or more pathways that can be substituted by SV40 T antigen.

6.6 Neuronal Differentiation

IGF-1 promotes neuronal development in vitro and in vivo (DUDEK et al. 1997; ZACKENFELS et al. 1995; TORRES-ALEMAN et al. 1994). Homozygous $IGF1^{(-/-)}$ mice display reduced brain size, hypomyelination and reduced number of hipocampal, granule and striatal neurons (BECK et al. 1995). Moreover, disruption of the IGF-1 receptor gene generally delays nervous system development (LIU et al. 1993). PI-3 kinase appears to play an important role, as PC12 cells and cerebellar neurons require it for neurite extension and inhibition of apoptosis (YAO and COOPER 1995; DUDEK et al. 1997). Although the 85-kDa regulatory subunits are expressed in developing neurons, $p55^{PIK}$ is highly expressed in neurite extensions where it binds to IRS-1 or IRS-2 during IGF-1/insulin stimulation (PONS et al., submitted). The expression of $p55^{PIK}$ remains very high during development of the nervous system and increases during differentiation of P19 cells into a neuronal phenotype. In both primary and P19 neuronal cultures $p55^{PIK}$ largely associates with IRS proteins and stimulates PI-3 kinase activity during stimulation with IGF-1. Using a combination of approaches, we propose that the regulation of PI-3 kinase by $p55^{PIK}$ is essential for some of the actions of IGF-1 on differentiation.

7 Future Perspectives

The expanding role of docking proteins raises new questions about the relation between insulin signaling and the action of other growth factors and cytokines. Clearly, IRS proteins represent an important multifunctional interface between many receptors and intracellular signaling pathways. One direction for the future is to complete the identification of more IRS protein family members, as well as members of other docking protein families. However, the most important direction will require a return to physiology and the analysis of mouse model systems lacking these proteins or expressing mutant proteins to resolve the contribution of the various pathways for normal development and resistance to disease.

Acknowledgements. This work was supported by NIH grants DK38712, DK43808 and DK48712 to M.F.W. L.Y. is supported by NIH grant DK07260.

References

Accili D, Drago J, Lee EJ, Johnson MD, Cool MH, Salvatore P, Asico LD, Jose PA, Taylor SI, Westphal H (1996) Early neonatal death in mice homozygous for a null alele of the insulin receptor gene. Nat Genet 12:106–109

Alessi DR, Andjelkovic M, Caudwell B, Cron P, Morrice N, Cohen P, Hemmings BA (1996) Mechanism of activation of protein kinase B by insulin and IGF-1. EMBO J 15:6541–6551

Alexandropoulos K, Baltimore D (1996) Coordinate activation of c-Src by SH3- and SH2-binding sites on a novel p130Cas-related protein, Sin. Genes Dev Genes Dev 10:1341–1355

Almind K, Inoue G, Pedersen O, Kahn CR (1996) A common amino acid polymorphism in insulin receptor substrate-1 causes impaired insulin signaling. Evidence from transfection studies. J Clin Invest 97:2569–2575

Ando A, Yonezawa K, Gout I, Nakata T, Ueda H, Hara K, Kitamura Y, Noda Y, Takenawa T, Hirokawa N, Waterfield MD, Kasuga M (1994) A complex of Grb2 dynamin binds to tyrosine-phosphorylated insulin receptor substrate-1 after insulin treatment. EMBO J 13:3033–3038

Antonetti DA, Algenstaedt P, Kahn CR (1996) Insulin receptor substrate 1 binds two novel splice variants of the regulatory subunit of phosphatidylinositol 3-kinase in muscle and brain. Mol Cell Biol 16:2195–2203

Araki E, Haag BL III, Kahn CR (1994) Cloning of the mouse insulin receptor substrate-1 (IRS-1) gene and complete sequence of mouse IRS-1. Biochim Biophys Acta 1221:353–356

Araki E, Lipes MA, Patti ME, Bruning JC, Haag BL III, Johnson RS, Kahn CR (1994) Alternative pathway of insulin signalling in mice with targeted disruption of the IRS-1 gene. Nature 372:186–190

Argetsinger LS, Norstedt G, Billestrup N, White MF, Carter-Su C (1996) Growth hormone interferon-gamma, and leukemia inhibitory factor utilize insulin receptor substrate-2 in intracellular signaling. J Biol Chem 271:29415–29421

Arrandale JM, Gore-Willse A, Rocks S, Ren JM, Zhu J, Davis A, Livingston JN, Rabin DU (1996) Insulin signaling in mice expressing reduced levels of Syp. J Biol Chem 271:21353–21358

Backer JM, Myers MG Jr, Shoelson SE, Chin DJ, Sun XJ, Miralpeix M, Hu P, Margolis B, Skolnik EY, Schlessinger J, White MF (1992a) Phosphatidylinositol 3′-kinase is activated by association with IRS-1 during insulin stimulation. EMBO J 11:3469–3479

Backer JM, Schroeder GG, Kahn CR, Myers MG Jr, Wilden PA, Cahill DA, White MF (1992b) Insulin stimulation of phosphatidylinositol 3-kinase activity maps to insulin receptor regions required for endogenous substrate phosphorylation. J Biol Chem 267:1367–1374

Backer JM, Shoelson SE, Weiss MA, Hua QX, Cheatham RB, Haring E, Cahill DC, White MF (1992c) The insulin receptor juxtamembrane region contains two independent tyrosine/beta-turn internalization signals. J Cell Biol 118:831–839

Baker J, Liu JP, Robertson EJ, Efstratiadis A (1993) Role of insulin-like growth factors in embryonic and postnatal growth. Cell 75:73–82
Bandyopadhyay G, Standaert ML, Zhao LM, Yu B, Avignon A, Galloway L, Karnam P, Moscat J, Farese RV (1997) Activation of protein kinase C (a, b and zeta) by insulin in 3T3/L1 cells. J Biol Chem 272:2551–2558
Baserga R (1994) Oncogenes and the strategy of growth factors. Cell 79:927–930
Baserga R (1996) Controlling IGF-receptor function: a possible strategy for tumor therapy. Tibtech 14:150–152
Beck KD, Powell-Braxton L, Widmer HR, Valverde J, Hefti F (1995) IGF-1 gene disruption results in reduced brain size, CNS hypomyelination, and loss of hippocampal granule and striatal parvalbumin-containing neurons. Neuron 14:717–730
Beitner-Johnson D, Blakesley VA, Shen-Orr Z, Jimenez M, Stannard B, Wang LM, Pierce JH, LeRoith D (1996) The proto-oncogene product c-Crk associates with insulin receptor substrate-1 and 4PS. J Biol Chem 271:9287–9292
Berlanga JJ, Gualillo O, Buteau H, Applanat M, Kelly PA, Edery M (1997) Prolactin activates tyrosyl phosphorylation of insulin receptor substrate 1 and phosphatidylinositol-3-OH kinase. J Biol Chem 272:2050–2052
Bruning JC, Winnay J, Bonner-Weir S, Taylor SI, Accili D, Kahn CR (1997a) Development of a novel polygenic model of NIDDM in mice heterozygous for IR and IRS-1 null alleles. Cell 88:561–572
Bruning JC, Winnay J, Cheatham B, Kahn CR (1997b) Differential signaling by IRS-1 and IRS-2 in IRS-1 deficient cells. Mol Cell Biol (in press)
Carpino N, Wisniewski D, Strife A, Marshak D, Kobayashi R, Stillman B, Clarkson B (1997) p62dok: a constitutively tyrosine-phosphorylated, GAP-associated protein in chronic myelogenous leukemia progenitor cells. Cell 88:197–204
Chen XH, Patel BKR, Wang LM, Frankel M, Ellmore N, Flavell RA, Larochelle WJ, Pierce JH (1997) Jak1 expression is required for mediating interleukin-4-induced tyrosine phosphorylation of insulin receptor substrate and stat6 signaling molecules. J Biol Chem 272:6556–6560
Cheney RE, Mooseker MS (1992) Unconventional myosins. Curr Opin Cell Biol 4:27–35
Clausen JO, Hansen T, Bjorbaek C, Echwald SM, Urhammer SA, Rasmussen S, Andersen CB, Hansen L, Almind K, Winther K, Haraldsdottir J, Borch-Johnsen K, Pedersen O (1995) Insulin resistance: interactions between obesity and a common variant of insulin receptor substrate-1. Lancet 346:397–402
Craparo A, Freund R, Gustafson TA (1997) 14-3-3 epsilon interacts with the insulin-like growth factor I receptor and insulin receptor substrate I in phosphotyrosine-independent a manner. J Biol Chem 272:11663–11670
Craparo A, O'Neill TJ, Gustafson TA (1995) Non-SH2 domains within the insulin receptor substrate-1 and SHC mediate their phosphotyrosine-dependent interaction with the NPEY motif of the insulin-like growth factor 1 receptor. J Biol Chem 270:15639–15643
Cross DAE, Alessi DR, Cohen P, Andjelkovich M, Hemmings BA (1996) Inhibition of glycogen synthase kinase-3 by insulin mediated protein kinase B. Nature 378:785–787
Cross DAE, Alessi DR, Cohen P, Andjelkovich M, Hemmings BA (1995) Inhibition of glycogen synthase kinase-3 by insulin mediated by protein kinase B. Nature 378:785–789
D'Ambrosio C, Keller SR, Morrione A, Lienhard GE, Baserga R, Surmacz E (1995) Transforming potential of the insulin receptor substrate 1. Cell Growth Differ 6:557–562
Dent P, Lavoinne A, Nakielny S, Caudwell FB, Watt P, Cohen P (1990) The molecular mechanisms by which insulin stimulates glycogen synthesis in mammalian skeletal muscle. Nature 348:302–307
Dhand R, Hara K, Hiles I, Bax B, Gout I, Panayotou G, Fry MJ, Yonezawa K, Kasuga M, Waterfield MD (1994) PI 3-kinase: structural and functional analysis of intersubunit interactions. EMBO J 13:511–521
Diaz-Meco MT, Lozano J, Municio MM, Berra E, Frutos S, Sanz L, Moscat J (1994) Evidence for the in vitro and in vivo interaction of Ras with protein kinase C zeta. J Biol Chem 269:31706–31710
Dudek H, Datta SR, Franke TF, Birnbaum MJ, Yao R, Cooper GM, Segal RA, Kaplan DR, Greenberg ME (1997) Regulation of neuronal survival by the serine-threonine protein kinase Akt. Science 275:661–665
Eck MJ, Dhe-Paganon S, Trub T, Nolte RT, Shoelson SE (1996a) Structure of the IRS-1 PTB domain bound to the juxtamembrane region of the insulin receptor. Cell 85:695–705
Eck MJ, Pluskey S, Trub T, Harrison SC, Shoelson SE (1996b) Spatial constraints on the recognition of phosphoproteins by the tandem SH2 domains of the phosphatase SH-PTP2. Nature 379:277–280

Ellis CA, Moran M, McCormick F, Pawson T (1990) Phosphorylation of GAP and GAP-associated proteins by transforming and mitogenic tyrosine kinases. Nature 343:377–381
Fei ZL, D'Ambrosio C, Li S, Surmacz E, Baserga R (1995) Association of insulin receptor substrate 1 with simian virus 40 large T antigen. Mol Cell Biol 15:4232–4239
Feng GS, Hui C-C, Pawson T (1993) SH-2 containing phosphotyrosine phosphatase as a target of protein-tyrosine kinase. Science 259:1607–1614
Fingar DC, Birnbaum MJ (1994) A role for raf-1 in the divergent signaling pathways mediating insulin-stimulated glucose transport. J Biol Chem 269:10127–10132
Franke TF, Yang S, Chan TO, Datta K, Kazlauskas A, Morrison DK, Kaplan DR, Tsichlis PN (1995) The protein kinase encoded by the Akt proto-oncogene is a target of the PDGF-activated phosphatidylinositol 3-kinase. Cell 81:727–736
Freeman RM Jr, Plutzky J, Neel BG (1992) Identification of a human src homology 2-containing protein-tyrosine-phosphatase: a putative homologue of Drosophila corkscrew. Proc Natl Acad Sci USA 89:11239–11243
Fruman DA, Cantley LC, Carpenter CL (1996) Structural organization and alternative splicing of the murine phosphoinositide 3-kinase p85a gene. Genomics 37:113–121
Gale NW, Kaplan S, Lowenstein EJ, Schlessinger J, Bar-Sagi D (1993) Grb2 mediates the EGF-dependent activation of guanine nucleotide exchange on Ras. Nature 363:88–92
Gibson TJ, Hyvonen M, Musacchio A, Saraste M (1994) PH domain: the first anniversary. TIBS 19:349–353
Gustafson TA, He W, Craparo A, Schaub CD, O'Neill TJ (1995) Phosphotyrosine-dependent interaction of Shc and IRS-1 with the NPEY motif of the insulin receptor via a novel non-SH2 domain. Mol Cell Biol 15:2500–2508
Hara K, Yonezawa K, Sakaue H, Ando A, Kotani K, Kitamura T, Kitamura Y, Ueda H, Stephens L, Jackson TR, Hawkins PT, Dhand R, Clark AE, Holman GD, Waterfield MD, Kasuga M (1994) 1-Phosphatidylinositol 3-kinase activity is required for insulin-stimulated glucose transport but not for ras activation in CHO cells. Proc Natl Acad Sci USA 91:7415–7419
Harrison SC (1996) Peptide-surface association: the case of PDZ and PTB domains. Cell 86:341–343
Haslam RJ, Koide HB, Hemmings BA (1993) Pleckstrin domain homology. Nature 363:309–310
He W, Craparo A, Zhu Y, O'Neill TJ, Wang LM, Pierce JH, Gustafson TA (1996) Interaction of insulin receptor substrate-2 (IRS-2) with the insulin and insulin-like growth factor I receptors. J Biol Chem 271:11641–11645
Heldin CH (1995) Dimerization of cell surface receptors in signal transduction. Cell 80:213–223
Hemmings BA (1997) Akt signaling Linked membrane events to life and death decisions. Science 275:628–630
Herbst R, Carroll PM, Allard JD, Schilling J, Raabe T, Simon MA (1996) Daughter of sevenless is a substrate of the phosphotyrosine phosphatase corkscrew and functions during sevenless signaling. Cell 85:899–909
Hirai S, Izawa M, Osada S, Spyrou G, Ohno S (1996) Activation of the JNK pathway by distantly related protein kinases, MEKK and MUK. Oncogene 12:641–650
Holgado-Madruga M, Emlet DR, Moscatello DK, Godwin AK, Wong AJ (1996) A Grb2-associated docking protein in EGF- and insulin-receptor signalling. Nature 379:560–563
Hosomi Y, Shii K, Ogawa W, Matsuba H, Yoshida M, Okada Y, Yokono K, Kasuga M, Baba S, Roth R (1994) Characterization of a 60-Kilodalton substrate of the insulin receptor kinase. J Biol Chem 269:11498–11502
Hotamisligil GS, Spiegelman BM (1994) Tumor necrosis factor alpha: a key component of the obesity-diabetes link. Diabetes 43:1271–1278
Hotamisligil GS, Arner P, Caro JF, Atkinson RL, Spiegelman BM (1995) Increased adipose tissue expression of tumor necrosis factor-a in human obesity and insulin resistance. J Clin Invest 95:2409–2415
Hotamisligil GS, Peraldi P, Budvari A, Ellis RW, White MF, Spiegelman BM (1996) IRS-1 mediated inhibition of insulin receptor tyrosine kinase activity in TNF-α- and obesity-induced insulin resistance. Science 271:665–668
Hunter T (1997) Oncoprotein networks. Cell 88:333–346
Inukai K, Anai M, van Breda E, Hosaka T, Katagiri H, Funaki M, Fukushima Y, Ogihara T, Yazaki Y, Kikuchi M, Oka Y, Asano T (1996) A novel 55-kDa regulatory subunit for phosphatidylinositol 3-kinase structurally similar to $p55^{PIK}$ is generated by alternative splicing of the p85a gene. J Biol Chem 271:5317–5320

Isakoff SJ, Yu YP, Su YC, Blaikie P, Yajnik V, Rose E, Weidner KM, Sachs M, Margolis B, Skolnik EY (1996) Interaction between the phosphotyrosine binding domain of Shc and the insulin receptor is required for Shc phosphorylation by insulin in vivo. J Biol Chem 271:3959–3962

Johnson J, Wang LM, Hanson EP, Sun XJ, White MF, Oakes SA, Pierce JH, O'Shea J (1995) Interleukins 2, 4, 7, and 15 stimulate tyrosine phosphorylation of insulin receptor substrates 1 and 2 in T cells. J Biol Chem 270:1–4

Jullien D, Tanti JF, Heydrick SJ, Gautier N, Gremeaux T, Van Obberghen E, Le Marchand-Brustel Y (1993) Differential effects of okadaic acid on insulin-stimulated glucose and amino acid uptake and phosphatidylinositol 3-kinase activity. J Biol Chem 268:15246–15251

Kanety H, Feinstein R, Papa MZ, Hemi R, Karasik A (1995) Tumor necrosis factor α-induced phosphorylation of insulin receptor substrate-1 (IRS-1). J Biol Chem 270:23780–23784

Kaplan DR, Morrison DK, Wong G, McCormick F, Williams LT (1990) PDGF beta-receptor stimulates tyrosine phosphorylation of GAP and association of GAP with a signaling complex. Cell 61:125–133

Kavanaugh WM, Turck CW, Williams LT (1995) PTB domain binding to signaling proteins through a sequence motif containing phosphotyrosine. Science 268:1177–1179

Keegan AD, Nelms K, White M, Wang LM, Pierce JH, Paul WE (1994) An IL-4 receptor region containing an insulin receptor motif is important for IL-4-mediated IRS-1 phosphorylation and cell growth. Cell 76:811–820

Kharitonenkov A, Schnekenburger J, Chen Z, Knyazev P, Ali S, Zwick E, White MF, Ullrich A (1995) Adapter function of PTP1D in insulin receptor/IRS-1 interaction. J Biol Chem 270:29189–29193

Kimura K, Hattori S, Kabuyama Y, Shizawa Y, Takayanagi J, Nakamura S, Toki S, Matsuda Y, Onodera K, Fukui Y (1994) Neurite outgrowth of PC12 cells is suppressed by wortmannin, a specific inhibitor of phosphatidylinositol 3-kinase. J Biol Chem 269:18961–18967

Kohn AD, Summers SA, Birnbaum MJ, Roth RA (1996) Expression of a constitutively active Akt Ser/Thr kinase in 3T3-L1 adipocytes stimulates glucose uptake and glucose transporter 4 translocation. J Biol Chem 271:31372–31378

Kuhne MR, Pawson T, Lienhard GE, Feng GS (1993) The insulin receptor substrate 1 associates with the SH2-containing phosphotyrosine phosphatase Syp. J Biol Chem 268:11479–11481

Kundra V, Escobedo JA, Kazlauskas A, Kim HK, Rhee SG, Williams LT, Zetter BR (1994) Regulation of chemotaxis by the platelet-derived growth factor receptor-beta. Nature 367:474–476

Lavan BE, Lane WS, Lienhard GE (1997) The 60-kDa phosphotyrosine protein in insulin-treated adipocytes is a new member of the insulin receptor substrate family. J Biol Chem 272:11439–11443

Lechleider RJ, Freeman RM Jr, Neel BG (1993a) Tyrosyl phosphorylation and growth factor receptor association of the human corkscrew homologue, SHPTP2. J Biol Chem 268:13434–13438

Lechleider RJ, Sugimoto S, Bennett AM, Kashishian AS, Cooper JA, Shoelson SE, Walsh CT, Neel BG (1993b) Activation of the SH2-containing phosphotyrosine phosphatase SH-PTP2 by its binding site, phosphotyrosine 1009, on the human platelet-derived growth factor receptor B. J Biol Chem 268:21478–21481

Lee CH, Li W, Nishimura R, Zhou M, Batzer AG, Myers MG Jr, White MF, Schlessinger J, Skolnik EY (1993) Nck associates with the SH2 domain docking proteins IRS-1 in insulin stimulated cells. Proc Natl Acad Sci USA 90:11713–11717

Lemmon MA, Ferguson KM, Schlessinger J (1996) PH domains: diverse sequences with a common fold recruit signaling molecules to the cell surface. Cell 85:621–624

Lienhard GE, Slot JW, James DE, Mueckler MM (1992) How cells absorb glucose. Sci Am 266:86–91

Liu JP, Baker J, Perkins JA, Robertson EJ, Efstratiadis A (1993) Mice carrying null mutations of the genes encoding insulin-like growth factor I (Igf-1) and type 1 IGF receptor (Igf1r). Cell 75:59–72

Luttrell LM, Hawes BE, Touhara K, van Biesen T, Kock WJ, Lefkowitz RJ (1995) Effect of cellular expression of pleckstrin homology domains on Gi-coupled receptor signaling. J Biol Chem 270:12984–12989

Mendez R, Myers MG Jr, White MF, Rhoads RE (1996) Stimulation of protein synthesis, eukaryotic translation initiation factor 4E phosphorylation, and PHAS-1 phosphorylation by insulin requires insulin receptor substrate-1 and phosphotidylinositol-3-kinase. Mol Cell Biol 16:2857–2864

Mendez R, White MF, Myers MG, Rhodes RE (1997) Requirement of protein kinase C zeta for stimulation of protein synthesis by insulin. Mol Cell Biol (in press)

Miralpeix M, Sun XJ, Backer JM, Myers MG, Jr., Araki E, White MF (1992) Insulin stimulates tyrosine phosphorylation of multiple high molecular weight substrates in FAO hepatoma cells. Biochemistry 31:9031–9039

Momomura K, Tobe K, Seyama Y, Takaku F, Kasuga M (1988) Insulin-induced tyrosine phosphorylation in intact rat adipocytes. Biochem Biophys Res Commun 155:1181–1186

Morrison D (1994) 14-3-3: modulators of signaling proteins? (Comment) Science 266:56–57

Morrison DK, Cutler RE (1997) The complexity of Raf-1 regulation. Curr Opin Cell Biol 9:174–179

Mothe I, Van Obberghen E (1996) Phosphorylation of insulin receptor substrate-1 on multiple serine residues, 612, 662, and 731, modulates insulin action. J Biol Chem 271:11222–11227

Munshi HG, Burks DJ, Joyal JL, White MF, Sacks DB (1996) Ca^{2+} regulates calmodulin binding to IQ motifs in IRS-1. Biochemistry 35:15883–15889

Muslin AJ, Tanner JW, Allen PM, Shaw AS (1996) Interaction of 14-3-3 with signaling proteins is mediated by the recognition of phosphoserine. Cell 84:889–897

Myers MG Jr, White MF (1995) New frontiers in insulin receptor substrate signaling. Trends Endocrinol Metab 6:209–215

Myers MG Jr, Wang LM, Sun XJ, Zhang Y, Yenush L, Schlessinger J, Pierce JH, White MF (1994) The role of IRS-1/GRB2 complexes in insulin signaling. Mol Cell Biol 14:3577–3587

Myers MG Jr, Grammer TC, Brooks J, Glasheen EM, Wang LM, Sun XJ, Blenis J, Pierce JH, White MF (1995) The pleckstrin homology domain in IRS-1 sensitizes insulin signaling. J Biol Chem 270:11715–11718

Myers MG Jr, Zhang Y, Aldaz GAI, Grammer TC, Glasheen EM, Yenush L, Wang LM, Sun XJ, Blenis J, Pierce JH, White MF (1996) YMXM motifs and signaling by an insulin receptor substrate 1 molecule without tyrosine phosphorylation sites. Mol Cell Biol 16:4147–4155

Noguchi T, Matozaki T, Horita K, Fujioka Y, Kasuga M (1994) Role of SH-PTP2, a protein-tyrosine phosphatase with Src homology 2 domains, in insulin-stimulated ras activation. Mol Cell Biol 14:6674–6682

O'Neill TJ, Craparo A, Gustafson TA (1994) Characterization of an interaction between insulin receptor substrate-1 and the insulin receptor by using the two-hybrid system. Mol Cell Biol 14:6433–6442

Ogawa W, Hosomi Y, Shii K, Roth RA (1994) Evidence for two distinct 60-kilodalton substrates of the SRC tyrosine kinase. J Biol Chem 269:29602–29608

Okada T, Sakuma L, Fukui Y, Hazeki O, Ui M (1994a) Blockage of chemotactic peptide-induced stimulation of neutrophils by wortmannin as a selective inhibitor of phosphatidylinositol 3-kinase. J Biochem 269:3563–3567

Okada T, Kawano Y, Sakakibara T, Hazeki O, Ui M (1994b) Essential role of phosphatidylinositol 3-kinase in insulin-induced glucose transport and antilipolysis in rat adipocytes. J Biochem 269:3568–3573

Ouwens DM, Van der Zon GC, Pronk GJ, Bos JL, Moller W, Cheatham B, Kahn CR, Maassen JA (1994) A mutant insulin receptor induces formation of a Shc-growth factor receptor bound protein 2 (Grb2) complex and $p21^{ras}$-GTP without detectable interaction of insulin receptor substrate 1 (IRS1) with Grb2. Evidence for IRS1-independent $p21^{ras}$-GTP formation. J Biol Chem 269:33116–33122

Patti ME, Sun XJ, Bruning JC, Araki E, Lipes MA, White MF, Kahn CR (1995) IRS-2/4PS is an alternative substrate of the insulin receptor in IRS-1 deficient transgenic mice (abstract). Diabetes 44:31A

Pawson T (1995) Protein modules and signalling networks. Nature 373:573–580

Pellici G, Lanfrancone L, Grignani F, McGlade J, Cavallo F, Forni G, Nicoletti I, Pawson T, Pelicci PG (1992) A novel transforming protein (SHC) with an SH2 domain is implicated in mitogenic signal transduction. Cell 70:93–104

Perkins LA, Larsen I, Perrimon N (1992) Corkscrew encodes a putative protein tyrosine phosphatase that functions to transduce the terminal signal from the receptor tyrosine kinase torso. Cell 12:225–236

Platanias LC, Uddin S, Yetter A, Sun XJ, White MF (1996) The type I interferon receptor mediates tyrosine phosphorylation of insulin receptor substrate 2. J Biol Chem 271:278–282

Pons S, Asano T, Glasheen EM, Miralpeix M, Zhang Y, Fisher TL, Myers MG Jr, Sun XJ, White MF (1995) The structure and function of $p55^{PIK}$ reveals a new regulatory subunit for the phosphatidylinositol-3 kinase. Mol Cell Biol 15:4453–4465

Pons S, Bernal MD, White MF (1997) (submitted)

Prisco M, Hongo A, Rizzo MG, Sacchi A, Baserga R (1997) The insulin-like growth factor-1 receptor as a physiologically relevant target of p53 in apoptosis caused by interleukin-3 withdrawal. Mol Cell Biol 17:1084–1092

Raabe T, Riesgo-Escovar J, Liu X, Bausenwein BS, Deak P, Maroy P, Hafen E (1996) DOS, a novel pleckstrin homology domain-containing protein required for signal transduction between sevenless and ras1 in Drosophila. Cell 85:911–20

Ridderstrale M, Tornqvist HE (1994) PI-3-kinase inhibitor wortmannin blocks the insulin-like effects of growth hormone in isolated rat adipocytes. Biochem Biophys Res Commun 203:306–10

Rordorf-Nikolic T, Van Horn DJ, Chen D, White MF, Backer JM (1995) Regulation of phosphatidylinositol 3-kinase by tyrosyl phosphoproteins. Full activation requires occupancy of both SH2 domains in the 85 kDa regulatory subunit. J Biol Chem 270:3662–3666

Roth RA, Zhang B, Chin JE, Kovacina K (1992) Substrates and signalling complexes: the tortured path to insulin action. J Cell Biochem 48:12–18

Rothenberg PL, Lane WS, Karasik A, Backer JM, White M, Kahn CR (1991) Purification and partial sequence analysis of pp185, the major cellular substrate of the insulin receptor tyrosine kinase. J Biol Chem 266:8302–8311

Ryan JJ, McReynolds LJ, Keegan A, Wang LH, Garfein E, Rothman P, Nelms K, Paul WE (1996) Growth and gene expression are predominantly controlled by distinct regions of the human IL-4 receptor. Immunity 4:123–132

Sakai R, Iwamatsu A, Hirano N, Ogawa S, Tanaka T, Mano H, Yazaki Y, Hirai H (1994) A novel signaling molecule, p130, forms stable complexes in vivo with v-Crk and v-Src in a tyrosine phosphorylation-dependent manner. EMBO J 13:3748–3756

Saklatvala J, Davis W, Guesdon F (1996) Interleukin 1 (IL1) and tumour necrosis factor (TNF) signal transduction. Philos Trans R Soc Lond Biol Sci 351:151–157

Sasaoka T, Draznin B, Leitner JW, Langlois WJ, Olefsky JM (1994a) Shc is the predominant signaling molecule coupling insulin receptors to activation of guanine nucleotide releasing factor and p21ras formation. J Biol Chem 269:10734–10738

Sasaoka T, Rose DW, Jhun BH, Saltiel AR, Draznin B, Olefsky JM (1994b) Evidence for a functional role of Shc proteins in mitogenic signaling induced by insulin, insulin-like growth factor-1, and epidermal growth factor. J Biol Chem 269:13689–13694

Sawka-Verhelle D, Tartare-Deckert S, White MF, Van Obberghen E (1996) IRS-2 binds to the insulin receptor through its PTB domain and through a newly identified domain comprising amino acids 591 to 786. J Biol Chem 271:5980–5983

Sawka-Verhelle D, Baron V, Mothe I, White MF, van Obberghen E (1997) (submitted)

Schlessinger J (1988) Signal transduction by allosteric receptor oligomerization. TIBS 13:443–447

Schlessinger J (1993) How receptor tyrosine kinases activate Ras. Trends Biochem Sci 18:273–275

Sell C, Baserga R, Rubin R (1995) Insulin-like growth factor I (IGF-I) and the IGF-I receptor prevent etoposide-induced apoptosis. Cancer Res 55:303–306

Shpetner HS, Herskovits JS, Vallee RB (1996) A binding site for SH3 domains targets dynamin to coated pits. J Biol Chem 271:13–16

Skolnik EY, Lee CH, Batzer AG, Vicentini LM, Zhou M, Daly RJ, Myers MG Jr, Backer JM, Ullrich A, White MF, Schlessinger J (1993a) The SH2/SH3 domain-containing protein GRB2 interacts with tyrosine-phosphorylated IRS-1 and Shc: implications for insulin control of ras signalling. EMBO J 12:1929–1936

Skolnik EY, Batzer AG, Li N, Lee CH, Lowenstein EJ, Mohammadi M, Margolis B, Schlessinger J (1993b) The function of GRB2 in linking the insulin receptor to ras signaling pathways. Science 260:1953–1955

Smith-Hall J, Pons S, Patti ME, Burks DJ, Yenush L, Sun XJ, Kahn CR, White MF (1997) The 60-kDa insulin receptor substrate functions like an IRS protein ($pp60^{IRS3}$) in adipocytes. Biochemistry (in press)

Sun XJ, Rothenberg PL, Kahn CR, Backer JM, Araki E, Wilden PA, Cahill DA, Goldstein BJ, White MF (1991) The structure of the insulin receptor substrate IRS-1 defines a unique signal transduction protein. Nature 352:73–77

Sun XJ, Crimmins DL, Myers MG Jr, Miralpeix M, White MF (1993) Pleiotropic insulin signals are engaged by multisite phosphorylation of IRS-1. Mol Cell Biol 13:7418–7428

Sun XJ, Wang LM, Zhang Y, Yenush L, Myers MG Jr, Glasheen EM, Lane WS, Pierce JH, White MF (1995) Role of IRS-2 in insulin and cytokine signalling. Nature 377:173–177

Sun XJ, Pons S, Asano T, Myers MG Jr, Glasheen EM, White MF (1996) The fyn tyrosine kinase binds IRS-1 and forms a distinct signaling complex during insulin stimulation. J Biol Chem 271:10583–10587

Sun XJ, Pons S, Wang LM, Zhang Y, Yenush L, Burks D, Myers MG Jr, Glasheen E, Copeland NG, Jenkins NA, Pierce JH, White MF (1997) The IRS-2 gene on murine chromosome 8 encodes a unique signaling adapter for insulin and cytokine action. Mol Endocrinol 11:251–262

Surmacz E, Sell C, Swantek J, Kato H, Roberts CT Jr, LeRoith D, Baserga R (1995) Dissociation of mitogenesis and transforming activity by C-terminal truncation of the insulin-like growth factor-I receptor. Exp Cell Res 218:370–380

Tamemoto H, Kadowaki T, Tobe K, Yagi T, Sakura H, Hayakawa T, Terauchi Y, Ueki K, Kaburagi Y, Satoh S, Sekihara H, Yoshioka S, Horikoshi H, Furuta Y, Ikawa Y, Kasuga M, Yazaki Y, Aizawa S (1994) Insulin resistance and growth retardation in mice lacking insulin receptor substrate-1. Nature 372:182–186

Tauchi T, Feng GS, Marshall MS, Shen R, Mantel C, Pawson T, Broxmeyer HE (1994) The ubiquitously expressed Syp phosphatase interacts with c-kit and Grb2 in hematopoietic cells. J Biol Chem 269:25206–25211

Tanti JF, Gremeaux T, Van Obberghen E, Le Marchand-Brustel Y (1994) Serine/threonine phosphorylation of insulin receptor substrate 1 modulates insulin receptor signaling. J Biol Chem 269:6051–6057

Tanti JF, Gremeaux T, Grillo S, Calleja V, Klippel A, Williams LT, Van Obberghen E, Le Marchand-Brustel Y (1996) Overexpression of a consititutively active form of phosphatidylinositol 3-kinase is sufficient to promote Glut 4 translocation adipocytes. J Biol Chem 271:25227–25232

Thies RS, Molina JM, Ciaraldi TP, Freidenberg GR, Olefsky JM (1990) Insulin-receptor autophosphorylation and endogenous substrate phosphorylation in human adipocytes from control, obese, and NIDDM subjects. Diabetes 39:250–259

Tobe K, Tamemoto H, Yamauchi T, Aizawa S, Yazaki Y, Kadowaki T (1995) Identification of a 190 kDa protein as a novel substrate for the insulin receptor kinase functionally similar to insulin receptor substrate-1. J Biol Chem 270:5698–5701

Torres-Aleman I, Pons S, Arevalo MA (1994) The insulin-like growth factor I system in the rat cerebellum: developmental regulation and role in neuronal survival and differentiation. J Neurosci Res 39:117–126

Tranisijevik M, Myers MG Jr, Thomas RS, Crimmons D, White MF, Sacks D (1993) Phosphorylation of the insulin receptor substrate IRS-1 by casein kinase II. J Biol Chem 268:18157–18166

Uddin S, Yenush L, Sun XJ, Sweet ME, White MF, Platanias LC (1995) Interferon-alpha engages IRS-signaling proteins to regulate the phosphatidylinositol 3-kinase. J Biol Chem 270:15938–15941

Valius M, Kazlauskas A (1993) Phospholipase C-gamma1 and phosphatidylinositol 3- kinase are the downstream mediators of the PDGF receptor's mitogenic signal. Cell 73:321–334

van der Geer P, Wiley S, Gish GD, Lai VKM, Stephens R, White MF, Pawson T (1996) Identification of residues that control specific binding of the Shc PTB domain to phosphotyrosine sites. Proc Natl Acad Sci USA 93:963–968

Velloso LA, Folli F, Sun XJ, White MF, Saad MJA, Kahn CR (1996) Cross-talk between the insulin and angiotensin signaling system. Proc Natl Acad Sci USA 93:12490–12495

Verheij M, Bose R, Lin XH, Yao B, Jarvis WD, Grant S, Birrer MJ, Szabo E, Son LI, Kyriakis JM, Haimovitz-Friedman A, Fuks Z, Kolesnick RN (1996) Requirement for ceramide-initiated SAPK/JNK signaling in stress-induced apoptosis. Nature 380:75–79

Vuori K, Ruoslahti E (1994) Association of insulin receptor substrate-1 with integrins. Science 266:1576–1578

Weidner KM, Di Cesare S, Sachs M, Brinkmann V, Behrens J, Birchmeier W (1996) Interaction between Gab1 and the c-Met receptor tyrosine kinase is responsible for epithelial morphogenesis. Nature 384:173–176

Wennstrom S, Hawkins P, Cooke F, Hara K, Yonezawa K, Kasuga M, Jackson T, Claessonwelsh L, Stephens L (1994) Activation of phosphoinositide 3-kinase is required for PDGF-stimulated membrane ruffling. Curr Biol 4:385–393

White MF, Maron R, Kahn CR (1985) Insulin rapidly stimulates tyrosine phosphorylation of a Mr 185,000 protein in intact cells. Nature 318:183–186

White MF, Livingston JN, Backer JM, Lauris V, Dull TJ, Ullrich A, Kahn CR (1988) Mutation of the insulin receptor at tyrosine 960 inhibits signal transmission but does not affect its tyrosine kinase activity. Cell 54:641–649

Wolf G, Trub T, Ottinger E, Groninga L, Lynch A, White MF, Miyazaki M, Lee J, Shoelson SE (1995) The PTB domains of IRS-1 and Shc have distinct but overlapping specificities. J Biol Chem 270:27407–27410

Xu X, Sonntag WE (1996) Growth hormone and aging: regulation, signal transduction and replacement therapy. TEM 7:145–150

Yamanashi Y, Baltimore D (1997) Identification of the Abl- and rasGAP-associated 62 kDa protein as a docking protein, Dok. Cell 88:205–211

Yamao T, Matozaki T, Amano K, Matsuda Y, Takahashi N, Ochi F, Fujioka Y, Kasuga M (1997) Mouse and human SHPS-1: molecular cloning of cDNAs and chromosomal localization of genes. Biochem Biophys Res Commun 231:61–67

Yamasaki H, Prager D, Gebremedhin S, Melmed S (1992) Human insulin-like growth factor I receptor ^{950}Tyrosine is required for somatroph growth factor signal transduction. J Biol Chem 267:20953–20958
Yamauchi K, Milarski KL, Saltiel AR, Pessin JE (1995) Protein-tyrosine-phosphatase SHPTP2 is a required positive effector for insulin downstream signaling. Proc Natl Acad Sci USA 92:664–668
Yamauchi T, Tobe K, Tamemoto H, Ueki K, Kaburagi Y, Yamamoto-Handa R, Takahadhi Y, Yoshizawa F, Aizawa S, Akanuma Y, Sonenberg N, Yazaki Y, Kadowaki T (1996) Insulin signaling and insulin actions in the muscles and livers of insulin-resistant, insulin receptor substrate 1-deficient mice. Mol Cell Biol 16:3074–3084
Yao R, Cooper GM (1995) Requirement for phosphatidylinositol-3 kinase in the prevention of apoptosis by nerve growth factor. Science 267:2003–2006
Yenush L, Makati KJ, Smith-Hall J, Ishibashi O, Myers MG Jr, White MF (1996) The pleckstrin homology domain is the principle link between the insulin receptor and IRS-1. J Biol Chem 271:24300–24306
Zackenfels K, Oppenheim RW, Rohrer H (1995) Evidence for an important role of IGF-I and IGF-II for the early development of chick sympathetic neurons. Neuron 14:731–741
Zhou MM, Harlan JE, Wade WS, Crosby S, Ravichandran KS, Burakoff SJ, Fesik SW (1995) Binding affinities of tyrosine-phosphorylated peptides to the COOH-terminal SH2 and NH_2-terminal phosphotyrosine binding domains of Shc. J Biol Chem 270:31119–31123
Zhou MM, Huang B, Olejniczak ET, Meadows RP, Shuker SB, Miyazaki M, Trub T, Shoelson SE, Fesik SW (1996) Structural basis for IL-4 receptor phosphopeptide recognition by the IRS-1 PTB domain. Nat Struct Biol 3:388–393

PDZ Domains and the Formation of Protein Networks at the Plasma Membrane

A.S. FANNING and J.M. ANDERSON

1 Introduction and Perspective

Protein-protein associations are the basis of much of the macromolecular organization underlying cellular structure and function. Two general mechanisms exist to achieve binding specificity: "surface-surface" and "modular domain-peptide" associations. The former is based on interacting surface chemistry unique to each pair of binding partners, e.g., oligomerization of hemoglobin monomers. Surface-surface interactions provide for highly specific binding, but the complex nature of these associations has not provided evolution with the opportunity to generate binding diversity. In contrast, "modular domain-peptide" associations are based on conserved domains which recognize variations of simple peptide motifs. Both the modular domains and peptide targets are found as motifs within multidomain proteins of diverse function, e.g., enzymes, receptors, cytoskeletal proteins. Examples of these domains include the well characterized SH3, SH2 and PTB domains.

Departments of Internal Medicine and Cell Biology, Yale University School of Medicine, PO Box 208019, New Haven, CT 06520-8019, USA

Modular domains have been used extensively to link proteins within signal transduction pathways and the cytoskeleton. The reason may be that once a set of interactions is established it can be adapted to create an independent parallel pathway by simple mutational changes in the domain and its cognate binding peptide. Another reason is that both the domains and peptide motifs can be shuffled among various proteins, by chromosomal translocations, without disrupting other functions of the protein, such as its enzymatic or other binding functions.

PDZ domains (aka GLGF or DHR repeats) are the most recently recognized class of modular surface-peptide binding domains. Much remains obscure about their distribution, function and regulation, yet recent data allow anticipation of their diverse potential. While similar in some respects to previously recognized modular binding domains, PDZ domains clearly have several unique properties. Unlike other well described modular domains, PDZ domains are ancient motifs found in organisms from bacteria to higher plants and animals (FANNING and ANDERSON 1996). In many cases PDZ domains recognize short peptide sequence at the extreme C-terminus of target proteins. These targets are often transmembrane receptors or channels (SHENG 1996). PDZ domains are frequently reiterated from two to six times within a protein and, unlike any other modular domain, some PDZ domains dimerize. Both features contribute to their utility in forming crosslinked protein networks and the organization of microdomains on the cell cortex at sites like synapses, and intercellular tight junctions (FANNING et al. 1996).

Because binding recognition may be based on a motif as short as 3–4 continuous residues, interactions with target proteins might be expected to be quite promiscuous. Indeed, some PDZ domains appear to bind many targets, and, conversely, any single target may bind a PDZ domain of several different proteins. While this too may contribute to creation of extensive and redundant protein networks, it is equally likely that PDZ domains are highly localized within a cell and thus never encountered all potential ligands. Like PTB and SH2 domain interactions, there is early evidence that PDZ domain binding is also regulated by phosphorylation of target motifs (COHEN et al. 1996).

In this review we highlight several aspects of the rapidly evolving knowledge of PDZ domains. Extensive references to the distribution of PDZ domains are provided along with discussion of the structural basis for sequence-specific peptide recognition, the likely function of PDZ domains in organizing cytoskeletal and signaling networks, and the possibility of regulation of PDZ-peptide binding. Although the protein binding function of PDZ domains was described only 3 years ago, the literature in this field has expanded enormously and we attempt to be inclusive, but not detailed. We refer the reader to other reviews on more focused aspects of PDZ domain biology (HARRISON 1996; FANNING and ANDERSON 1996; ANDERSON 1996; SHENG 1996).

2 The PDZ Domain is a Modular Protein-Binding Motif

2.1 PDZ Domains are Conserved Sequence Elements Found in Diverse Proteins in Animals, Plants, and Bacteria

The PDZ domain was originally identified as a region of repeated sequence homology between the synaptic protein PSD-95 (aka SAP90) and the *Drosophila* tumor suppressor Dlg (CHO et al. 1992). This 80–90 amino acid motif has subsequently been identified in over 50 distinct proteins (Table 1), with the number of domains in each protein varying from one to as many as six. Although sequence homology between any pair of PDZ domains can be as low as 20% (Fig. 1), all PDZ domains appear to maintain a core consensus which presumably underlies a common tertiary structural motif (DOYLE et al. 1996). From sequence comparisons it is also apparent that PDZ domains of different proteins fall into distinct classes (Fig. 2). Often the PDZ domain of one protein (e.g., Dlg/hDlg) is more similar to the PDZ domain in a second protein (e.g., PSD-95) than it is to others within the same protein. Given the very high identity among some PDZ motifs in different proteins (the second PDZ domains of hDlg and PSD-95 are 88% identical), it seems likely that domains of the same class in different proteins have a similar function. As discussed below, this speculation is supported by recent biochemical analysis.

The cellular function of proteins which contain PDZ domains varys considerably. However, analysis of invertebrate mutations suggests the majority of these proteins are involved in some aspect of signal transduction. Many examples are observed in *Drosophila*. For example, the Dlg protein, which contains three PDZ domains, has been identified as the product of a tumor suppressor locus (WOODS and BRYANT 1989, 1991). Mutations in Dlg result in imaginal disk overgrowth and loss of cell polarity. The *Drosophila dishevelled* (Dsh) protein is a component of the wingless signaling pathway, which is involved in the establishment of segment polarity (THEISEN et al. 1994). Several human and mouse homologs of this segment polarity gene have already been identified (SUSSMAN et al. 1994; PIZZUTI et al. 1996). The *Drosophila* cno protein is a component of the notch signal transduction pathway (MIYAMOTO et al. 1995), and is required for normal photoreceptor development. A human homolog of cno, AF6, has been identified as a gene at the translocation breakpoint in several patients with acute myelogenous leukemia (PRASAD et al. 1993). The SIF protein, encoded by the *Drosophila still life* gene, has high homology to guanine nucleotide exchange factors for the rho family of small GTPases (SONE et al. 1997). Mutations in SIF result in abnormal neuronal morphology. Another potential signaling protein is encoded by the *Drosophila* InaD gene. InaD binds to and appears to regulate the activity of the TRP Ca^{2+} channel, which is involved in phototransduction (SHIEH and NIEMEYER 1995; SHIEH and ZHU 1996). Mutations in PDZ domain-containing proteins have also been described in *C. elegans*; for example the *lin2* and *lin7* genes encode components of a ras signaling pathway (HOSKINS et al. 1995; SIMSKE et al. 1996). Mutations in either gene result in failure to differentiate by a class of cells known as vulval precursor cells. The

Table 1. PDZ Domain-containing proteins

Protein (species)		Acc. #	Amino acid boundaries	References
MAGuKs				
dlg-like				
dlg	(Dm)	U47110	40–126, 154–244, 486–567	Woods and Bryant (1991)
hdlg/SAP97	(h/r)	U13897/U14950	222–309, 318–404, 458–546	Lue et al. (1994); Müller et al. (1995)
PSD93/chapsyn110	(r/h)	U32376/U50717	98–184, 193–279, 421–500	Kim et al. (1996); Brenman et al. (1996)
PSD95/SAP90	(r)	A45436	65–151, 160–245, 313–393	Cho et al. (1992); Kistner et al. (1993)
SAP102	(r)	U50147	149–235, 244–329, 404–484	Müller et al. (1996)
p55-like				
Camguk	(Dm)	U53190	515–598	Dimitratos et al. (1997)
CASK	(r)	U47110	489–572	Hata et al. (1996)
DLG2	(h)	X82898	163–244	Mazoyer et al. (1995)
DLG3	(h)	U37707	137–215	Smith et al. (1996)
hlin2	(h)	Unpublished	489–572	A. Brecher, personal communication
lin2	(ce)	X92564	545–620	Hoskins et al. (1995)
p55	(h)	A39599	71–152	Ruff et al. (1991)
tight junction				
ZO-1	(h)	A47747	11–97, 177–251, 412–490	Willot et al. (1993)
ZO-2	(h)	L27476	33–119, 307–384, 513–601	Duclos et al. (1994)
p130 (“ZO-3”)	(c)	Unpublished	Unknown (3 PDZ domains)	B. Stevenson, personal communication
tamou	(Dm)	D83477	23–110, 167–251, 408–486	Takahisa et al. (1996)

Syntrophins				
α1-syntrophin	(h)	U40571	87–169	Ahn et al. (1996)
β1-syntrophin	(h)	L31529	115–193	Ahn et al. (1994)
β2-syntrophin	(h)	U40572	118–196	Ahn et al. (1996)
Tyrosine phosphatases				
PTPH1	(h)	A41109	510–596	Yang and Tonks (1991)
PTPmeg	(h)	A41105	517–603	Gu et al. (1991)
PTPbas	(h)	D21209	1093–1178, 1368–1452,	Maekawa et al. (1994)
(aka fap-1, PTPL1)			1501–1589, 1788–1869, 1883–1966	Saras et al. (1994)
dsh homologs				
dsh	(Dm)	L26974	252–337	Theisen et al. (1994)
Dvl-1/DVL1	(m/h)	U46461	252–337	Sussman et al. (1994)
Dvl-2	(m)	U24160	267–353	Yang et al. (1996)
Dvl-3	(m)	U41285	249–335	Kligensmith et al. (1996)
DVL/RACK8	(h)	U49262	245–333	Pizutti et al. (1996); Kuroda et al., unpublished
LIM Domain proteins				
CLP36	(r)	U23769	4–81	Wang et al. (1995)
Enigma	(h)	L35240	6–79	Wu and Gill (1994)
	(r)	U48247	6–82	Kuroda et al., unpublished
ril	(r)	X76454	3–80	Keiss et al. (1995)
Ser/Thr Protein kinases				
LIMK (aka Kiz-1)	(h)	D26309	164–259	Mizuno et al. (1994)
MAST205	(m)	U02313	1042–1130	Walden and Cowan (1993)

Table 1. (*contd.*)

Protein (species)		Acc. #	Amino acid boundaries	References
Other				
apxl	(h)	X83543	29–107	Schiaffino et al. (1995)
AF6	(h)	U02478	993–1077	Prasad et al. (1993)
cno	(Dm)	D49534	844–938	Miyamoto et al. (1995)
densin-180	(r)	U66707	1401–1492	Apperson et al. (1996)
IL-16 (aka LCF)	(h)	M90391	174–258, 294–378	Cruikshank et al. (1994); Bannert et al. (1996)
inaD	(Dm)	U15803	18–103, 361–447	Shieh and Niemeyer (1995)
KIAA147	(h)	D63481	663–743, 798–881, 939–1024, 1035–1122	Nagase et al. (1995)
lin7	(ce)	U78092	199–280	Simske et al. (1996)
NHE-RF	(r)	U19815	15–93, 155–233	Weinman et al. (1995)
nNOS	(h)	S28878	17–99	Nakane et al. (1993)
par3	(ce)	U25032	523–608	Etemed-Moghadam et al. (1995)
PICK1	(m)	Z46720	22–99	Staudinger et al. (1995)
Rhophilin	(m)	U43154	500–573	Watanabe et al. (1996)
SIF (*still life*)	(Dm)	D86546	1204–1293	Sone et al. (1997)
TKA-1	(h)	Z50150	13–88, 153–226	Seedorf and Ullrich, unpublished
Tiam1	(h)	U16296	845–929	Habets et al. (1994)
X11	(h)	A47176	526–614	Duclos et al. (1993)

Bacterial, cyanobacteria, and plants				
htrA and htrA-like				
Ba_htrA		L09274	275–390	TATUM et al. (1994)
Ba_htrA2		U07351	238–354	TATUM et al. (1994)
Cj_htrA		U27271	139–252	TAYLOR and HIRATSUKA (1990)
Rm_degP		U31512	302–391	GLAZEBROOK et al. (1996)
Ec_htrA (aka degP)		S01899	254–369	LIPINSKA et al. (1988)
Ec_hhoA		P39099	232–347	BASS et al. (1996)
Hi_hhoB		P44947	219–323	FLEISCHMANN et al. (1995)
Ml_htrA		U15180	437–517	Smith et al., unpublished
Pa_mucD		U32853	235–350	BOUCHER et al. (1996)
Bb_htrA		L20127	265–379	ANDERSON et al. (1994)
St_htrA		A18802	255–370	Anonymous submission (patent)
ctpA/ctpB homologs				
Bb_ctp		L37094	87–165	Mitchell and Minnick, unpublished
Hv_ctpA		X90929	47–136	Oelmuller et al., unpublished
So_ctpA		X90558	225–308	INAGAKI et al. (1996)
Ss_ctpA		A53964	105–183	SHESTAKOV et al. (1994)
Ss_ctpB		X96490	132–213	Sidoruk et al., unpublished
Sc_synp2	P42784 sp	103–183	BRAND et al. (1992)	
Ps_ctp (aka vicilin)	S02355 pir	18–75 (part)	GATEHOUSE et al. (1982)	

Bacteria: Ba, *Brucella abortis*; Bb, *Bartonella baciliformis*; Cj, *Campylobacter jejuni*; Ec, *Escherichia coli*; Hi, *Haemophilus influenza*; Ml, *Mycobacterium leprae*; Pa, *Pseudomonas aeruginosa*; Rm, *Rhizobium meliloti*; St, *Salmonella typhimurium*. Cyanobacteria: Ss, *Synechocystis* sp.; Sc, *Synechococcus* sp. Plants: Hv, *Hordeum vulgare* (barley); Ps, *Pisum sativum* (garden pea); So, *Spinacia oleracea* (spinach). Other: Dm, *Drosophila melanogaster*; c, canine; ce, *Ceanorhabditis elegans*; h, human; m, mouse; r, rat. See text for full description of these proteins.

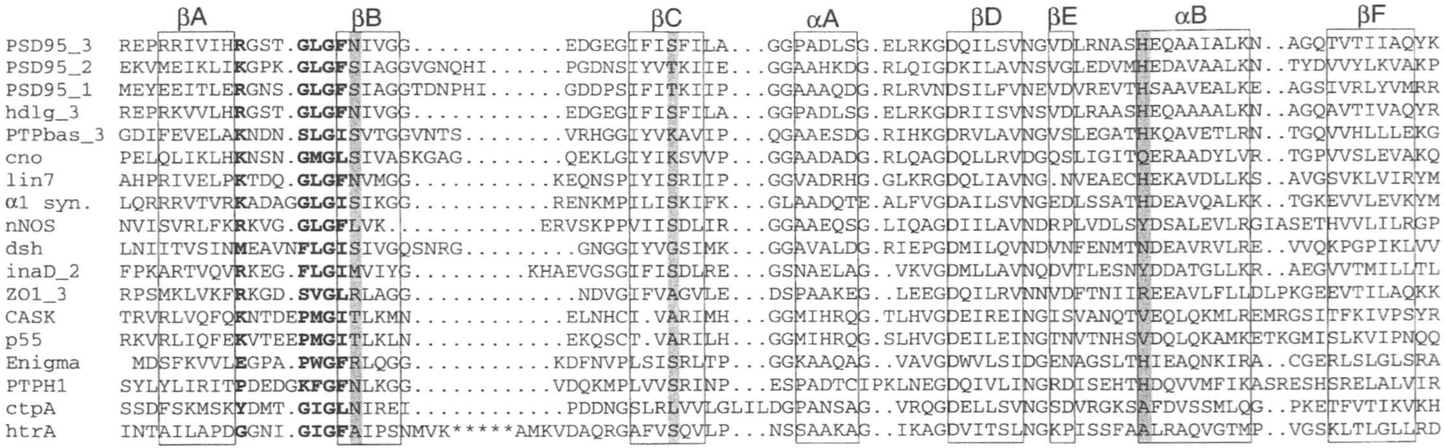

Fig. 1. Sequence alignment of selected PDZ domains with PSD-95 comparing presumed structural motifs and domains involved in ligand binding. The sequence alignment was created using the PILEUP program (DEVEREUX et al. 1984). The positions of the secondary structural elements identified in PSD95 are boxed. Amino acid residues within the carboxylate loop of PSD-95 that hydrogen bond with the Val (0) carboxylate group of the ligand are indicated *in bold*. Residues which directly contact the amino acid side chains of the peptide ligand are *highlighted in gray*. The corresponding residues in the other PDZ domains are similarly highlighted for comparison. The ctpA and htrA polypeptides represented here are proteases from barley and *E. coli*, respectively. "***" represents a 30 amino acid insertion in the βB-βC loop of htrA. See Table 1 for amino acid boundaries and references associated with each polypeptide

identification of PDZ domains in so many proteins with signaling capacities led to the speculation early on that PDZ domains were somehow involved in assembling proteins involved in signal transduction pathways.

In other cases, known signaling proteins have been found to contain PDZ domains. For example, the neuronal isoform of nitric oxide synthase (nNOS), which generates the intracellular second messenger NO, contains a PDZ domain (NAKANE et al. 1993). The PTPH1, PTPmeg, and PTPbas proteins have all been biochemically identified as protein tyrosine phosphatases (YANG and TONKS 1991; GU et al. 1991; MAEKAWA et al. 1994; SARAS et al. 1994). PTPbas binds directly to the Fas receptor, which is involved in a signal transduction pathway that regulates apoptosis (SATO et al. 1995). At least two PDZ domain-containing proteins, LIMK and MAST205, have been identified as probable serine threonine kinases by sequence homology and biochemistry (MIZUNO et al. 1994; WALDEN and COWAN 1993). The protein Rhophilin was originally identified as a substrate for the small GTP binding protein rho (WATANABE et al. 1996). Similarly, the protein PICK1 was originally identified as a substrate for protein kinase C (STAUDINGER et al. 1995). A regulatory factor which may bind the Na-H exchanger through a PDZ domain, NHE-RF, has been identified and shown to be required for protein kinase A-mediated regulation of the Na-H exchanger (WEINMAN et al. 1995).

Many other PDZ domain-containing proteins are implicated in signal transduction based on the presence of other motifs commonly found in signaling proteins. For instance, the proteins CLP36, Enigma, ril, and LIMK all contain LIM domains (MIZUNO et al. 1994; WU and GILL 1994; KIESS et al. 1995; WANG et al. 1995), and the polypeptide encoded by the gene KIAA147 has a region of homology to adenylate kinase (NAGASE et al. 1995).

PDZ domains are also one of the signature characteristics of a family of membrane-associated signaling proteins known as the membrane-associated guanylate kinase homologs (MAGUKs). This family of proteins has been one of the most intensely studied classes of PDZ-containing proteins in terms of PDZ function (FANNING et al. 1996; GOMPERTS 1996; EHLERS et al. 1996). All MAGUKs are distinguished by a core of homologous protein domains which include a region homologous to the enzyme guanylate kinase (GuK), a src-homology region 3 (SH3) domain, and one to three PDZ domains. Presumably this core of domains serves a conserved coordinated binding and/or signaling function. The MAGUKs can be subdivided into three classes based on the number of PDZ domains and expression of other sequence motifs. The first class is characterized by the prototypic member Dlg (see Table 1), and comprises several related proteins which have three PDZ domains. The tissue specificity and subcellular distribution of this class of proteins varies considerably, although the majority of them have been localized to synapses (SHENG 1996). The second class of MAGUKs also have three PDZ domains, but differ significantly from the Dlg-like MAGUKs in terms of sequence (FANNING et al. 1996). These proteins were initially characterized as components of the vertebrate tight junction, and include the proteins ZO-1 (WILLOTT et al. 1993), ZO-2 (JESAITIS and GOODENOUGH 1994; DUCLOS et al. 1994), and a 130-kDa protein provisionally identified as "ZO-3" (B. Stevenson, University of Alberta,

Edmonton, personal communication). More recently the *tamou* gene, an invertebrate homolog of ZO-1 and ZO-2, has been identified in *Drosophila* (TAKAHISA et al. 1996). Mutations in the *tamou* gene cause a reduction in expression of the *extramacrochaetae* transcription factor, suggesting that the tight junction MAGUKs may be directly involved in signal transduction at the level of transcription. The third class of MAGUKs contains only one PDZ domain, and is characterized by the p55 protein (RUFF et al. 1991). A subset of this class of proteins (CASK, CAMGUK, and LIN2, sometimes collectively referred to as the CAMGUKs) also contains a region of homology to the enzyme calcium-calmodulin protein kinase II (HOSKINS et al. 1995; DIMITRATOS et al. 1997; HATA et al. 1996a, b). As described above, analysis of the mutant phenotypes of the invertebrate MAGUK genes *dlg*, *lin2*, and *tamou* strongly suggests they also function in signal transduction.

Database homology searches indicate that PDZ domains are even found in two protease families in both plants and bacteria (see Table 1, Fig. 1). The first class is represented by the *E. coli htrA* gene. The htrA protein, also known as degP, is a serine protease found in the periplasmic space and appears to be involved in the proteolytic degradation of aberrant proteins (LIPINSKA et al. 1988, 1989). It is a heat shock protein, essential for the growth of *E. coli* at elevated temperatures (LIPINSKA et al. 1990), and is also required for virulence in a number of bacterial pathogens such as *Salmonella typhimurium*, *Pseudomonas aeruginosa*, and *Brucella abortis* (JOHNSON et al. 1991; BAUMLER et al. 1994; BOUCHER et al. 1996). The second class is found in plants, bacteria and even cyanobacteria (SHESTAKOV et al. 1994; INAGAKI et al. 1996). The most extensively characterized example of this class is encoded by the spinach *ctpA* gene. The ctpA protein is a novel protease associated with the thylakoid membrane which cleaves the D1 precursor protein of the photosystem II reaction center (ANBUDURAI et al. 1994; TAGUCHI et al. 1995). In each case it appears the PDZ domain interacts with C-terminal sequences to bring the protease domain and its substrate into proximity. The presence of this conserved sequence element with a conserved binding function in bacterial, plant and animal proteins clearly establishes its early evolution and maintained function. Not surprisingly, a search of the yeast database reveals that PDZ domains are also present in the fungi (data not shown).

2.2 PDZ Domains Bind to a C-Terminal Motif

There is now an overwhelming accumulation of experimental evidence supporting the hypothesis that PDZ domains are protein-binding motifs. Biochemical analysis using both *in vivo* and *in vitro* binding assays indicates that PDZ domains bind to

Fig. 2. Sequence relationships between PDZ domains found in different proteins. This dendrogram was assembled using the PILEUP program (DEVEREUX et al. 1984). Note that sequences that cluster in a distinct branch of the tree share ligands that are distinct from ligands for PDZ domains in other branches of the tree (e.g., distinguish PSD95_2 and hdlg_2 from SAP102_l)

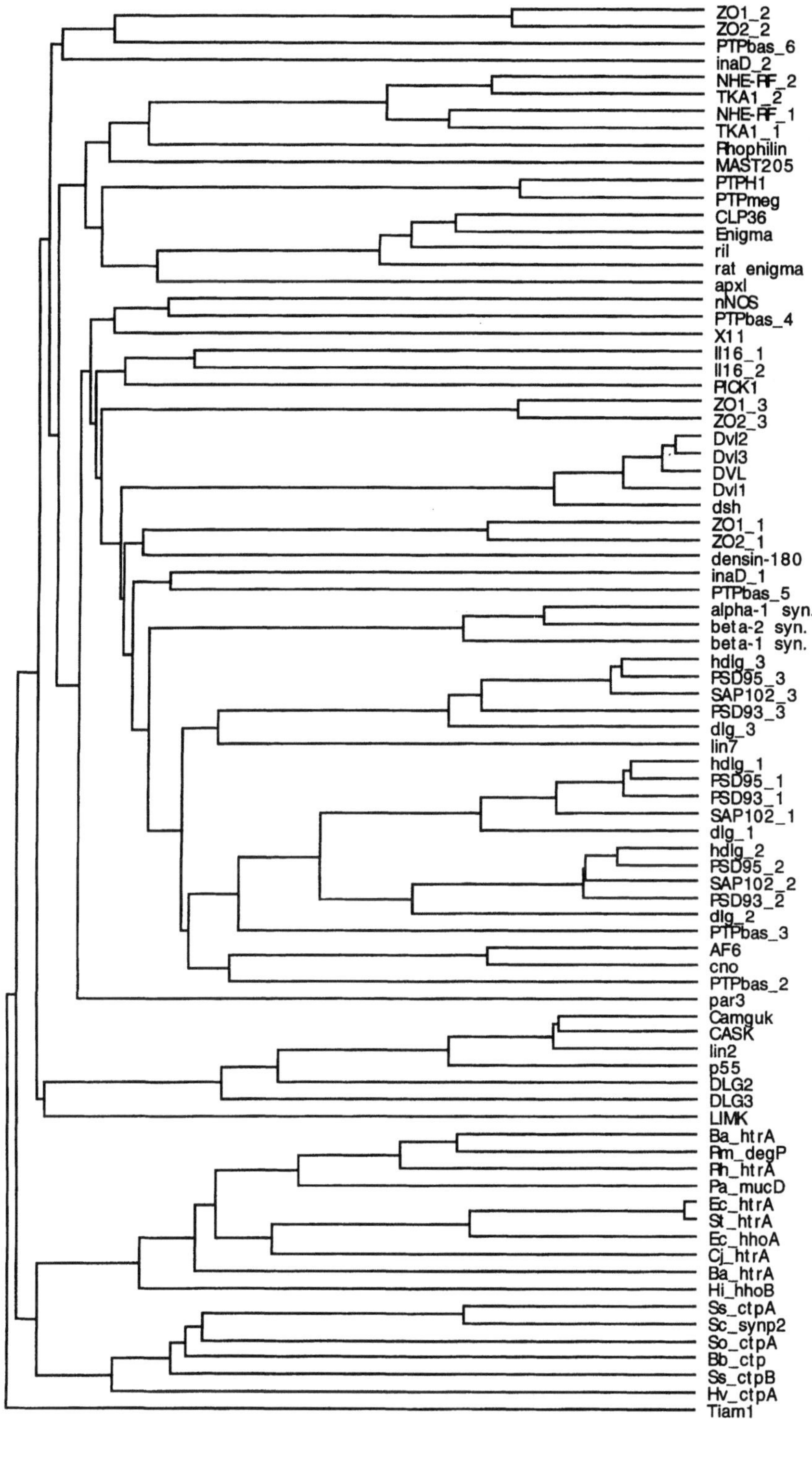
ZO1_2
ZO2_2
PTPbas_6
inaD_2
NHE-RF_2
TKA1_2
NHE-RF_1
TKA1_1
Rhophilin
MAST205
PTPH1
PTPmeg
CLP36
Enigma
ril
rat enigma
apxl
nNOS
PTPbas_4
X11
Il16_1
Il16_2
PICK1
ZO1_3
ZO2_3
Dvl2
Dvl3
DVL
Dvl1
dsh
ZO1_1
ZO2_1
densin-180
inaD_1
PTPbas_5
alpha-1 syn.
beta-2 syn.
beta-1 syn.
hdlg_3
PSD95_3
SAP102_3
PSD93_3
dlg_3
lin7
hdlg_1
PSD95_1
PSD93_1
SAP102_1
dlg_1
hdlg_2
PSD95_2
SAP102_2
PSD93_2
dlg_2
PTPbas_3
AF6
cno
PTPbas_2
par3
Camguk
CASK
lin2
p55
DLG2
DLG3
LIMK
Ba_htrA
Rm_degP
Rh_htrA
Pa_mucD
Ec_htrA
St_htrA
Ec_hhoA
Cj_htrA
Ba_htrA
Hi_hhoB
Ss_ctpA
Sc_synp2
So_ctpA
Bb_ctp
Ss_ctpB
Hv_ctpA
Tiam1

specific sequence motifs at the C-termini of their target proteins (see Table 2). For example, the C-termini of both the *N*-methyl D-aspartate (NMDA) receptor and the Shaker-type potassium channel Kv1.4 have been identified as ligands for the first (PDZ1) and second (PDZ2) PDZ domains of three related proteins the synapse associated proteins PSD-95 and chapsyn 110 (aka KAP-5 or PSD-93), and the human homolog of the *Drosophila* dlg protein, hdlg (Kim et al. 1995; Kornau et al. 1995; Niethammer et al. 1996). The measured K_d of the interaction of the K^+ channel with hdlg PDZ2 ranges from 42 to 160 n*M*, suggesting that these interactions are of relatively high affinity, (Marfatia et al. 1996). The NMDA receptor and the K^+ channel also bind to PDZ2 of another synaptic protein, SAP102 (Müller et al. 1996). The C-terminus of the Fas receptor, a transmembrane protein involved in apoptosis, has been shown to bind to one of the six PDZ domains of the protein tyrosine phosphatase FAP-1 (Sato et al., 1995) and the C-terminus of the

Table 2. PDZ domain containing proteins and their cellular targets

PDZ protein	Target	C-terminal Motif
A. PDZ/C-terminal interactions		
PSD-95 (SAP90)	NMDA receptor (NR2 subunit)	ESDV
	Shaker K^+channel	ETDV
	Kir2.3	ESAI
PSD-93 (chapsyn-110)	NMDA receptor	See above
	Shaker K^+ channel	
SAP97	NMDA receptor	See above
SAP102	NMDA receptor	See above
hdlg	Shaker K^+channel	See above
	APC	VTSV
FAP-1 (PTPbas)	Fas receptor	QSLV
inaD	TRP Ca^{2+} channel	rgKSTVtgrmisgwl
CASK	Neurexin	EYYV
p55	Glycophorin C	EYFI
hlin2	Syndecan-2	EFYA
lin7[a]	let23	ETCL
dsh[a]	Dfz2, fz	ASHV
B. PDZ/PDZ interactions		
nNOS	PSD-95, PSD-93	
	α1-syntrophin	
PSD-95[a]	PSD-93	

[a] Direct interaction not yet established. See text for details and references.

APC tumor suppressor gene has also been identified as a ligand for PDZ2 of hdlg (Matsumine et al. 1996). Again, FAP-1 has a relatively high affinity for its ligand, with a measured K_d of 154 nM (Songyang et al. 1997). All of the ligands for these PDZ domains contain the consensus sequence (S/T)XV at the extreme C-terminus, in which S/T represents serine or threonine, X is any amino acid, and V is a valine residue. In all cases mutational analysis has demonstrated that this motif is both necessary and sufficient to mediate the interaction with PDZ domains.

Other classes of PDZ domains seem to recognize distinct C-terminal sequences. For example, p55 binds directly to the erythrocyte transmembrane protein glycophorin C, which ends in the sequence EYFI (Songyang et al. 1997). The synapse-associated protein CASK binds to the C-terminus of the transmembrane protein neurexin, a putative adhesion/signaling protein of neurons (Hata et al. 1996a). Neurexin ends in the C-terminal sequence EYYV. Finally, the human homolog of the *C. elegans* protein LIN2, hLIN2, has been shown to interact with the transmembrane heparin sulfate proteoglycan syndecan-2 (A. Brecher, Yale University, personal communication). Syndecans all contain the C-terminal sequence EFYA. The chemical basis for the differences in PDZ-target peptide specificity is discussed in greater detail below.

Still other PDZ interactions with C-termini can be inferred from genetic studies, even though direct interactions have yet to be established. The *C. elegans* protein LIN7, for example, has been shown to interact genetically with the transmembrane receptor tyrosine kinase LET23, which contains the C-terminal sequence ETCL (Simske et al. 1996). A possible ligand for the PDZ domain within the Dsh protein has also been identified. The product of the *Dfz2* gene has been identified as a receptor for the wingless protein, a secreted signaling protein (Bhanot et al. 1996). Dfz2 protein contains the C-terminal sequence ASHV. The Dsh protein, a cytosolic component of the wingless signaling pathway, is genetically downstream of this wingless receptor, and thus may bind via its PDZ domain to the cytoplasmic tail of the receptor (Theisen et al. 1994).

2.3 PDZ Domain Interactions with Other Motifs

Not all PDZ domains are restricted to C-terminal binding; PDZ domains have also been shown to bind to other PDZ domains and, in at least one case, to an internal (T/S)XV motif (Table 2). For example, the PDZ2 domain of PSD-95 and PSD-93 has been shown to bind directly to PDZ domains in both nNOS and α1-syntrophin (Brenman et al. 1996b), suggesting that PDZ domains also mediate heterotypic dimerization. Perhaps of even greater interest is the fact that an individual PDZ domain can bind to both C-terminal sequences and to other PDZ domains, suggesting another level of versatility in forming protein complexes. More recently, the second PDZ domain in the *Drosophila* inaD photoreceptor protein was shown to bind the TRP calcium channel at an STV motif which is nine residues from the C-terminus (Shieh and Zhu 1996). This observation suggests a third possible binding modality in which a (S/T)XV consensus sequence is located internally within the target protein. It remains to be determined whether any single PDZ domain can participate in all three types of binding interactions, i.e., dimerization and binding to internal as well as C-terminal peptide motifs. Conceivably different classes of PDZ domains are capable of one, two or all three binding modalities. It is also unclear whether dimerization and C-terminal binding can occur simultaneously, or whether these interactions are competitive events. Either possibility has interesting implications for how they participate in assembly of macromolecular complexes.

2.4 The Binding Interactions Mediated by PDZ Domains are Selective

In vitro and *in vivo* binding assays suggest that the interactions between PDZ domains and their ligands are somewhat selective. For example, only one of the six PDZ domains in FAP-1 (PDZ3) will bind to Fas (Sato et al. 1995), and only PDZ1 and PDZ2 of PSD-95 will bind to the NMDA receptor and Kv1.4 (Kornau et al. 1995; Kim et al. 1995). Although all three of the PDZ domains in SAP102 will bind to the NMDA receptor *in vitro*, the binding affinity of PDZ2 (K_d = 6 nM) is more than 100 times tighter than that of PDZ3 (K_d = 1.0 μM) (Müller et al. 1996). The same degree of specificity exists for dimerization between PDZ domains. The PDZ domain in nNOS will bind only to the second PDZ domain in PSD-95 and PSD-93 (Brenman et al. 1996a), and although it does bind to PSD-95, it will not bind to the C-terminus of the NMDA receptor (Kornau et al. 1995).

These results are not entirely unexpected. The organization of PDZ domains by sequence similarity using the PILEUP program (Fig. 2) illustrates that while some PDZ domains fall into groups with high sequence similarity, and thus presumably have similar cellular targets (e.g., PSD-95 and hdlg), there is actually a great overall diversity, suggesting that many domains have novel binding specificities. This idea has been extended and confirmed by the work of Songyang et al. (1997), who used the oriented peptide libraries to isolate peptide ligands for PDZ domains with no known binding partners. These studies confirmed that divergent PDZ domains each had distinct optimal binding partners. It is probable that further analysis of the ligands for different classes of PDZ domains will reveal much greater complexity, perhaps analogous to that of SH3 and SH2 binding recognition motifs.

3 Structural Basis for PDZ Interactions

The structural basis of PDZ binding has been elucidated from the crystal structure of the third PDZ domain of PSD-95 complexed to a peptide ligand (Doyle et al. 1996). These results suggest a mechanism for the specificity in binding C-termini containing the (T/S)XV motif, although it remains to be determined how general this model will be for other ligand specificities and to what extent sequences outside this short motif contribute to specificity.

The PDZ3 in PSD-95 was solved with a bound peptide ligand which ends in the residues Gln (at position −3 from the C-terminus), Thr (−2), Ser (−1), and a terminal Val (0) (DOYLE et al. 1996). The PDZ domain in this complex (Fig. 3) forms a globular structure composed of 5–6 β-strands (βA-F, see also Fig. 1) and 2 α-helices (αA and αB) arranged into what has been described as a 'β sandwich' or 'up-and-down β barrel' (DOYLE et al. 1996; MORIAS-CABRAL et al. 1996). In the co-crystal structure, the peptide ligand binds within a hydrophobic pocket created by the βB-strand, the αA-helix, and the loop connecting the βA- and βB-strands designated the "carboxylate binding loop" (DOYLE et al. 1996). The peptide ligand is oriented antiparallel to βB, with its C-terminal valine side chain within a hydrophobic cavity, and peptide binding is stabilized by interactions between the backbone of the peptide and the βB-strand. The interaction between the PDZ domain and the peptide ligand is also promoted by hydrogen bonding between the Val (0) carboxylate group and residues of the so-called "carboxylate-binding loop" (R and GLGF; Fig. 1, bold residues, and Fig. 3), and between the peptide side chains of Gln (−3), Thr (−2), and Val (0) and side chains of residues within the PDZ domain (Fig. 1, shaded residues). The side chain of Ser (−1) is oriented into solution and does not interact with the PDZ domain.

These results lead to a speculative model in which differences in specificity of distinct PDZ domains are the result of changes in amino acid residues that mediate the interactions with the (−3), (−2), and (0) residues of the target peptide. Because the (−1) side chain makes no contact with the PDZ domain in PSD-95, it would not be predicted to contribute to specificity. As shown in Fig. 1, there is a considerable diversity in primary sequence of different PDZ domains at positions which presumably directly contact the ligand. PDZ domains which bind to the (T/S)XV motif actually comprise a family of very closely related domains (Fig. 2). Less closely related sequences would be predicted to have novel specificities because of differences at residues which interact with the amino acid side chains of the peptide ligand.

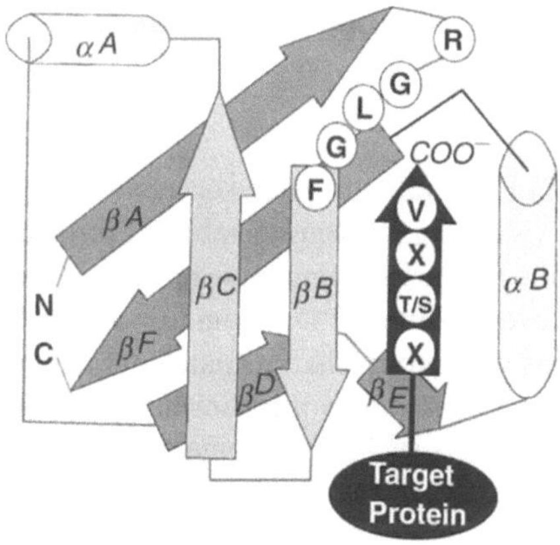

Fig. 3. Schematic diagram of PDZ domain structure. β-Strands are indicated by *arrows* and α-helices are represented by *cylinders*. Secondary structure elements are labeled as in Fig. 1. The C-terminal ligand (*black*) lies in a groove created by the βB-strand and αC-helix, the C-terminus of the target protein is embedded in the hydrophobic cavity and the carboxylate group is coordinated by a loop created by βA-βB elements or "carboxylate binding loop" (residues R, GLGF)

Although the PDZ of the CASK protein has not been crystallized, its unique specificity can be rationalized by the model described above. CASK binds to the cell surface protein neurexin, which contains the C-terminal motif EYYV (HATA et al. 1996b). Instead of the Asn (in the βB strand) and Ser (in βC) residues found in PSD-95 that normally interact with the Gln (−3) residue in the peptide ligand, the CASK PDZ contains Thr and Ala, respectively (see Fig. 1). Similarly, the His residue (in helix αB) in PSD-95 that binds to Thr (−2) in the peptide ligand is replaced by a Val residue. This hydrophobic side chain would be predicted to be positioned to interact with the hydrophobic side chain at position −2 of neurexin. Although this model is simple and supported by the limited experimental data, it remains to be determined whether residues outside of the immediate area of the C-terminus are also involved in mediating binding specificity.

The structural basis for heterodimerization between PDZ domains, such as occurs between nNOS and PSD-95, is currently unknown. The same is true for the interactions with internal (T/S)XV motifs such as occur between of inaD and TRP. Like many other PDZ domains, these PDZs also differ from PSD-95 at residues that are proposed to directly contact the peptide ligand (see Fig. 1). In addition, the carboxylate binding loop in inaD contains the motif FLGI instead of the GLGF motif seen in PSD-95 (Table 2). It is possible that alterations within residues of the carboxylate binding loop might relax the constraints for a C-terminal residue, and thus allow binding at an internal (T/S)XV motif such as seen with inaD. HARRISON (1996) has suggested that the binding of the peptide motif to the PDZ domain can be described as the addition of a β-strand to the B-sheet of the PDZ domain. One implication of this model is that the supplementing β-strand need not necessarily come from the C-terminus of a protein, but could be contributed by an internal sequence, such as in inaD, or by a β-strand from another PDZ domain. An understanding of these interactions awaits resolution of the crystal structures of other complexed PDZ domains.

4 PDZ Domain Interactions Create Crosslinked Networks at the Plasma Membrane

The great majority of proteins which contain PDZ domains are associated with the plasma membrane. Interaction with the C-termini of transmembrane proteins probably provides one mechanism for targeting or maintaining these proteins at the plasma membrane. However, the presence of multiple PDZ domains within a protein, and the association of these domains with motifs that mediate interactions with the cortical cytoskeleton, also appears to provide a mechanism for cross-linking and organizing transmembrane proteins. As such, PDZ domains appear to be critical for the formation of organized structures at the plasma membrane such as synapses and cell-cell junctions.

4.1 PDZ Domains Promote the Clustering of Transmembrane Proteins

The *in vivo* distribution of many PDZ proteins overlaps with that of the transmembrane targets that have been identified by *in vitro* biochemistry. This has been particularly well established with the MAGUK proteins present in vertebrate and invertebrate synapses. PSD-95, PSD-93, hdlg, and SAP102 all colocalize with the NMDA receptor and the K^+ channel within distinct tissues and cell types in the nervous system (Kornau et al. 1995; Kim et al. 1995, 1996; Müller et al. 1996). There are several observations, however, which suggest that the organization of these transmembrane proteins within synapses is a direct result of the interaction of these proteins with the PDZ domains in different MAGUKs. For example, coexpression of PSD-95 with the NMDA receptor or K^+ channel in cultured fibrobroblasts results in the formation of discrete clusters of these proteins on the cell surface. If expressed individually, these proteins are diffusely distributed (Kim et al. 1995). Similar results were obtained with PSD-93 (Kim et al. 1996). In a similar set of experiments, Ehlers et al. (1995) demonstrated that splice variants of the NMDA receptor which contain the conserved C-terminal T/SXV motif cluster into discrete patches at the plasma membrane, while those that lack this motif are diffusely distributed. Attaching the C-terminal motif to a receptor protein that is normally diffusely distributed was sufficient to mediate the clustering of the chimeric receptor. Both studies suggest that MAGUK proteins are directly involved in the crosslinking and organization of transmembrane proteins in the plasma membrane via interactions mediated by PDZ domains.

The idea that PDZ domains are required for channel clustering and synaptic organization is supported by experimental evidence from *Drosophila* Dlg mutants (Tejedor et al. 1997). In *Drosophila* larvae the Dlg protein normally colocalizes with Shaker K^+ channels at glutamatergic neuromuscular junctions. The cytoplasmic C-terminal tail of the *Drosophila* K^+ channel binds directly to PDZ1-2 of Dlg *in vitro*, and Dlg is required for receptor clustering in heterologous cell transfection experiments such as those described above. In Dlg null mutant larvae the K^+ channels fail to cluster at neuromuscular junctions. Similarly, synaptic clustering of the K^+ channel is abolished by simple deletion of the EDTV PDZ-binding motif in the K^+ channel. Finally, expression of a truncated Dlg protein encoding only PDZ1-2 results in ectopic coclustering of the truncated protein with the channel. The implication of this study is that PDZ1-2 can cluster channels but lacks information to recruit clusters to synapes. This additional information must reside in the binding functions of some other domain.

In many cases the PDZ domain may be less important for organizing a crosslinked domain than it is for targeting a particular protein to a subcellular domain or promoting the interaction with a cellular substrate. For example, nNOS normally colocalizes with PSD-95 in neurons and with α1-syntrophin in the skeletal muscle sarcolemma. In mutant mice expressing only a nNOS isoform lacking the PDZ domain, nNOS does not interact with PSD-95 and syntrophins, nor does it interact with plasma membranes, in the brain and skeletal muscle sarcolemma

(Brenman et al. 1996). These results suggest that the PDZ domain is necessary to recruit nNOS to specific protein complexes at the plasma membrane. The APC tumor suppressor, which is found at synapses and at the lateral membranes of colonic epithelial cells, may also be recruited in an analogous fashion (Matsumine et al. 1996). In fact, there are several potential signaling molecules listed in Table 1 that may be recruited to potential targets or substrates in this way. The implication is that PDZ domains are involved in recruiting signaling proteins to protein complexes at the plasma membrane, and that this localization may be required for proper signal transduction.

4.2 Possible Mechanisms of Transmembrane Protein Crosslinking by PDZ Domains

There are several potential mechanisms by which PDZ-containing proteins like the MAGUKs might cluster their transmembrane targets (Fig. 4). One possibility (Fig. 4A) is that a protein with several PDZ domains could provide multiple binding sites for a particular transmembrane protein or combination of proteins. There are several examples of proteins with multiple PDZ domains (Table 1). These include many of the MAGUKs, which have three PDZ domains, and the protein tyrosine phosphatase PTPBas, which contains six PDZ domains. However, only one of the six PDZ domains in PTPbas binds the Fas receptor *in vitro* (Sato et al. 1995), and only one of the PDZ domains in MAGUKs like PSD-95 appears to have a high affinity for a particular ligand (Kim et al. 1995; Kornau et al. 1995; Müller et al. 1996). These observations make it difficult to imagine how a single protein like PSD-95 could crosslink two transmembrane proteins, unless crosslinking involved interactions of the other PDZ domains within the molecule with distinct transmembrane proteins within the same complex. Another possibility is that proteins with PDZ domains form dimers. Each dimer would have two identical PDZ domains, making it possible to crosslink two transmembrane proteins. There is genetic evidence that the *Drosophila* Dlg protein forms dimers (Woods and Bryant 1991), and immunoprecipitation assays suggest that PSD-95 dimerizes with PSD-93 (Kim et al. 1996) and that ZO-1 forms dimers with ZO-2 (Gumbiner et al. 1991). When forming heterodimers new combinations of additional binding functions can be brought together. Such interactions can be quite specific, since yet another synaptic MAGUK called SAP97 does not dimerize with PSD-93 but can bind K^+ channels (Kim et al. 1996). Presumably this specificity is important in the molecular architecture of synaptic complexes.

A third possibility is that transmembrane proteins are crosslinked via linkages to the cortical cytoskeleton (Fig. 4C). This would be accomplished by coupling PDZ domains to cytoskeletal binding motifs within the same polypeptide. In this way the PDZ domain would serve as the transmembrane linkage within a protein that can also interact with the cortical cytoskeleton. Several examples of this type protein already exist. For example, the erythrocyte membrane protein p55 and hdlg have been shown to bind directly to the actin-binding protein band 4.1 (Marfatia

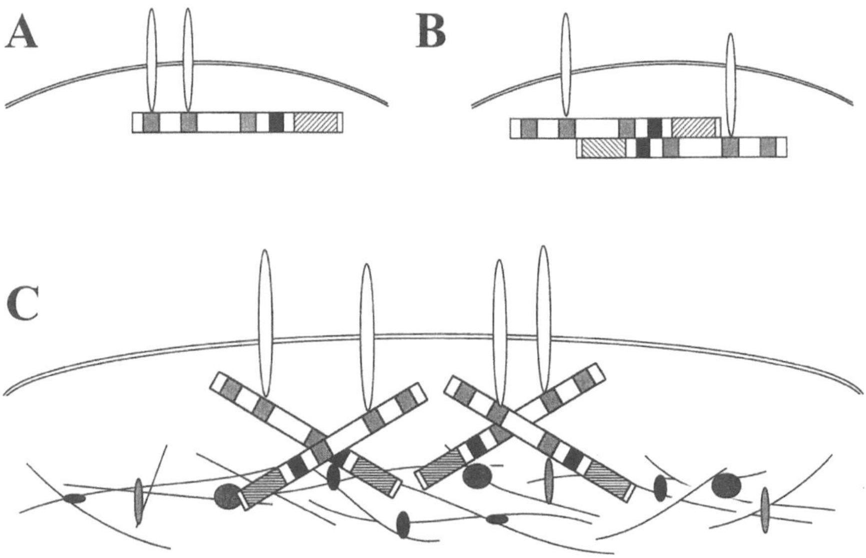

Fig. 4A–C. Three models for PDZ domain mediated organization of transmembrane proteins at the plasma membrane. In model **A** there are multiple functional PDZ domains in each MAGUK (*box*) which can bind to and crosslink one or more transmembrane proteins (*ovals*). In model **B** the oligomerization of a MAGUK enables crosslinking of a transmembrane protein that binds to only one specific PDZ domain within a MAGUK. In model **C** transmembrane proteins are crosslinked to the cortical cytoskeleton (*lines and shaded ovals*) via MAGUKs. Experimental evidence suggests that a combination of all three models is involved in the organization of subcortical domains. *Shaded box*, PDZ domain; *black box*, SH3 domain; *diagonal hatching*, GuK domain (figure used with permission from FANNING et al. 1996)

et al. 1996; LUE et al. 1996). Protein 4.1 is a member of the ERM family of tyrosine kinase substrates, which includes ezrin, radixin, moesin, and the product of the neurofibromatosis type 2 gene (TROFATTER et al. 1993). Protein 4.1 binds directly to spectrin-actin complexes and links these proteins to the plasma membrane via an association with the transmembrane protein glycophorin C. Since both p55 and hdlg also interact with transmembrane proteins via their PDZ domains, their interaction with band 4.1 would link these transmembrane proteins to the cortical cytoskeleton. PTPH1 and PTPMeg, tyrosine phosphatases that contain PDZ domains, also possess a domain similar to the region in the ERM proteins that mediates interactions with the cortical cytoskeleton, suggesting that they also may associate with the cytoskeleton (YANG and TONKS 1991; GU et al. 1991). Another group of PDZ containing proteins, the syntrophins, bind directly to the cytoskeletal protein dystrophin (AHN et al. 1996). Dystrophins interact with actin to form a specialized cytoskeletal array at the plasma membrane of muscles and synapses.

The reality, however, is that all three models are probably relevant. The formation of a complex cortical structure like a synapse or a tight junction probably involves linkages between many different transmembrane proteins as well as proteins of the cortical cytoskeleton. Not infrequently, PDZ proteins contain several other domains which mediate protein-protein interactions, such as SH3 domains, the GuK domian (KIM et al. 1997), and LIM domains. It is likely that these domains also contribute to the organization of plama membrane domains by creating linkages to other components of the cell cortex.

4.3 Regulation of PDZ-Mediated Interactions

Recent speculation has surrounded the question of whether the interactions of PDZ domains with their ligands can be regulated. Several PDZ domains have been shown to bind ATP, but this interaction does not appear to affect the *in vitro* binding of these domains to their peptide ligands (KISTNER et al. 1995; MARFATIA et al. 1996). Binding of many modular domains, i.e., PTB and SH2 domains, is promoted by tyrosine phosphorylation of their targets. However, one example already suggests phosphorylation interrupts binding to PDZ domains (COHEN et al. 1996). These investigators demonstrated that the inwardly rectifying K^+ channel Kir 2.3, which ends in the residues RRESAI, binds *in vitro* and *in vivo* to the PDZ2 domain of PSD-95. The Ser-440 residue within the C-terminal PDZ-binding is also a consensus site (RRXS) for phosphorylation by protein kinase A. In Kir-transfected fibroblasts activation of protein kinase A results in phosphorytaion of Ser-440, dissociation of the channel from PSD95, and inhibition of K^+ conductance. Mutation of Ser-440 to Ala eliminates the effects of PKA. These results imply binding of Kir 2.3 to PSD-95 not only clusters the channel but bestows PKA sensitivity to channel conductance. Given that many PDZ domain targets include a Ser, Thr or Tyr residue which could be phosphorylated, this may turn out to be a general mechansim to regulate binding.

5 Conclusions and Speculations

PDZ domains are an evolutionarily ancient form of protein-binding module. An important use of their binding function has been to organize signaling and cytoskeletal networks on the cytoplasmic surface of the plasma membrane. The frequent concatenation of PDZ domains and occasional ability to dimerize both contribute to their ability to cluster target proteins into functional domains. Their occasional coexpression with motifs which bind cytoskeletal proteins further enhances this function. An area for further research will be to define if PDZ domains are more than just a form of protein glue. Does binding ever induce an allosteric regulatory effect on the target protein? Is binding regulated? These seem reasonable questions given that the binding targets often express enzymatic or ion channel activity. These speculations are encouraged by the observation that PKA regulates conductance of the Kir channels through binding the PDZ domain of PSD95. Oddly, all PDZ interactions so far identified involve binding to either an integral protein on the plasma membrane or recruitment of a soluble protein to the plasma membrane. It is not obvious why PDZ domains could not be used to organize cytosolic or organellar complexes and this remains an area for investigation.

PDZ domains bind to distinct C-terminal peptide motifs, and we can begin to rationalize the basis for specificity. For example a hydrophic side chain at the extreme C-terminus of the target motif appears a universal requirement. Already two general subclasses of PDZ domain can be defined, namely those recognizing either Ser/Thr or a hydrophobic side chain at position −2 from the C-terminus. However, given the great diversity in primary sequences among PDZ domains it seems likely that many more subclasses exist whose target motifs remain to be defined.

We predict extensive use of PDZ domain-containing proteins organizing proteins on the plasma membrane . They may even be involved in establishing cell polarity. Searching the protein database for potential C-terminal ligands has already revealed an extensive list of potential targets for the defined specificities. In addition, many PDZ domains probably have highly divergent target peptides whose sequence we cannot even guess at. Much work remains to identify these new motifs and the structural basis for their recognition, as well as the structural basis for recognizing internal motifs and dimerization to other PDZ domains.

Acknowledgements. The authors would like to thank Bruce Stevenson, Peter Bryant, Stuart Kim, and Athar Chisti for helpful discussions and sharing information prior to publication and Christina Van Itallie for reviewing the manuscript. A.S.F. was supported in part by NRSA DK09261 from NIDDK and the Irwin M. Arias Postdoctoral Research Fellowship from the American Liver Foundation. This work was supported by NIH grants DK34898 and CA66263.

References

Ahn AH, Yoshida M, Anderson MS, Feener CA, Selig S, Hagiwara Y, Ozawa E, Kunkel LM (1994) Cloning of human basic A1, a distinct 59-kDa dystrophin-associated protein encoded on chromosome 8q23-24. Proc Natl Acad Sci USA 91:4446–4450

Ahn AH, Freener CA, Gussoni E, Yoshida M, Oxawa E, Kunkel LM (1996) The three human syntrophin genes are expressed in diverse tissues, have distinct chromosomal locations, and each bind to dystrophin and its relatives. J Biol Chem 271:2724–2730

Anbudurai PR, Mor TS, Ohad I, Shestakov SV, Pakrasi HB (1994) The ctpA gene encodes the C-terminal processing protease for the D1 protein of the photosystem II reaction center complex. Proc Natl Acad Sci USA 91:8082–8086

Anderson JM (1996) MAGUK magic. Curr Biol 6:326–329

Anderson B, Sims K, Regnery R, Robinson L, Schmidt MJ, Goral S, Hager C, Edwards K (1994) Detection of Rochalimaea henselae DNA in specimens from cat scratch disease patients by PCR. J Clin Microbiol 32:942–948

Apperson ML, Moon IS, Kennedy MB (1996) Characterization of densin-180, a new brain-specific synaptic protein of the O-sialoglycoprotein family. J Neurosci 16:6839–6852

Bannert N, Baier M, Werner A, Kurth R (1996) Interleukin-16 or not? Nature 381:30

Bass S, Gu Q, Christen A (1996) Multicopy suppressors of prc mutant *Escherichia coli* include two HtrA (DegP) protease homologs (HhoAB), DksA, and a truncated RlpA. J Bacteriol 178:1154–1161

Baumler AJ, Kusters JG, Stojiljkovic I, Heffron F (1994) Salmonella typhimurium loci involved in survival within macrophages. Infect Immun 62:1623–1630

Bhanot P, Brink M, Samos CH, Hsieh J, Wang Y, Macke JP, Andrew D, Nathans J, Nusse R (1996) A new member of the frizzled family from Drosophila functions as a Wingless receptor. Nature 382:225–230
Boucher JC, Martinez-Salazar J, Schurr MJ, Mudd MH, Yu H, Deretic V (1996) Two distinct loci affecting conversion to mucoidy in Pseudomonas aeruginosa in cystic fibrosis encode homologs of the serine protease HtrA. J Bacteriol 178:511–523
Brand SN, Tan X, Widger WR (1992) Cloning and sequencing of the petBD operon from the cyanobacterium Synechococcus sp. PCC 7002. Plan Mol Biol 20:481–491
Brenman JE, Chao DS, Gee SH, McGee AW, Craven SE, Santillano DR, Wu Z, Huang F, Xia H, Peters MF et al (1996) Interaction of nitric oxide synthase with the postsynaptic density protein PSD-95 and a1-syntrophin mediated by PDZ domains. Cell 84:757–767
Cho K, Hunt CA, Kennedy MB (1992) The rat brain postsynaptic density fraction contains a homolog of the *Drosophila* discs-large tumor suppressor protein. Neuron 9:929–942
Cohen NA, Brenman JE, Snyder SH, Bredt DS (1996) Binding of the inward rectifier K^+ channel Kir 2.3 to PSD-95 is regulated by protein kinase A phosphorylation. Neuron 17:759–767
Cruikshank WW, Center DM, Nisar N, Wu M, Natke BC, Theodore AC, Kornfeld H (1994) Molecular and functional analysis of a lymphocyte chemoattractant factor: association of biologic function with CD4 expression. Proc Natl Acad Sci USA 91:5109–5113
Devereux J, Haeberli P, Smithies O (1984) A comprehensive set of sequence analysis programs for the VAX. Nucleic Acids Res 12:387–395
Dimitratos Sd, Woods DF, Bryant PJ (1997) Camguk, Lin-2 and CASK: novel membrane-associated guanylate kinase homologs that also contain CaM kinase domains. Mech Dev (in press)
Doyle DA, Lee A, Lewis J, Kim E, Sheng M, MacKinnon R (1996) Crystal structure of a complexed and peptide-free membrane protein-binding domain: molecular basis of peptide recognition by PDZ. Cell 85:1067–1076
Duclos F, Boschert U, Sirugo G, Mandel JL, Hen R, Koenig M (1993) Gene in the region of the Friedreich ataxia locus encodes a putative transmembrane protein expressed in the nervous system. Proc Natl Acad Sci USA 90:109–113
Duclos F, Rodius F, Wrogemann K, Mandel JL, Koenig M (1994) The Friedreich Ataxia region: characterization of two novel genes and reduction of the critical region to 300kb. Hum Mol Genet 3:909–914
Ehlers MD, Tingley WG, Huganir RL (1995) Regulated subcellular distribution of the NR1 subunit of the NMDA receptor. Nature 269:1734–1737
Ehlers MD, Mammen AL, Lau L, Huganir RL (1996) Synaptic targeting of glutamate receptors. Curr Opin Cell Biol 8:484–489
Etemad-Moghadam B, Guo S, Kemphues KJ (1995) Asymmetrically distrubuted PAR-3 protein contributes to cell polarity and pindle alignment in early *C. elegans* embryos. Cell 83:743–752
Fanning AS, Anderson JM (1996) Protein-protein interactions: PDZ domain networks. Curr Biol 6:1385–1388
Fanning AS, Lapierre LA, Brecher AR, Van Itallie CM, Anderson JM (1996) Protein interactions in the tight junction: the role of MAGUK proteins in regulating tight junction organization and function. Curr Top Membr 43:211–235
Fleischmann RD, Adams MD, White O, Clayton RA, Kirkness EF, Kerlavage AR, Bult CJ, Tomb J-F, Dougherty BA, Merrick JM et al (1995) Whole-genome random sequencing and assembly of Haemophilus influenzae Rd. Science 269 (5223):496–512
Gatehouse JA, Lycett GW, Croy RR, Boulter D (1982) The post-translational proteolysis of the subunits of vicilin from pea (Pisum sativum L.). Biochem J 207:629–632
Glazebrook J, Ichige A, Walker GC (1996) Genetic analysis of *Rhizobium meliloti* bacA-phoA fusion results in identification of degP: two loci required for symbiosis are closely linked to degP. J Bacteriol 178:745–752
Gomperts SN (1996) Clustering membrane proteins: it's all coming together with the PSD-95/SAP90 protein family. Cell 84:659–662
Gu M, York JD, Warshawsky I, Majerus PW (1991) Identification, cloning and expression of a cytosolic megakaryocyte protein-tyrosine-phosphatase with sequence homology to cytoskeletal protein 4.1. Proc Natl Acad Sci USA 88:5867–5871
Gumbiner BT, Lowenkopf, Apatira D (1991) Identification of 160-kDa polypeptide that binds to the tight junction protein ZO-1. Proc Natl Acad Sci USA 88:3460–3464

Habets GG, Scholtes EH, Zuydgeest D, van der Kammen RA, Stam JC, Berns A, Collard JG (1994) Identification of an invasion-inducing gene, Tiam-1 that encodes a protein with homology to GDP-GTP exchangers for Rho-like proteins. Cell 77:537–549

Harrison SC (1996) Peptide-surface association: the case of PDZ and PTB domains. Cell 86:341–343

Hata Y, Butz S, Sudhof TC (1996a) CASK: a novel dlg/PDZ95 homolog with an N-terminal calmodulin-dependent protein kinase domain identified by interaction with neurexins. J Neurosci 16:2488–2494

Hata Y, Butz S, Sudhof TC (1996b) CASK: a novel dlg/PSD95 homologue with an N-terminal CaM kinase domain identified by interaction with neurexins. J Neurosci 16:2488–2494

Hoskins R, Hajinal A, Harp S, Kim SK (1995) The *C. elegans* vulval induction gene lin-2 encodes a member of the MAGUK family of cell junction proteins. Development 122:97–111

Inagaki N, Yamamoto Y, Mori H, Satoh K (1996) Carboxyl-terminal processing protease for the D1 precursor protein: cloning and sequencing of the spinach cDNA. Plant Mol Biol 30:39–50

Jesaitis LA, Goodenough DA (1994) Molecular characterization and tissue distribution of ZO-2, a tight junction protein homologous to ZO-1 and the *Drosophila* discs-large tumor suppressor protein. J Cell Biol 124:949–961

Johnson K, Charles I, Dougan G, Pickard D, O'Gaora P, Costa G, Ali T, Miller I, Hormaeche C (1991) The role of a stress-response protein in *Salmonella typhimurium* virulence. Mol Microbiol 5:401–407

Kiess M, Scharm B, Aguzzi A, Hajnal A, Klemenz R, Schwarte-Waldhoff I, Schafer R (1995) Expression of ril, a novel LIM domain gene, is down-regulated in Hras-transformed cells and restored in phenotypic revertants. Oncogene 10:61–68

Kim E, Niethammer M, Rothschild A, Jan YN, Sheng S (1995) Clustering of the Shaker-type K^+ channels by direct interaction with the PSD-95/SAP90 family of membrane-associated guanylate kinases. Nature 378:85–88

Kim E, Cho KO, Rothschild A, Sheng M (1996) Heteromultimerization and NMDA receptor clustering activity of chapsyn-110, a novel member of the PSD-95 family of synaptic proteins. Neuron 17:103–113

Kim E, Naisbitt S, Hseuh Y-P, Rao A, Rothschild A, Craig AM, Sheng M (1997) GKA, a novel synaptic protein that interacts with the guanylate Kinase-like domain of the PSD95/SAP90 family of channel clustering molecules. J Cell Biol 136:669–678

Kistner U, Wenzel BM, Veh RW, Cases-Langhoff C, Garner AM, Appeltauer U, Voss B, Gundelfinger ED, Garner CC (1993) SAP90, a rat presynaptic protein related to the product of the *Drosophila* tumor suppressor gene dlg-A. J Biol Chem 268:4580–4583

Kistner U, Garner CC, Linial M (1995) Nucleotide binding by the synapse associated protein SAP90. FEBS Let 359:159–163

Kligensmith J, Yang Y, Axelrod JD, Beier DR, Perrimon N, Sussman DJ (1996) Conservation of dishevelled structure and function between flies and mice: isolation and characterization of Dvl2. Mech Dev 58:15–26

Kornau HC, Schenker LT, Kennedy MB, Seeburg PH (1995) Domain interaction between NMDA receptor subunits and the postsynaptic density protein PSD-95. Science 269:1737–1740

Lipinska B, Sharma S, Georgopoulos C (1988) Sequence analysis and regulation of the htrA gene of *Escherichia coli*: a sigma 32-independent mechanism of heat-inducible transcription. Nucleic Acids Res 16:10053–10067

Lipinska B, Fayet O, Baird L, Georgopoulos C (1989) Identification, characterization, and mapping of the *Escherichia coli* htrA gene, whose product is essential for bacterial growth only at elevated temperatures. J Bacteriol 171:1574–1584

Lipinska B, Zylicz M, Georgopoulos C (1990) The HtrA (DegP) protein, essential for *Escherichia coli* survival at high temperatures, is an endopeptidase. J Bacteriol 172:1791–1797

Lue RA, Marfatia SM, Branton D, Chishti AH (1994) Cloning and characterization of hdlg: the human homologue of the *Drosophila* discs large tumor suppressor binds to protein 4.1. Proc Natl Acad Sci USA 91:9818–9822

Lue RA, Brandin E, Chan EP, Branton D (1996) Two independent domains of hDlg are sufficient for subcellular targeting: the PDZ1-2 conformational unit and an alternatively spliced domain. J Cell Biol 135:1125–1137

Maekawa K, Imagawa N, Nagamatsu M, Harada S (1994) Molecular cloning of a novel protein-tyrosine phosphatase containing a membrane-binding domain and GLGF repeats. FEBS Lett 337:200–206

Marfatia SM, Morais Cabral JH, Lin L, Hough C, Bryant PJ, Stolz L, Chishti AH (1996) Modular organization of the PDZ domains in the human discs-large protein suggests a mechanism for coupling PDZ domain-binding proteins to ATP and the membrane cytoskeleton. J Cell Biol 135:753–766

Matsumine A, Ogai A, Senda T, Okumura N, Satoh K, Baeg GH, Kawahara T, Kobayashi S, Okada M, Toyoshima K et al (1996) Binding of APC to the human homolog of the *Drosophila discs large* tumor suppressor protein. Science 272:1020–1023

Mazoyer S, Gayther SA, Nagai MA, Smith SA, Dunning A, vanRensburg EJ, Albertsen H, White R, Ponder BAJ (1995) A gene (DLG2) located at 17q12-q21 encodes a new homologue of the *Drosophila* tumor suppressor dlg-A. Genomics 28:25–31

Miyamoto H, Nihonmatsu I, Kondo S, Ueda R, Togashi S, Hirata K, Ikegami Y, Yamamoto D (1995) Canoe encodes a novel protein containing a GLGF/DHR motif and functions with Notch and scabrous in common developmental pathways in *Drosophila*. Genes Dev 9:612–625

Mizuno K, Okano I, Ohashi K, Nunoue K, Kuma K, Miyata T, Nakamura T.(1994) Identification of a human cDNA encoding a novel protein kinase with two repeats of the LIM/double zinc finger motif. Oncogene 9:1605–1612

Morias-Cabral JH, Petosa C, Sutcliffe MJ, Raza S, Byron O, Poy F, Marfatia SM, Chishti AH, Liddington RC (1996) Crystal structure of a PDZ domain from the human homolog of discs-large protein. Nature 384:649–652

Müller BM, Kistner U, Veh RW, Cases-Langhoff C, Becker B, Gundelfinger ED, Garner CC (1995) Molecular characterization and spatial distribution of SAP97, a novel presynaptic protein homologous to SAP90 and the *Drosophila discs-large* tumor suppressor protein. J Neurosci 15:2354–2366

Müller BM, Bistner U, Kindler S, Chung WK, Kuhlendahl S, Fenster SD, Lau L, Veh RW, Huganir RL, Gundelfinger ED et al. (1996) SAP102, a novel postsynaptic protein that interacts with NMDA receptor complexes *in vivo*. Neuron 17:255–265

Nagase T, Seki N, Tanaka A, Ishikawa K, Nomura N (1995) Prediction of the coding sequences of unidentified human genes. IV. The coding sequences of 40 new genes (KIAA0121-KIAA0160) deduced by analysis of cDNA clones from human cell line KG-1. DNA Res 2:167–174

Nakane M, Schmidt HHHW, Pollock JS, Forstermann U, Murad F (1993) Cloned human brain nitric oxide synthase is highly expressed in skeletal muscle. FEBS Lett 316:175–180

Niethammer M, Kim E, Sheng M (1996) Interaction between the C-terminus of NMDA receptor subunits and multiple members of the PSD-95 family of membrane-associated guanylate kinases. J Neurosci 16:2157–2163

Pizzuti A, Novelli G, Mari A, Ratti A, Colosimo A, Amati F, Penso D, Sangiuolo F, Calabrese G, Palka G et al. (1996) Human homologue sequences to the *Drosophila dishevelled* segment-polarity gene are deleted in the DiGeorge syndrome. Am J Hum Genet 58:722–729

Prasad R, Gu Y, Alder H, Nakamura T, Canaani O, Saito H, Huebner K, Gale RP, Nowell PC, Kuriyama K et al (1993) Cloning of the ALL-1 fusion partner, the AF-6 gene, involved in acute myeloid leukemias with the t(6;11) chromosome translocation. Cancer Res 53:5624–5628

Ruff P, Speicher DW, Husain-Chishti A (1991) Molecular identification of a major palmitoylated erythrocyte membrane protein containing the src homology 3 motif. Proc Natl Acad Sci USA 88:6595–6599

Saras J, Claesson-Welsh L, Heldin C, Gonez LJ (1994) Cloning and characterization of PTPL1, a protein tyrosine phosphatase with similarities to cytoskeletal-associated proteins. J Biol Chem 269:24082–24089

Sato T, Irie S, Kitada S, Reed JC (1995) Fap-1: a protein tyrosine phosphatase that associates with Fas. Science 268:411–415

Schiaffino MV, Bassi MT, Rugarli EI, Renieri A, Galli L, Ballabio A (1995) Cloning of a human homologue of the *Xenopus laevis* APX gene from the ocular albinism type 1 critical region. Hum Mol Genet 4:373–382

Sheng M (1996) PDZs and receptor/channel clustering: rounding up the latest suspects. Neuron 17:575–578

Shestakov SV, Anbudurai PR, Stanbekova GE, Gadzhiev A, Lind LK, Pakrasi HB (1994) Molecular cloning and characterization of the ctpA gene encoding a carboxyl-terminal processing protease. Analysis of a spontaneous photosystem II-deficient mutant strain of the Cyanobacterium *Synechocystis* sp. PCC 6803. J Biol Chem 269:19354–19359

Shieh B, Niemeyer B (1995) A novel protein encoded by the InaD gene regulates recovery of visual transduction in *Drosophila*. Neuron 14:201–210

Shieh BH, Zhu MY (1996) Regulation of the TRP channel by INAD in *Drosophila* photoreceptors. Neuron 16:991–998

Simske JS, Kaech SM, Harp SA, Kim SK (1996) LET-23 receptor localization by the cell junction protein LIN-7 during *C. elegans* vulval induction. Cell 85:195–204

Smith SA, Holik P, Stevens J, Mazoyer S, Melis R, Williams B, White R, Albertsen H (1996) Isolation of a gene (DLG3) encoding a second member of the discs-large family on chromosome 17q12-q21. Genomics 31:145–150

Sone M, Hoshino M, Suzuki E, Kuroda S, Kaibuchi K, Nakagoshi H, Saigo K, Nabeshima Y, Hama C (1997) Still life, a protein in synaptic terminals of drosophila homologous to GDP-GTP exchangers. Science 275:543–547

Songyang Z, Fanning AS, Fu C, Xu J, Marfatia SM, Chishti AH, Crompton A, Chan AC, Anderson JM, Cantley LC (1997) Recognition of unique carboxyl-terminal motifs by distinct PDZ domains. Science 275:73–77

Staudinger J, Zhou J, Burgess R, Elledge SJ, Olson EN (1995) Pick1: a perinuclear binding protein and substrate for protein kinase C isolated by the yeast two-hybrid system. J Cell Biol 128:263–271

Sussman DJ, Klingensmith J, Salinas P, Adams PS, Nusse R, Perrimon N (1994) Isolation and characterization of a mouse homolog of the *Drosophila* segment polarity gene dishevelled. Dev Biol 166:73–86

Taguchi F, Yamamoto Y, Satoh K (1995) Recognition of the structure around the site of cleavage by the carboxyl-terminal processing protease for D1 precursor protein of the photosystem II reaction center. J Biol Chem 270:10711–10716

Takahisa M, Togashi S, Suzuki T, Kobayashi M, Murayama A, Kondo K, Miyake T, Ueda R (1996) The *Drosophila tamou* gene, a component of the activating pathway of extramacrochaetae expression, encodes a protein homologous to mammalian cell-cell junction-associated protein ZO-1. Genes Dev 10:1783–1795

Tatum FM, Cheville NF, Morfitt D (1994) Cloning, characterization and construction of htrA and htrA-like mutants of *Brucella abortus* and their survival in BALB/c mice. Microb Pathog 17:23–36

Taylor DE, Hiratsuka K (1990) Use of non-radioactive DNA probes for detection of *Campylobacter jejuni* and *Campylobacter coli* in stool specimens. Mol Cell Probes 4:261–271

Tejedor FJ, Bokari A, Rogero O, Gorczyca M, Zhang J, Kim E, Sheng M, Budnik V (1997) Essential role for dlg in synaptic clustering of Shaker K^+ channels *in vivo*. J Neurosci 17:152–159

Theisen H, Purcell J, Bennett M, Kansagara D, Syed A, Marsh JL (1994) Dishevelled is required during wingless signaling to establish both cell polarity and cell identity. Development 120:347–360

Trofatter JA, MacCollin MM, Rutter JL, Murrell JR, Duyao MP, Parry DM, Eldredge R, Kley N, Menon AG, Pulaski K (1993) A novel moesin, ezrin-, radixin-like gene is a candidate for the neurofibromatosis 2 tumor suppressor. Cell 72:791–800

Walden PD, Cowan NJ (1993) A novel 205-kilodalton testis-specific serine/threonine protein kinase associated with microtubules of the spermatid manchette. Mol Cell Biol 13:7625–7635

Wang H, Harrison-Shostak DC, Lemasters JJ, Herman B (1995) Cloning of a rat cDNA encoding a novel LIM domain protein with high homology to rat RIL. Gene 165:267–271

Watanabe G, Saito Y, Madaule P, Ishizaki T, Fujisawa K, Morii N, Mukai H, Ono Y, Kakizuka A, Narumiya S (1996) Protein kinase N (PKN) and PKN-related protein rhophilin as targets of small GTPase Rho. Science 271:645–648

Weinman EJ, Steplock D, Wang Y, Shenolikar S (1995) Characterization of a protein cofactor that mediates protein kinase A regulation of the renal brush border membrane $Na^{(+)}$-H^+ exchanger. J Clin Invest 95:2143–2149

Willott E, Balda MS, Fanning AS, Jameson B, Van Itallie C, Anderson JM (1993) The tight junction protein ZO-1 is homologous to the *Drosophila discs-large* tumor suppressor protein of septate junctions. Proc Natl Acad Sci USA 90:7834–7838

Woods DF, Bryant PJ (1989) Molecular cloning of the Lethal(1) Discs Large-1 oncogene of *Drosophila*. Dev Biol 134:222–235

Woods DF, Bryant PJ (1991) The *discs-large* tumor suppressor gene of *Drosophila* encodes a guanylate kinase homolog localized at septate junctions. Cell 66:451–464

Wu RY, Gill GN (1994) LIM domain recognition of a tyrosine-containing tight turn. J Biol Chem 269:25085–25090

Yang Q, Lijam N, Sussman DJ, Tsang M (1996) Genomic organization of mouse *Dishevelled* genes. Gene 180:121–123

Yang Q, Tonks NK (1991) Isolation of a cDNA clone encoding a human protein-tyrosine phosphatase with homology to the cytoskeletal-associated proteins band 4.1, ezrin, and talin. Proc Natl Acad Sci USA 88:5949–5953

Mechanism and Function of Signaling by the TGFβ Superfamily

P.A. Hoodless[1] and J.L. Wrana[1,2]

1 Introduction

Transforming growth factor β (TGFβ) was initially identified based on its ability to induce the anchorage-independent growth of normal rat kidney fibroblasts (Roberts et al. 1981). It is now apparent that TGFβ is the founding member of a superfamily of growth and differentiation factors that includes almost 40 members from animals as diverse as *C. elegans*, *Drosophila* and humans. The superfamily is

[1]Program in Developmental Biology and Division of Gastroenterology, The Hospital for Sick Children, 555 University Avenue, Toronto, ON, Canada
[2]Department of Medical Genetics and Microbiology, University of Toronto, Toronto, ON, Canada

generally subdivided into three groups, the prototypic TGFβs, the activins and the bone morphogenetic proteins (BMPs). However, with new members constantly being identified these divisions are becoming increasingly difficult to define. Nevertheless, the study of the biology of this large family has provided us with some interesting and surprising insights into how these factors can regulate a staggering array of diverse developmental and physiological processes. In addition, major advances have been made in elucidating the mechanism of signaling by TGFβ-like factors. In this review we will focus on recent developments in our understanding of TGFβ signaling and on the multiple roles these factors play in controlling inductive interactions and patterning during development.

2 Mechanisms of Signal Transduction

2.1 The Ligands

The mature, biologically active forms of TGFβ superfamily members are typically homodimers of two cysteine-rich, 12–15 kDa subunits, linked by a single disulfide bond. These ligands are synthesized as part of a larger inactive precursor. Release of the biologically active C-terminal portion of this precursor occurs at a characteristic tetrabasic cleavage site and is mediated by furin proteases (Gentry et al. 1988). The amino terminal proregion is not required for the biological activity of the factors, but plays a role in mediating the correct folding and dimerization of the mature regions and is required for efficient proteolytic processing at the tetrabasic site. Some members of the TGFβ superfamily are constitutively and efficiently processed in most cells and are secreted in a biologically active form. However, there are exceptions to this, and many of these factors are processed very poorly. A notable example is the *Xenopus* ligand, Vg1, whose inefficient processing has led to the suggestion that specific, developmentally regulated pathways are required for its secretion in an active form (Dale et al. 1993; Thomsen and Melton 1993). Since correct processing of TGFβ factors is necessary for their biological activity, regulation at this level could provide an important point for controlling activity of the secreted factors.

2.2 The Receptors

Members of the TGFβ superfamily signal through a conserved family of transmembrane ser/thr kinase receptors, all of which have a characteristic structure that includes a short, cysteine rich extracellular domain, a single transmembrane domain and an intracellular serine/threonine kinase domain (reviewed in Attisano and Wrana 1996; Massagué and Weis-Garcia 1996; Miyazono et al. 1994). Many of these receptors have now been identified in a wide variety of vertebrates and invertebrates including humans, mouse, *Xenopus*, *C. elegans* and *Drosophila*. Functional characterization and comparison of their primary amino acid sequence indicates that these receptors can be divided into two classes, the type I and the type II receptors. The type I receptor class is distinguished by a highly conserved sequence known as the "GS domain," which contains a repetitive glycine-serine motif and is located between the transmembrane and kinase domains. The kinase domains of type I receptors are relatively well conserved, while the type II receptors possess more distantly related kinase domains. In addition, the type II receptor has a C-terminal extension which is rich in serine and threonine and, in the case of the mammalian BMPRII and *C. elegans* DAF4, can be quite large (Liu et al. 1995; Nohno et al. 1995; Rosenzweig et al. 1995). It is not clear what role this region plays, since it does not seem to be required for type II receptor function (Wieser et al. 1993). In addition to these signaling receptors, TGFβ binds to two other transmembrane proteins, betaglycan (or type III receptor) and endoglin (reviewed in Attisano et al. 1994). In the case of betaglycan, this binding may play a role in controlling access of the TGFβ2 isoform to the signaling receptors (Lin et al. 1995; López-Casillas et al. 1993). The role of endoglin in TGFβ signaling is less clear. This protein bears some sequence similarity with betaglycan, is expressed at high levels on endothelial cells and when overexpressed in a monocyte cell line can modulate TGFβ-responsiveness (Lastres et al. 1996). However, both endoglin and the orphan receptor, TSRI (ALK1), are mutated in hereditary hemorrhagic telangiectasia (Johnson et al. 1996; McAllister et al. 1994), suggesting that endoglin may fulfill functions in a TGFβ-independent pathway.

Considerable biochemical and genetic data have shown that signaling by TGFβ-like factors occurs through heteromeric complexes of type II and type I ser/thr kinase receptors. For instance, in mink lung epithelial cells lacking either component of the TGFβ heteromeric receptor complex, signaling is abolished (Cárcamo et al. 1994; Wrana et al. 1992). Furthermore, studies in *Drosophila* show that signaling by DPP (an ortholog of mammalian BMP2) requires both receptor II (PUNT) and receptor I (TKV) (Brummel et al. 1994; Letsou et al. 1995; Nellen et al. 1994; Penton et al. 1994; Ruberte et al. 1995) and in mammalian cells, BMP signaling similarly occurs through a complex of type II and type I receptors (Hoodless et al. 1996; Liu et al. 1995; Yamashita et al. 1995). Thus, heteromeric complexes of type II and type I receptors appear to be a general requirement for signaling by the TGFβ superfamily.

Studies on TGFβ and activin signaling have provided a mechanistic understanding of how the type II and type I kinase domains interact to initiate signaling (Attisano et al. 1996; Chen and Weinberg 1995; Chen et al. 1995; Souchelnytskyi et al. 1996; Wrana et al. 1994a). In the absence of ligand, both type II and type I receptors appear to exist as independent homomers on the cell surface (reviewed in ten Dijke et al. 1996). Heterotetrameric complex formation is initiated when ligand binds to receptor II, which then leads to recruitment of receptor I. Receptor II, which is a constitutively active kinase, then phosphorylates receptor I on serine and threonine residues within the 'GS domain,' which is found in all type I receptors (Souchelnytskyi et al. 1996; Wieser et al. 1995; Wrana et al. 1994a). Once phosphorylated, receptor I is activated to signal to downstream targets (Fig. 1A). The recent demonstration that MADR2 is a substrate of the TGFβ receptor (see below), and that phosphorylation is mediated by the type I kinase, highlights the critical role this receptor plays in activating intracellular signaling pathways.

The downstream position of receptor I has important implications when the response of cells to TGFβ-like factors is considered. For example, the TGFβ and activin type I receptors (TβRI and ActRIB, respectively), which have kinase domains that are over 90% identical, mediate common biological responses to their respective ligands (CÁRCAMO et al. 1994). Furthermore, BMP receptor complexes

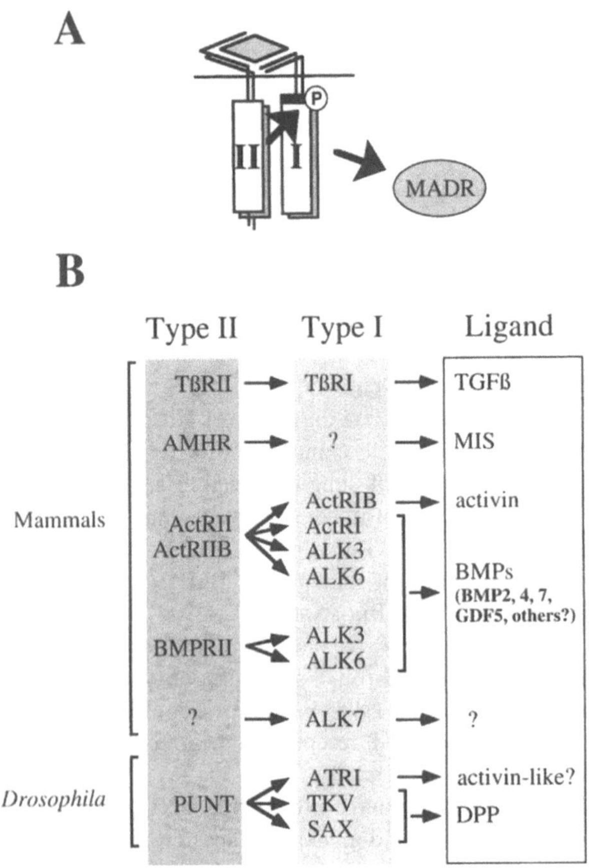

Fig. 1A, B. TGFβ superfamily signaling through ser/thr kinase receptors. **A** Schematic representation of the general mechanism of activation of ser/thr kinase receptors. Ligand (*diamond*) induces formation of a heterotetrameric complex of type II receptors (*II*) and type I receptors (*I*). The kinase of type II phosphorylates the GS domain (*black box*), activating the type I receptor kinase, which signals to downstream targets (MADR proteins). **B** Combinatorial nature of heteromeric ser/thr kinase receptor complexes. Mammalian and *Drosophila* type II receptors (*dark box*) can interact with and activate a number of type I receptors (*light box*) to generate a variety of heteromeric complexes. Ligands known to interact with these receptor complexes are shown (*white box*). The ligand and type II receptor for the recently identified type I receptor, ALK7, are unknown. (RYDEN et al. 1996; TSUCHIDA et al. 1996)

that contain either ALK3 or ALK6, which also have highly related kinase domains, similarly activate a common BMP-specific signaling pathway (Hoodless et al. 1996). Thus, the type I receptor plays a central role in defining the nature of the biological response to ligand.

Although some ser/thr kinase receptors appear to bind a limited subset of ligands, others can bind multiple members of the TGFβ superfamily. TβRII and TβRI appear to be quite specific, interacting with and transducing signals only for TGFβ1, 2 and 3 (Wrana et al. 1992). In contrast, the type II receptors ActRII and ActRIIB appear to be capable of binding activin, BMP2, BMP7, GDF5 and likely many other members of the superfamily (reviewed in Attisano and Wrana 1996). Moreover, depending on the ligand that is involved, recruitment of different sets of type I receptors leads to the formation of distinct receptor complexes that mediate distinct biological responses. Thus, BMP2-induced formation of complexes between ActRIIB and either ALK3 or ALK6 mediates BMP signaling (Hoodless et al. 1996), while activin induces the same type II receptor to associate with ActRIB to initiate activin signaling (Attisano et al. 1996; Tsuchida et al. 1995). Surprisingly, this feature is conserved in *Drosophila* receptors where PUNT, a type II receptor, can associate with ATRI, TKV and SAX (Wrana et al. 1994b; Letsou et al. 1995), suggesting that this characteristic may play a functional role in TGFβ signaling. This mixing and matching of different type II and type I receptors leads to a combinatorial model for ligand-receptor interaction (Fig. 1B). It will be interesting to see whether any of the opposing effects often observed for activins and BMPs during early development are mediated in part through their interaction with shared receptors.

2.3 GDNF

Glial-derived neurotrophic factor (GDNF) is a distant relative of TGFβ and is a potent survival factor for central and peripheral neurons. GDNF provides a notable exception to the general observation that members of the TGFβ superfamily signal through serine/threonine kinase receptors. In the course of identifying receptors for GDNF, several groups identified a novel glycolipid-anchored receptor known as GDNFR-α (Jing et al. 1996; Treanor et al. 1996), which, together with the orphan tyrosine kinase receptor Ret, appears to form a functional GDNF receptor complex (Durbec et al. 1996; Jing et al. 1996; Treanor et al. 1996; Trupp et al. 1996). These findings were particularly surprising on two accounts. Not only is this the first example of a TGFβ-like factor signaling through a tyrosine kinase receptor, but it is also the first example of a tyrosine kinase receptor that requires a glycolipid-anchored receptor component (GDNFR-α) to bind ligand. It would certainly be intriguing if other members of the TGFβ superfamily, which are known to signal through ser/thr kinase receptors, were also found to signal through tyrosine kinase receptors. Alternatively, GDNF and the recently identified ligand neurturin (Kotzbauer et al. 1996) may represent the prototypic members of a new family of factors that signal through a distinct receptor system.

2.4 The Signaling Pathways

In the search for intracellular components of the signal transduction pathway, many proteins have been identified by yeast two-hybrid screens using the receptors as bait. These include TGFβ-receptor interacting protein-1 (TRIP-1), farnesyl transferase-α (FT-α), apolipoprotein J/clusterin and FK506/rapamycin binding protein (FKBP12) (reviewed in ATTISANO and WRANA 1996). However, while these proteins may play a regulatory role in receptor function, there is little evidence to indicate that they are required for signaling by TGFβ ligands. In constrast, using genetic screens, two components have been identified which may be directly involved in signal transduction: a MAP kinase kinase kinase, TAK-1, and its putative regulator, TAB-1, and the family of proteins related to the *Drosophila* gene *Mothers against dpp* (MAD).

2.4.1 TAK-1/TAB-1

TAK-1 (TGFβ-activated kinase) was identified in a yeast screen for novel mammalian components of mitogen-activated protein kinase (MAPK) cascades (YAMAGUCHI et al. 1995). The potential role of TAK-1 as a mediator of TGFβ signaling is suggested by the observation that an activated form of TAK-1, in which the N-terminal 22 amino acids are deleted, induces expression of a TGFβ-responsive promoter, while a kinase-deficient version of TAK-1 blocks TGFβ-dependent responses. In addition, the kinase activity of TAK-1, measured in vitro using an exogenous substrate, is activated by TGFβ but not EGF, further supporting a putative role for TAK-1 in TGFβ signal transduction. However, since TAK-1 activity is also modulated by BMP4, the role of this factor in maintaining specificity in response to different ligands is unclear. More recently, a TAK-1 associated protein, TAB-1, was cloned using a yeast two-hybrid system (SHIBUYA et al. 1996). This protein activates TAK-1 by directly binding to its catalytic domain and can also affect gene induction by TGFβ. Together, TAK-1 and TAB-1 may function in TGFβ signaling.

2.4.2 MAD-Related Proteins

While the role of TAK-1 in TGFβ signal transduction remains to be fully elucidated, it is now evident that MAD-related (MADR) proteins (also known as Smad proteins) play a critical role in the signal transduction pathways of the TGFβ superfamily. The first family member, the *Drosophila Mothers against dpp* gene (*Mad*), was identified, along with a second locus *Medea*, in a screen for maternal effect enhancers of mutations in *dpp* (RAFTERY et al. 1995). Dorsal-ventral patterning and imaginal disk defects resulting from mutations in *dpp* were enhanced in *Mad* or *Medea* loss of function mutations, suggesting that these genes encode rate-limiting components required for DPP signaling. *Mad* homozygous mutants exhibit defects in dorsal-ventral patterning, imaginal disk development and midgut morphogenesis reminiscent of *dpp* mutant phenotypes (SEKELSKY et al. 1995). Moreover, MAD is required for DPP function specifically in the cells which respond to DPP, indicating that MAD is a component of the DPP signal transduction pathway (Newfeld et al. 1996). Further evidence in *Drosophila* was supplied by studies involving a constitutively active DPP type I receptor, TKV. Ectopic expression of activated TKV in wing imaginal disks yielded a mutant wing phenotype (Hoodless et al. 1996). This mutant phenotype could be suppressed in flies which were heterozygous for null alleles of *Mad*, conclusively placing MAD downstream of the receptor. This was supported by similar experiments in which mutations of *Mad* were able to rescue eye defects resulting from expression of activated TKV using an eye specific promoter (Wiersdorff et al. 1996).

Three related homologues of MAD, *sma-2*, *sma-3* and *sma-4*, have been identified in *C. elegans* (Savage et al. 1996). Mutations in these three genes yield a phenotype similar to that observed in mutants of the type II receptor gene *daf*-4. The DAF-4 receptor, which likely binds a BMP-like ligand, functions in dauer larvae formation and in the development and morphogenesis of the male tail rays. Identification of mutants in male tail morphology led to the cloning of the three *sma* genes all of which are essential for tail ray development, suggesting that these genes are components of the DAF-4 signaling pathway. This observation is bolstered by mosaic analysis which indicates that *sma-2* is required in the same cell as DAF-4 (Savage et al. 1996).

Several vertebrate homologues of Mad have now been identified in human, mouse, rat and *Xenopus* (reviewed in Derynck and Zhang 1996, Massagué 1996, Wrana and Attisano 1996). The first human homologue, DPC4 (or MADR4), was identified as a candidate tumor suppressor gene involved in pancreatic cancer (Hahn et al. 1996), while mouse MADR2 was cloned in a functional screen in *Xenopus* to identify mesoderm inducers (Baker and Harland 1996). At present, at least six distinct MADR proteins (MADR1 through 6) have been reported in vertebrates, although other members are likely to exist (Derynck and Zhang 1996; Massagué 1996; Wrana and Attisano 1996). Sequence comparisons of this family indicates the presence of three distinct domains, a highly conserved amino-terminal MH1 domain (for MAD homology 1), a highly conserved carboxy-terminal MH2 domain and a central, divergent, nonconserved region (Fig. 2A). Analysis of the sequence of MADR proteins reveals that there are no known structural motifs. However, all mutations identified in *Drosophilia*, *C. elegans* and human cancers map to highly conserved residues within the MH domains, indicating that these regions are critical to the function and regulation of the protein (Eppert et al. 1996; Hahn et al. 1996; Riggins et al. 1996; Savage et al. 1996; Sekelsky et al. 1995). Closer analysis indicates that MADR proteins in vertebrates can be paired, based on amino acid sequence homology, with MADR1 and MADR5 (Dwarfin C) comprising a pair which are 90% identical and MADR2 and MADR3 forming a second pair which are 91.5% identical. DPC4 is more distantly related to the other family members, sharing only 42% amino acid identity with MADR1 and 45% with MADR2. In addition, MADR2 possesses two inserts in the MH1 domain while DPC4 contains an insert in the MH2 domain (Fig. 2A).

2.4.3 Ligand Specific Regulation of MADR Proteins

While the amino acid sequence of MADR proteins offers no clues as to how these proteins function in mediating intracellular signaling, biochemical studies have shown that MADR proteins are rapidly phosphorylated in response to ligand (Eppert et al. 1996; Hoodless et al. 1996; Nakao et al. 1997; Yingling et al. 1996; Zhang et al. 1996). This phosphorylation occurs primarily on serine residues in a single tryptic peptide (Eppert et al. 1996; Hoodless et al. 1996; Macías-Silva et al. 1996). Remarkably, this phosphorylation is highly specific. Thus, phosphorylation of MADR1 has only been observed in response to BMP signaling through ALK3 and ALK6 (Hoodless et al. 1996), while MADR2 is phosphorylated by TGFβ or activin signaling through TβRI and ActRIB, respectively (Fig. 2B, Eppert et al.

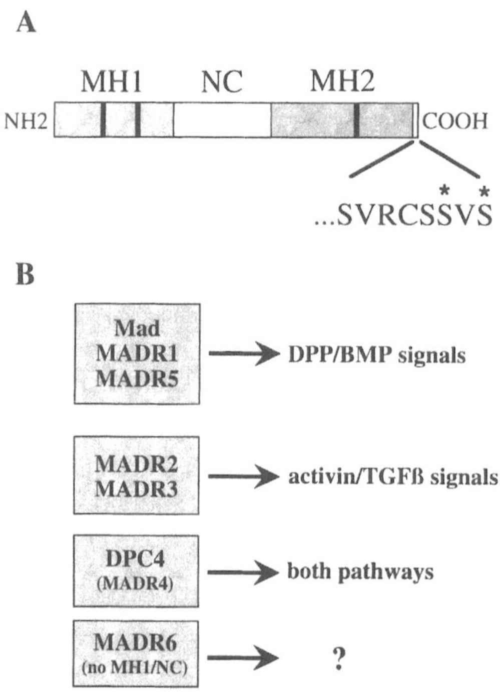

Fig. 2A, B. MAD-related (MADR) proteins. **A** Schematic representation of the structure of MADR proteins. The position of the MH1 (*light box*), MH2 domain (*dark box*) and the nonconserved linker domain (*NC, white box*) is shown. The presence of amino acid inserts in the MH1 domain of MADR2 and the MH2 domain of DPC4, relative to other MADR proteins, are indicated (*vertical black lines*). The amino acid sequence of the C-terminus of MADR2 and position of the phosphorylated serine residues within this sequence (*asterisks*) is shown. **B** Specific function of MADR proteins in TGFβ signaling. MADR proteins are grouped according to structural and functional characteristics. The prototypic *Drosophila* Mad which functions in DPP signaling and the highly related mammalian proteins MADR1 and MADR5 all signal in BMP pathways. In contrast, the highly related mammalian MADR2 and MADR3 proteins function in activin/TGFβ signaling pathways. DPC4 is more distantly related and may function in both pathways. MADR6, which has no MH1 or nonconserved region, was identified in an EST search and its function in signaling is not known

1996, LAGNA et al. 1996, NAKAO et al. 1997). The observation that both activin and TGFβ were able to induce phosphorylation of MADR2 supports previous observations that suggested these two ligands signal through highly related type I receptors to activate common intracellular pathways and common biological responses (ATTISANO et al. 1996; CÁRCAMO et al. 1994). The studies on the regulation of MADR protein phosphorylation also showed that a point mutation in MADR1 that corresponds to a null allele of *Mad* in *Drosophila* is not phosphorylated in response to BMP2 (HOODLESS et al. 1996). Furthermore, nonfunctional mutants of MADR2 identified in colorectal carcinomas also fail to be phosphorylated (EPPERT et al. 1996). Collectively, these results indicate that phosphorylation of MADR1 or MADR2 in response to ligand is critical for protein function and signal transduction.

The specific role of MADR1 in BMP and MADR2 in activin/TGFβ signaling is supported by a number of functional studies. In *Xenopus*, overexpression of different MADR proteins is sufficient to initiate developmental signals that recapitulate the responses observed for their respective ligands. For example, expression of XMAD1 (the *Xenopus* homologue of MADR1) induces ventral mesoderm in ectodermal explant cultures similar to the effects of the ligands BMP2 and 4 (GRAFF et al. 1996; THOMSEN 1996). Furthermore, expression of XMAD1 can rescue phenotypes generated by a dominant-negative BMP2/4 receptor, consistent with XMAD1 functioning in BMP2/4 signaling (GRAFF et al. 1996; THOMSEN 1996). In contrast, expression of XMAD2 (the *Xenopus* homologue of MADR2) induces dorsal mesoderm, similar to that observed in response to the ligands activin, Vg1 or nodal (BAKER and HARLAND 1996; GRAFF et al. 1996). Studies like these also showed that *Drosophila* Mad and human homologues of XMAD1 and XMAD2 function similarly to their *Xenopus* counterparts, emphasizing the high degree of conservation observed in this pathway across diverse species (BAKER and HARLAND 1996; EPPERT et al. 1996; LIU et al. 1996; NEWFELD et al. 1996).

In mammalian cells, differential regulation of transcription by MADR proteins has also been observed using the TGFβ and activin responsive promoter, 3TP, which combines portions of the plasminogen activator inhibitor (PAI-1) promoter with 3 TPA responsive elements. This promoter is highly responsive to TGFβ and activin, but responds poorly to BMPs. Consistent with this, 3TP is strongly activated by MADR2 or MADR3, which are closely related to each other, but not MADR1 (Y. CHEN et al. 1996; LAGNA et al. 1996; ZHANG et al. 1996). Interestingly, DPC4 seems to be able to synergize with MADR2 or MADR3 in activating this promoter, suggesting that MADR proteins may cooperate in signaling biological responses. However, this synergistic effect in some cell types is not ligand-dependent and may be due to overexpression. Nevertheless, in support of their functional cooperativity, MADR proteins have the potential to physically interact, forming homo- and heteromers. MADR3 and DPC4 exhibit strong hetero- and homomeric interactions in yeast two hybrid assays (ZHANG et al. 1996). In mammalian cells, DPC4 forms homomers and, upon stimulation by BMP4 or TGFβ/activin, forms heteromeric complexes with MADR1 or MADR2, respectively (LAGNA et al. 1996). Furthermore, expression of a dominant-negative form of DPC4, which has a small

C-terminal deletion, in *Xenopus* ectodermal explant cultures inhibits the induction of dorsal mesoderm by MADR2 and ventral mesoderm by MADR1 (LAGNA et al. 1996), suggesting that DPC4 is required for both activin and BMP signaling (Fig. 2B). The functional role of DPC4 in these pathways and the significance of these physical interactions needs to be established. However, since DPC4 is structurally more distantly related, it may play a more general role in signal transduction by the TGFβ superfamily, while specificity is provided by the MADR protein with which it interacts.

2.4.4 MADR Proteins are Substrates of the Activated Type I Receptor

The regulation of MADR1 and MADR2 phosphorylation by BMP and TGFβ/activin signaling pathways, respectively, is tightly regulated. Recent studies have revealed that MADR proteins are direct substrates of ser/thr kinase receptor complexes. In mammalian cells, MADR2 and MADR3 interact directly with the TGFβ receptor complex (MACÍAS-SILVA et al. 1996; ZHANG et al. 1996). The interaction of MADR2 is transient but is stabilized by a kinase-deficient version of the type I receptor and requires activation of TβRI by TβRII, indicating that association occurs through interaction with the type I kinase domain (MACÍAS-SILVA et al. 1996). Furthermore, in vitro kinase assays using purified receptors demonstrated that these MADR proteins are direct substrates of the TGFβ receptor complex and that phosphorylation is mediated by the activated type I receptor. Mapping studies have localized the ligand-dependent phosphorylation site to within a 5 amino acid tryptic peptide at the C-terminal tail of MADR2 which contains three serine residues (Fig. 2A, MACÍAS-SILVA et al. 1996). Mutation of these serines to alanine residues blocks phosphorylation and yields a mutant form of MADR2 that interacts stably with the receptor and functions in a dominant-negative manner in TGFβ signaling. Interestingly, DPC4 does not interact with the TGFβ receptor complex and is not phosphorylated in vitro (MACÍAS-SILVA et al. 1996; ZHANG et al. 1996). These results are consistent with the suggestion by LAGNA et al. (1996) that the function of DPC4 in TGFβ signaling is distinct from that fulfilled by receptor-targeted MADR proteins.

2.4.5 MADR Proteins Function in the Nucleus

MADR proteins do not contain any known protein motifs that may indicate their function in signal transduction. However, immunohistochemical localization of the proteins has indicated that MADR1 and MADR2 are dispersed throughout the cells but accumulate in the nucleus in response to BMP and TGFβ signaling, respectively (HOODLESS et al. 1996; LIU et al. 1996; MACÍAS-SILVA et al. 1996; BAKER and HARLAND 1996). This role is supported by immunolocalization studies in *Drosophila* which indicate that in the absence of signaling, MAD is localized to the cytoplasm (NEWFELD et al. 1996). Furthermore, studies in *Xenopus* demonstrate that in vivo a lacZ/MADR2 fusion protein is localized to the nucleus at the anterior portion of the axis, in a region in which activin, Vg1 and/or nodal are thought to signal dorsal mesoderm induction (BAKER and HARLAND 1996). Receptor-dependent phosphorylation on the C-terminal serines is required for this nuclear localization since a phosphorylation site mutant of MADR2 fails to accumulate in response to TGFβ signaling (MACÍAS-SILVA et al. 1996). Moreover, an N-terminal truncation of MADR2 that contains the MH2 domain and a small portion of the nonconserved domain is more active than full length MADR2 and is constitutively nuclear, suggesting that these proteins function in the nucleus. These latter studies further indicate that the MH1 domain may play an inhibitory function by maintaining MADR proteins in the cytoplasm, possibly by association with cytosolic components (BAKER and HARLAND 1996). Saturation of this cytosolic inhibitor could lead to spurious entry of MADR proteins into the nucleus upon overexpression, providing one possible explanation for the observation that overexpressed MADR proteins often signal in the absence of ligand. Interestingly, in contrast to MADR1 and MADR2, overexpression of MADR3 results in predominantly nuclear localization and constitutive signaling in the absence of ligand (Y. CHEN et al. 1996).

Once in the nucleus MADR proteins may act as transcriptional activators. A fusion protein between MADR1 and the DNA-binding domain of GAL4 is able to induce a GAL4 promoter construct in a ligand-dependent manner, suggesting that MADR1 can act as a transcriptional activator (LIU et al. 1996). GAL4 fusions with the C-terminal half of MADR1 or the C-terminal portion of DPC4 are able to constitutively activate transcription (LIU et al. 1996) consistent with the observation that the downstream functions of MADR proteins are localized within the C-terminal domain and that removal of the inhibitory N-terminal domain can unmask these activities.

2.5 The Nuclear Targets

The search for downstream effectors molecules in the signal transduction pathways of TGFβ superfamily members has led to the identification of two potential nuclear transcription factor targets, schnurri (SHN) and FAST-1. The *schnurri* gene was identified in *Drosophilia* screens for zygotic embryonic lethal mutations (NÜSSLEIN-VOLHARD et al. 1984). The phenotypes of *shn* mutants are strikingly similar to that observed in mutants of *dpp* and of the DPP receptors *tkv* and *punt*, suggesting that these genes are involved in a common developmental pathway (ARORA et al. 1995; GRIEDER et al. 1995). Furthermore, mutations in *shn* affect many, although not all, aspects of DPP signaling including dorsal ectoderm and mesoderm patterning, morphogenesis of the midgut and dorsal closure (ARORA et al. 1995; GRIEDER et al. 1995; STAEHLING-HAMPTON et al. 1995). Expression of *shn* is not dependent on *dpp*, *tkv* or *punt*, nor is the expression of *tkv* or *punt* dependent on *shn*, indicating that SHN does not act by regulating expression of these signaling components (ARORA et al. 1995). Cloning of *shn* has shown that the SHN protein is a member of the zinc finger family of transcription factors with seven zinc fingers and an associated acidic domain which could function as a transcriptional activation domain

(Arora et al. 1995; Grieder et al. 1995; Staehling-Hampton et al. 1995). SHN is most closely related to the human major histocompatibility complex-binding proteins 1 and 2 (MBP1, also known as HIV-EP1 or PRDII-BF1, and MBP2, which is closely related to HIV-EP2) and kappa binding protein 1 (KBP1). Currently, little is known about the precise role that SHN plays in controlling DPP-responsiveness in *Drosophila* and the relationship, if any, that the related mammalian proteins have in signaling by TGFβ superfamily members.

The second nuclear factor, FAST-1 (forkhead activin signal transducer-1), was identified in *Xenopus* by its ability to bind to an activin responsive element (ARE) found in the promoter region of the gene *Mix.2* (X. Chen et al. 1996; Huang et al. 1995). The ARE mediates an immediate early response to activin and comprises a 50-bp segment which can bind FAST-1 in electrophoretic mobility shift assays. FAST-1 contains a winged helix DNA binding domain but otherwise has no sequence similarity to any other proteins. In the absence of signaling, FAST-1 binds constitutively to the ARE, but in the presence of activin it assembles into a higher order complex, termed the ARF (activin responsive factor). Significantly, this ARF contains MADR2 in addition to FAST-1 and possibly other components, but does not contain MADR1. Moreover, FAST-1 is constitutively nuclear, convincingly demonstrating that MADR2 can function in the nucleus by interacting with nuclear targets to potentially regulate transcriptional activation. Thus, in the case of *Mix.2* regulation, FAST-1 provides a link between activation of the receptor and transcriptional events in the nucleus. Since no homologues of FAST-1 have yet been identified, whether all signaling by TGFβ superfamily members occurs through FAST-1 and FAST-1-like molecules remains to be determined.

2.6 A Model for TGFβ Signaling

The wealth of data that has recently emerged on TGFβ and BMP signaling has allowed the synthesis of a general model for MADR proteins in ser/thr kinase receptor signaling (Fig. 3). This model is most completely described for the case of TGFβ signaling through MADR2. Ligand-dependent activation of the TGFβ receptor results in activation of receptor I. Activated receptor I is recognized by MADR2, which then associates with the receptor. MADR2 serves as a direct substrate of the type I receptor kinase and is phosphorylated on the C-terminal serines. This leads to dissociation of MADR2, ligand-dependent complex formation with DPC4, nuclear accumulation of the complex and association with targets, such as FAST-1, to ultimately activate expression of target genes. It is important to note that several aspects of this model remain speculative. In particular, the subcellular localization of DPC4, its association with FAST-1 and the determinants that mediate association with MADR2 have not been reported. Mammalian versions of FAST-1 have yet to be reported. In addition, direct interaction between the BMP receptors and MADR1 or MADR5 has not been demonstrated thus far and nuclear targets for these proteins have not been identified. Nevertheless, it seems likely that the example of MADR2 provides a general mechanism for under-

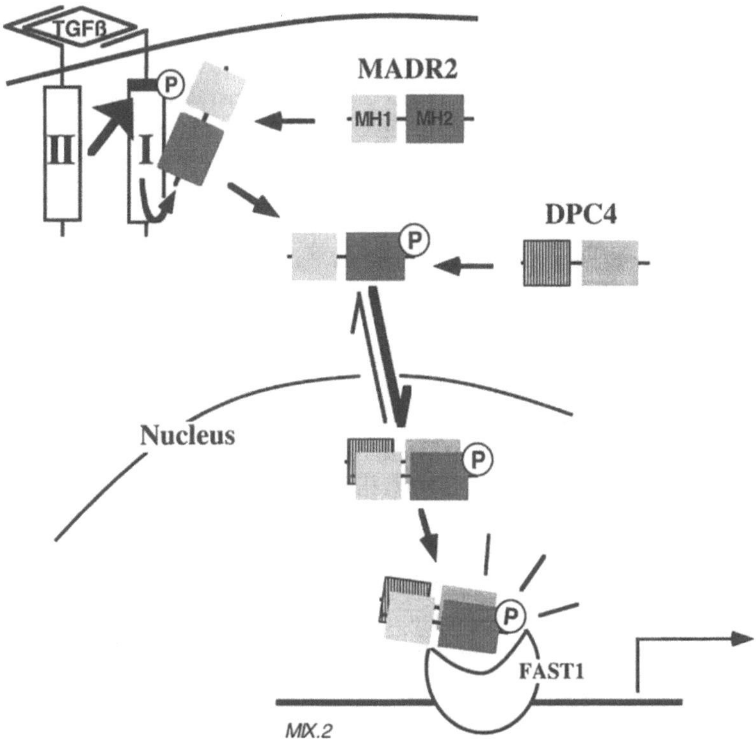

Fig. 3. A model of TGFβ signal transduction. Activation of the receptors by TGFβ leads to recruitment of MADR2, which is phosphorylated by the type I receptor on the C-terminal serine residues. This leads to dissociation of phosphorylated MADR2 from the receptor complex, followed by association with DPC4 and nuclear accumulation. In the nucleus, heteromeric complexes of MADR2 and DPC4 then associate with DNA binding proteins, such as FAST-1, to activate the transcription of target genes

standing how the TGFβ superfamily signals biological responses through MADR proteins.

At our current level of understanding, the role of MADR proteins in signaling directly from the receptor to the nucleus presents some significant differences compared to other mechanisms of transmitting cell surface signals. In particular, the apparent lack of an enzymatic activity for MADR proteins provides little opportunity for the cell to amplify TGFβ signals. Combined with the short length of the pathway and the high degree of specificity this could function to provide the nucleus with a precise and sensitive readout of the concentration of ligand present in the extracellular environment. Since small changes in the concentration of TGFβ-like factors such as activin can alter cell fate decisions, the precision afforded by this type of pathway may be crucial in allowing cells to respond appropriately within a morphogenetic gradient during development.

3 Biology of TGFβ Signaling

The TGFβ superfamily of growth factors has a staggering array of diverse effects on cell biology depending on the type of cell, its environment and its state of differentiation. These responses range from cell proliferation, cell cycle inhibition, expression of extracellular matrix proteins, morphogenetic movements, apoptosis and induction of differentiation. The role of these factors in development is one of the most rapidly expanding areas in TGFβ research and provides insight into the distinct functions of this family of growth factors. It is now evident that members of this family are involved in the very early proliferative and inductive events that occur at gastrulation to establish the basic body plan of the embryo (reviewed in Harland 1994, Hogan 1996, Slack 1994). In addition, later during embryogenesis, various TGFβ superfamily ligands have been implicated in the epithelial-mesenchymal interactions required for the development of many organs and tissues of the embryo. This research has revealed that the same members of this family are commonly "reused" at different stages and locations during embryogenesis and in many cases induce very different responses.

While many of the advances in vertebrate development have come from studies in *Xenopus laevis*, advances in mouse genetics, in particular analysis of "knockout" mice, has recently provided insight into the role of TGFβ superfamily members in mammalian development. Table 1 provides a list of mutations generated in ligands or in components of the signaling pathways of the TGFβ superfamily. In comparisons with the expression patterns, the resulting mutant phenotype has, in many cases, provided some surprises. For example, BMP7 is expressed during the early stages of gastrulation in the node and notochord (Lyons et al. 1995). However, mice mutant in BMP7 have no defects in gastrulation but die as a result of abnormal kidney development (Dudley et al. 1995; Luo et al. 1995). This disparity between expression pattern and mutant phenotypes is commonly observed for the TGFβ superfamily and is possibly due to the redundancy in many TGFβ-like factors. In addition, mutations in the ligand often do not have the same phenotype as mutants in the corresponding receptor. This is likely due to the capacity of some receptors to bind multiple TGFβ-like factors (Fig. 1). The Müllerian-inhibiting substance (MIS) ligand-receptor system stands in stark contrast to these general observations. MIS is expressed in the fetal testis and is responsible for the regression of the Müllerian ducts, the precursor of the female reproductive organs. Male mice deficient in MIS develop both male and female reproductive organs (Behringer et al. 1994). Furthermore, mice deficient in the MIS type II receptor are phenotypically identical to the MIS ligand-deficient mice and the MIS ligand/MIS receptor double mutants are indistinguishable from the single mutants (Mishina et al. 1995). These experiments indicate that MIS receptor may be the only functional type II receptor for MIS in vivo and that this ligand/receptor combination functions in a pathway required for Müllerian duct regression.

Research into the role of TGFβ family members and their signaling pathways has contributed to a greater understanding of many developmental processes. A

Table 1. Expression patterns and phenotypes of mouse mutations in TGFβ superfamily member ligands and in the signaling components

Gene	Function	Normal embryonic expression	Mutant phenotype	Gastrulation	Refs.
Activin A	Ligand	Present in oocyte and preimplantation embryo e5.5–e8.5: decidual expression only surrounding the whole embryo e8.5: weak expression in the heart Later expression in mesenchyme of developing face, whiskers, hair follicles, tooth bud, heart and digestive tract	Die within 24 h of birth. Lack whiskers and incisors and have abnormalities in the mouth	Normal	1, 2, 3, 4
Activin B	Ligand	Present in oocyte and preimplantation embryo e5.5–e8.5: weak expression around ectoplacental cone e8.5: weak expression in the heart	Viable with defective eyelid development and females are unable to rear offspring	Normal	1, 2, 3, 5
α-Inhibin	Ligand	Present in oocyte e7.5: no embryonic or decidual expression e10.5: in developing gonads and placenta	Both males and females are normal and fertile but develop gonadal stromal tumors	Normal	1, 2, 3, 6
Nodal	Ligand	e5.5: primitive ectoderm e6.5: proximal posterior ectoderm and primitive endoderm overlying streak e7.5: restricted to node	Arrests shortly after gastrulation with no evidence of primitive streak formation	Mesoderm evident in 25%. 10% have expression of *brachyury*. Embryonic ectoderm proliferates but then degenerates.	7,8

Table 1. (*Contd.*)

BMP2	Ligand	e6.5: detected by PCR analysis in embryo: present in decidua e7.5: mesoderm of amnion/chorion visceral endoderm and presumptive precardiac mesoderm e9.5: limb buds and myocardial layer of the heart and edges of neural folds later stages in tooth buds, whisker folicles and craniofacial mesenchyme	Arrests between e7.5 and e9.5. Malformation of amnion and chorion. Abnormal cardiac development	Normal except for persistence of open proamniotic canal in 69% of null mutants	9, 10, 11
BMP4	Ligand	e6.5: low levels in posterior primitive streak e7.5: amnion, allantois and posterior primitive streak e9.5: posterior primitive streak, posterior ventral mesoderm, foregut, branchial arches, heart and limb buds	Two primary phenotypes observed. Early phenotype: arrests at gastrulation with no primitive streak formation. Late phenotype: Truncated or disorganized posterior structures	Early phenotype shows extra embryonic mesoderm but no *brachyury* expression. Possible defects in cell survival and proliferation in epiblast layer. Late phenotype lacks posterior mesoderm	10
BMP5 (short ear)	Ligand	Expressed at early stages of skeletal development. Also in soft tissues including lungs, liver, ureter, bladder, and intestines	Viable and fertile but have effects in cartilage and bone including altered sternum and the loss of one pair of ribs. Bones are smaller and weaker. Also have reduced size of external ear	Normal	12, 13
BMP7	Ligand	e6.5: anterior primitive streak and node e7.5: allantois, axial mesoderm, low level in vicseral endoderm e8.5: notochord and definitive endoderm, nonneural ectoderm, myocardial cells and somatopleure e9.5: developing brain, limb bud e10.5: developing kidney, limb buds	Severe defects in kidney and eye development. Die at birth. Polydactyly in hind limbs, minor skeletal defects in ribs and skull	Normal	9, 14, 15

BMP-8B	Ligand	Male germ cells of testis and trophoblast cells of the placenta	Viable. Females are normal while males display germ cell deficiencies or infertility	Normal	16
ALK3 (BMPRI)	Type I receptor	e7.0: epiblast and mesoderm. Later stages: ubiquitous but absent from liver	Die between e7.0 and e9.5 with no primitive streak formation	No mesoderm induction occurs and no *brachyury* expression. Defects in epiblast cell proliferation	17
ActRII	Type II receptor	Present in oocytes. e5.5–e8.5: weak expression in embryonic ectoderm. Also in decidua	Viable but females are sterile. A few had skeletal and facial abnormalities	Normal	18, 19
TGFβ1	Ligand	Extraembryonic mesoderm, cardiac mesoderm, endothelial cells of blood vessels, fetal liver and surrounding decidua	Appear normal until approximately 3 weeks after birth, than exhibit "wasting" and develop multifocal inflamatory disease	Normal	1, 20
MIS (Müllerian-inhibiting substance)	Ligand	e12.5: in fetal tesis. Expression persists after birth in testis but at lower levels in Sertoli cells. Also in ovaries 6 days after birth	Mice are viable but males are infertile since they develop both male and female reproductive organs	Normal	21
MIS type II receptor (AMHR)	Type II receptor	In the mesenchymal cells adjacent to the Müllerian duct epithelium. Also in Sertoli cells and granulosa cells of fetal and adult testes and ovaries	Identical to above MIS deficient mice	Normal	22
GDF-5 (brachypodism)	Ligand		Viable with defects in the number and length of bones in the limb	Normal	23
GDF-9	Ligand	Expression is restricted to oocyte from primary one-layer follicle stage until ovulation	Females are viable but infertile. Follicular development arrests at one-layer stage	Normal	24

References: [1]Manova et al. 1992, [2]Albano et al. 1993, [3]Albano et al. 1994, [4]Matzuk et al. 1995c, [5]Vassali et al. 1994, [6]Matzuk et al. 1992, [7]Conlon et al. 1994, [8]Zhou et al. 1993, [9]Lyons et al. 1995, [10]Winnier et al. 1995, [11]Zhang and Bradley 1996, [12]Kingsley et al. 1992, [13]King et al. 1994, [14]Dudley et al. 1995, [15]Luo et al. 1995, [16]Zhao et al. 1996, [17]Mishina et al. 1995, [18]Manova et al. 1994, [19]Matzuk et al. 1995b, [20]Kulkarni et al. 1993, [21]Behringer et al. 1994, [22]Mishina et al. 1996, [23]Storm et al. 1994, [24]Dong et al. 1996

complete discussion of the role of these ligands in development is beyond the scope of this review. However, recent investigations in the areas of gastrulation and limb development have led to major advances in our understanding of the in vivo relationships between TGFβ family members and serve to illustrate the diverse and important functions these ligands fulfill during development.

3.1 Gastrulation

Gastrulation involves the process by which the three primary germ layers, the ectoderm, the mesoderm and the endoderm, are formed and patterned to establish the basic body plan of the embryo. The ectodermal derivatives give rise to the neural tube and epidermis, the mesoderm forms the muscular, skeletal, circulatory, and excretory systems, and the endoderm forms the gut tube. It is now evident that members of the TGFβ family are essential in patterning all three embryonic axes established during gastrulation, anterior-posterior (A-P), dorsal-ventral (D-V) and left-right (L-R).

3.1.1 TGFβ Superfamily in *Xenopus* Gastrulation

For many years, biologists have searched for the primary signals required for mesoderm induction and formation of the "organizer." The organizer is a region of dorsal mesoderm which can drive gastrulation movements and induction of nervous tissue when explanted on the ventral side of the embryo, resulting in the formation of a second axis (reviewed in LEMAIRE and KODJABACHIAN 1996). Many factors such as Wnts and FGFs have been implicated in the induction of mesoderm. However, a critical role for signaling by TGFβ family members during *Xenopus* gastrulation was illustrated by experiments using a dominant-negative activin type II receptor, tAR (the *Xenopus* homologue of ActRIIB), which contains a transmembrane domain but lacks a kinase domain. Overexpression of this receptor, which binds multiple members of the TGFβ superfamily, blocks signaling by BMPs, Vg1 and possibly other ligands (HEMMATI-BRIVANLOU and MELTON 1992; HEMMATI-BRIVANLOU and THOMSEN 1995; SCHULTE-MERKER et al. 1994). Significantly, expression of tAR in *Xenopus* blocks mesoderm induction, indicating the importance of TGFβ-like factors in early develoment. Studies using specific members of the TGFβ superfamily such as activins, BMPs, Xnrs (nodal-related) and Vg1 have shown more directly that these proteins can regulate the early induction and patterning of the mesoderm and neural tube which occurs during gastrulation (reviewed in JONES et al. 1995, SLACK 1994).

The first class of TGFβ family members extensively studied in mesoderm induction was the activins. *Xenopus* ectodermal explants, which will normally form epidermis when cultured in vitro, can be induced to form mesoderm and elongate into tube-like structures in response to activins. These induced explants contain dorsal mesoderm (notochord and muscle) and neural tissue, suggesting that activin has organizer activity (THOMSEN et al. 1990). Moreover, injection of activin mRNA into the ventral side of a *Xenopus* embryo can induce a secondary body axis. Similar axial induction by activin has also been shown in chicks (MITRANI et al. 1990). While these observations demonstrate that activin can function in mesoderm induction and axial patterning, the role of activins as the endogenous primary signal for mesoderm induction is questioned. In *Xenopus*, only a small amount of maternally deposited activin protein is present prior to the start of zygotic transcription, after the time that a primary mesoderm inducer is required. In addition, activin does not appear to be localized (reviewed in SLACK 1994). Despite these findings, expression of the extracellular domain of a type II activin receptor, which presumably can still bind ligand but cannot transduce signals, delays mesoderm induction (DYSON and GURDON 1997). This form of the receptor is likely to be secreted and blocks activin signaling but not Vg1, Xnr1, Xnr2 or BMPs, indicating that activin is required for normal mesodermal development. The ability of the embryos to recover later in development suggests that other factors can compensate for the early lack of activin signaling. Reconciliation of these contradictory data may be provided by the identification of two new members of the activin family, activin C and D (HÖTTEN et al. 1995; ODA et al. 1995). In particular, activin D is maternally expressed and is present at low levels when mesoderm induction takes place but is not localized within the embryo (ODA et al. 1995). Interestingly, a similar delay in mesoderm induction was observed in *Xenopus* embryos injected with a dominant-negative form of DPC4, possibly reflecting a disruption of activin signaling (LAGNA et al. 1996).

Currently, the most likely candidate for the endogenous mesoderm inducer is Vg1. Vg1 is expressed maternally and is localized to the vegetal pole of the embryo, which is the appropriate time and place expected for a mesoderm inducer (WEEKS and MELTON 1987). Vg1 is also interesting since in *Xenopus* it appears to be present in an inactive form in the embryo due to inefficient processing of the ligand. This creates the possibility that Vg1 is regulated by post-translational mechanisms in vivo and that activation of the processing machinery permits specific activation of Vg1. Using a chimera between BMP4 and Vg1 (BVg1) which is efficiently processed, Vg1 can induce dorsal mesoderm in *Xenopus* ectodermal explants and can efficiently induce an ectopic secondary axis (DALE et al. 1993; THOMSEN and MELTON 1993). The recent identification of chicken Vg1 (cVg1) has supported the role of Vg1 as a mesoderm and axial inducer (SELEIRO et al. 1996). cVg1 is expressed in a region of the chick embryo known to contain the axial organizing activity. Moreover, grafts of cell pellets expressing cVg1 can induce the ectopic formation of a complete primitive streak at the graft location, indicating that the functions of Vg1 may be conserved in vertebrate development.

Other possible mesoderm inducers from the TGFβ superfamily include the Xnrs. Xnr1 and Xnr2, similar to Vg1, are localized in the vegetal pole in the blastula and during gastrulation become asymmetrically restricted to the marginal zone where mesoderm formation occurs. In addition, they have been shown to induce both ventral and dorsal mesoderm in ectodermal explant experiments and can dorsalize the ventral marginal zone (JONES et al. 1995; LUSTIG et al. 1996).

In contrast to activin, Vg1 and Xnrs, all of which can induce dorsal mesoderm, the BMPs are able to induce and pattern ventral mesoderm (reviewed in HARLAND 1994, HOGAN 1996). In the *Xenopus* embryo, ectopic overexpression of BMP4 can suppress dorsal axis formation and in ectodermal explants BMPs ventralize mesoderm induced by factors such as activin. Interestingly, while homodimeric BMPs can pattern mesoderm, they are themselves weak mesoderm inducers. In contrast, an alternative form of BMP, composed of a dimer of BMP4 and BMP7, strongly induces mesoderm, suggesting that heterodimeric BMPs may also play a role in inducing mesoderm in vivo (S. Nishimatsu and G. Thomsen, personal communication). It will certainly be intriguing to determine how the signaling pathways activated by homodimeric vs. heterodimeric ligands might differ to elicit these distinct biological activities. Interestingly, another TGFβ superfamily member may play a role in the early patterning of the mesoderm. In embryos, anti-dorsalizing morphogenetic protein (ADMP) is restricted to the organizer region and the posterior neural floor plate (Moos et al. 1995). Paradoxically, ADMP expression is induced by dorsalizing factors, yet the protein appears to suppress dorsal and anterior structures. Thus, ADMP may function as a moderating factor within the organizer.

Investigations of mesoderm induction in *Xenopus* have shown that activins, Vg1, Xnrs and BMPs have the capacity to induce and pattern mesoderm to generate the complete range of dorsal and ventral cell fates that are typically assumed during embryogenesis. In addition to their role in induction and patterning of the mesoderm, the TGFβ superfamily may play a role in patterning the ectoderm. The first indication of this was provided by the observation that overexpression of the dominant-negative ser/thr kinase receptor, tAR, in ectodermal explant cultures induces neural tissue (HEMMATI-BRIVANLOU and MELTON 1994). As mentioned above, this form of the receptor disrupts signaling by multiple members of the superfamily, thus demonstrating that TGFβ-like factors might function in the ectoderm to block neural differentiation. This idea is supported by the observation that neural cell fates are induced when cells in the embryo are dissociated, a treatment that presumably disrupts signaling by endogenous TGFβ-like factors (reviewed in HEMMATI-BRIVANLOU and MELTON 1997). Consistent with this conclusion, treatment of dissociated ectodermal cells with BMP4 can inhibit this neural fate (WILSON and HEMMATI-BRIVANLOU 1995) and introduction of antisense RNA for BMP4 into ectodermal explants can induce neural markers (SASAI et al. 1995). Moreover, dominant-negative forms of BMP4 and BMP7, in which the cleavage site between the pro-region and the mature region of the ligands is destroyed, induce neural cells in ectodermal explants, supporting a specific function for endogenous BMPs in blocking neural induction (HAWLEY et al. 1995).

Previous models of ectodermal patterning suggested that neural induction occurred secondary to mesoderm and was mediated by specific signals emanating from the organizer region of the embryo. However, the ability of neural tissue to form in the absence of mesoderm has led to the suggestion of an alternative model in which the neural cell fate is actually a default state for ectodermal cells (reviewed in HEMMATI-BRIVANLOU and MELTON 1997). The discovery of how organizer molecules that induce neural differentiation function has resulted in a dramatic reconciliation of these apparently contradictory models (reviewed in THOMSEN 1997). In particular, studies on noggin and chordin have shown that these two proteins, which are secreted by the organizer and are potent neural inducers, function as antagonists of BMP signaling by directly interacting with the ligands to prevent binding to the receptors (PICCOLO et al. 1996; ZIMMERMAN et al. 1996). In addition, follistatin which binds activin, and possibly other TGFβ family members, may also act in a similar way (HEMMATI-BRIVANLOU et al. 1994). Thus, the organizer region may function to induce neural tissue in vivo by providing a source of secreted molecules that block BMP signaling in the adjacent ectoderm, thereby allowing neural differentiation to occur. These studies have also pointed to an emerging aspect of the organizer in controlling mesodermal patterning. Since the secreted factors described above are derived from the organizer, it seems likely that they may also function to protect this region of the embryo from the influence of ventralizing factors such as BMPs. Interestingly, genetic studies in *Drosophila* have shown that the *short gastrulation* gene (*Sog*), which is structurally related to chordin, functions as an antagonist of DPP in a manner analogous to that described for noggin and chordin in *Xenopus* (HOLLEY et al. 1996). Together, these studies suggest that the use of TGFβ antagonists may be a common mechanism for establishing dorsal-ventral body patterning in both vertebrates and invertebrates.

3.1.2 TGFβ Superfamily in Mammalian Gastrulation

It appears that ectodermal and mesodermal patterning in the *Xenopus* embryo results from a complex balance between ventralizing signals such as BMPs, dorsalizing signals such as activins and Vg1, and inhibitors of TGFβ ligands, such as noggin, chordin and follistatin. While the models of gastrulation appear to be better understood in *Xenopus*, it is important to determine whether similar models hold true for mammalian development. During gastrulation, cells in the ectodermal layer delaminate, ingress through the primitive streak and differentiate into mesoderm. In recent years, knockouts of ligands and other components of the signaling pathways in mice have begun to provide insights into the role of TGFβ ligands in mammalian gastrulation.

As in *Xenopus*, the importance of activin as a primary mesodermal inducer in early mammalian gastrulation is in question. Both activin A and B are not expressed in the early mouse embryo until after gastrulation. Homologous recombination was used to generate a null mutation of activin B in mice. These animals were viable, although defects in eyelid development were present (VASSALLI et al. 1994). Furthermore, homozygous mutant females were able to bear live young which died within 24 h of birth. Together these results indicte that activin B is not required either zygotically or maternally for mesoderm formation. Activin A deficient mice die within 24 h of birth with abnormalities in mouth formation and a lack of whiskers, similarly indicating that a zygotic source of activin A is not required for mesoderm induction or gastrulation (MATZUK et al. 1995c). Double null mutants in both activin A and B display a phenotype which is the combination of the two single mutants, eliminating the possibility that zygotic sources of activin A and B can compensate for each other during gastrulation.

In order to potentially avoid the problem of redundant ligands during development, null mutants have also been produced for the activin type II receptor, ActRII (MATZUK et al. 1995b). Mutants are viable with minor skeletal and facial abnormalities, although the females are sterile. However, like the activin ligand mutants, mesoderm and neural tube formation is unaffected. Interestingly, mice deficient in ActRII did not have the identical phenotype as mice deficient in either of the activin ligands, suggesting that in certain tissues other receptors, such as ActRIIB, may be involved in the embryonic functions of activin and that ActRII functions as a receptor for other ligands (MATZUK et al. 1995).

Taken together, these mutants strongly suggest that activins are not the primary inducer of mesoderm in mammals (SMITH 1995). However, the possibility cannot be completely eliminated since activin A derived from the decidua could provide inducing activity (ALBANO et al. 1994; MANOVA et al. 1992). Studies with TGFβ1 have shown that this ligand may be able to cross the placenta and rescue the embryo, supporting the possibility that activins might display a similar property (LETTERIO et al. 1994). In addition, two new activins, C and D, have been identified from human liver and *Xenopus*, respectively (HÖTTEN et al. 1995; ODA et al. 1995). While the mammalian homologue of activin D has not been reported and the embryonic expression pattern of these new activins is unknown, their identification raises the possibility that they could function during early embryogenesis.

In mammals, the BMP pathway appears to be important for mesoderm induction and gastrulation. Embryos deficient in the BMP2/4 receptor, ALK3 (BMPRIA), exhibit a complete absence of mesoderm (MISHINA et al. 1995). The interpretation of this phenotype is somewhat complicated since ectoderm proliferation was reduced in mutant embryos. Since the onset of gastrulation is thought to require the accumulation of 1400–1500 cells in the ectodermal layer of the embryo, mesoderm induction and cell proliferation are highly coupled (HOGAN et al. 1986). Thus, the lack of mesoderm in the mutants could be secondary to proliferative defects. However, teratomas derived from mutant embryos were impaired in mesoderm formation, suggesting that some BMP signaling is important for mesoderm induction.

A similar phenotype was observed in mice deficient in BMP4, a known ligand for ALK3. BMP4 mRNA is expressed in 6.5-day-old embryos in the posterior end of the primitive streak, the time and location required for a mesoderm inducer (WINNIER et al. 1995). BMP4 mutant embryos could be separated into two main phenotypes, suggesting that there are two critical phases in development during which BMP4 is required. The early phenotype was similar to the ALK3 mutants in that they arrested at gastrulation. The embryos displayed some extraembryonic mesoderm but no embryonic mesoderm as indicated by the lack of expression of *brachyury*. Furthermore, the BMP4 mutants, like the ALK3 mutants, displayed defects in cell survival and proliferation in the ectodermal layer. The second BMP4 deficient phenotype arrested later in development with embryos reaching beyond the head fold stage. These embryos were truncated or had disorganized posterior structures as well as a reduction of extraembryonic mesoderm. This later phenotype may be due to a rescue of some embryos by another ligand. BMP2, a second ALK3 ligand, is a potential candidate since it is present at very low levels in the embryo at gastrulation (WINNIER et al. 1995). However, while BMP2 might compensate for BMP4, it is not critical for normal gastrulation since BMP2 deficient mice arrest later in development with defects in heart formation and in closure of the exocoelomic cavity (ZHANG and BRADLEY 1996). Taken together, these studies suggest a role of BMPs in cell proliferation and survival in the ectoderm and in mesoderm formation. In addition, they emphasize that BMP2 and BMP4, which are highly related and interact with common receptors, fulfill unique functions during cardiac development and mesoderm formation, respectively.

Another strong candidate for a primary mesoderm inducer in mice is nodal, which has been implicated in mesoderm induction and gastrulation (CONLON et al. 1994; ZHOU et al. 1993). Nodal is expressed throughout the ectoderm prior to formation of the primitive streak and becomes localized to the prospective posterior ectoderm in the region where streak formation will occur. Expression is also observed transiently in the visceral endoderm prior to and during early streak formation. Later, expression becomes restricted to the node. Nodal mutants arrest shortly after the onset of gastrulation with little or no mesoderm and no evidence of primitive streak formation. Mesoderm is observed in 25% of the embryos with 10% exhibiting *brachyury* expression, suggesting that nodal is important but not absolutely required for mesoderm formation. In contrast to the ALK3 and BMP4 knockouts, mice deficient in nodal have normal proliferation of the ectoderm so the lack of cell numbers cannot account for the deficiencies in mesoderm. Evidence suggests that nodal may be involved in cell migration and the cell-cell interactions required for proper streak formation and patterning. In addition, using embryo chimeras between nodal mutant and wild-type cells it has been demonstrated that nodal expression in the ectoderm is required for primitive streak formation. However, visceral endoderm expression is required for proper anterior neural development and embryos lacking nodal in this region lack forebrain structures (VARLET et al. 1997).

Thus, in mammals, TGFβ family members are required during gastrulation in many capacities and at least two of these, BMP4 and nodal, are required prior to streak formation. In the gastrulating embryo, although both ligands are involved in mesoderm formation, BMP4 appears to play a role in proliferation and survival of the ectodermal layer while nodal is important for mesoderm and primitive streak formation. However, neither of these ligands on its own fulfills all the requirements for a primary mesodermal inducer. The variability observed in mesoderm induction in both BMP4 and nodal mice suggests that TGFβ family members may be able to compensate for each other. Thus, there may be no single TGFβ ligand which is required for mesoderm formation at all times. Compensatory mechanisms have been suggested to operate in *Xenopus* mesoderm induction as discussed earlier and similar mechanisms may function in mammals. Characterization of other TGFβ-like ligands in early development may provide some further insights into how mammalian gastrulation is controlled. For instance, Vg1 remains a good candidate in *Xenopus* and chicken as a mesoderm inducer but a mammalian homologue has not yet been identified. In addition, introduction of mutations in other components of the signaling pathways, such as the type I activin receptors or MADR proteins, may be informative by addressing the requirement of specific pathways during development, thereby avoiding the problems of ligand redundancy. Moreover, the identification of gene targets which mediate specific downstream events required for gastrulation may allow us to define mechanistically how these ligands control early embryogenesis. In addition, other factors, such as Wnts and FGFs, may act in concert with TGFβ ligands to regulate cell proliferation, mesoderm induction and cell migration during gastrulation. A greater understanding of the signal transduction pathways of TGFβ ligands may reveal how these interactions may occur.

3.1.3 Left-Right Axis

During gastrulation, formation of the left-right (L-R) body axis is linked to the dorsoventral and anterioposterior axes. In vertebrates, the morphological markers of L-R asymmetry are the position and shape of the heart, gut, liver, lungs and other organs and the direction of embryo turning. In the chick embryo, one of the earliest molecular indications of L-R asymmetry is the expression of the activin type II receptor, ActRII (Levin et al. 1995). *ActRII* is expressed more strongly on the right side of the primitive streak and later is restricted to the right side of Henson's node (the chick equivalent to the mouse node), suggesting that an activin-like factor is involved in early L-R patterning. In fact, this activin-like factor has been implicated in a signaling cascade which affects L-R patterning involving Sonic hedgehog (Shh) and a chicken nodal-related factor, cNR-1. *Shh*, a secreted growth factor not related to TGFβ, is initially expressed symmetrically in the node but becomes restricted to ectoderm on the left side of the node opposite to *ActRII* expression. Following *Shh* induction, *cNR-1* is expressed in the endoderm on the left side of the node adjacent to the domain of *Shh* expression and later is also observed in a large patch in the lateral plate mesoderm on the left side (Levin et al. 1995). Interestingly, this pathway can be manipulated experimentally. Activin coated beads placed on the left side of the node can induce ectopic *ActRII* expression with concomitant downregulation of *Shh* and cNR-1. Furthermore, ectopic expression of SHH on the right side of the node, where it is not normally expressed, induces ectopic expression of *cNR-1*. Moreover, ectopic implants of activin or Shh in the node causes loss of L-R asymmetry and randomization of the sidedness of the heart (Levin et al. 1995). Thus, these studies suggest a putative pathway in which signaling by an activin-like ligand through ActRII downregulates *Shh* and subsequently *cNR-1*, initiating L-R patterning in the chick embryo.

A general role for nodal in L-R patterning is supported by experiments in the mouse. Asymmetrical expression of *nodal* is first observed in the node at the early somite stages with a more intense expression on the left side (Collignon et al. 1996; Lowe et al. 1996). In addition, *nodal* is expressed in a region of the lateral plate mesoderm on the left side of the embryo. Interestingly, in the mouse mutant *inv* in which homozygotes develop a mirror-image L-R axis (situs invertus), *nodal* expression in the lateral plate mesoderm is restricted to the right side of the embryo and expression in the node is either symmetrical or enhanced on the right side (COLLIGNON et al. 1996; LOWE et al. 1996). In another mouse mutant, *iv*, in which the L-R axis is randomized, nodal expression was often bilaterally symmetrical or not expressed at all (LOWE et al. 1996). Taken together, these results indicate that there is a strong relationship between sidedness of the embryo and *nodal* expression and that misexpression of *nodal* can disrupt L-R axis formation.

Another TGFβ family member, Lefty, which is expressed in the left half of the gastrulating embryo, has also been implicated in L-R axis patterning (MENO et al. 1996). Lefty contains the cysteine-knot motif characteristic of the TGFβ superfamily but only shares 20%–25% similarity to other known members. The early expression of *lefty* is symmetrical with expression in the primitive streak mesoderm beginning at 7 days of embryogenesis. However, at the early somite stage, lefty is expressed transiently and exclusively on the left side in the region of the lateral plate mesoderm and in the floorplate of the developing neural tube. In about 50% of *iv* mutant mice, and in all homozygous *inv* mutant mice, *lefty* expression is observed on the right side indicating that, like nodal, lefty is downstream of these mutations and is likely involved in L-R patterning (MENO et al. 1996).

Although nodal and lefty expression is correlated with L-R asymmetry, neither factor appears to be involved in the early establishment of L-R axis formation. Nevertheless, as described above, the studies in chick using recombinant activin protein suggest that an early activin-like factor may be involved. The endogenous gene responsible for this has not yet been identified in chick or mouse. However, in *Xenopus*, recent studies have implicated Vgl in the initiation of L-R patterning (HYATT et al. 1996). As mentioned above, although Vgl mRNA is localized throughout the vegetal pole, tight regulation of the processing of the proform may control the expression of functional protein in specific regions. HYATT et al. (1996) circumvented this by using a chimeric protein that contains the pro-region and processing site from BMP-2 and the mature region of Vgl (BVgl). Expression of BVgl in the dorsovegetal cell on the right side of the embryo disrupted the L-R axis, causing randomization of cardiac and gut orientation (HYATT et al. 1996). In contrast, expression of BVgl in the left side had no significant effect on development. In addition, *Xnr-1*, which like chick and mouse nodal displays asymmetry in *Xenopus*, was bilaterally expressed in embryos injected with BVgl on the right side (HYATT et al. 1996). Thus, differential regulation of the processing of Vgl on the left side of the embryo may initiate L-R axis formation and control asymmetrical expression of nodals. Whether a Vgl homologue plays the same role in chick and mouse awaits further analysis.

3.2 Limb Development

Limb development provides a good example of the inductive abilities of TGFβ family members in epithelial-mesenchymal interactions. Development of the limb bud has been extensively studied and involves a complex orchestration of events coordinated by many growth factors including BMPs, TGFβ, Wnts, sonic hedgehog (SHH), and FGFs (reviewed in Cohn and Tickle 1996, Tickle 1996). The early limb bud is composed of mesoderm covered by ectoderm. The rim of the limb bud contains a thickened ectodermal region known as the apical ectodermal ridge (AER), which regulates the proximal-distal outgrowth of the limb and maintains the underlying mesoderm in a proliferative and undifferentiated state, called the progress zone. BMP2, BMP4 and BMP7 are all expressed in the AER and in the underlying mesoderm of the progress zone and are thought to play a role in signaling between these two regions. Expression of BMP2 is activated in the limb by sonic hedgehog (SHH), a regulator of patterning in the limb bud. However, BMP2 itself does not supply a polarizing signal and BMPs have been postulated to control cell proliferation in the progress zone and may be involved in controlling overall limb bud shape which is linked to limb patterning (Hogan 1996; Tickle 1996).

Recently, exciting progress has been made in understanding the role of BMPs and TGFβs in the formation of digits. As the limb extends and cells move away from the progress zone and the influences of the AER, they either undergo chondrogenic differentiation to form the cartilaginous skeleton of the digits or the cells remain undifferentiated in the interdigital regions. Eventually the cells in the interdigital regions undergo massive cell death by apoptosis. BMP2, BMP4 and BMP7 display expression patterns which correlate with the interdigital spaces, suggesting that BMPs may regulate apoptosis in this region (Zou and Niswander 1996). A direct role for BMPs in signaling apoptosis was demonstrated in recent experiments in which a dominant-negative BMP type I receptor (BMPRIB) capable of blocking BMP2 and BMP4 signals was injected into chick limb buds (Zou and Niswander 1996). This resulted in extensive webbing between the digits, suggesting that endogenous BMP signaling induces apoptosis between the digits. These results correlate with studies which have shown that beads soaked with BMP4 can accelerate apoptosis when implanted in the interdigital space and, when placed at the tip of growing digits, can cause bifurcation of the digit and establish an ectopic region of cell death (Gañan et al. 1996). Finally, it was shown that in duck feet, which are naturally webbed, there is little or no expression of BMP2, 4 or 7 in the interdigital spaces (Zhou and Niswander 1996). A developmental role for BMPs in apoptosis may not be unique since BMP4 is also implicated in programmed cell death of neural crest cells (Graham et al. 1994). In the mouse the role of BMPs in digit formation is supported by the observations that BMP7 deficient mice often have extra digits in the hind legs (Dudley et al. 1995; Luo et al. 1995). Unfortunately, the effects of BMP2 and 4 deficiency in limb formation cannot be assessed due to the early lethality in these mutants. In addition to the BMPs, growth and differentiation factor 5 (GDF5) has also been implicated in proper limb development, since the gene is mutated in the natural mouse mutation *brachypodism* (*bp*) (Storm et al. 1994). *Bp* null mutants have shortened limb bones and alterations in segmentation of the digits which are detectable as early as day 12 of embryogenesis, thereby suggesting that GDF5 is involved in regulating condensation of mesoderm in the limb.

In contrast to BMPs, TGFβs are present in the limb bud in the precartilaginous cells which will form the digits. Implantation of beads soaked with either

TGFβ1 or TGFβ2 into the interdigital region of chicks leads to inhibition of cell death and formation of extra digits in the interdigital regions (GAÑAN et al. 1996). Moreover, TGFβ is chondrogenic and exogenous TGFβ can induce chondrogenesis in cultures of limb mesenchyme. BMPs are also implicated in formation of the digit skeleton after the chondrogenic aggregates are established by promoting radial growth of the cartilage and controlling joint location and formation (Macias et al. 1997).

Studies in the limb bud highlight two major themes in TGFβ biology. First, a single ligand has the ability to elicit multiple biological responses and second, distinct ligands, such as TGFβ or BMPs, often evoke opposite effects in cells that are competent to respond to both. This balance between different TGFβ factors is a reiteration of the competition between BMPs and activin, Xnr and Vg1 that occurs during mesoderm induction in *Xenopus*. The mechanisms dictating which ligand prevails to exert its effects may to some extent be controlled by restricted expression of inhibitors or functional ligands, but may also involve regulation by signaling receptors. In this context, it is intriguing that BMP and activin/TGFβ utilize distinct MADR proteins to signal their respective biological responses, with no crosstalk observed between these pathways.

3.3 TGFβ Ligands as Morphogens

The question of whether TGFβ ligands act at a distance by diffusion, or by signaling through relay mechanisms, is important to answer since it allows us to define how cell fate decisions are controlled during development. As mentioned above, the balance between competing ligands and the presence or absence of inhibitors can influence the pathway of differentiation that a cell takes. For example, activins appear to act in a concentration-dependent manner in which high concentrations of activin can induce dorsal mesoderm (notochord) while lower levels induce mesoderm with more ventral characteristics (muscle). In fact, changes of as little as 1.5-fold can alter cell fate (reviewed in GREEN and SMITH 1991). This differential response to the same ligand supports the hypothesis that diffusible morphogens can differentially control cell fate determination by creating concentration gradients that radiate from a source. Cells could potentially respond to variations in concentration and differentiate into different lineages depending on their proximity to the morphogen and the thresholds required to induce a given response. Studies in *Drosophila* suggest that DPP can function as a true morphogen (reviewed in SMITH 1996). Expression of a constitutively activated receptor for DPP, which recapitulates downstream responses in the absence of ligand, induces transcription of two DPP-responsive genes, *spalt* and *omb*, in the wing imaginal disk. Importantly, these genes are only induced in cells that express the activated receptor and not in neighboring cells, indicating that a relay mechanism is not functioning (LECUIT et al. 1996; NELLEN et al. 1996). In contrast, cells ectopically expressing DPP induce expression of these genes in a halo of surrounding cells, in a manner consistent with DPP acting at a distance.

In vertebrates, TGFβ-like factors also appear to act as diffusible morphogens. Using activin-coated beads in explant cultures of *Xenopus* ectoderm, it has been demonstrated that the distance of a cell from an activin source can affect the set of genes expressed by the cell within a few hours (GURDON et al. 1994). Furthermore, cells respond directly to changes in the concentration of activin by expressing genes characteristic of the highest concentration of the morphogen, in a "ratchet-like" process (GURDON et al. 1995). These studies also demonstrated that passive diffusion of activin can occur by using explant cultures in which cells incapable of responding to activin were sandwiched between activin expressing cells and cells competent to respond to activin. Although direct contact with the activin source was prevented, the induction of *brachyury* was observed in the competent cells (GURDON et al. 1994). Consistent with these findings, ectodermal explants expressing activin, when juxtaposed with lineage-traced explants, induce *brachyury* expression in the latter explant and the position of expression relative to the activin source depends on the concentration of activin, indicating that activin can have long range effects (JONES et al. 1996). In addition, expression of constitutively active type I activin receptors induces gene expression exclusively cell autonomously, suggesting that the long range effects of activin are not due to a relay mechanism. In contrast to the observations with activin, the ligands, Xnr-2 and BMP4, appear to act only locally, suggesting that interaction of these proteins with the extracellular matrix can act to limit diffusion (JONES et al. 1996). Recently, by overexpressing epitope-tagged versions of the ligands and using lineage tracers to follow the injected cells in *Xenopus* embryos, activin and Vg1 appeared to have limited diffusion, acting only on adjacent cells (REILLY and MELTON 1996). Despite these latter studies, the preponderance of evidence indicates that TGFβ ligands can act at long distances to establish concentration gradients, thereby functioning as diffusible morphogens to differentially regulate cell fate decisions.

3.4 Disruption of TGFβ Signaling in Human Disease

In addition to their normal functions as regulators of developmental processes and homeostasis, mutation in members of this superfamily and their signaling components have recently been implicated in a diverse array of human disorders. Recent investigations in human cancer have suggested that components of the signaling pathway for TGFβ can act as tumor suppressors. Furthermore, hereditary mutations in ligands, receptors and other regulatory components have been described in three human syndromes, persistent Müllerian duct syndrome (PMDS), Hunter-Thompson type acromesomelic chondrodysplasia and hereditary hemorrhagic telangiectasia.

3.4.1 Cancer and TGFβ

Due to its roles in cell cycle inhibition, apoptosis and the expression of cell adhesion molecules, the involvement of the TGFβ pathway in cancer and tumor progression has long been suspected. Many epithelial cancer cell lines are resistant to TGFβ-mediated growth inhibition, including lines derived from squamous cell cancer, breast cancer, bladder cancer, lymphomas and, in particular, small-cell lung, gastric and colon cancers (reviewed in MARKOWITZ and ROBERTS 1996). Increased resistance to TGFβ in colorectal carcinoma cell lines correlates with biological aggressiveness, suggesting that release from TGFβ growth inhibition is an important step in tumor progression. In addition, restoration of TGFβ-responsiveness has been correlated with reversion of transformation in the colon cancer cell line SW480. However, loss of autocrine expression of the TGFβ ligand itself does not correlate with malignancy. Moreover, high levels of plasma TGFβ in colorectal cancer patients have been associated with an increase of TGFβ mRNA in the tumor and with tumor progression (TSUSHIMA et al. 1996). Thus, the resistance to TGFβ growth inhibition in tumors likely results from disruption of the ability of the cell to respond to TGFβ and suggests that components of the TGFβ signaling pathway may function as tumor suppressors. Exciting evidence is now emerging to provide a molecular basis for understanding the development of TGFβ resistance and its role in cancer progression.

Direct evidence that resistance to TGFβ in human cancers could be a result of mutations in components of the signaling pathways was first suggested by studies on the receptors TβRII and TβRI. In some gastric carcinoma cell lines, rearrangement of the TβRII gene has been detected and correlates with loss of sensitivity to TGFβ (PARK et al. 1994). Perhaps more significant was the finding that approximately 90% of colon cancers exhibiting microsatellite instability (RER+ for "replication errors") contain mutations causing inactivation of TβRII (MARKOWITZ et al. 1995). Microsatellite instability can be attributed to mutations in the mismatch repair genes, which in turn cause mutations in repetitive regions of the genome. Approximately 15%–17% of all colorectal carcinomas exhibit microsatellite instability either as a hereditary form of cancer, termed HNPCC (hereditary nonpolyposis colorectal cancer), or as sporadic cases (reviewed in KINZLER and VOGELSTEIN 1996). In RER+ cancers, mutations in TβRII accumulate and are primarily manifested as the addition or deletion of one or two adenine residues in a 10 base pair polyadenine region residing at the junction between the extracellular and transmembrane domains of the receptor. This introduces frameshifts and premature stops codons. Interestingly, mutations of TβRII are rare in endometrial cancers with microsatellite instability, emphasizing the importance that escape from TGFβ signaling may play in colorectal cancer (MYEROFF et al. 1995). In addition to RER+ colorectal cancer, mutations in the ser/thr kinase domain of TβRII have been identified in squamous head and neck carcinomas (GARRIGUE-ANTAR et al. 1995). These findings implicate TβRII as a tumor suppressor and present an important mechanistic link between defects in DNA repair mechanisms and tumor progression that can result from escape from TGFβ-mediated growth control.

In addition to defects in TβRII, alterations in the type I receptor, TβRI, have been found in prostate, colon and gastric cancer cells which are insensitive to TGFβ. Furthermore, a lack of TβRI protein has been observed in the AIDS-related Kaposi's sarcoma although the molecular basis has not been investigated (reviewed in Markowitz and Roberts 1996). While the functional significance of these alterations in TβRI expression has yet to be determined, these cases suggest that TβRI is a target for inactivating mutations.

Similar to the receptors, intracellular components of the TGFβ signaling pathways may also function as tumor suppressors (reviewed in Rustgi 1996). Both DPC4 and MADR2 map to chromosome 18q21, a region which is often deleted in pancreatic, lung and colorectal carcinomas. Loss of heterozygosity (LOH) of this region occurs in about 90% of pancreatic tumors and in over 60% of colorectal cancers, pointing to the importance of this chromosomal region in tumor progression. Evidence is now accumulating to implicate the MADR proteins as the candidate tumor suppressor genes in that region. Somatic missense mutations of DPC4 have been identified in 6 out of 27 (22%) pancreatic carcinomas (Hahn et al. 1996) and 5 out of 31 (16%) colorectal cancers (Takagi et al. 1996). Two missense mutations and a 2 base pair frameshift have also been identified in a sampling of 42 lung cancers (Nagatake et al. 1996). Further evidence that mutations in MADR genes in this region can contribute to tumor progression was obtained in two studies that reported genetic alterations in MADR2 (JV-18) (Eppert et al. 1996; Riggins et al. 1996). Interestingly, MADR2 functions in the TGFβ signaling pathway and in one study, missense mutations in MADR2 were identified in 4 out of 66 (6%) sporadic colorectal carcinomas while none were found in 101 breast cancers and 76 sarcomas (Eppert et al. 1996). Moreover, the proteins resulting from these missense mutations are nonfunctional. Together, these results strongly suggest that these mutations can account for a disruption in TGFβ signaling. The bias for mutations in MADR2 in colorectal cancers parallels similar observations in the TGFβ receptors. This may indicate that this pathway is more susceptible to mutation in colorectal cancer and suggests that TGFβ may be particularly important in controlling the normal proliferation of gastrointestinal epithelium. Thus, the essential molecules of the TGFβ signal transduction pathway can serve as tumor suppressors which, when mutated, can contribute to tumor progression due to an inability to respond to TGFβ.

The involvement of other TGFβ superfamily members in cancer has been suggested by studies on the α-inhibin deficient mice (Matzuk et al. 1992). Both male and female mice develop mixed granulosa/Sertoli cell tumors in 100% of the cases, indicating that α-inhibin is a potent negative regulator of gonadal stromal cell proliferation. Thus, α-inhibin is the first secreted protein demonstrated to have tumor-suppressor activity. Furthermore, mice which are deficient in both MIS and α-inhibin develop tumors earlier and faster than the α-inhibin deficient controls, suggesting that α-inhibin and MIS have synergistic effects in tumor development (Matzuk et al. 1995a). It is yet to be determined if alterations in either of these ligands participates in the development of gonadal tumors in humans.

3.4.2 Hereditary Diseases

The finding that mutations in Müllerian inhibitory substance (MIS) could result in persistent Müllerian duct syndrome (PMDS) was the first confirmation of an association between a human syndrome and a member of the TGFβ superfamily (IMBEAUD et al. 1994). PMDS is characterized by the presence of Müllerian duct derivatives in otherwise normal males. Approximately 50% of patients with PMDS have a normal *MIS* gene and normal serum levels of MIS, indicating that organ insensitivity may also contribute to PMDS. Supporting this observation is the recent finding of a mutation in the MIS type II receptor gene of a PMDS patient (IMBEAUD et al. 1995). At present, a type I receptor for MIS or any of the intracellular components in this signaling pathway have not been identified, but these proteins could also be potential targets for mutations resulting in PMDS syndrome.

Another member of the TGFβ superfamily, cartilage-derived morphogenetic protein 1 (CDMP-1), the human homologue of murine GDF-5, is the primary defect in a human chondrodysplasia (THOMAS et al. 1996). CDMP-1 is predominantly expressed in the long bones during human development and mutations in this gene have been described in Hunter-Thompson type acromesomelic chondrodysplasia. This autosomal recessive disorder is characterized by a shortening of the long bones in the limbs, with the proximal bones of the hands and feet being most severely affected. Remarkably, this syndrome is phenotypically similar to the murine mutant *brachpodism* (*bp*), which, as mentioned above, is due to mutations in the mouse homologue GDF-5 (STORM et al. 1994).

A third syndrome, hereditary hemorrhagic telangiectasia or Osler-Rendu-Weber (ORW) syndrome, is characterized by multisystemic vascular dysplasia. ORW can be linked to two distinct loci. One loci has been shown to correspond to the endoglin gene, a TGFβ binding protein expressed in endothelial cells and placenta (MCALLISTER et al. 1994). Most mutations in endoglin linked to ORW result in protein truncations and it is currently unclear how these changes may affect endothelial cell function. The second locus was shown to be the activin-like type I receptor, ALK1 (also known as TSR1 or ACVRLK1) (JOHNSON et al. 1996), which can bind TGFβ or activin when overexpressed with TβRII or ActRIIB respectively, although the in vivo ligand remains uncertain. ALK1, similar to endoglin, is expressed in endothelial cells and other highly vascularized tissues such as lung and placenta (ATTISANO et al. 1993). Hereditary mutations in ALK1 identified in three ORW families affect critical residues in the kinase domain, suggesting that the resulting protein is nonfunctional (JOHNSON et al. 1996). Thus, endoglin and ALK1 may play a critical role in regulating vascular endothelium structure during development or repair and disruption of this function may lead to the vascular dysplasia associated with this disease.

4 Conclusions

The TGFβ superfamily of growth factors can regulate diverse biological processes from the earliest stages of development into adulthood. Our understanding of how these factors function has progressed rapidly and is uniquely propelled by an amalgamation of molecular and developmental studies. Thus, molecular studies of TGFβ signal transduction are providing mechanistic insights into the biology of these factors while studies of the role of these factors in development continues to contribute mechanistic insights into the signaling pathways. Continuing investigations into both the biochemistry and the biology of TGFβ signaling will undoubtedly be crucial for us to gain a complete understanding of the actions of this diverse family of growth factors. With this knowledge, we can then begin to assess how mutations within components of the signaling pathways can affect human diseases, such as cancer, and potentially how the TGFβ superfamily can be used as a therapeutic treatment.

Acknowledgements. We would like to thank Dr. Liliana Attisano for critical reading of this manuscript and Dr. G. Thomsen for communicating results prior to publication. P.A.H. is the recipient of a Canadian Association of Gastroenterology fellowship and J.L.W. is a scholar of the Medical Research Council of Canada.

Note added in proof. The currently accepted nomenclature for vertebrate members of the MAD-related family of proteins is Smad1 through Smad6 (Derynck et al. (1996) Nomenclature: Vertebrate mediators of TGFβ family signals. Cell 87:173).

References

Albano RM, Groome N, Smith JC (1993) Activins are expressed in preimplantation mouse embryos and in ES and EC cells and are regulated on their differentiation. Development 117:711–723

Albano RM, Arkell R, Beddington RSP, Smith JC (1994) Expression of inhibin subunits and follistatin during postimplantation mouse development: decidual expression of activin and expression of follistatin in primitive streak, somites and hindbrain. Development 120:803–813

Arora K, Dai H, Kazuko SG, Jamal J, O'Connor MB, Letsou A, Warrior R (1995) The Drosophila schnurri gene acts in the dpp/TGF-β signaling pathway and encodes a transcription factor homologous to the human MBP family. Cell 81:781–790

Attisano L, Wrana JL (1996) Signal transduction by members of the transforming growth factor-β superfamily. Cytokine Growth Fact Rev 7:327–339

Attisano L, Cárcamo J, Ventura F, Weis FMB, Massagué J, Wrana JL (1993) Identification of human Activin and TGF-β type I receptors that form heterodimeric kinase complexes with type II receptors. Cell 75:671–680

Attisano L, Wrana JL, López-Casillas F, Massagué J (1994) TGF-β receptors and actions. Biochim Biophys Acta 1222:71–80

Attisano L, Wrana JL, Montalvo E, Massagué J (1996) Activation of signalling by the activin receptor complex. Mol Cell Biol 16:1066–1073

Baker JC, Harland R (1996) A novel mesoderm inducer, Madr2, functions in the activin signal transduction pathway. Genes Dev 10:1880–1889

Behringer RR, Finegold MJ, Cate RL (1994) Müllerian-inhibiting substance function during mammalian sexual development. Cell 79:415–425

Brummel T, Twombly V, Marques G, Wrana JL, Newfeld S, Attisano L, Massagué J, O'Connor MB, Gelbart WM (1994) Characterization and relationship of DPP receptors encoded by the saxophone and thick veins genes in Drosophila. Cell 78:251–261

Cárcamo J, Weis FMB, Ventura F, Wieser R, Wrana JL, Attisano L, Massagué J (1994) Type I receptors specify growth inhibitory and transcriptional responses to TGF-β and activin. Mol Cell Biol 14:3810–3821

Chen F, Weinberg RA (1995) Biochemical evidence for the autophosphorylation and transphosphorylation of transforming growth factor β receptor kinases. Proc Natl Acad Sci USA 92:1565–1569

Chen R-H, Moses HL, Maruoka EM, Derynck R, Kawabata M (1995) Phosphorylation-dependent interaction of the cytoplasmic domains of the type I and type II transforming growth factor-β receptors. J Biol Chem 270:12235–12241
Chen X, Rubock MJ, Whitman M (1996) A transcriptional partner for MAD proteins in TGF-β signalling. Nature 383:691–696
Chen Y, Lebrun J-J, Vale W (1996) Regulation of transforming growth factor β- and activin-induced transcription by mammalian Mad proteins. Proc Natl Acad Sci USA 93:12992–12997
Cohn MJ, Tickle C (1996) Limbs: a model for pattern formation within the vertebrate body plan. Trends Genet 12:253–257
Collignon J, Varlet I, Robertson EJ (1996) Relationship between asymmetric nodal expression and the direction of embryonic turning. Nature 381:155–158
Conlon FL, Lyons KM, Takaesu N, Barth KS, Kispert A, Herrmann B, Robertson EJ (1994) A primary requirement for nodal in the formation and maintenance of the primitive streak in the mouse. Development 120:1919–1928
Dale L, Matthews G, Colman A (1993) Secretion and mesoderm-inducing activity of the TGF-β-related domain of xenopus-Vg1. EMBO J 12:4471–4480
Derynck R, Zhang Y (1996) Intracellular signalling: the mad way to do it. Curr Biol 6:1226–1229
Dong J, Albertini DF, Nishimori K, Kumar TR, Lu N, Matzuk MM (1996) Growth differentiation factor-9 is required during early ovarian folliculogenesis. Nature 383:531–535
Dudley AT, Lyons KM, Robertson EJ (1995) A requirement for bone morphogenetic protein-7 during development of the mammalian kidney and eye. Genes Dev 9:2795–2807
Durbec P, Marcos-Gutierrez CV, Kilkenny C, Grigoriou M, Wartiowaara K, Suvanto P, Smith D, Ponder B, Costantini F, Saarma M, Sariola H, Pachnis V (1996) GDNF signalling through the Ret receptor tyrosine kinase. Nature 381:798–793
Dyson S, Gurdon JB (1997) Activin signalling has a necessary function in Xenopus early development. Curr Biol 7:81–84
Eppert K, Scherer SW, Ozcelik H, Pirone R, Hoodless P, Kim H, Tsui L-C, Bapat B, Gallinger S, Andrulis I, Thomsen G, Wrana JL, Attisano L (1996) MADR2 maps to 18q21 and encodes a TGFβ-regulated MAD-related protein that is functionally mutated in colorectal carcinoma. Cell 86:543–552
Gañan Y, Macias D, Duterque-Coquillaud M, Ros MA, Hurle JM (1996) Role of TGFβs as signals controlling the position of the digits and the areas of interdigital cell death in the developing chick limb autopod. Development 122:2349–2357
Garrigue-Antar L, Muñoz-Antonia T, Antonia SJ, Gesmonde J, Vellucci VF, Reiss M (1995) Missense mutations of the Transforming growth factor β type II receptor in human head and neck squamous carcinoma cells. Cancer Res 55:3982–3987
Gentry LE, Liobin MN, Purchio AF, Marquardt H (1988) Molecular events in the processing of recombinant type 1 pre-pro-transforming growth factor beta to the mature polypeptide. Mol Cell Biol 8:4162–4168
Graff JM, Bansal A, Melton DA (1996) Xenopus Mad proteins transduce distinct subsets of signals for the TGFβ superfamily. Cell 85:479–487
Graham A, Francis-West P, Brickell P, Lumsden A (1994) The signalling molecule BMP4 mediates apoptosis in the rhombencephalic neural crest. Nature 372:684–686
Green JBA, Smith JC (1991) Growth factors as morphogens. Trends Genet 7:245–250
Grieder NC, Nellen D, Burke R, Basler K, Affolter M (1995) schnurri is required for Drosophila dpp signalling and encodes a zinc finger protein similar to the mammalian transcription factor PRDII-BF1. Cell 81:791–800
Gurdon JB, Harger P, Mitchell A, Lemaire P (1994) Activin signalling and response to a morphogen gradient. Nature 371:487–492
Gurdon JB, Mitchell A, Mahony D (1995) Direct and continuous assessment by cells of their position in a morphogen gradient. Nature 376:520–521
Hahn SA, Schutte M, Shamsul Hoque ATM, Moskaluk CA, da Costa LT, Rozenblum E, Weinstein CL, Fischer A, Yeo CJ, Hruban RH, Kern SE (1996) DPC4, a candidate tumor suppressor gene at human chromosome 18q21.1. Science 271:350–353
Harland RM (1994) The transforming growth factor β family and induction of vertebrate mesoderm: bone morphogenetic proteins are ventral inducers. Proc Natl Acad Sci USA 91:10243–10246
Hawley SHB, Wünnenberg-Stapleton K, Hashimoto C, Laurent MN, Watabe T, Blumberg BW, Cho KWY (1995) Disruption of BMP signals in embryonic Xenopus ectoderm leads to direct neural induction. Genes Dev 9:2923–2935

Hemmati-Brivanlou A, Kelly OG, Melton DA (1994) Follistatin, an antagonist of activin is expressed in the Spemann organizer and displays direct neuralizing activity. Cell 77:283–295
Hemmati-Brivanlou A, Melton DA (1992) A truncated activin receptor inhibits mesoderm induction and formation of axial structures in Xenopus embryos. Nature 359:609–614
Hemmati-Brivanlou A, Melton DA (1994) Inhibition of activin receptor signalling promotes neuralization in Xenopus. Cell 77:273–281
Hemmati-Brivanlou A, Melton D (1997) Vertebrate embryonic cells will become nerve cells unless told otherwise. Cell 88:13–17
Hemmati-Brivanlou A, Thomsen GH (1995) Ventral mesodermal patterning in Xenopus embryos: expression patterns and activities of BMP-2 and BMP-4. Dev Genet 17:78–89
Hogan BLM (1996) Bone morphogenic proteins: multifunctional regulators of vertebrate development. Genes Dev 10:1580–1594
Hogan BLM, Constantini F, Lacy E (1986) Manipulating the mouse embryo: a laboratory manual. Cold Spring Harbor Laboratory Press, Cold Spring Harbor, New York
Holley SA, Neul JL, Attisano L, Wrana JL, Sasai Y, O'Connor MB, De Robertis EM, Ferguson EL (1996) The Xenopus dorsalizing factor noggin ventralizes Drosophila embryos by preventing DPP from activating its receptor. Cell 86:607–617
Hoodless PA, Haerry T, Abdollah S, Stapleton M, O'Connor MB, Attisano L, Wrana JL (1996) MADR1, a MAD-related protein that functions in BMP2 signalling pathways. Cell 85:489–500
Hötten G, Neidhardt H, Schneider C, Pohl J (1995) Cloning of a new member of the TGF-β family: a putative new activin βC chain. Biochem Biophys Res Commun 206:608–613
Huang H-C, Murtaugh LC, Vize PD, Whitman M (1995) Identification of a potential regulator of early transcriptional responses to mesoderm inducer in the frog embryo. EMBO J 14:5965–5973
Hyatt BA, Lohr JL, Yost HJ (1996) Initiation of vertebrate left-right axis formation by maternal Vg1. Nature 384:62–65
Imbeaud S, Carré-Eusèbe D, Rey R, Belville C, Josso N, Picard JY (1994) Molecular genetics of the persistent Müllerian duct syndrome: a study of 19 families. Hum Mol Genet 3:125–131
Imbeaud S, Faure E, Lamarre I, Mattei MG, di Clemente N, Tizard R, Carré-Eusèbe D, Belville C, Tragethon L, Tonkin C, Nelson J, McAuliffe M, Bidart J-M, Lababidi A, Josso N, Cate RL, Picard J-Y (1995) Insensitivity to anti-Müllerian hormone due to a mutation in the human anti-Müllerian hormone receptor. Nature Genet 11:382–388
Jing S, Wen D, Yanbin Y, Hoist PL, Luo Y, Fang M, Tamir R, Antonio L, Hu Z, Cupples R, Louis J-C, Hu S, Altrock BW, Fox GM (1996) GDNF-induced activation of the Ret protein tyrosine kinase is mediated by GDNFR-α, a novel receptor for GDNF. Cell 85:1113–1124
Johnson DW, Berg JN, Baldwin MA, Gallione C, Marondel I, Stenzil T, Speer M, Pericak-Vance M, Qumsiyeh WA, Schwartz C, Diamond A, Guttmacher AE, Jackson CE, Attisano L, Kucherlapati R, Porteous MEM, Marchuk DA (1996) Mutations in the activin receptor-like kinase 1 gene in hereditary hemorrhagic telangiectasia type 2. Nature Genet 13:189–196
Jones CM, Kuehn MR, Hogan BLM, Smith JC, Wright CVE (1995) Nodal-related signals induce axial mesoderm and dorsalize mesoderm during gastrulation. Development 121:3651–3662
Jones CM, Armes N, Smith JC (1996) Signalling by TGF-β family members: short range effects of Xnr-2 and BMP-4 contrast with the long-range effects of activin. Curr Biol 6:1468–1475
King JA, Marker PC, Seung KJ, Kingsley DM (1994) BMP5 and the molecular, skeletal, and soft-tissue alterations in short ear mice. Dev Biol 166:112–122
Kingsley DM, Bland AE, Grubber JM, Marker PC, Russell LB, Copeland NG, Jenkins NA (1992) The mouse short ear skeletal morphogenesis locus is associated with defects in a bone morphogenetic member of the TGF-β superfamily. Cell 71:399–410
Kinzler KW, Vogelstein B (1996) Lessons from hereditary colorectal cancer. Cell 87:159–170
Kotzbauer PT, Lampe PA, Heuckeroth RO, Golden JP, Creedon DJ, Johnson EMJ, Milbrandt J (1996) Neurturin, a relative of glial-cell-line-derived neurotrophic factor. Nature 384:467–470
Kulkarni AB, Huh C-G, Becker D, Geiser A, Lyght M, Flanders KC, Roberts AB, Sporn MB, Ward JM, Karlsson S (1993) Transforming growth factor β1 null mutation in mice causes excessive inflammatory response and early death. Proc Natl Acad Sci USA 90:770–774
Lagna G, Hata A, Hemmati-Brivanlou A, Massagué J (1996) Partnership between DPC4 and SMAD proteins in TGF-β signalling pathways. Nature 383:832–836
Lastres P, Letamend'a A, Zhang H, Rius C, Almendro N, Raab U, López LA, Langa C, Fabra A, Letarte M, Bernabéu C (1996) Endoglin modulates cellular responses to TGF-β1. J Cell Biol 133:1109–1121

Lecuit T, Brook WJ, Ng M, Calleja M, Sun H, Cohen SM (1996) Two distinct mechanisms for long-range patterning by decapentaplegic in the Drosophila wing. Nature 381:387–393
Lemaire P, Kodjabachian L (1996) The vertebrate organizer: structure and molecules. Trends Genet 12:525–531
Letsou A, Arora K, Wrana J, Simin K, Twombly V, Jamal J, Staehling-Hampton K, Hoffmann FM, Gelbart WM, Massagué J, O'Connor MB (1995) Dpp signaling in Drosophila is mediated by the punt gene product: a dual ligand binding type II receptor of the TGF-β receptor family. Cell 80:899–908
Letterio JJ, Geiser AG, Kulkarni AB, Roche NS, Sporn MB, Roberts AB (1994) Maternal rescue of transforming growth factor-β1 null mice. Science 264:1936–1938
Levin M, Johnson RL, Stern CD, Kuehn M, Tabin C (1995) A molecular pathway determining left-right asymmetry in chick embryogenesis. Cell 82:803–814
Lin HY, Moustakas A, Knaus P, Wells RG, Henis YI, Lodish HF (1995) The soluble exoplasmic domain of the type II transforming growth factor (TGF)-β receptor: a heterogeneously glycosylated protein with high affinity and selectivity for TGF-β ligands. J Biol Chem 270:2747–2754
Liu F, Ventura F, Doody J, Massagué J (1995) Human type II receptor for bone morphogenetic proteins (BMPs): extension of the two-kinase receptor model to the BMPs. Mol Cell Biol 15:3479–3486
Liu F, Hata A, Baker J, Doody J, Cárcamo J, Harland R, Massagué J (1996) A human Mad protein acting as a BMP-regulated transcriptional activator. Nature 381:620–623
López-Casillas F, Wrana JL, Massagué J (1993) Betaglycan presents ligand to the TGF-β signaling receptor. Cell 73:1435–1444
Lowe LA, Supp DM, Sampath K, Yokoyama T, Wright CVE, Potter SS, Overbeek P, Kuehn MR (1996) Conserved left-right asymmetry of nodal expression and alterations in murine situs inversus. Nature 381:158–161
Luo G, Hofmann C, Bronckers ALJJ, Sohocki M, Bradley A, Karsenty G (1995) BMP-7 is an inducer of nephrogenesis, and is also required for eye development and skeletal patterning. Genes Dev 9:2808–2820
Lustig KD, Kroll K, Sun E, Ramos R, Elmendorf H, Kirschner MW (1996) A Xenopus nodal-related gene that acts in synergy with noggin to induce complete secondary axis and notochord formation. Development 122:3275–3282
Lyons KM, Hogan BLM, Robertson EJ (1995) Colocalization of BMP 7 and BMP 2 RNAs suggests that these factors cooperatively mediate tissue interactions during murine development. Mech Dev 50:71–83
Macias D, Gañan Y, Sampath TK, Piedra ME, Ros MA, Hurle JM (1997) Role of BMP-2 and OP-1 (BMP-7) in programmed cell death and skeletogenesis during chick limb development. Development 124:1109–1117
Macías-Silva M, Abdollah S, Hoodless PA, Pirone R, Attisano L, Wrana JL (1996) MADR2 is a substrate of the TGFβ receptor and its phosphorylation is required for nuclear accumulation and signalling. Cell 87:1215–1224
Manova K, Paynton BV, Bachvarova RF (1992) Expression of activins and TGFβ1 and β2 RNAs in early postimplantation mouse embryos and uterine decidua. Mech Dev 36:141–152
Manova K, DeLeon V, Angeles M, Kalantry S, Giarre M, Attisano L, Wrana JL, Bachvarova RF (1994) mRNAs for activin receptors II and IIB are expressed in mouse oocytes and in the epiblast of pregastrula and gastrula stage mouse embryos. Mech Dev 49:3–11
Markowitz S, Wang J, Myeroff L, Parsons R, Sun LZ, Lutterbaugh J, Fan RS, Zborowska E, Kinzler KW, Vogelstein B, Brattain M, Willson JKV (1995) Inactivation of the Type II TGF-β receptor in colon cancer cells with microsatellite instability. Science 268:1336–1338
Markowitz SD, Roberts AB (1996) Tumor suppressor activity of the TGF-β pathway in human cancers. Cytokine Growth Fact Rev 7:93–102
Massagué J (1996) TGFβ signalling: receptors, transducers and MAD proteins. Cell 85:947–950
Massagué J, Weis-Garcia F (1996) Serine/threonine kinase receptors: mediators of transforming growth factor beta family signals. In: Pawson T, Parker P (eds) Cancer surveys cell signalling, vol 27. ICRF Press, London, pp 41–64
Matzuk MM, Finegold MJ, Su J-GJ, Hsueh AJW, Bradley A (1992) α-Inhibin is a tumour-suppressor gene with gonadal specificity in mice. Nature 360:313–319
Matzuk MM, Finegold MJ, Mishina Y, Bradley A, Behringer RR (1995a) Synergistic effects of inhibins and Müllerian-inhibiting substance on testicular tumorigenesis. Mol Endocrinol 9:1337–1345
Matzuk MM, Kumar TR, Bradley A (1995b) Different phenotypes for mice deficient in either activins or activin receptor type II. Nature 374:356–359

Matzuk MM, Kumar TR, Vassalli A, Bickenbach JR, Roop DR, Jaenisch R, Bradley A (1995c) Functional analysis of activins during mammalian development. Nature 374:354–356

McAllister KA, Grogg KM, Johnson DW, Gallione CJ, Baldwin MA, Jackson CE, Helmbold EA, Markel DS, McKinnon WC, Murell J, McCormick MD, Pericak-Vance MA, Heutink P, Oostra BA, Haitjema T, Westerman CJ, Porteous ME, Guttmacher AE, Letarte M, Marchuk DA (1994) Endoglin, a TGF-β binding protein of endothelial cells is the gene for hereditary haemorrhagic telangiectasia type I. Nature Genet 8:345–351

Meno C, Saijoh Y, Fujii H, Ikeda M, Yokoyama T, Yokoyama M, Toyoda Y, Hamada H (1996) Left-right asymmetric expression of the TGFβ-family member Lefty in mouse embryos. Nature 381:151–155

Mishina Y, Suzuki A, Ueno N, Behringer RR (1995) Bmpr encodes a type I bone morphogenetic protein receptor that is essential for gastrulation during mouse embryogenesis. Genes Dev 9:3027–3037

Mishina Y, Rey R, Finegold MJ, Matzuk MM, Josso N, Cate RL, Behringer RR (1996) Genetic analysis of the Müllerian-inhibiting substance signal transduction pathway in mammalian sexual differentiation. Genes Dev 10:2577–2587

Mitrani E, Ziv T, Thomsen G, Shimoni Y, Melton DA, Bril A (1990) Activin can induce the formation of axial structures and is expressed in the hypoblast of the chick. Cell 63:495–501

Miyazono K, ten Dijke P, Ichijo H, Heldin C-H (1994) Receptors for transforming growth factor-β. Adv Immunol 55:181–220

Moos M Jr, Wang S, Krinks M (1995) Anti-dorsalizing morphogenetic protein is a novel TGF-β homolog expressed in the Spemann organizer. Development 121:4293–4301

Myeroff LL, Parsons R, Kim S-J, Hedrick L, Cho KR, Orth K, Mathis M, Kinzler KW, Lutterbaugh J, Park K, Bang Y-J, Lee HY, Park J-G, Lynch HT, Roberts AB, Vogelstein B, Markowitz SD (1995) A transforming growth factor β receptor type II gene mutation common in colon and gastric but rare in endometrial cancers with microsatellite instability. Cancer Res 55:5545–5547

Nagatake M, Takagi Y, Osada H, Uchida K, Mitsudomi T, Saji S, Shimokata K, Takahashi T, Takahashi T (1996) Somatic in vivo alteration of the DPC4 gene at 18q21 in human lung cancers. Cancer Res 56:2718–2720

Nakao A, Röijer E, Imamura T, Souchelnytskyi S, Stenman G, Heldin C-H, ten Dijke P (1997) Identification of Smad2, a human Mad-related protein in the transforming growth factor β signaling pathway. J Biol Chem 272:2896–2900

Nellen D, Affolter M, Basler K (1994) Receptor ser/thr kinases implicated in the control of Drosophila body pattern by decapentaplegic. Cell 78:225–237

Nellen D, Burke R, Struhl G, Basler K (1996) Direct and long-range action of a DPP morphogen gradient. Cell 85:357–368

Newfeld SJ, Chartoff EH, Graff JM, Melton DA, Gelbart WM (1996) Mothers against dpp encodes a conserved cytoplasmic protein required in DPP/TGFβ responsive cells. Development 122:2099–2108

Nohno T, Ishikawa T, Saito T, Hosokawa K, Noji S, Wolsing DH, Rosenbaum JS (1995) Identification of a human type II receptor for bone morphogenetic protein-4 that forms differential heteromeric complexes with bone morphogenetic protein type I receptors. J Biol Chem 270:22522–22526

Nüsslein-Volhard C, Wieschaus E, Kluding H (1984) Mutations affecting the pattern of the larval cuticle in Drosophila melanogaster. I. Zygotic loci on the second chromosome. Rouxs Arch Dev Biol 183:267–282

Oda S, Nishimatsu S-I, Murakami K, Ueno N (1995) Molecular cloning and functional analysis of a new activin b subunit: a dorsal mesoderm-inducing activity in Xenopus. Biochem Biophys Res Commun 210:581–588

Park K, Kim S-J, Bang Y-J, Park J-G, Kim NK, Roberts AB, Sporn MB (1994) Genetic changes in the transforming growth factor β (TGF-β) type II receptor gene in human gastric cancer cells: correlation with sensitivity to growth inhibition by TGF-β. Proc Natl Acad Sci USA 91:8772–8776

Penton A, Chen Y, Staehling-Hampton K, Wrana JL, Attisano L, Szidonya J, Cassill A, Massagué J, Hoffmann FM (1994) Identification of two bone morphogenetic protein type I receptors in drosophila and evidence that Brk25D is a decapentaplegic receptor. Cell 78:239–250

Piccolo S, Sasai Y, Lu B, De Robertis EM (1996) Dorsoventral patterning in xenopus: inhibition of ventral signals by direct binding of chordin to BMP-4. Cell 86:589–598

Raftery LA, Twombly V, Wharton K, Gelbart WM (1995) Genetic screens to identify elements of the decapentaplegic signaling pathway in drosophila. Genetics 139:241–254

Reilly KM, Melton DA (1996) Short-range signaling by candidate morphogens of the TGFβ family and evidence for a relay mechanism of induction. Cell 86:743–754

Riggins GJ, Thiagalingam S, Rozenblum E, Weinstein CL, Kern SE, Hamilton SR, Willson JKV, Markowitz SD, Kinzler KW, Vogelstein B (1996) MAD-related genes in the human. Nature Genet 13:347–349
Roberts AB, Anzano MA, Lamb LC, Smith JM, Sporn MB (1981) New class of transforming growth factors potentiated by epidermal growth factor. Proc Natl Acad Sci USA 78:5339–5343
Rosenzweig BL, Imamura T, Okadome T, Cox GN, Yamashita H, ten Dijke P, Heldin CH, Miyazono K (1995) Cloning and characterization of a human type II receptor for bone morphogenetic proteins. Proc Natl Acad Sci USA 92:7632–7636
Ruberte E, Marty T, Nellen D, Affolter M, Basler K (1995) An absolute requirement for both the type II and type I receptors, punt and thick veins for dpp signalling in vivo. Cell 80:889–897
Rustgi AK (1996) MAD about colorectal cancer. Gastroenterology 111:1387–1389
Ryden M, Imamura T, Jornvall H, Neveu I, Trupp M, Belluardo N, Okadome T, ten Dijke P, Ibanez CF (1996) A novel type I receptor serine-threonine kinase predominantly expressed in the adult central nervous system. J Biol Chem 271:30603–30609
Sasai Y, Lu B, Steinbeisser H, De Robertis EM (1995) Regulation of neural induction by the Chd and BMP-4 antagonistic patterning signals in Xenopus. Nature 376:333–336
Savage C, Das P, Finelli A, Townsend S, Sun C, Baird S, Padgett R (1996) Caenorhabditis elegans genes sma-2, sma-3 and sma-4 define a conserved family of transforming growth factor β pathway components. Proc Natl Acad Sci USA 93:790–794
Schulte-Merker S, Smith JC, Dale L (1994) Effects of truncated activin and FGF receptors and of follistatin on the inducing activities of BVg1 and activin: does activin play a role in mesoderm induction? EMBO J 13:3533–3541
Sekelsky JJ, Newfeld SJ, Raftery LA, Chartoff EH, Gelbart WM (1995) Genetic characterization and cloning of Mothers against dpp, a gene required for decapentaplegic function in Drosophila melanogaster. Genetics 139:1347–1358
Seleiro EAP, Connolly DJ, Cooke J (1996) Early development expression and experimental axis determination by the chicken Vg1 gene. Curr Biol 6:1476–1486
Shibuya H, Yamaguchi K, Shirakabe K, Tonegawa A, Gotoh Y, Ueno N, Irie K, Nishida E, Matsumoto K (1996) TAB1: an activator of the TAK1 MAPKKK in TGF-β signal transduction. Science 272:1179–1182
Slack JMW (1994) Inducing factors in Xenopus early embryos. Curr Biol 4:116–126
Smith J (1995) Angles on activin's absence. Nature 374:311–312
Smith J (1996) How to tell a cell where it is. Nature 381:367–368
Souchelnytskyi S, ten Dijke P, Miyazono K, Heldin C-H (1996) Phosphorylation of Ser165 in TGF-β type I receptor modulates TGF-β1-induced cellular responses. EMBO J 15:6231–6240
Staehling-Hampton K, Laughon AS, Hoffmann FM (1995) A Drosophila protein related to the human zinc-finger transcription factor PRDII/MBPI/HIV-EP1 is required for dpp signaling. Development 121:3393–3403
Storm EE, Huynh TV, Copeland NG, Jenkins NA, Kingsley DM, Lee S-J (1994) Limb alterations in brachypodism mice due to mutations in a new member of the TGFβ-superfamily. Nature 368:639–642
Takagi Y, Kohmura H, Futamura M, Kida H, Tanemura H, Shimokawa K, Saji S (1996) Somatic alterations of the DPC4 gene in human colorectal cancers in vivo. Gastroenterology 111:1369–1372
ten Dijke P, Miyazono K, Heldin C-H (1996) Signaling via hetero-oligomeric complexes of type I and type II serine/threonine kinase receptors. Curr Opin Cell Biol 8:139–145
Thomas JT, Lin K, Nandedkar M, Camargo M, Cervanka J, Luyten FP (1996) A human chondrodysplasia due to a mutation in a TGF-β superfamily member. Nature Genet 12:315–317
Thomsen G, Woolf T, Whitman M, Sokol S, Vaughan J, Vale W, Melton DA (1990) Activins are expressed early in Xenopus embryogenesis and can induce axial mesoderm and anterior structures. Cell 63:485–493
Thomsen GH (1996) Xenopus Mothers against decapentaplegic is an embryonic ventralizing agent that acts downstream of the BMP2/4 receptor. Development 122:2359–2366
Thomsen GH (1997) Antagonism within and around the Spemann Organizer: BMP inhibitors in vertebrate body patterning. Trends Genet 13:209–211
Thomsen GH, Melton DA (1993) Processed Vg1 protein is an axial mesoderm inducer in xenopus. Cell 74:433–441
Tickle C (1996) Vertebrate limb development. Cell Dev Biol 7:137–143
Treanor JJS, Goodman L, de Sauvage F, Stone DM, Poulsen KT, Beck CD, Gray C, Armanini MP, Pollock RA, Hefti F, Phillips HS, Goddard A, Moore MW, Buj-Bello A, Davies AM, Asai N,

Takahashi M, Vandlen R, Henderson CE, Rosenthal A (1996) Characterization of a multicomponent receptor for GDNF. Nature 382:80–83

Trupp M, Arenas E, Fainzilber M, Nilsson A-S, Sieber B-A, Grigoriou M, Kilkenny C, Salazar-Grueso E, Pachnis V, Arumäe U, Sariola H, Saarma M, Ibáñez CF (1996) Functional receptor for GDNF encoded by the c-ret proto-oncogene. Nature 381:785–789

Tsuchida K, Vaughan JM, Wiater E, Gaddy-Kurten D, Vale WW (1995) Inactivation of activin-dependent transcription by kinase-deficient activin receptors. Endocrinology 136:5493–5503

Tsuchida K, Sawchenko PE, Nishikawa S, Vale WW (1996) Molecular cloning of a novel type I receptor serine/threonine kinase for the TGFβ superfamily from rat brain. Mol Cell Neurosci 7:467–478

Tsushima H, Kawata S, Tamura S, Ito N, Shirai Y, Kiso S, Imai Y, Shimomukai H, Nomura Y, Matsuda Y, Matsuzawa Y (1996) High levels of transforming growth factor β in patients with colorectal cancer: association with disease progression. Gastroenterology 110:375–382

Varlet I, Collingnon J, Robertson EJ (1997) Nodal expression in the primitive endoderm is required for specification of the anterior axis during mouse gastrulation. Development 124:1033–1044

Vassalli A, Matzuk MM, Gardner HAR, Lee K-F, Jaenish R (1994) Activin/inhibin βB subunit gene disruption leads to defects in eyelid development and female reproduction. Genes Dev 8:414–427

Weeks DL, Melton DA (1987) A maternal mRNA localized to the vegetal hemisphere in xenopus eggs codes for a growth factor related to TGF-β. Cell 51:861–867

Wiersdorff V, Lecuit T, Cohen SM, Mlodzik M (1996) Mad acts downstream of Dpp receptors, revealing a differential requirement for dpp signalling and propagation of morphogenesis in the drosophila eye. Development 122:2153–2162

Wieser R, Attisano L, Wrana JL, Massagué J (1993) Signalling activity of TGF-β type II receptors lacking specific domains in the cytoplasmic region. Mol Cell Biol 13:7239–7247

Wieser R, Wrana JL, Massague J (1995) GS domain mutations that constitutively activate TβR-I, the downstream signalling component in the TGF-β receptor complex. EMBO J 14:2199–2208

Wilson PA, Hemmati-Brivanlou A (1995) Induction of epidermis and inhibition of neural fate by BMP-4. Nature 376:331–333

Winnier G, Blessing M, Labosky PA, Hogan BLM (1995) Bone morphogenetic protein-4 is required for mesoderm formation and patterning in the mouse. Genes Dev 9:2105–2116

Wrana JL, Attisano L (1996) MAD-related proteins in TGFβ signalling. Trends Genet 12:493–496

Wrana JL, Attisano L, Carcamo J, Zentella A, Doody J, Laiho M, Wang X-F, Massagué J (1992) TGF-β signals through a heteromeric protein kinase receptor complex. Cell 71:1003–1014

Wrana JL, Attisano L, Wieser R, Ventura F, Massagué J (1994a) Mechanism of activation of the TGF-β receptor. Nature 370:341–347

Wrana JL, Tran H, Attisano L, Arora K, Childs SR, Massagué J, O'Connor MB (1994b) Two distinct transmembrane serine/threonine kinases from drosophila form an activin receptor complex. Mol Cell Biol 14:944–950

Yamaguchi K, Shirakabe K, Shibuya H, Irie K, Oishi I, Ueno N, Taniguchi T, Nishida E, Matsumoto K (1995) Identification of a member of the MAPKKK family as a potential mediator of the TGF-β signal transduction. Science 270:2008–2011

Yamashita H, ten Dijke P, Huylebroeck D, Sampath TK, Andries M, Smith JC, Heldin C-H, Miyazono K (1995) Osteogenic protein-1 binds to activin type II receptors and induces certain activin-like effects. J Cell Biol 130:217–226

Yingling JM, Das P, Savage C, Zhang M, Padgett RW, Wang X-F (1996) Mammalian dwarfins are phosphorylated in response to TGF-β and are implicated in control of cell growth. Proc Natl Acad Sci USA 93:8940–8944

Zhang H, Bradley A (1996) Mice deficient for BMP2 are nonviable and have defects in amnion/chorion and cardiac development. Development 122:2977–2986

Zhang Y, Feng X-H, Wu R-Y, Derynck R (1996) Receptor-associated Mad homologues synergize as effectors of the TGF-β response. Nature 383:168–172

Zhao GQ, Deng K, Labosky PA, Liaw L, Hogan BLM (1996) The gene encoding bone morphogenetic protein 8B is required for the initiation and maintenance of spermatogenesis in the mouse. Genes Dev 10:1657–1669

Zhou X, Sasaki H, Lowe L, Hogan BLM, Kuehn MR (1993) Nodal is a novel TGF-β-like gene expressed in the mouse node during gastrulation. Nature 361:543–547

Zimmerman LB, De Jesús-Escobar JM, Harland R (1996) The Spemann organizer signal noggin binds and inactivates bone morphogenetic protein 4. Cell 86:599–606

Zou H, Niswander L (1996) Requirement for BMP signaling in the interdigital apoptosis and scale formation. Science 272:738–741

Notch Receptors, Partners and Regulators: From Conserved Domains to Powerful Functions

S.E. Egan, B. St-Pierre and C.C. Leow

Division of Immunology and Cancer Research and the Program in Developmental Biology, The Hospital for Sick Children, 555 University Avenue, Toronto, Ontario, M5G 1X8, Canada
The Department of Molecular and Medical Genetics, The University of Toronto, Canada

1 The Notch Family of Receptors

The Notch gene was first identified in *Drosophila* as a locus required for development of many structures (Poulson 1937). The diversity of biological processes regulated by Notch in *Drosophila* and its varied relatives from *C. elegans* to man is only now becoming clear. Indeed the Notch receptor system may be used to regulate development and homeostasis of every tissue in complex animals. Some of these processes have been analyzed in great detail and we refer readers to several excellent recent reviews focused on more biological aspects of Notch function (Artavanis-Tsakonas et al. 1995; Kopan and Turner 1996; Lardelli et al. 1995). Genetic analysis of invertebrate Notch family receptors has highlighted a role of Notch proteins as transmembrane transcription factors which require proteolytic activation and nuclear translocation for activity. Numerous proteins which participate in the Notch signaling pathway have been discovered. These varied proteins include ligands for Notch, regulators of Notch ligand-binding specificity, regulators of Notch signaling efficiency, and transcription factors which cooperate with Notch to control gene expression. In this article we will highlight the structural features of Notch receptors and the proteins with which they interact. These structures are conserved in all Notch-family receptors and their partners. Through the analysis of structure and function together, we hope to highlight some important features of these signaling proteins which may help to explain their role in integrating many extracellular signals to control cell fate determination, differentiation, proliferation and cell survival.

1.1 Invertebrate Notch Genes and Signaling Pathway Components

In *Drosophila melanogaster*, individuals heterozygous for loss of function mutations in *Notch* develop with wing *notches*. Homozygous loss of function mutations in *Notch* results in a lethal expansion of neural tissues at the expense of epidermal tissue in the developing embryo (Artavanis-Tsakonas et al. 1995). This phenotype is also seen in flies with mutations in several other genes, collectively known as the "neurogenic genes" (Lehmann et al. 1983). The *Notch* locus codes for a large receptor protein (Kidd et al. 1986; Wharton et al. 1985). Many mutations in *Notch* have been identified which give rise to distinct phenotypes in wings, eyes and external sensory organs. Gene products which participate in the signaling pathway upstream and downstream of Notch have been identified through cloning of other neurogenic genes, including Delta, a ligand for Notch (Kopczynski et al. 1988; Vassin et al. 1987). In addition, mutations which enhance or suppress Notch-dependent phenotypes have been generated and the mutant gene(s) in many cases have been characterized. Alleles of the *Suppressor of Hairless* (*Su*(*H*)) gene have been identified which enhance the glossy eye phenotype in sensitized N^{ts1}/fa^{g2} *Notch* mutant flies (Fortini and Artavanis-Tsakonas 1994; Honjo 1996). Also, a complex locus termed *Enhancer of split* (*E*(*spl*)) was identified which, when mutated, enhanced or worsened the small and rough eye phenotype in flies with the *split* allele of *Notch* (Campos-Ortega 1991). Su(H) and E(spl) gene products are critical components of the Notch-pathway in many tissues (see below). In contrast to these enhancers of Notch-dependent phenotypes, mutations in the *Drosophila deltex* gene can suppress the lethality associated with a specific combination of *Notch* gain of function *Abruptex* (*Ax*) alleles (Busseau et al. 1994). Thus, genetic experiments in *Drosophila* have led to the isolation of Notch and many Notch-pathway components.

The *C. elegans lin-12* and *glp-1* genes are highly related in both sequence and function (Greenwald 1994; Greenwald et al. 1983; Lambie and Kimble 1991; Yochem and Greenwald 1989). Like their distant relative Notch, LIN-12 and GLP-1 control diverse cellular functions such as cell fate induction, lateral specification and cell division (Austin and Kimble 1987; Greenwald et al. 1983; Lambie and Kimble 1991). LIN-12 and GLP-1 are redundant for many but not all developmental functions and dramatic phenotypes are observed in a large number of tissues when both Notch-like genes are mutated. The phenotypes observed in *LIN-12/GLP-1* compound mutants are collectively referred to as the "lag phenotype" (for *lin-12* and *glp-1*) (Lambie and Kimble 1991). Genetic analysis in *C. elegans* has led to the isolation of novel genes which, when mutated, induce the lag phenotype. These lag genes, termed *lag-1* and *lag-2*, encode homologues of *Drosophila Su(H)* and *delta* respectively (Christensen et al. 1996; Henderson et al. 1994; Tax et al. 1994). Isolation of genes which induce similar mutant phenotypes to mutations of Notch-family receptors in flies (Neurogenic) and worms (Lag) has thus led to the discovery of Notch or LIN-12/GLP-1 pathway components in both organisms. Enhancer/suppressor screens have also been used to isolate components of the LIN-12/GLP-1 pathway in *C. elegans* (Grant and Greenwald 1996; Levitan and Greenwald 1995; Sundaram and Greenwald 1993a, b). Mutations in the *sel-12* gene (*s*uppressors and/or *e*nhancers of *lin-12*), for example, can suppress phenotypes induced by a hyperactive LIN-12 protein (Levitan and Greenwald 1995). Finally, an additional component of the worm LIN-12/GLP-1 pathway was identified in a yeast two hybrid screen with the cytoplasmic domain of LIN-12 as bait. The LIN-12-binding protein identified in this screen, EMB-5, was shown to function downstream of LIN-12 and GLP-1 in genetic experiments (Hubbard et al. 1996).

1.2 Vertebrate Notch Genes and Oncogenes

Coffman et al. isolated the first vertebrate Notch-family gene by low stringency hybridization screening of a *Xenopus* cDNA library using the *Drosophila Notch* cDNA as probe (Coffman et al. 1990). The *Xenopus* Notch gene sequence was then used by many labs to obtain mammalian homologues. Low stringency hybridization approach and degenerate PCR led to the isolation of three mammalian Notch genes: Notchl, 2 and 3 (Franco Del Amo et al. 1992; Kopan and Weintraub 1993; Lardelli et al. 1994; Reaume et al. 1992; Stifani et al. 1992; Weinmaster et al. 1991, 1992). Mammalian Notch genes have been independently isolated as oncogenes responsible for cancerous transformation in human T-cell leukemias (Tan1/Notch1) (Ellisen et al. 1991) and mouse mammary tumor virus-induced breast cancer (int-3/Notch4) (Robbins et al. 1992). The discovery of human and mouse Notch oncogenes in addition to the well studied role for these proteins in invertebrate development has heightened interest in these proteins as regulators of vertebrate development, cell division and differentiation. The study of Notch signaling in vertebrate development has been pursued, primarily through ectopic expression studies in *Xenopus* and gene targeting in mice (de la Pompa et al. 1997; Dorsky et al. 1997; Wettstein et al. 1997).

1.3 Conserved Structure of Notch Family Receptors

Notch-family receptors are very large transmembrane glycoproteins which are present in all complex animals. These proteins contain several copies of specific domains which have been implicated in protein-protein interactions critical for activation, regulation, or signal transduction. The modular nature of Notch proteins is revealed by comparing the distinct organization of multiple Notch-family receptors within a species as well as by comparing Notch proteins between species. For example, gene duplication and mutation has resulted in the generation of at least four Notch-family proteins in mammals which subtly differ from each other in specific subdomains. In contrast, comparison of *Drosophila* Notch with *C. elegans* Notch-like proteins, LIN-12 and GLP-1, reveals that these gene products are dramatically different in size, structure, and primary amino acid sequence from Notch. Features which are shared by the *Drosophila* and *C. elegans* proteins are likely to be fundamental to signal integration and transduction by all Notch-like receptors. Genetic and biochemical studies of the Notch proteins in several systems have helped to define the fundamental structural features of these receptors as well as Notch-pathway signaling proteins.

1.4 Notch Receptor Proteins in *Drosophila* and Vertebrates

Notch family receptors are highly conserved between *Drosophila* and vertebrates (Fig. 1). The extracellular domain of *Drosophila* Notch is made up of an N-terminal signal peptide, 36 tandem cysteine-rich EGF-like repeats (EGFL repeats), and 3 copies of a distinct conserved cysteine-rich repeat termed the LIN-12/Notch repeat (LNR), which is followed by a cysteine-poor domain (Artavanis-Tsakonas et al. 1995; Kidd et al. 1986; Wharton et al. 1985). The signal peptide is a universal feature of all Notch-related receptor proteins as each must pass through the secretory pathway on its way to the plasma membrane. EGFL repeats are protein-protein interaction motifs, found in epidermal growth factor (EGF) and dozens of other extracellular proteins (Campbell and Bork 1993). Six of the EGFL repeats in Notch are also predicted to bind calcium (Rao et al. 1995). Recently, EGFL repeats

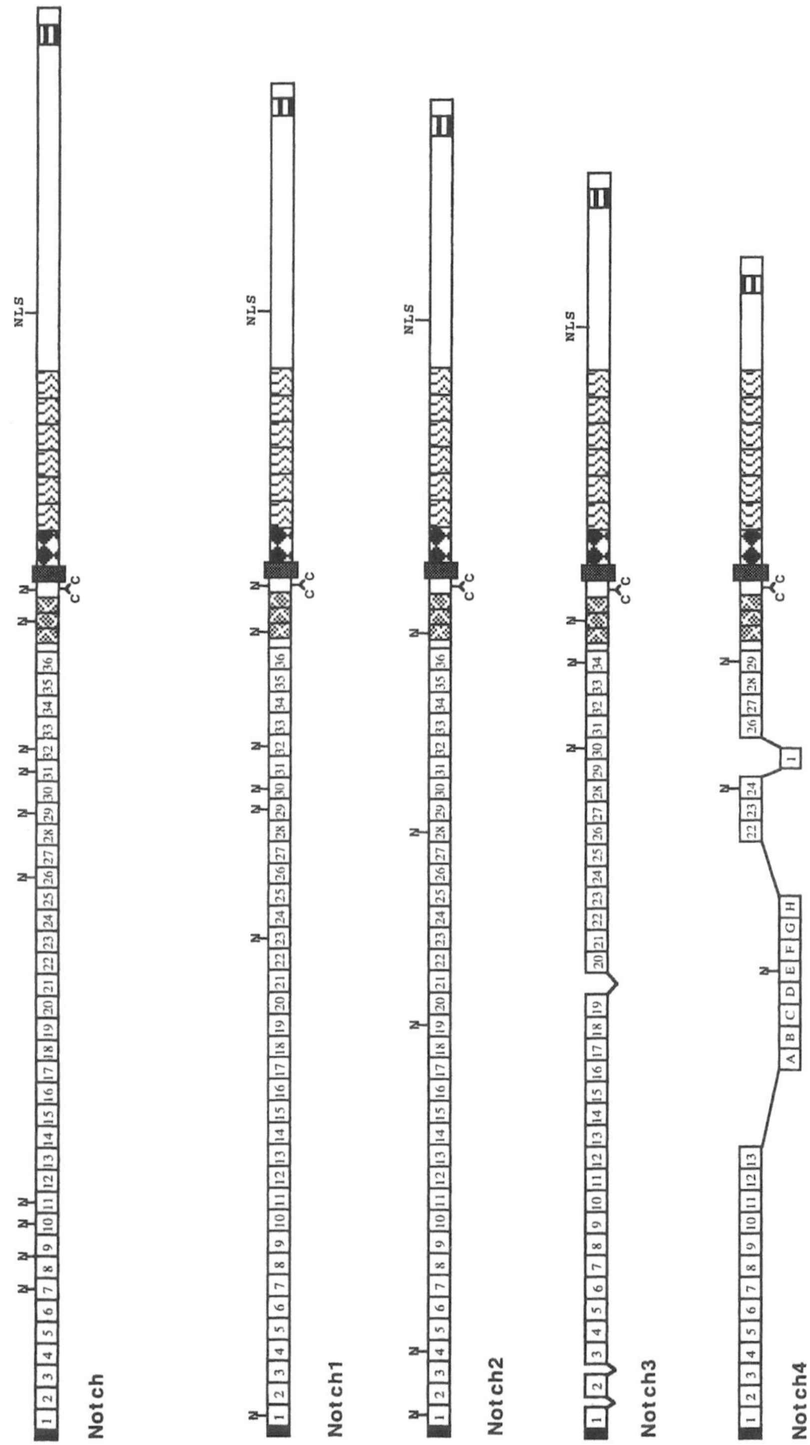

Fig. 1.

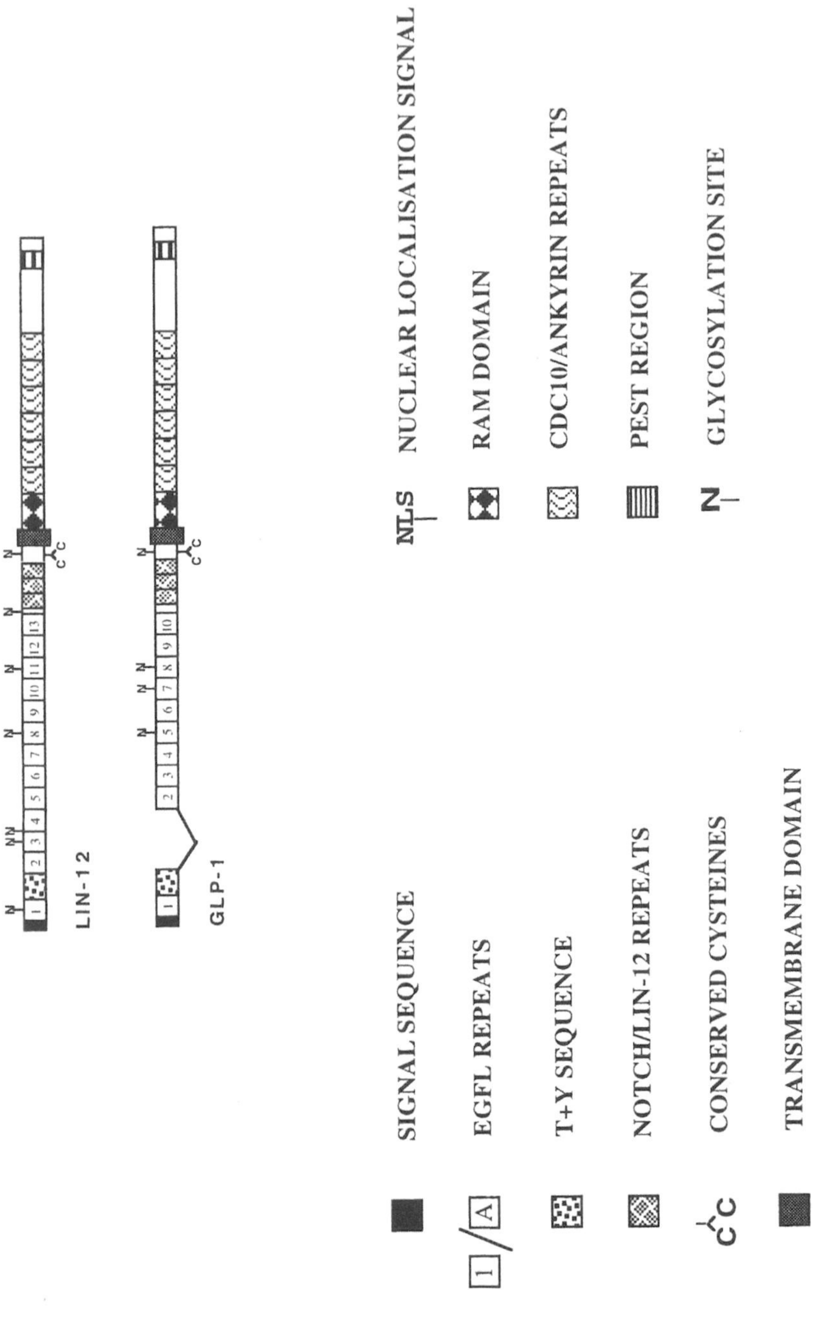

Fig. 1. Continued

have even been detected in two transcription factor proteins which are predicted to reside in the nucleus (Gomez-Skarmeta et al. 1996). Each repeat is approximately 40 amino acids long and contains 6 invariant cysteine residues whose spacing is conserved to form 3 disulfide bridges (Fig. 2). The importance of these residues is further reinforced by the lethality of two dominant alleles of Notch which involve changes in an EGFL repeat cysteine (Kelley et al. 1987).

One *Xenopus*, one zebrafish and four mammalian Notch homologues have been identified to date. The four mammalian Notch proteins, Notch-1 through -4, share strong overall homology with each other as well as with the *Xenopus* and zebrafish proteins (Bierkamp and Campos-Ortega 1993; Coffman et al. 1990; Franco Del Amo et al. 1992; Gallahan and Callahan 1997; Lardelli et al. 1994; Reaume et al. 1992; Stifani et al. 1992; Uyttendaele et al. 1996; Weinmaster et al. 1991, 1992). Comparison of EGFL repeat domains between vertebrate and *Drosophila* proteins indicates that the number of EGFL repeats can vary through loss of individual repeats (Fig. 1). *Drosophila* Notch, mammalian Notch1, mammalian Notch2 and the identified Notch proteins in *Xenopus* and zebrafish each possess 36 EGFL repeats, whereas mammalian Notch3 and Notch4 only encode 34 and 29 EGFL repeats respectively. Typically, each repeat is more homologous to its equivalent repeat in other Notch-family proteins than to its adjacent EGFL repeats or other EGFL repeats within the same protein (Fig. 2). For example, EGFL repeats 11 and 12 from *Drosophila* Notch are more similar to EGFL repeats 11 and 12 from *Xenopus* Notch1 than they are to any of the remaining 34 EGFL repeats within the *Drosophila* protein (Coffman et al. 1990; Rebay et al. 1991). This suggests that individual EGFL repeats may have specific conserved ligands, or conserved binding surfaces within ligands which contact several EGFL repeats. The specific repeats lost in mammalian Notch 3 are an EGFL-repeat sized segment from EGFL repeats 2 and 3 (which results in the fusion of repeats 2 and 3 into a novel EGFL repeat), and EGFL repeat 21 (Lardelli et al. 1994). Deletion of these repeats, which are highly conserved in the typical 36 repeat containing Notch receptor proteins, may be an important feature of Notch 3 which is required for its specific biological functions. The Notch 4 protein has lost and rearranged several EGFL repeats with the result that repeats 14–21 in this protein cannot be easily lined up with specific repeats from Notch 1, 2 or 3 (Fig. 1) (Uyttendaele et al. 1996; Gallahan and Callahan 1997). The biological significance of these deletions and rearrangements in Notch 4 is unclear, although it is tempting to speculate that Notch 4 may interact with some distinct ligands or partners as compared to Notch 1, 2 and 3.

Fig. 1. Schematic representation of Notch family proteins. Consensus sequence for identifying N-linked glycosylation site is NXS/T (where X is any amino acid except Proline). PEST regions were identified based on the consensus published in Rogers et al. (1986). The conserved EGFL repeats are aligned and numbered. In Notch-4, EGFL repeats A-H are unique, and EGFL repeat I is a hybrid of EGFL repeats 31 and 32 in Notch-1/2 (Uyttendaele et al. 1996). Consensus nuclear localization sequences are indicated. Notch4, Lin-12 and Glp-1 do not contain consensus NL sequences. It is not clear whether these proteins are targeted to the nucleus through nonconsensus NLS or whether they are cotransported into the nucleus with transcription factors including the CSL proteins. Alternatively, it remains possible that Notch4 does not function in the nucleus

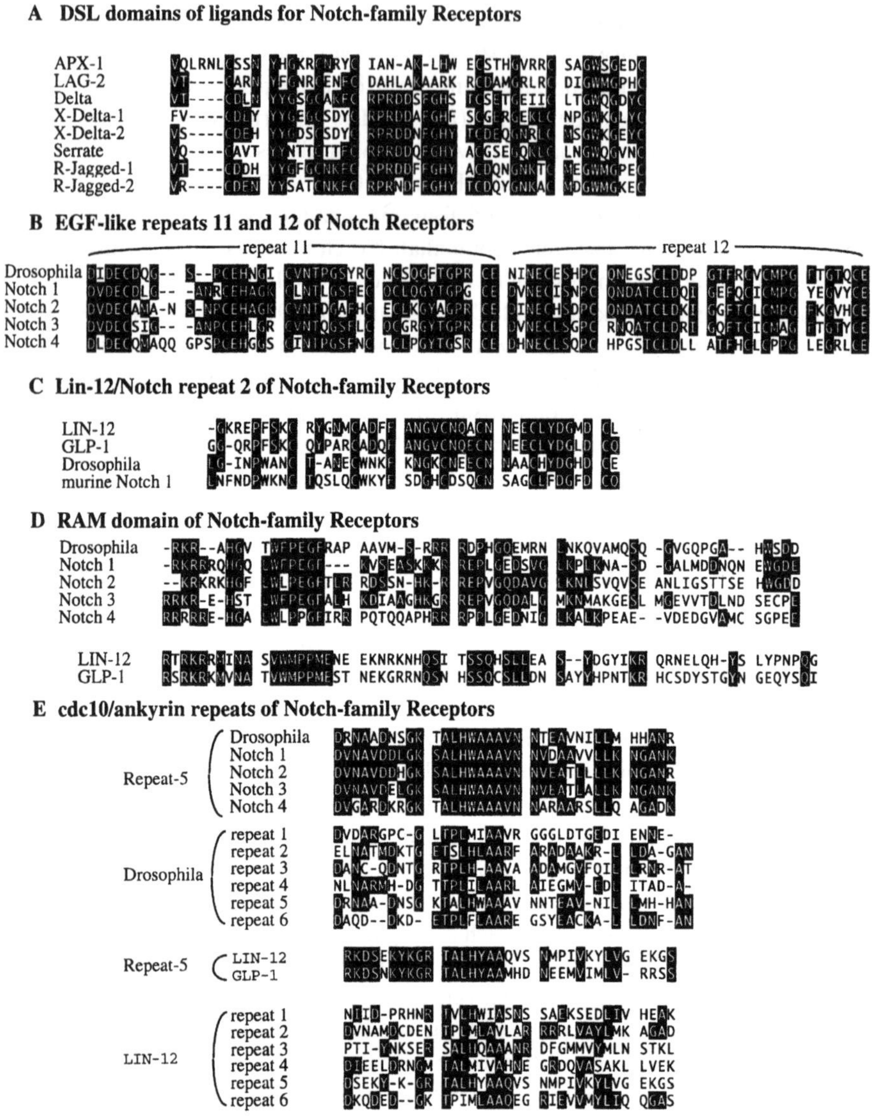

Fig. 2. Alignment of specific DSL ligand and Notch-family receptor domains which have been implicated in ligand binding or in signal transduction. Note, only representative LNR domains and cdc10 domains are shown in order to indicate conservation across species (LNR) or conservation across species and between domains (cdc10)

Conservation of the 36 individual EGFL repeats between *Drosophila* and the vertebrate Notch1 and 2 proteins, as well as deletion and alteration of specific repeats in Notch3 and 4, together highlight the modular nature of the large extracellular domain of the Notch proteins. It is likely that specific combinations of EGFL repeats are retained by each Notch protein to facilitate interaction with many different ligands, co-receptors or regulators.

The LIN-12/Notch repeat (LNR) is also a cysteine-rich domain with specific conserved spacing (Fig. 2) (YOCHEM et al. 1988). All Notch-family proteins contain exactly three LIN-12/Notch repeats. It has been suggested that Notch receptors are activated through dimerization and that LIN-12/Notch repeats prevent dimerization prior to the activation of Notch by one of its ligands. Deletion of this domain can generate an activated form of Notch which no longer depends on ligand for stimulation (GREENWALD 1994; LIEBER et al. 1993). C-terminal to the LNR domain in Notch, and its relatives, are two cysteines within a cysteine-poor region, which also function to maintain Notch in an inactive state. Mutation of either or both cysteines to serine residues produces an activated Notch protein (KIDD et al. 1989; LIEBER et al. 1993). In contrast to Notch proteins with deletions of the LNR, Notch proteins which have cysteine to serine mutations in this region are still dependent on ligand for function (LIEBER et al. 1993). It is interesting to note that cysteine-rich sequences in the extracellular domains of the Ret and Neu tyrosine kinase receptors inhibit dimerization and, further, that mutations which disrupt these domains result in constitutive receptor activation (ASAI et al. 1995; MULLIGAN et al. 1993; SIEGEL et al. 1994). The cysteine residues in the LNR domain and the two cysteine residues C-terminal to the LNR may together regulate the dimerization of Notch. Ligand activation of Notch-family proteins could disrupt or exchange the specific disulfide bonds which are formed between cysteine residues in this extracellular region, inducing dimerization and signal transduction.

All Notch family proteins also contain a transmembrane domain immediately downstream of these cysteine-rich domains. On the cytoplasmic side of the transmembrane domain, Notch proteins contain a Ram domain, six copies of a cdc10/SWI6/ankyrin-related repeat (hereafter referred to as cdc10 repeats) (BREEDEN and NASMYTH 1987), and PEST sequences which are thought to be important in protein turnover (Fig. 1). The cytoplasmic domains of Notch proteins are variable in length and contain the least conserved regions of these receptors. The cytoplasmic domains of the Notch-family proteins will be discussed in greater detail below.

1.5 The *C. elegans* Notch-Related Receptors: LIN-12 and GLP-1

The *C. elegans* LIN-12 and GLP-1 receptors are highly related to each other and are related in structure to the *Drosophila* and vertebrate Notch proteins (Fig. 1) (YOCHEM and GREENWALD 1989; YOCHEM et al. 1988). Both LIN-12 and GLP-1 contain signal sequences and membrane spanning domains for insertion into the plasma membrane. The LIN-12 protein contains 13 EGFL repeats whereas GLP-1 contains 10. A small sequence is present near the N-termini of both proteins. This sequence, termed T + Y, is present between the first and second EGFL repeat in either protein and has no known function. The T + Y domains could simply reflect the common ancestry of LIN-12 and GLP-1. Like the Notch proteins discussed above, the specific EGFL repeats in LIN-12 and GLP-1 are highly related to their equivalent repeats in each protein. The fact that LIN-12 has 13 repeats and GLP-1 has 10 is due to the presence of 3 additional EGFL repeats which immediately follow the T + Y sequence in LIN-12 but not in GLP-1. Just C-terminal to the last EGFL repeat in both LIN-12 and GLP-1 are three LNR repeats followed by a cysteine-poor sequence containing two critical cysteines as discussed above (YOCHEM et al. 1988). Analysis of dominant gain of function *LIN-12(d)* alleles reveals that mutations in amino acids which are within the LNR region or very close to one of the two conserved cysteine residues (downstream of the LNR) result in hyperactivation of LIN-12 (GREENWALD and SEYDOUX 1990). Thus, related domains just outside of the plasma membrane are required to repress signal transduction by Notch and these Notch-like receptors.

The cytoplasmic domains of LIN-12 and GLP-1 are also highly related to each other and similar in overall structure to the Notch proteins. These proteins each contain a Ram domain, six cdc10 repeats, and PEST sequences. All of these structural features are conserved between the Notch proteins and the LIN-12/GLP-1 receptors which have been characterized in *C. elegans*. Genetic analysis in *C. elegans* and in *Drosophila* has suggested that conservation of overall domain structure from Notch-like proteins in both species may be related to conservation of mechanisms of receptor activation, regulation, and signal transduction by Notch, LIN-12 and GLP-1.

2 Notch Receptors as Integrators of Multiple Extracellular Signals

Notch receptors possess very large extracellular domains, yet the minimal region which is necessary and sufficient for interaction of Notch with the Delta ligand represents less than 5% of these extracellular sequences (REBAY et al. 1991). Analysis of Notch mutants in *Drosophila* and humans has revealed that specific point mutations in many of the conserved EGFL repeats can cause alterations in Notch signaling with associated biological consequences. The extracellular domains of Notch-family proteins are thought to participate in numerous protein-protein interactions. The multiple EGFL repeats of these receptors may be used to integrate a myriad of extracellular signals. It is therefore important to characterize the interaction between Notch and specific Notch ligands as well as to understand how Notch:Notch ligand interactions are spatially restricted. Through the analysis of individual interactions between the Notch extracellular domain and its partners or regulators, we may begin to understand how integration of multiple signals can occur.

2.1 DSL-Ligands for Notch and LIN-12/GLP-1 Receptors

Ligands for Notch-family proteins have been identified in *Drosophila*, *C. elegans* and several vertebrate species. There are two known ligands for Notch in

→

Fig. 3. Structural comparison of Notch ligands. The cysteine rich region of Serrate/Jagged is drawn according to LINDSELL et al. (1995)

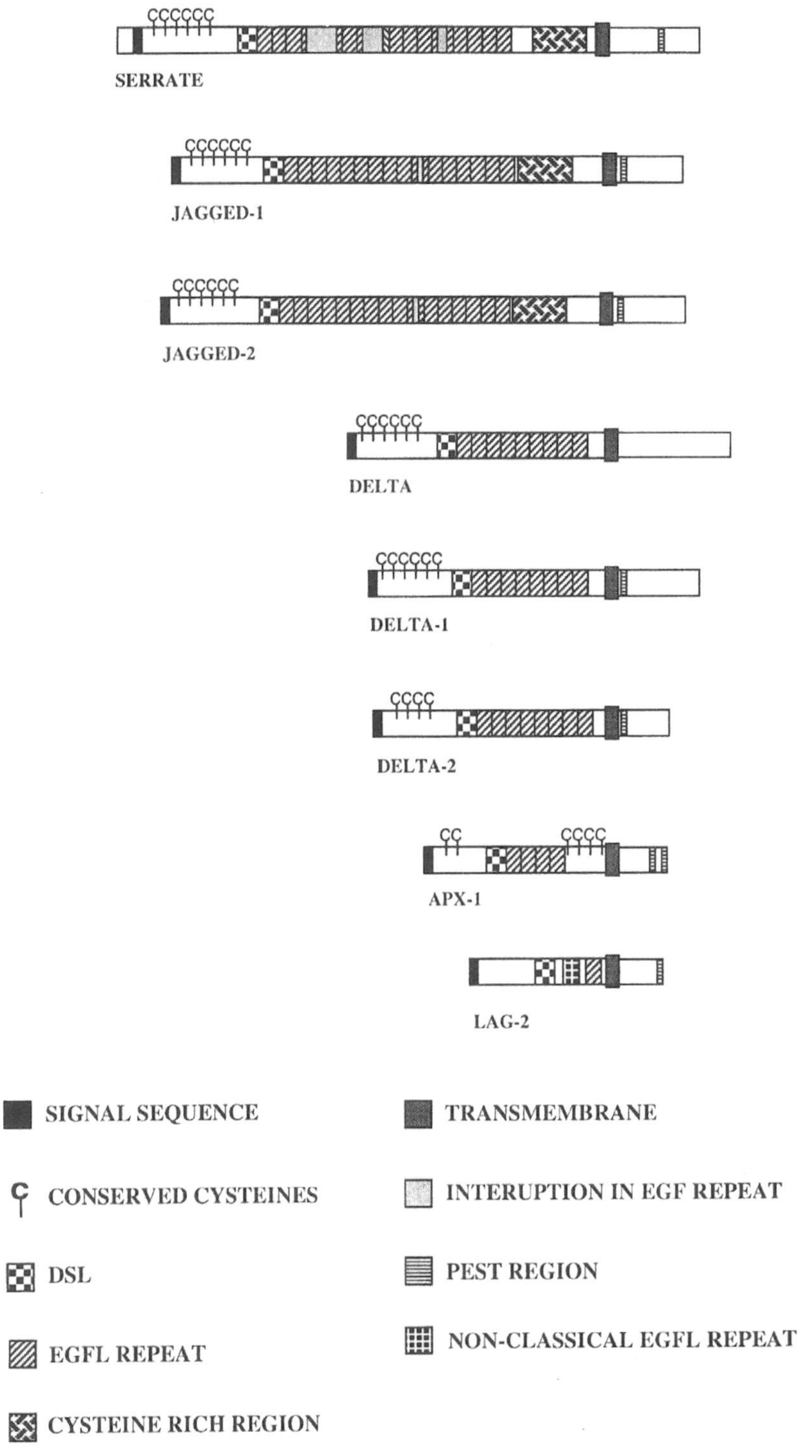
CCCCCC
SERRATE
CCCCCC
JAGGED-1
CCCCCC
JAGGED-2
CCCCCC
DELTA
CCCCCC
DELTA-1
CCCC
DELTA-2
CC
CCCC
APX-1
LAG-2
SIGNAL SEQUENCE
TRANSMEMBRANE
CONSERVED CYSTEINES
INTERUPTION IN EGF REPEAT
DSL
PEST REGION
EGFL REPEAT
NON-CLASSICAL EGFL REPEAT
CYSTEINE RICH REGION

Drosophila, Delta and Serrate (Fig. 3) (FLEMING et al. 1990; KOPCZYNSKI et al. 1988; THOMAS et al. 1991; VASSIN et al. 1987). Mutations in the gene for Delta cause lethal overproduction of neuronal cell types in the *Drosophila* embryo (MUSKAVITCH 1994). This "neurogenic" phenotype also occurs in flies which are mutant for Notch and several other genes which participate in the Notch signal transduction pathway as discussed above. Serrate is known to be required for proliferation and patterning of imaginal disks (SPEICHER et al. 1994). Delta and Serrate proteins both contain an N-terminal region which is highly related (FLEMING et al. 1990; THOMAS et al. 1991). This region of approximately 200 amino acids contains a conserved cysteine rich domain termed the DSL domain (*D*elta/*S*errate/*L*ag-2 domain) and the first three EGFL repeats of each protein. The DSL domain has been named after several Notch ligands (HENDERSON et al. 1994; TAX et al. 1994) and is required for binding of these ligands to their respective Notch or LIN-12/GLP-1 receptors (Fig. 2) (FITZGERALD and GREENWALD 1995; MUSKAVITCH 1994). In fact, physical interaction between *Drosophila* Notch and its two ligands has been demonstrated by means of an in vitro cell adhesion assay (FEHON et al. 1990; REBAY et al. 1991). In this assay, EGF-like repeats 11 and 12 of Notch are necessary and sufficient for its interaction with Delta or Serrate (DE CELIS et al. 1993; FEHON et al. 1990; REBAY et al. 1991). The high affinity Delta-binding site likely includes additional EGFL repeats within Notch, as mutations have been identified in EGFL repeats 14 and 29 of Notch which weaken its interaction with Delta (LIEBER et al. 1992). Like Notch, Delta and Serrate contain multiple EGFL repeats. Indeed, mutations in several of the Delta EGFL repeats can also decrease the affinity of Delta:Notch interaction in vitro, suggesting that EGFL repeat regions in Notch-ligands may also participate in high affinity binding (LIEBER et al. 1992). The Delta protein contains a total of 9 EGFL repeats and Serrate contains 14. In Serrate, the fourth, sixth and tenth EGFL repeats contain small insertions which disrupt the typical EGFL cysteine spacing pattern. Serrate also contains an additional cysteine-rich domain between the EGFL repeats and the transmembrane domain which is not present in Delta (Fig. 3). The Delta and Serrate cytoplasmic domains are unrelated to each other or to other proteins.

Vertebrate homologues of the Delta and Serrate proteins have been reported by several labs in the past 2 years (Fig. 3). To date, two genes with homology to Delta have been described in *Xenopus*, *X-Delta1* and *X-Delta2* (CHITNIS et al. 1995; JEN et al. 1997). One of these has also been described in chickens, mice, rats and zebrafish (*Delta1* in chickens and rats = *Dll1* in mice and *DeltaD* in zebrafish) (BETTENHAUSEN et al. 1995; HENRIQUE et al. 1995) (Rat Delta1 submitted to Genebank by G. DiSibio, L. Hebshi, J. Boulter and G. Weinmaster and zebrafish DeltaD submitted to Genebank by J.A. Campos-Ortega). Two Serrate homologues have been described in rats and chickens; these genes have been named *Jagged1* and *2* (LINDSELL et al. 1995; SHAWBER et al. 1996a) or *Serrate1* and *2* respectively (HAYASHI et al. 1996; MYAT et al. 1996). The human *Jagged1/Serrate1* orthologue has also been described (ZIMRIN et al. 1996; Human Jagged1 submitted directly to Genebank by L. Li, Y. Deng, A.B. Banta and L. Hood as well as G.E. Gray, R.S. Mann, E. Mitsiadis, D. Henrique, M. Caracangiu, D. Ish-Horowicz and S. Artavanis-

Tsakonas). Additional vertebrate members of the Delta and Serrate families exist, although they have yet to be published (NYE and KOPAN 1995). The vertebrate Delta proteins contain the N-terminal Delta-Serrate homology region, which includes a DSL domain and three EGFL repeats. Delta1 and Delta2 each contain a total of eight EGFL repeats which are followed by transmembrane domains and short cytoplasmic domains with limited homology to the cytoplasmic domain of the *Drosophila* Delta protein. The Vertebrate Serrate-like proteins also contain the N-terminal Delta-Serrate homology region (including DSL domain and 3 EGFL repeats), 13 additional EGFL repeats (total of 16 EGFL repeats where repeat 10 contains an insertion), the conserved Serrate cysteine-rich domain, a transmembrane domain, and a cytoplasmic domain which is conserved between Jagged1 and Jagged2 but which is unrelated to the *Drosophila* Serrate cytoplasmic domain.

The conservation of DSL domains and EGFL repeats in the vertebrate Delta/Serrate homologues indicates that ligand:receptor interaction is likely to involve the DSL domain in ligands and EGFL repeats 11 and 12 in Notch proteins. The conservation of domains which are unique to Delta proteins (cytoplasmic domain) or to Serrate proteins (cysteine-rich domain) indicates that regulation or presentation of these two ligand families may differ in some important way (see below).

In *C. elegans*, three DSL ligands have been identified (Fig. 3). These three proteins are named Lag-2, Apx-1 and Arg-1 (HENDERSON et al. 1994; MELLO et al. 1994; TAX et al. 1994) (Apx-1: Mello and Priess, unpublished: see FITZGERALD and GREENWALD 1995). Like the Delta and Serrate proteins in *Drosophila* (and vertebrates), the *C. elegans* ligands each contain sequences which are similar to the region in common between Delta and Serrate discussed above. In Lag-2, this region contains the DSL domain and an EGFL repeat which is most similar to the third Delta or Serrate repeat. This ligand is missing sequences corresponding to EGFL repeats one and two from Delta and Serrate as repeats one and two may have fused during evolution to create a nonclassical EGFL repeat in their place. Thus, the small Lag-2 protein contains only one consensus EGFL repeat. The larger Apx-1 protein contains the same Delta-Serrate homology region (with DSL domain followed by three EGFL repeats) and one additional EGFL repeat (for a total of four EGFL repeats). Both proteins contain transmembrane domains and short cytoplasmic domains with no obvious domain structure. It has been suggested that PEST sequences in the cytoplasmic domains of both Lag-2 and Apx-1 may induce rapid degradation of these proteins (ROGERS et al. 1986). The Lag-2 cytoplasmic domain is not homologous to either Delta or Serrate cytoplasmic domains, and no cysteine-residues are present between the EGFL repeat and transmembrane domain, indicating that this ligand cannot be assigned to either the Delta or Serrate subfamilies of Notch ligands (HENDERSON et al. 1994; TAX et al. 1994). Similarly, Apx-1 cannot be easily assigned to either Delta or Serrate subfamilies on the basis of homology. It is interesting to note, however, that Apx-1 contains two cysteine residues between the signal peptide and the DSL domain and also contains four cysteine residues between the fourth and final EGFL repeat and the transmembrane domain (MELLO et al. 1994). The significance of four cysteine residues downstream of the EGFL repeats is unknown but cysteine residues are present in an equivalent region of Serrate proteins as noted above. The detailed structure of Arg-1 has not yet been described in the literature.

The Notch family receptors may be "activated," through dimerization, stabilization or internalization in response to specific ligand stimulation. DSL-family ligands in all species encode transmembrane proteins which are likely presented on the surface of signaling cells. Surprisingly, Delta and Serrate in *Drosophila*, as well as Lag-2 in *C. elegans*, have been detected in cells which are separated from their sites of synthesis (Couso et al. 1995; Henderson et al. 1994; Kooh et al. 1993; Thomas et al. 1991). These results suggest that DSL proteins may be endocytosed by Notch-receptor bearing cells or may be proteolytically cleaved from their membrane spanning precursors. The Lag-2 protein which is expressed in the distal tip cell of the *C. elegans* gonad is taken up by GLP-1 expressing germ cells in such a way that the cytoplasmic domain of Lag-2 remains intact and the endocytosed protein colocalizes with internalized GLP-1 (Henderson et al. 1994). In this case, proteolytic cleavage of Lag-2 is not responsible for transfer of Lag-2 from signaling cell to receiving cell. The process of DSL ligand release and endocytosis may be similar to the mechanism by which the transmembrane ligand Boss is taken up by cells expressing the Boss receptor, Sevenless (Cagan et al. 1992). Artificially generated secreted forms of Lag-2 and Apx-1 can fully substitute for the endogenous Lag-2 in several developmental contexts. Indeed, secreted DSL domains of either Lag-2 or Apx-1 are even sufficient to enhance GLP-1 activation by endogenously expressed Lag-2 protein (Fitzgerald and Greenwald 1995). These data suggest that the DSL domains of either protein may be able to induce dimerization, stabilization or endocytosis of GLP-1. Analogous studies with secreted forms of Delta or Serrate proteins have yet to be reported.

A dominant allele of *Xenopus Delta1* (*X-Delta1*^{STU}) has been described which can inhibit activation of endogenous Notch when ectopically expressed in *Xenopus* embryos (Chitnis et al. 1995). This mutant was created through deletion of C-terminal sequences from *Delta1*. The effect of *X-Delta1*^{STU} can be overcome through co-injection of excess wild type *X-Delta1* cRNA, which restores Notch activation. It is not clear whether truncated Delta1 protein inhibits normal Delta function (perhaps by forming inactive oligomers with wild type Delta1 protein), or whether it binds to Notch in a nonproductive fashion (and therefore blocks endogenous Delta from "activating" Notch). Similar experiments have also been reported in *Drosophila* where Delta or Serrate proteins with C-terminal truncations (DeltaTM or SerrateTM) behave as dominant inhibitors to block Notch activation in neighboring cells (Sun and Artavanis-Tsakonas 1996). These mutant DSL ligands can still bind to Notch on the surface of S2 tissue culture cells in vitro, indicating that adhesion between ligand and receptor is not dependent on the cytoplasmic domain of either ligand.

Delta and Serrate proteins with C-terminal truncations behave as dominant negative alleles which block activation of Notch as discussed above. In contrast, Lag-2 protein with B-galactosidase in place of its cytoplasmic domain can still function to suppress the lethal effects of severe *lag-2(0)* loss of function mutations in *C. elegans* (Henderson et al. 1994; Wilkinson et al. 1994). Thus, the cytoplasmic domains of these proteins are dispensable, at least in as much as they can be replaced by β-galactosidase. Do the cytoplasmic domains of Lag-2 and Apx-1 perform different and dispensable functions than the cytoplasmic domains of Delta and Serrate proteins in *Xenopus*, or in *Drosophila*? Oligomerized Delta may be required to activate Notch and the intracellular domain may facilitate oligomerization (MUSKAVITCH 1994). β-Galactosidase can form tetramers and may, therefore, fortuitously generate higher order complexes of DSL ligands which are capable of activating LIN-12 and GLP-1 receptors. It will be interesting to see if Delta and Serrate β-galactosidase fusions can stimulate Notch in *Drosophila*. Alternatively, as DSL ligands can be endocytosed, it is possible that X-Delta1^{STU}, DeltaTM and SerrateTM mutants are endocytosis impaired. If endocytosis of Notch:Notch ligand complexes is required for activation of Notch, and the truncated DSL ligands cannot be endocytosed, then Notch may be locked into non-productive complexes at the cell surface by truncated ligands.

How do DSL ligands "activate" Notch-family receptors? Biochemical analysis of Notch signaling complexes before and after binding of Delta, DeltaTM or Delta + DeltaTM should yield the answer. As discussed above, ligand induced Notch oligomerization and/or endocytosis may be critical.

2.2 DSL Ligands Encode Redundant and Specific Functions

In *Drosophila*, the Serrate protein can substitute for Delta during neuroblast segregation (GU et al. 1995). Similarly, Apx-1 and Arg-1 can suppress the phenotype of *lag-2* loss of function mutations in several cell fate decisions (FITZGERALD and GREENWALD 1995). These experiments reveal that distinct DSL ligands in both *Drosophila* and *C. elegans* share overlapping biochemical functions. In *Drosophila* wing development, however, the Delta and Serrate proteins have very distinct and nonoverlapping signaling capabilities (DIAZ-BENJUMEA and COHEN 1995; DOHERTY et al. 1996; KIM et al. 1995; SPEICHER et al. 1994). Ectopic expression studies have shown that an activated *Notch* allele can stimulate transcription of wing margin specific genes in both dorsal and ventral compartments of the developing wing disk (DE CELIS et al. 1996; DOHERTY et al. 1996). The Delta protein can only stimulate margin genes in the dorsal compartment and therefore can only activate Notch dorsally (DOHERTY et al. 1996). In contrast, Serrate cannot stimulate margin genes in the dorsal compartment but only in the ventral wing disk (COUSO et al. 1995; JONSSON and KNUST 1996; KIM et al. 1995; SPEICHER et al. 1994). Indeed the Serrate protein may even function to antagonize Notch under specific circumstances (JONSSON and KNUST 1996). In the fly wing then, Delta and Serrate proteins can only activate the Notch proteins in distinct compartments whereas their receptor, Notch, can function in either. One or more spatially restricted components must be limiting the specificity or distribution of Notch such that it can only respond to one specific DSL ligand in each compartment.

It has recently been demonstrated that the secreted protein Wingless inhibits Notch activation by Delta and yet may facilitate the activation of Notch by Serrate

(Axelrod et al. 1996; Couso et al. 1995; Jonsson and Knust 1996). In contrast, the secreted protein Fringe inhibits Notch activation by Serrate but may facilitate Notch activation by Delta (see below) (Irvine and Wieschaus 1994). Through complex spatial regulation of Delta and Serrate presentation in the developing wing, as well as Wingless and Fringe mediated control of Notch sensitivity for these ligands, the Notch protein can integrate multiple signals to precisely control induction of wing margin specific gene expression.

The conservation of 36 EGFL repeats between *Drosophila* and vertebrate Notch proteins, together with the fact that EGFL repeats 11 and 12 are sufficient for interaction between Notch and DSL ligands, suggests that there may be other important protein-protein interactions which occur through the extracellular domain of this large protein. Perhaps Wingless and Fringe can bind to distinct EGFL repeats and control the activation of Notch by Delta or Serrate. It has also been speculated that Brainiac and Egghead may bind to the extracellular domain of Notch (Goode et al. 1996a). These and other interactions could be envisioned to occur through Notch or through the formation of a multisubunit receptor which includes Notch. Indeed, the Wingless protein cannot bind to Notch on the surface of tissue culture cells. Perhaps Wingless can bind to a D-Frizzled2:Notch complex (Bhanot et al. 1996) or to a complex between Notch and Serrate (Couso et al. 1995). Interestingly a number of point mutations have been identified in the extracellular domain of Notch which can produce very specific phenotypes, suggesting that this receptor may interact with several different proteins in the secretory pathway or extracellular space.

2.3 Subcellular Localization of Notch

In the very early *Drosophila* embryo, Notch protein is found on apical, lateral and basal cell surfaces (Fehon et al. 1991; Kidd et al. 1989). As development proceeds, Notch becomes highly localized to specific membrane subcompartments in cells of several tissues. In developing foregut, hindgut and salivary gland, the Notch protein is apical (Fehon et al. 1991). In the developing wing and eye imaginal disks, Notch is specifically localized to an apicolateral ring in each cell of the columnar epithelial layer. Notch is also highly concentrated at the interface of dividing neuroblasts and their progeny. In some developing neurons, Notch is detected in axonal bundles. In the vertebrate neural tube, Notch1 protein is concentrated on the basal membrane of differentiating neurons which have just divided from their self-renewing siblings (Chenn and McConnell 1995). The very specific subcellular localization of Notch protein in each case is likely essential for spatial control of Notch activation since the DSL ligands are transmembrane proteins as discussed above (Kooh et al. 1993). In many cases DSL ligands are presented on specific subdomains of signaling cells (Kooh et al. 1993; Sun and Artavanis-Tsakonas 1996), and therefore access to Notch will be dependent on the direct juxtaposition of Notch and Delta in neighboring cells. In order to understand Notch activation, the mechanism by which Notch is localized within a polarized cell must also be determined. Structure/

function analysis of Notch proteins will be required to decipher which domains are responsible for localization of Notch in each cellular context. Are the same domains of Notch required for colocalization of Notch with intercellular junctions in the wing imaginal disk as are required for Notch1 basal localization during differentiation of neurons in the neural tube (CHENN and MCCONNELL 1995)? Are the proteins (or other molecules) which tether Notch to the adherence junction in imaginal disk cells also controlling ligand binding or signaling properties of Notch in this context? In addition, there is evidence to suggest that Notch proteins may perform distinct functions in apical vs. lateral membranes of the same cell (GOODE et al. 1996a). It will be necessary to determine which domains of Notch are responsible for its localization to specific membrane subcompartments in each developmental context in order to understand the signaling pathway and biological functions which are mediated downstream from this receptor in each case.

2.4 Notch Proteins as Regulators of Adhesion, Polarity and Cell Signaling

Historically, Notch proteins have been considered as having one of two distinct roles. In one case, Notch proteins are receptors which receive important signals from neighboring cells, and directly alter specific gene expression programs in response. The other view of Notch is as a regulator of cell adhesion and polarity, and that control of these processes by Notch indirectly controls the efficiency by which other signals can be received from neighboring cells. Both of these views may actually be correct.

Truncated Notch-family proteins have been constructed and analyzed in *Drosophila* and *C. elegans* which function like "activated receptors" (FORTINI and ARTAVANIS-TSAKONAS 1993; GREENWALD 1994). These cytoplasmic-domain miniproteins are typically found in the nucleus and have been shown to induce Notch-dependent or LIN-12-dependent events, even in the absence of their respective wild type Notch-family protein. Thus, many functions of Notch or LIN-12 are not dependent on adhesion mediated by the extracellular domains but can be mimicked through activation of signal transduction by the cytoplasmic domain in isolation. Consequently, Notch proteins are signaling receptors.

Drosophila Notch and other Neurogenic genes in this organism are required for the development of specific cell lineages in all three germ layers. In order to identify "key characteristics" of processes which depend on Notch in these diverse tissues, HARTENSTEIN et al. (1992) used molecular markers to analyze the transformations associated with loss of Notch function in several tissues. Notch and other neurogenic genes were found to control the process by which many tissues acquire or maintain epithelial properties. These studies have been extended to show that proneural proteins, which are antagonized by Notch (see Sect. 4.2 below), promote mesenchymal fates whereas neurogenic proteins including Delta, Notch and downstream targets of Notch (the E(spl) proteins) are required for epithelial phenotypes in endodermal cells (TEPASS and HARTENSTEIN 1995). The distinction between epithelial and mesenchymal is important for cell-cell communication. Communication between polarized epithelial cells often involves ligands and receptors which are localized to specific membrane subcompartments. Indeed this is the case with Notch and its ligands as discussed above. Epidermal growth factor receptor family of tyrosine kinases are also highly localized and require proper cell polarity to function (Maratos-Flier et al. 1987). In *C. elegans*, for example, the EGFR related Let-23 protein can only perceive the presence of its ligand Lin-3, if Let-23 is concentrated in the lateral membranes at cell junctions by the Lin-2, Lin-7 and Lin-10 proteins (Lambie 1996; Simske et al. 1996). By localizing receptors to specific membrane subcompartments, these receptors can become highly responsive to a ligand which is localized in the correct subcompartment of a neighboring cell. Alternatively, a localized receptor can become completely insensitive to signals which are localized to the opposite side of a neighboring cell. Thus, if Notch induces or maintains the epithelial character of cells then it will control the sensitivity of cells to other specific extracellular signals.

Cell polarity and the epithelial behavior of a cell may be *directly* regulated by Notch or may be *indirectly* regulated as a result of Notch-induced epithelial cell fate specification. In the *Drosophila* follicular epithelium, extracellular Brainiac and Egghead proteins function together with Notch to control the epithelial morphogenesis and polarity (Goode et al. 1996a, b). This Notch-dependent control of epithelial polarity occurs independent of Notch-induced cell fate transformations. Notably the adhesion between oocyte and follicular epithelium is dependent on brainiac and egghead from the germ line cells (oocyte and nurse cells) and Notch in the epithelium (Goode et al. 1996a). Furthermore, loss of the oocyte-follicular epithelium adhesion system results in overgrowth and disorganization of the follicular epithelium. In this case, the loss of *brainiac* or *egghead* in the germ line leads to a decrease in the sensitivity of follicular epithelial cells to other signals including the TGF-α homologue, Gurken, which stimulates the *Drosophila* EGF receptor (DER). Not only is DER signal efficiency dramatically lowered but Notch normally expressed in the lateral membrane of the follicle cell becomes much less sensitive to Delta expressed by its neighbor. Goode et al. have therefore proposed that Notch in a follicle cell can participate in one type of cell-cell signaling system at the apical membrane (together with Brainiac and Egghead) to control the adhesion, epithelial organization and polarity and that another pool of Notch in the lateral membrane of the same follicle cell can respond to Delta through a distinct signaling system to control cell fate.

2.5 Notch Glycoproteins; Fringe and Brainiac Control Activation

All Notch-family proteins which have been described to date are predicted to be glycosylated on one or more asparagine residues. For example, the *Drosophila* Notch extracellular domain contains ten sequon targets for potential glycosylation (sequon = Asparagine followed by any amino acid, followed by either serine or threonine). The functional significance of glycosylation on transmembrane or secreted proteins may vary for specific proteins and so it is difficult to predict what the function of glycosylation will be for the Notch-family receptors. It is known that at least some of the potential glycosylation sites are used, since Notch immunoprecipitated from *Drosophila* embryos or tissue culture cells can bind to the lectins concanavalin-A and wheat germ agglutinin (JOHANSEN et al. 1989; KIDD et al. 1989). The number and location of predicted glycosylation sites are different between individual Notch-family proteins (Fig. 1). The DSL ligands are also predicted to be glycosylated on asparagine residues, which vary dramatically in both number and position between individual ligands. *Drosophila* Serrate, for example, has 14 potential glycosylation sites which are dispersed throughout the entire extracellular domain (FLEMING et al. 1990; THOMAS et al. 1991). In contrast, mouse Dll1 has only one potential N-linked glycosylation site (BETTENHAUSEN et al. 1995).

Genetic analysis has highlighted the importance of Fringe in controlling Notch activation in the *Drosophila* wing disk (IRVINE and WIESCHAUS 1994; KIM et al. 1995). We and others have recently cloned mammalian Fringe proteins (Lunatic Fringe, Manic Fringe and Radical Fringe; WU et al. 1996) and expressed them in *Drosophila* (COHEN et al. 1997; JOHNSTON et al. 1997). Interestingly, mouse Manic Fringe and Radical Fringe alter Notch dependent development in several fly tissues. These data led us to propose that Fringes control the sensitivity of Notch for its specific ligands (COHEN et al. 1997), by inhibiting Serrate:Notch interaction while facilitating Delta:Notch. It has recently been demonstrated that Fringe does inhibit activation of Notch by Serrate in *Drosophila* (FLEMING et al. 1997). This effect requires the N-terminal domain of Serrate, since Delta-Serrate chimeric ligands which contain the N-terminus of Delta are resistant to inhibition by Fringe. As discussed above, the Brainiac protein also effects the efficiency of Notch signaling in *Drosophila.* Recent analysis of the Fringe and Brainiac amino acid sequences has led to the suggestion that these secreted proteins may encode either galactosyl transferases or lectins (YUAN et al. 1997). Fringe and Brainiac proteins both possess signal sequences and therefore pass through the secretory pathway (GOODE et al. 1996b; IRVINE and WIESCHAUS 1994). N-linked glycosylation and galactosyl transfer reactions normally occur in the endoplasmic reticulum and golgi complex where nucleotide sugar substrates are present. If Fringe and Brainiac function within the secretory pathway then they may control transfer of galactose residues onto specific substrates. If Fringe and Brainiac function in the extracellular space then they are likely to be lectins.

These data suggest that Fringe may inhibit Serrate function by directly altering the glycosylation of either Serrate or Notch, in such a way as to prevent their association. Alternatively, Fringe could function in the extracellular milieu, which does not contain the nucleotide sugar substrates for galactosyl transfer. In this case, secreted Fringe proteins may function as lectins which bind to Notch or Serrate in a galactose-dependent and peptide sequence specific manner. Fringe binding to either Notch or Serrate would prevent Serrate-Notch complex formation. These possibilities can explain how Fringe blocks Serrate from activating Notch and can also explain how Fringe facilitates the interaction of Delta and Notch. Fringe mediated glycosylation of Notch (or Delta) may strengthen the interaction between receptor and ligand. Alternatively, Fringe as a lectin could bind to both Delta and Notch, creating a highly stable complex. Similar models can be used to explain the effect of the neurogenic protein Brainiac on Notch signaling.

3 Notch Signaling

Numerous genetic and in vitro cell culture studies have demonstrated a crucial role for Notch proteins as receptors of extracellular signals. Elements of the biochemical pathway(s) which transduce these signals from the membrane to the nucleus have been identified. Below we summarize these recent results. The cytoplasmic domain of Notch, acting as a molecular scaffold, is now believed to facilitate assembly of a transcription regulating complex on specific gene promoters which in turn regulate differentiation and proliferation in many tissues.

3.1 The Intracellular Domain of Notch Functions as an Activated Notch Receptor

Dominantly transforming mutants of human Notch1, feline Notch2 and mouse Notch4 have been isolated which are activated through deletion of most or all extracellular sequences (Ellisen et al. 1991; Robbins et al. 1992; Rohn et al. 1996). In addition, mutants of *Drosophila* Notch with deleted cytoplasmic domains behave as dominant inhibitory alleles (Lieber et al. 1993; Lyman and Young 1993; Rebay et al. 1993; Struhl et al. 1993). Together these data suggest that Notch proteins are receptors for extracellular signals and that "Notch-signaling" can be activated simply by removing the extracellular domain. Experiments in *Drosophila* and *C. elegans* have revealed that Notch or LIN-12 cytoplasmic domain alleles can induce cell fates which are normally dependent on wild type Notch or LIN-12 activity, even in the absence of the wild type protein (Fortini and Artavanis-Tsakonas 1993; Greenwald 1994; Lieber et al. 1993; Struhl et al. 1993). Thus, wild type Notch or LIN-12 functions can be supplied simply by activating the signaling pathway(s) which function downstream from the cytoplasmic domains.

The intracellular domains of Notch and LIN-12 behave phenotypically as do their ligand-activated wild-type receptor counterparts, yet there are no obvious enzymatic domains within the cytoplasmic portion to account for the signaling potential of Notch-family proteins. Amino acid sequence comparisons and deletion mapping of the regions involved in protein-protein interactions have led to the identification of several distinct domains within the Notch/LIN-12/GLP-1 intracellular regions (Figs. 1, 2). At the very amino-terminal end of the cytoplasmic region, just inside the transmembrane domain, is the Ram domain (Hsieh et al. 1996; Tamura et al. 1995). This domain is poorly conserved between Notch proteins and the *C. elegans* Notch-like proteins, but binds to orthologous partner proteins in each species (see below) (ROEHL et al. 1996; TAMURA et al. 1995). The next structurally distinct region following the Ram domain is composed of six tandemly arranged cdc10 repeats.These repeats are protein-protein interaction motifs found in a wide variety of transcription factors and cytoskeletal proteins (for review see BLANK et al. 1992). In Notch proteins the cdc10 repeats are followed by a small cluster of conserved residues which are unrelated to any previously identified domain. LIN-12 and GLP-1 also are highly related to each other and share a low level of homology to the Notch proteins in this region. Finally, at the carboxy-terminal end of the *Drosophila* Notch receptor and mammalian Notch1 are two domains thought to be important for regulating protein stability: an opa region (polyglutamine sequence) and PEST sequences (ROGERS et al. 1986). The C-termini of all Notch and Notch-related receptors contain PEST sequences which are likely responsible for rapid degradation of these proteins. The C-terminal regions of LIN-12 and GLP-1, downstream of the cdc10 repeats, are significantly shorter than the equivalent regions in the Notch proteins (LIN-12 has a 163 amino acid tail which is C-terminal to the cdc10 repeats, the GLP-1 C-terminal tail has 162 amino acids, whereas Notch has 537, Notch1 has 456, Notch2 has 436, Notch3 has 320, and Notch4 has 177 amino acids in this region respectively). In addition to the Ram domain, cdc10 repeat domain, the short region of homology C-terminal to the cdc10 repeats, and PEST sequences, one or two nuclear localization signals have been defined in several Notch-like proteins; (Fig. 1) (STIFANI et al. 1992; LIEBER et al. 1993).

3.2 Interaction of the Notch Intracellular Domain with Nuclear Factors: CSL Proteins

The Notch intracellular domain, whether expressed as a dominantly activated mutant protein or proteolytic cleavage product from wild type Notch, can become localized to the nucleus (AHMAD et al. 1995; KOPAN et al. 1996). This nuclear localization is important for Notch function, since the removal of both nuclear localization signals reduces the ability of the intracellular domain to inhibit differentiation of MyoD-expressing 3T3 cells (KOPAN et al. 1994). Genetic analysis in flies and worms has led to the characterization of several nuclear proteins which interact with the cytoplasmic domains of Notch-family proteins. These nuclear partners of Notch/LIN-12/GLP-1 are essential for signal transduction downstream from Notch-family receptors in many developmental contexts.

In *Drosophila*, the first nuclear Notch-partner protein to be identified was Suppressor of Hairless (Su(H)) (FORTINI and ARTAVANIS-TSAKONAS 1994). Su(H) is a highly conserved DNA-binding protein which had been previously identified in mammals (variously known in mammals as RBPJκ/KBF2/CBF1: we will refer to the mammalian protein as CBF1 for simplicity) (FURUKAWA et al. 1992; HONJO 1996; MATSUNAMI et al. 1989; SCHWEISGUTH and POSAKONY 1992). Loss of function mutations in *Su(H)* can dramatically impair Notch signal transduction (FORTINI and ARTAVANIS-TSAKONAS 1994). In *C. elegans*, the homologous protein, Lag-1, functions genetically downstream of LIN-12 and GLP-1 (CHRISTENSEN et al. 1996). Vertebrate CBF1, fly Su(H) and worm Lag-1 proteins share striking sequence conservation over a central region of 412 amino acids. Together these proteins are termed CSL proteins (for CBF1, Su(H), and LAG-1) (CHRISTENSEN et al. 1996). CSL proteins interact strongly with the Ram domains, and with low affinity to the cdcd10 repeat domains, of their respective Notch-protein cytoplasmic domains (HONJO 1996; ROEHL et al. 1996). Amino acids 179–361 of human CBF1 (numbering for the aPCR-3 spliced variant which corresponds to protein isoform RBP3) are necessary and sufficient for interaction with the Ram domain of Notch1 (HSIEH et al. 1996). The DNA-binding domain of CSL proteins has proved difficult to define. The minimal region of the human CBF-1 isoform RBP3 required for specific DNA-binding (see below) is from amino acid 7 to 435, whereas the RBP3 isoform is only 486 amino acids in total (BROU et al. 1994). Mutational analysis has shown that residues 178–187 and 251–283 (RBP3 numbering) are critical for DNA-binding in transfection studies (CHUNG et al. 1994). Mammalian CBF1 has been shown to repress expression of several viral and cellular enhancers or promoters which contain RTGGGAA elements (HSIEH et al. 1996). In contrast, when complexed to the cytoplasmic domain of Notch-family proteins, CSL proteins can activate expression of cellular genes which contain this sequence (BAILEY and POSAKONY 1995; HSIEH et al. 1996; JARRIAULT et al. 1995; LECOURTOIS and SCHWEISGUTH 1995; LU and LUX 1996). The cytoplasmic domain of Notch proteins can be found in the cytoplasm and nucleus. Similarly, the CSL proteins can be found in the cytoplasm or nucleus. Like Notch-family proteins, CSL proteins contain nuclear localization signals (residues 67–70 in RBP3) (HONJO 1996). It has been speculated that unstimulated Notch can sequester CSL proteins in the cytoplasm (FORTINI and ARTAVANIS-TSAKONAS 1994). However, the detailed mechanisms responsible for controlling accumulation of either protein in the cytoplasm or in the nucleus have yet to be defined. Interestingly, the Su(H) protein can be found in both cytoplasm and nucleus of "socket cells" in *Drosophila*. In this case, the cytoplasmic accumulation of Su(H) is controlled by Notch signaling, yet cytoplasmic Su(H) does not colocalize with Notch at epithelial cell junctions (GHO et al. 1996). Each CSL protein also contains a P X S/T P sequence which corresponds to the consensus Mapk phosphorylation site (residues 336–339 in RBP3). It has not been determined whether this site is phosphorylated in vivo and, if so, whether phosphorylation regulates CSL location or function (CHRISTENSEN et al. 1996). Recently, Honjo and colleagues have identified a novel CSL family protein named RBP-L which binds to CGTGGGAA DNA elements but surprisingly does not bind to Ram domain sequences from Notch 1, 2, 3 or 4 (MINOGUCHI et al. 1997).

3.3 Interaction of the Notch Intracellular Domain with Nuclear Factors: EMB-5 Proteins

The cdc10 repeats of LIN-12 and GLP-1 bind to the *C. elegans* EMB-5 protein (HUBBARD et al. 1996). Furthermore, genetic analysis has revealed that EMB-5 is required downstream of LIN-12 and GLP-1. EMB-5 is highly related to the *Saccharomyces cerevisiae* Spt6p regulator of chromatin structure and shares significant homology to a hypothetical protein in *E. coli* (sp:P46837) (Bortvin and Winston 1996; Swanson et al. 1990). Mammalian EMB-5 proteins have also been identified (Supt6H) (Chiang et al. 1996; Nishiwaki et al. 1993). Homology between yeast, worm and mammalian EMB-5-like proteins extends over most of the coding region in each case (Chiang et al. 1996; Nishiwaki et al. 1993; Swanson et al. 1990). These proteins are speculated to encode an SH2 domain (at amino acid 1301–1396 in *C. elegans* EMB-5) and an S1 RNA-binding domain (amino acids 1127 to 1259 of *C. elegans* EMB-5) (Bycroft et al. 1997; MacLennan and Shaw 1993); however, the putative SH2 and S1 domains diverge significantly from consensus SH2 and S1 domains when all EMB-5 homologues are aligned, suggesting that these homologies may not be meaningful. The two hybrid interaction between cdc10 repeats of LIN-12 or GLP-1 and EMB-5 requires only the C-terminal 627 residues out of the 1521 amino acid EMB-5 protein (Hubbard et al. 1996). Perhaps the cytoplasmic domain of all Notch-family proteins forms a molecular scaffold to connect DNA-binding CSL proteins with EMB-5 proteins.

3.4 Interaction of the Notch Intracellular Domain with Nuclear Factors: NFkB Proteins

In a search for protein partners of the cdc-10 repeats of human Notch1, Aster et al. (1994) found that the cdc-10 repeat containing Bcl-3 protein can bind to Notch1 in vitro and in vivo. Bcl-3 is a member of the IkB family of inhibitors which regulate the subcellular localization and DNA-binding activity of NFkB/rel transcription factors. The interaction between Notch and Bcl-3 required cdc-10 repeat regions of both proteins. It has also been shown that the cytoplasmic domain of human Notch1 can block DNA binding of NFkB (p50:p65 dimers) in vitro (Guan et al. 1996). This is due to a direct interaction between the cdc-10 repeats of Notch1 and p50. Therefore, the cytoplasmic domain of Notch1 both mimics Bcl-3 (Bcl-3 can block p50 but not p65 from binding to DNA) and as noted above can bind to Bcl-3. Genetic analysis of Su(H) in *Drosophila* and Lag-1 and EMB-5 in *C. elegans* has established the physiological significance of these proteins in Notch signaling. In contrast, the physiological significance of Notch1:Bcl-3 or Notch1:p50 complex formation has yet to be defined. It is interesting to note, however, that CBF-1 regulates several promoters which contain NFkB binding sites in close proximity to RTGGGAA CBF-1 binding sequences (Kannabiran et al. 1997). Perhaps the cytoplasmic domain of Notch-family proteins can simultaneously bind to CSL and NFkB/rel proteins in order to regulate expression from this class of promoter.

4 Transcriptional Targets of the Notch Signaling Pathway

Several genes have been identified which are transcriptionally regulated by Notch in *Drosophila* or LIN-12/GLP-1 in *C. elegans*. These genes facilitate lateral specification, inhibition of differentiation or proliferation in response to Notch signaling.

4.1 A Positive Feedback Loop: Notch and CSL Genes are Transcriptional Targets of Notch

Many transmembrane receptors either possess or associate with cytoplasmic enzyme modules. Signal transduction from such receptors is therefore accompanied by enzymatic amplification of signal. Such is the case with Receptor tyrosine kinases, where one activated receptor molecule can phosphorylate and activate multiple cytoplasmic signaling proteins. In contrast, the cytoplasmic domain of Notch-family proteins perform a structural, rather than enzymatic, role in transcription regulating complexes (with CSL and EMB-5 proteins). This structural role does not allow for rapid signal amplification. The nonenzymatic nature of Notch-family proteins may be an important feature of this signaling system. Notch or LIN-12/GLP-1 signaling is often used to establish two alternative cell fates from a group of cells with the potential to adopt either fate (Campos-Ortega 1988; Greenwald and Rubin 1992; Heitzler and Simpson 1991; Kopan and Turner 1996; Sternberg 1993). This process is termed lateral specification. In *C. elegans*, two cells (Z1.ppp and Z4.aaa) each have the potential to become either an anchor cell (AC) or a ventral uterine cell (VU) (Kimble 1981; Kimble and Hirsh 1979). A strong LIN-12 signal will promote VU cell formation (Struhl et al. 1993). Both of the precursor cells express the LIN-12 receptor and the Lag-2 ligand (Wilkinson et al. 1994). In any given animal, one of the two precursor cells will randomly express more LIN-12 (or less Lag-2) than the other. This cell will therefore receive a slightly stronger LIN-12 signal than its neighbor. The LIN-12 receptor gene is then strongly induced by LIN-12 signal (Wilkinson et al. 1994). The LIN-12 gene contains 20 Lag-1 binding sites (RTGGGAA) (Christensen et al. 1996). These sites presumably bind Lag-1:LIN-12 (cytoplasmic domain):EMB-5 complexes which increase LIN-12 expression in response to a LIN-12 signal. In addition, the Lag-1 gene itself contains 18 Lag-1 binding sites and is therefore upregulated in response to a LIN-12 or GLP-1 signal (Christensen et al. 1996). The cell, which receives a slightly stronger LIN-12 signal initially, goes on to amplify the signal by dramatically upregulating expression of both Lag-1 and LIN-12. This cell will become a VU cell in response to its strong LIN-12 signal. The VU cell decreases expression of Lag-2 ligand to ensure that its neighbor will not receive a LIN-12 signal, and therefore to ensure that its neighbor will become an AC (Wilkinson et al. 1994). In this way, LIN-12 and Lag-1 positive autoregulation is used to ensure the specification of two distinct cell types based on a random fluctuation in LIN-12 or Lag-2 expression in one precursor cell. By strongly increasing the expression of

Notch-family receptor and CSL DNA-binding subunit and by decreasing the expression of DSL ligand in response to a Notch signal, this system is ideally suited to control lateral specification. This process of lateral specification was first described in grasshoppers, where it is used to ensure the proper segregation of neuronal and epidermal cell fates from a field of proneural precursor cells (Doe and Goodman 1985).

The positive feedback loop between Notch signaling and Notch receptor gene expression has also been observed in mammalian systems. The endogenous Notch4 gene is upregulated by activated int-3/Notch4 in MMTV-int-3 transgenic mice (Smith et al. 1995). Similarly, expression of an activated Notch1 gene in the thymus of transgenic mice results in upregulation of the endogenous Notch1 gene in this tissue (Robey et al. 1996). It is not known whether int-3 upregulates Notch1, 2 or 3 in mammary epithelium; or if activated Notch1 also upregulates Notch2, 3 or 4 in the thymus. Each Notch locus may upregulate specifically and only in response to activation of its corresponding protein, but not in response to other Notch signals. Alternatively, an activated Notch protein may upregulate expression of any Notch gene which is present within accessible or open chromatin.

4.2 bHLH Transcription Factor Networks: E(spl)/Groucho and Mastermind

The Notch signaling pathway regulates expression of basic Helix-Loop-Helix (bHLH) transcription factors (Campuzano and Modolell 1992; Olson and Klein 1994; Weintraub 1993). Defining features of the bHLH proteins are a basic domain for specific DNA binding, and a Helix-Loop-Helix domain for dimerization. In *Drosophila*, Notch can induce N-box bHLH proteins encoded by the *Enhancer of split* [*E(spl)*] gene complex (Bailey and Posakony 1995; Jennings et al. 1994, 1995; Lecourtois and Schweisguth 1995).

The *E(spl)* locus was identified in a screen to isolate genes, which, when mutated, would enhance the small eye phenotype of *Notch*^{split}mutant flies (Tietze et al. 1992; Welshons 1956). Loss of function mutations at this locus were later found to induce neurogenic phenotypes (Lehmann et al. 1983). Thus, the *E(spl)* locus is required genetically for Notch-dependent development. This locus encodes two distinct types of transcription regulating proteins, N-box bHLH proteins and Groucho, which interact to create a transcription repressor complex (Hartley et al. 1988; Klambt et al. 1989; Knust et al. 1987; Paroush et al. 1994). The E(spl) bHLH proteins are directly induced by Notch and Su(H) through RTGGGAA promoter sequences (Bailey and Posakony 1995; Brou et al. 1994; Furukawa et al. 1995; Lecourtois and Schweisguth 1995; Takebayashi et al. 1994). Once E(spl) bHLH DNA-binding proteins are induced, they bind to the widely expressed Groucho transcription repressor protein. E(spl)bHLH:E(spl)Groucho complexes, in turn, target specific promoter N-box sequences (CACNAG) for transcriptional repression (Tietze et al. 1992). In this way, Notch can simultaneously induce gene expression from CSL-binding sites and repress gene expression at N-box sites. In vertebrates, the Notch1 gene can also induce N-box bHLH proteins including Hes1 and the *E*nhancer of *S*plit *R*elated protein-1 (ESR-1) (JARRIAULT et al. 1995; WETTSTEIN et al. 1997).

The *Drosophila* E(spl)-C codes for 7 bHLH proteins of approximately 200 amino acids each (DELIDAKIS and ARTAVANIS-TSAKONAS 1992; KNUST et al. 1992). These proteins share three domains in common, including a conserved 57 amino acid basic helix-loop-helix domain at their N-termini. A proline residue is found in the basic region of E(spl) and Hairy bHLH domains, which is responsible for specific binding to N-box DNA recognition sites (BIER et al. 1992; FEDER et al. 1993; ISHIBASHI et al. 1993; RUSHLOW et al. 1989; SASAI et al. 1992). The bHLH domain is followed by a 37 amino acid *Orange* domain which imparts specific repressive functions on each protein (Dawson et al. 1995). For example, the Orange domain of Hairy is specifically required for Hairy-mediated inhibition of the Scute E-box bHLH protein. In other words, the Orange domain of E(spl)m8 cannot substitute for the Orange domain of Hairy. In contrast, the bHLH domain of Hairy can be replaced by the bHLH domain of E(spl) in this assay. The C-terminus of E(spl) and other N-box bHLH proteins codes for a tetrapeptide, WRPW, which binds to Groucho (PAROUSH et al. 1994). Recently, it has been speculated that the Orange domain facilitates repression of a specific set of transcription factors by each N-box bHLH protein (DAWSON et al. 1995). In contrast, the C-terminal WRPW:Groucho-binding domain is used by all N-box bHLH protein to inhibit more general transcriptional activation.

N-box bHLH proteins induced by Notch typically antagonize E-box bHLH transcription factors including the Achaete/Scute-Complex proteins during neuronal development (HEITZLER et al. 1996; SASAI et al. 1992). How do E(spl) bHLH proteins antagonize E-box bHLH proteins? One possibility is that the Orange domain of E(spl) proteins can bind directly to specific E-box proteins in such a way as to inhibit their transcriptional activation properties. Alternatively, the N-box element in the promoter of Achaete/Scute-Complex genes may bind the E(spl) bHLH:Groucho repressor complex during lateral specification (OHSAKO et al. 1994; PAROUSH et al. 1994; VAN DOREN et al. 1994). By repressing Achaete/Scute function or expression, the Notch-induced E(spl) N-box proteins can block neuronal differentiation. Furthermore, E-box proteins can regulate the expression of Delta (KUNISCH et al. 1994). By activating E(spl) expression to counteract the E-box bHLH protein responsible for DSL ligand expression, Notch signaling will repress ligand expression and ensure lateral specification.

Groucho is a 719 amino acid nuclear protein with 4 conserved domains (HARTLEY et al. 1988). The N-terminal 130 amino acid region of Groucho is highly related to its mammalian TLE/Groucho relatives (TLE1-4) (STIFANI et al. 1992) (numerous other homologues have been identified in other mammalian species: see Genebank). This is followed by a nuclear localization and putative phosphorylation sequences (HUSAIN et al. 1996), an N-box bHLH protein binding domain (from amino acid 251–414; GRBAVEC and STIFANI 1996; PAROUSH et al. 1994) and four WD40 repeats. Interestingly, Groucho related proteins have been identified in mammals which lack the WD40 repeats (SCHMIDT and SLADEK 1994). As the

Groucho proteins dimerize, it has been suggested that these shorter proteins may function to inhibit functions of full sized Groucho proteins (PINTO and LOBE 1996). Mastermind is also one of the neurogenic genes which cause lethal hypertrophy of neural tissues in the *Drosophila* embryo (SMOLLER et al. 1990). Alleles of *mastermind* have been identified as enhancers or suppressors of Notch dependent phenotypes in many genetic screens. In *Drosophila melanogaster*, *mastermind* encodes an unusual 1596 residue nuclear protein which is extremely rich in amino acid homopolymers. For example, almost 50% of this protein is made from glutamine, glycine and asparagine, which are organized into numerous homopolymeric runs of five or more residues. These stretches surround charged clusters of amino acids which are highly conserved in Mastermind from *Drosophila virilis* (NEWFELD et al. 1994). Recently the Mastermind and Groucho proteins have been found to colocalize at specific chromosomal regions (BETTLER et al. 1996), suggesting that, perhaps a complex containing E(spl) bHLH proteins Groucho and Mastermind may function to repress gene expression in response to Notch activation.

4.3 Notch Regulates the Expression of Vestigial

In *Drosophila*, the Vestigial protein is required for specification and growth of wing tissue (KIM et al. 1996). Expression of *vestigial* (*vg*) is highly regulated through activation and repression of specific enhancer sequences (KIM et al. 1996; WILLIAMS et al. 1994). At the dorsal/ventral (D/V) compartment boundary, Vestigial is expressed in response to Notch activation. A D/V-specific enhancer has been identified which resides on a 750-bp *Eco*RI fragment of the *vg* second intron. This enhancer is extremely conserved between *Drosophila melanogaster* and *D. virilis* and contains a Su(H) binding site in both species. Mutation of this Su(H) site completely obliterates D/V boundary expression of the enhancer, indicating that Notch and Su(H) function together to activate *vg* expression at the D/V boundary. It will be interesting to characterize other transcription factors which interact with the D/V enhancer as some may also bind to the Notch cytoplasmic domain or to the Notch:Su(H) complex.

4.4 CSL-Independent Signaling

N-Box bHLH proteins have been shown to block differentiation of neuronal cells (ISHIBASHI et al. 1994). In C2C12 myoblasts, Jagged1 induced activation of Notch1 blocks expression of muscle specific genes and therefore blocks differentiation (LINDSELL et al. 1995). The cytoplasmic domain of Notch1 can also inhibit muscle cell differentiation (KOPAN et al. 1994). In C2C12s, the Ram domain of Notch1 is not required for this effect (SHAWBER et al. 1996b), as expression of the cdc10 repeat region of Notch in C2C12 myoblasts is sufficient to prevent muscle cell differentiation. The cdc10 domain of Notch is unable to form a stable complex with CBF-1, is unable to activate expression from promoters containing CBF-1 binding sites and is unable to upregulate the expression of the N-box bHLH protein HES-1. In fact, exogenous expression of HES-1 in C2C12 myoblasts does not even block myogenesis, indicating that the pathway by which Notch and CSL proteins induce N-box bHLH proteins to block neuronal differentiation is not responsible for Notch-mediated inhibition of C2C12 muscle cell differentiation (SHAWBER et al. 1996b). Surprisingly, expression of HES-1 in 10T1/2 fibroblasts blocks MyoD-induced myogenic conversion, indicating that this N-box bHLH protein has cell type specific effects on the myogenic differentiation pathway (SASAI et al. 1992).

Experiments in *C. elegans* with GLP-1 reveal that the cdc10 repeat domain (with surrounding amino acids) behaves as an activated allele by inducing the multivulval phenotype (Muv) (ROEHL et al. 1996; ROEHL and KIMBLE 1993). The "specific activity" of this minigene is increased through addition of Ram domain sequences. Thus, interaction between the GLP-1 cytoplasmic domain and the CSL protein Lag-1 is not required for induction of the Muv phenotype but simply enhances GLP-1 signaling strength in this system. In *Drosophila*, expression of a fragment of Notch, from amino acid 1790 to the C-terminus, rescues the cuticle phenotypes in *Notch* null flies (LIEBER et al. 1993). This same fragment of Notch can also induce "activated" Notch phenotypes (antineurogenic phenotypes) in wild type flies. Surprisingly this fragment is missing part of the Ram domain (amino acids 1769–1880 constitute the full Ram domain although 1769–1825 can also bind to Su(H) with reduced affinity: Fig. 2). This activated Notch protein from amino acid 1790 to the C-terminus is expected to associate poorly with Su(H), yet it functions in the absence of wild type Notch, and can inhibit neurogenesis in its presence. While CSL proteins are known to be very important components of the Notch family of receptors, there are clearly CSL-independent signals downstream of Notch proteins in several systems (BAILEY and POSAKONY 1995; LECOURTOIS and SCHWEISGUTH 1995).

5 Regulators of Notch

There are numerous mechanisms by which Notch proteins are regulated. These include regulation of cell surface expression, regulation of subcellular localization, regulation of proteolytic activation, regulation of nuclear translocation and regulation of proteolytic degradation. In several cases a protein has been identified which may regulate these processes.

5.1 Proteolytic "Activation" of Notch-Family Proteins

The large transmembrane Notch protein in *Drosophila* and in mammals is processed to a smaller 100–110 kDa protein fragment, which reacts with antiserum raised against cytoplasmic domain sequences (AHMAD et al. 1995; ASTER et al. 1994; KIDD et al. 1989; ZAGOURAS et al. 1995). Very recently it has been shown that Notch proteins are cleaved by the disintegrin metalloprotease Kuzbanian (and its orthologues) (Pan and Rubin 1997; Rooke et al. 1996). This processing event is thought to occur either between the EGFL repeats region and the LNR region (Blaumueller et al. 1997) or between the LNR region and the transmembrane domain (Crittenden et al. 1994), in the Trans-Golgi network (Blaumueller et al. 1997). Thus, cell surface Notch is a heterdimeric receptor complex, much like the Receptor for Insulin. The two Notch peptides which are generated by Kuzbanian proteins remain associated at the cell surface, likely through disulfide bond formation (Blaumueller et al. 1997). It will be interesting to determine whether some activating mutations in Notch protein EGFL repeats or in the LNR domain actually disrupt association of N- and C-terminal peptides at the cell surface.

Cytoplasmic domain fragments of Notch or Notch-family proteins may in some cases represent an "*activated form*" of the protein. Activation through proteolysis may be an important mechanism to ensure irreversible commitment in a cellular signaling cascade. Very little is known about proteolytic "activation" of Notch proteins. All Notch family proteins, including LIN-12 and GLP-1, contain a short stretch of basic amino acids just inside the transmembrane domain which may be a site of cleavage (Fig. 1). This site forms the N-terminal boundary of the Ram domain and is therefore expected to encode the N-terminus of activated subfragments of Notch (Fig. 2). The extracellular LNR domain is likely involved in regulating processing, since inclusion of this domain in a truncated Notch1 miniprotein (lacking the EGF repeat region of the extracellular domain and the C-terminal PEST domain) blocks proteolytic cleavage in 3T3 cells (Kopan et al. 1996). The LNR domain, together with two conserved juxtamembrane cysteine residues, forms an inhibitory extracellular domain which is thought to prevent dimerization of Notch proteins in the absence of ligand stimulation. It is therefore possible that dimerization of Notch-receptors may be a prerequisite for proteolytic release of the cytoplasmic domain. Specific proteases are involved in this process, since cleavage of the Notch receptor miniprotein (see above) is sensitive to a peptidyl aldehyde inhibitor (N-Cbz-L-Leu-L-Leu-Leu-H or MG132), but not to other protease inhibitors such as the calpain I and calpain II inhibitors (Kopan et al. 1996). The protease responsible for this process has yet to be identified.

It is interesting to note that other transmembrane transcription factors can be activated through regulated proteolytic processing (Wang et al. 1994). In the cholesterol feedback system, for example, the *S*terol *R*egulatory *E*lement *B*inding *P*roteins (SREBP-1 and SREBP-2) each contain a basic-helix-loop-helix-leucine zipper (bHLH-ZIP) transcription factor domain attached to two membrane-spanning regions and a regulatory domain. In sterol depleted cells, a subfragment of one or both SREBPs is released by a two-step proteolytic process, after which the bHLH-ZIP domain translocates to the nucleus and activates expression of the LDL receptor gene and several cholesterol biosynthetic enzyme genes. For Notch proteins, ligand induced proteolytic activation has not been described and therefore it remains to be determined how this proteolysis reaction is controlled. Perhaps one or more of the proteins involved in SREBP activation or related proteins may participate in activation of Notch proteins (Hua et al. 1996). Activation of Notch may even require both ligand-induced activation and the presence of additional signals.

The biological function of endogenous Notch cytoplasmic domain subfragments also remains to be determined. Some of these fragments have even been detected in the nucleus of differentiated retinal neurons in the rat and in cervical epithelium and cervical carcinoma of humans (AHMAD et al. 1995; ZAGOURAS et al. 1995). It is also important to note that nuclear staining of Notch proteins has not been detected in either *Drosophila* or *C. elegans*. It is possible that nuclear forms of the Notch receptors have a very short half-life and are therefore difficult to detect.

5.2 Deltex is a Positive Regulator of Notch Signaling

The Deltex protein is a positive regulator of Notch signaling in *Drosophila*. Mutations in *deltex* suppress lethality observed in some heteroallelic combinations of activated *Notch* mutations (*Abruptex* alleles), and strongly enhance heterozygous loss of function mutations in *Notch*, *delta* and other Notch pathway components (Busseau et al. 1994). Overexpression of Deltex yields phenotypes which are similar to those observed in response to inappropriate activation of Notch (MATSUNO et al. 1995). Deltex binds to Notch cdc10 repeats 1 through 5 and co-localizes with Notch in transfected S2 tissue culture cells (DIEDERICH et al. 1994; MATSUNO et al. 1995). This protein is 737 amino acids and contains three domains, designated I, II and III, separated by the presence of two opa repeat regions (polyglutamine sequences commonly found in *Drosophila* proteins including Notch). Domain I, which lacks significant homology to known structural motifs, is sufficient for Notch-binding (MATSUNO et al. 1995). This domain is also sufficient to induce Deltex overexpression phenotypes in *Drosophila* or to suppress the lethality associated with *deltex* severe loss of function mutations. Domains II and III contain a putative SH3-binding domain and a RING-H2-finger domain respectively and are expected to interact with other signaling proteins or pathways (FREEMONT 1993; REN et al. 1993). The Deltex protein functions upstream of the "cytoplasmic domain" of Notch, as overexpression of an extracellular domain deletion mutant of Notch can rescue deltex null mutants (MATSUNO et al. 1995). Interestingly, Deltex and Su(H) bind to Notch in a mutually exclusive manner. Domains I, II and III each homodimerize in two hybrid assays (MATSUNO et al. 1995), suggesting that Deltex might regulate oligomerization or endocytosis of Notch in response to ligand stimulation.

A mammalian homologue of Deltex, FXI-T1, has recently been identified in a screen for cDNAs which were highly expressed in radiation-induced thymomas in comparison to normal thymic tissue (PAMPENO and MERUELO 1996). In addition, the sequences of two distinct but highly related genes with homology to Deltex have been deposited in the expressed sequence tag database. Sequence comparison between FXI-T1 and *Drosophila* Deltex reveals that all three domains are conserved. Domain I, the Notch-binding domain of Deltex, is similar in both proteins (29% identity over 136 amino acids) but significantly shorter in the mammalian protein. Domain II is unrelated in primary amino acid sequence between FXI-T1 and Deltex but contains putative SH3-binding domains in both proteins. Finally, domain III is highly conserved between FXI-T1 and Deltex, and contains a RING-H2-finger motif in each case. As there are four mammalian Notch proteins and at least three Deltex proteins in mammals, it seems possible that each Deltex protein may bind to a distinct Notch or to a partially overlapping set of the Notch proteins.

5.3 SEL-12 and the Presenilins; Positive Regulators of Notch Protein Signaling

Loss of function mutations in the *C. elegans sel-12* gene can suppress phenotypes associated with a gain-of-function *lin-12* point mutant (in the extracellular juxtamembrane region) and can enhance the phenotypes associated with loss of function mutations in either *lin-12* or *glp-1* (Levitan and Greenwald 1995). It is also known that SEL-12 is required in cells which are receiving a LIN-12 signal and not in their signaling neighbors. Thus, the SEL-12 protein is required for LIN-12 and GLP-1 signaling in *C. elegans*. Interestingly SEL-12 is highly related to human Presenilin-1 and -2, which, when mutated, cause early onset Alzheimer's disease (Rogaev et al. 1995; Sherrington et al. 1995). Human Presenilin proteins can even rescue loss of function *sel-12* mutant worms, indicating that SEL-12 and the Presenilins interact with similar targets or partners in their respective organisms (Levitan et al. 1996). Human Presenilin-1 protein (PS1) is localized to the endoplasmic reticulum and golgi complex when ectopically expressed in Cos or H4 neuroglioma cells (Cook et al. 1996; Kovacs et al. 1996). Human PS1 and *C. elegans* SEL-12 are both serpentine proteins, which contain cytosolic N- and C-termini, separated by seven to nine transmembrane domains (likely eight) (Doan et al. 1996; Li and Greenwald 1996). These proteins are cleaved in vivo after the sixth transmembrane domain (Thinakaran et al. 1996). Endoproteolysis of PS1 results in the accumulation of an ~27-kDa N-terminal fragment and a ~17-kDa C-terminal fragment. Accumulation of these fragments in vivo is limited, either by a rate limiting proteolytic process or by rapid clearance of N- and C-terminal fragments when they are produced in excess of a threshold concentration. Many of the Presenilin mutants which have been isolated from early onset Alzheimer's disease patients (or carriers) contain mutations near the natural site of cleavage and these proteins can accumulate as uncleaved precursors in vivo.

How do mutant Presenilins cause Alzheimer's disease (Tanzi et al. 1996; Yanker 1996)? In the brains of Alzheimer's disease patients, the amyloid β-peptide (Aβ) accumulates in plaques which ultimately disrupt neuronal function and survival. These fragments are generated through cleavage of the β-amyloid precursor protein at specific sites. This cleavage is believed to occur either within the golgi complex or early endosomes. Indeed mutant Presenilin proteins induce accumulation of elevated levels of pathological Aβ1–42 fragments in vitro and in vivo (Borchelt et al. 1996; Duff et al. 1996; Lemere et al. 1996; Scheuner et al. 1996). Recently it was discovered that Presenilin-2 can bind directly to unprocessed BAPP, suggesting that it controls processing and/or transport of BAPP (Weidemann et al. 1997).

What do these Presenilin proteins have to do with Notch signaling in worms or in other organisms? The above scheme could help to explain how PS proteins facilitate the proteolytic cleavage of the β-amyloid precursor protein, but what about Notch? Perhaps the presenilin proteins target LIN-12, GLP-1 and Notch proteins for specific cleavage as part of a Notch processing or "activation" pathway. Alternatively, Presenilins may facilitate the transport of Notch from its site of synthesis to the cell surface. Genetic analysis of the Presenilin proteins in *C. elegans* (Levitan and Greenwald 1995), and now in *Drosophila* (Boulliane et al. 1997) will help to define their role in Notch signaling.

5.4 Numb and Inhibition of Notch Signaling

During *Drosophila* neurogenesis, asymmetric division of neuroblasts and sensory organ precursor cells causes segregation of critical cell fate determinants to one daughter cell at the expense of the other (Posakony 1994). One intrinsic determinant is the Numb protein (Rhyu et al. 1994; Uemura et al. 1989). Numb is a 557 amino acid adapter protein that inhibits activation of Notch by Delta (Delta is expressed in adjacent mesodermal and neuroectodermal cells) (Campos-Ortega 1996; Frise et al. 1996; Guo et al. 1996; Spana and Doe 1996). Due to selective partitioning of the ~70-kDa Numb protein into one daughter cell, Notch is inhibited in this cell but not in its sibling. In the sensory organ precursor lineage, Notch is required to activate expression of the Tramtrack zinc finger transcription factors, which in turn are necessary to induce sheath cell fate (Guo et al. 1996).

Recently a mammalian homologue of *Drosophila* Numb has been identified (Verdi et al. 1996; Zhong et al. 1996). Both D-Numb and mNumb contain two identifiable domains: an N-terminal *P*hospho*t*yrosine-*b*inding-domain (PTB-domain) and a C-terminal domain which includes SH3-binding sequence(s). The PTB domain is contained within a larger conserved N-terminal region of approximately 300 amino acids (over 60% identity between D-Numb and mNumb in this region). The PTB domain of *Drosophila* and mouse Numb can bind directly to Notch and Notch1 respectively (Guo et al. 1996; Zhong et al. 1996). This interaction does not require tyrosine phosphorylation of Notch, as Numb can bind to a GST Notch cytoplasmic domain fusion protein which has been produced in bacteria. Numb-binding sequences of Notch have been defined to include both Ram and C-terminal domains. Surprisingly, Numb does not disrupt the association of Su(H) and Notch in tissue culture cells (Frise et al. 1996). Several mNumb proteins are generated through alternative splicing of mNumb mRNA (Verdi et al. 1996; Zhong et al. 1996). Indeed the cloned mouse and rat mNumb proteins are nearly identical, with the exception of an additional 11 amino acid insert in the rat Numb PTB domain with respect to the published mouse protein. It has not been determined whether this long PTB isoform from the published Rat mNumb is capable of binding directly to Notch proteins or whether it binds to distinct signaling molecules.

The C-termini of D-Numb and mNumb both contain multiple PXXP consensus SH3-binding domains. The natural SH3-containing partner(s) for Numb proteins have not been determined, but the mNumb C-terminal PXXP-containing region has selective affinity for the SH3 domain of p60^{src} in comparison to the SH3 domains of ABL, v-CRK, GAP, p85 and FYN (a distinct SRC-family kinase), suggesting that the src protein may bind to mNumb in vivo (VERDI et al. 1996).

In addition to PTB and SH3-binding domains, D-Numb possess a targeting sequence which partitions it to one of two daughter cells following division of the neuroblast or sensory organ precursor. The sequences responsible for partitioning D-Numb have yet to be defined but must be conserved since mNumb colocalizes with D-Numb in *Drosophila*, partitions with D-Numb in *Drosophila* (ZHONG et al. 1996), and rescues D-Numb loss of function phenotypes in *Drosophila* (VERDI et al. 1996; ZHONG et al. 1996). The mechanism for targeting Numb to a specific membrane subdomain and the mechanism for selecting the plane of cell division are coordinated in *Drosophila* (KNOBLICH et al. 1995; KRAUT and CAMPOS-ORTEGA 1996; KRAUT et al. 1996; LIN and SCHAGAT 1997; RHYU et al. 1994) in order to segregate D-Numb to one of the two daughter cells. In contrast, mNumb is apically localized within the mammalian neural tube independent of the cell division plane and can therefore be partitioned to one or both daughter cells depending on the plane of cell division (ZHONG et al. 1996). The partner proteins or molecules which regulate Numb localization or function await to be discovered. Perhaps Numb localization or function is regulated by phosphotyrosine containing proteins which copurify with the PTB domain of Rat mNumb in vitro or the highly conserved Seven-in-absentia proteins which can associate with mNumb in two hybrid assays (VERDI et al. 1996).

5.5 Dishevelled and Regulation of Notch Signaling by PDZ Containing Proteins

As mentioned in Sect. 2.2 above, the Wingless protein can inhibit Notch activation in response to Delta but may facilitate Notch activation in response to Serrate (AXELROD et al. 1996; COUSO et al. 1995; JONSSON and KNUST 1996). The mechanism by which Wingless controls the specificity of Notch for its ligands remains to be elucidated. However, a component of the "Wingless signaling pathway" is capable of binding to the Notch cytoplasmic domain and, when overexpressed, this protein can inhibit Notch function (AXELROD et al. 1996). Dishevelled is required for reception of a Wingless signal in *Drosophila* and is phosphorylated and membrane localized in response to Wingless stimulation (YANAGAWA et al. 1995). Dishevelled is a 623 amino acid adapter protein with three conserved domains (KLINGENSMITH et al. 1994; SUSSMAN et al. 1994; THEISEN et al. 1994). The first domain at the N-terminus is unique to Dishevelled proteins which have been identified in several species. The second domain is a PDZ domain which has been identified in a number of signaling proteins including PSD-95, *Drosophila Disk* Large and ZO-1 (between residues 257 and 338) (FANNING and ANDERSON 1996; HARRISON 1996), and the third domain is a domain found in Regulators of G-protein Signaling proteins or RGS proteins (located between residues 394 and 474) (IYENGAR 1997). *Drosophila* Dishevelled can bind directly to the C-terminal domain of *Drosophila* Notch (C-terminal to the cdc10 repeat region from amino acid 2109–2703) (AXELROD et al. 1996). The N-terminus of Dishevelled, from amino acid 1–393, is sufficient for Notch-binding in yeast two hybrid experiments. Interestingly, this region includes the PDZ domain, which in the context of many other proteins has been implicated in binding to the C-terminus of specific receptors or signaling proteins. The Dishevelled PDZ domain is predicted to bind C-terminal sequences in Frizzled receptor proteins (BHANOT et al. 1996; WANG et al. 1996).

The C-terminus of Notch (KGSEAIYI) is a candidate PDZ-binding domain. Originally the consensus PDZ-binding domain was a C-terminal S/TXV sequence. Recent analysis of several PDZ domains, however, indicates that additional C-terminal or even internal peptide motifs can bind to PDZ domains (SONGYANG et al. 1997). Indeed, most if not all Frizzled family and Notch-family receptors contain C-terminal sequences which potentially bind to PDZ domains. The C-termini of Notch proteins are conserved. For example Notch1 C-termini from humans, mice, rats, *Xenopus* and zebrafish are nearly identical (T/N*HIPEAFK*). The Notch2 and Notch3 C-termini are conserved and related (HSNM*QVY**A* and TPKR*QVM**A* respectively). The LIN-12 and GLP-1 C-termini may also bind to PDZ domains (LIN-12: YSEPAHYF and GLP-1: QMNGSFYC).

Dishevelled proteins are believed to regulate the organization and polarity of membrane subcompartments in response to multiple extracellular signals (KLINGENSMITH and NUSSE 1994; KRASNOW et al. 1995). *Drosophila* Dishevelled is an important mediator of Wingless signal transduction but also of transduction from the D-Frizzled1 receptor which regulates tissue polarity in flies. How does Dishevelled inhibit Notch? Five mechanisms can be easily imagined: (a) Dishevelled inhibits glycogen synthase kinase 3 (GSK3) family kinases, which may positively regulate Notch signaling (NOORDERMEER et al. 1994; RUEL et al. 1993; SIEGFRIED et al. 1994), (b) Dishevelled PDZ domain may target Notch to a specific subcellular compartment where Delta cannot stimulate Notch, (c) the Dishevelled PDZ domain may compete with PDZ domains of other proteins which are required for correct targeting of Notch to a specific subcellular compartment where Delta functions (MIYAMOTO et al. 1995), (d) Dishevelled may block transcriptional activation by nuclear Notch, or (e) Dishevelled may target Notch for degradation. It is interesting to note that a truncated Notch protein which has lost sequences downstream of the cdc10 repeat domain is resistant to inhibition by Dishevelled (AXELROD et al. 1996). This Notch mutant is therefore "activated" and dominant over Dishevelled when both proteins are ectopically expressed. Similarly truncated forms of GLP-1 are also hyperactive (MANGO et al. 1991). Interpretation of these results is complicated by the fact that deletion of C-terminal PEST sequences in addition to deletion of the Dishevelled-binding domain, may contribute to the activity of truncated Notch or GLP-1 proteins by increasing their respective half-lives. Mammals have at least three Dishevelled homologues which likely respond to distinct Frizzled receptors and may inhibit distinct Notch proteins.

Another PDZ containing protein has been implicated in Notch signaling. The Canoe protein is required for activated *abruptex* alleles of *Notch* to block vein tissue formation in *Drosophila* (MIYAMOTO et al. 1995). In addition, loss of function mutations in the *canoe* gene result in enhancement of wing notching observed in female flies heterozygous for *Notch* loss of function alleles. These and other data suggest that Canoe is required for normal Notch activation or signal transduction in several fruit fly tissues. The Canoe PDZ domain is centrally located between residues 843 and 926 in this 1893 amino acid protein. Canoe is related to the 1612 amino acid AF-6 protein in humans (Prasad et al. 1993). AF-6 was discovered as a fusion partner of the ALL-1 gene, a human homologue of *Drosophila trithorax* located on chromosome 11, which has been implicated in a large number of acute leukemias. Canoe and AF-6 share homology throughout most of their respective coding regions *particularly* in the PDZ domains and the N-terminal domains which bind to GTP-loaded Ras (Kuriyama et al. 1996). The exciting discovery that Canoe/AF-6 interacts genetically with Notch and physically with Ras raises the possibility that this protein represents a mechanism by which these two ancient and conserved signaling pathways are crossregulated (Matsuo et al. 1997) (see below). The PDZ domain of AF-6 selects [Y-EorA-F, I, MorY-Y or F-V, F, I or L] C-terminal peptides from a degenerate peptide library which are related to the C-terminus of Notch1 at two positions (P*E*A*F* K) (Songyang et al. 1997). Together, these results suggest that Ras may control the subcellular localization (or activity) of Notch.

5.6 SEL-1 and Proteolysis of Notch-Family Receptors

Loss of function mutations in the *C. elegans SEL-1* gene suppress the phenotypes associated with reduced LIN-12/GLP-1 activity but cannot suppress the phenotypes associated with complete loss of LIN-12 or GLP-1 (Grant and Greenwald 1996). Mutations in *SEL-1* do not, however, suppress phenotypes associated with reduced function of other receptor proteins including receptor tyrosine kinases, receptor serine/threonine kinases and serpentine receptors. Therefore, SEL-1 is a specific negative regulator of LIN-12 and GLP-1. Cell ablation studies have shown that SEL-1 functions within cells receiving LIN-12 or GLP-1 signals, as opposed to functioning within cells which presents ligands for these receptors (Grant and Greenwald 1997). SEL-1 is a widely expressed 685 amino acid protein with a functional signal sequence and is found primarily within intracellular vesicles. SEL-1 is related in structure to proteins in humans (predicted by the overlap of the IBD2 cDNA and multiple expressed sequence tag cDNA clones in the public databases) and yeast (HRD3p) (Grant and Greenwald 1996, 1997). These proteins contain a signal sequence, similar amino acid composition, multiple copies of a LXYY motif, and a novel 68 amino acid motif near their short hydrophobic C-termini. The yeast HRD3p protein regulates degradation of an endoplasmic reticulum membrane protein, HMG-CoA reductase, which is a critical cholesterol biosynthetic enzyme (Hampton et al. 1996). How might a *C. elegans* protein which is related to a yeast protein responsible for destruction of HMG CoA reductase negatively regulate Notch signaling in this organism? The Notch family proteins including LIN-12 and GLP-1 all contain PEST sequences which are typically associated with rapid turnover. Perhaps SEL-1 specifically targets Notch family proteins for destruction. Rapid turnover may be an important feature of Notch proteins since they can be used for cell fate decisions within two or more successive cell division cycles. Consequently, active fragments of Notch may need to be immediately cleared following each cell fate decision. Alternatively, the connection between cholesterol metabolism and Notch receptor signaling may be significant. As the SCAP protein (required for proteolytic activation of SREBP 1 and 2) contains a region of homlogy to HMG CoA reductase, perhaps SEL-1 and SCAP regulate activation of Notch and SREBP transcription factors (Hua et al. 1996; Pahl and Baeuerle 1996).

5.7 Hairless, Bearded and E(spl)m4

A serine-rich protein termed Hairless has been identified in *Drosophila* which inhibits the ability of Su(H) to bind to its RTGGGAA recognition sequence (Bang et al. 1995; Bang and Posakony 1992; Brou et al. 1994; Maier et al. 1992). The Hairless interaction domain on CSL proteins has been localized to a small and highly conserved region (between amino acid 330 and 438 in human CBF-1 isoform RBP3). Interestingly the region of CSL proteins which interact with Hairless overlaps the region required for DNA binding. This overlap explains how Hairless can block the association of CSL proteins with their target DNA sequences. The site of the Hairless interaction domain in CSL proteins also includes the putative Mapk phosphorylation site (Christensen et al. 1996), as discussed above (see Sect. 3.2), suggesting that phosphorylation of CSL proteins could even regulate the interaction between CSL proteins and Hairless proteins. Such regulation would constitute a mechanism by which the Ras/Mapk pathway could cross-regulate the Notch pathway. The Hairless protein, which is 1076 amino acids in length, only requires a small domain from amino acid 212 to 293 in order to interact with *Drosophila* Su(H) or human CBF-1 in vitro (Brou et al. 1994). Hairless orthologues have not yet been described in *C. elegans* or in vertebrates. Conservation of the Hairless interaction domain in CSL proteins from worms to humans, as well as the binding of *Drosophila* Hairless to human CBF-1, together suggest that such proteins exist.

Bearded and E(spl)m4 are related proteins which function to inhibit Notch signaling (Klambt et al. 1989; Leviten and Posakony 1996). Overexpression of Bearded in flies from gain-of-function Brd^1 or Brd^3 alleles results in increased SOP specification within several proneural regions (Leviten and Posakony 1996). In some cases these SOP cells divide to produce two pIIb progeny, instead of one pIIa and one pIIb. The generation of extra SOP cells and the transformation of pIIa into pIIb cells both suggest a failure to activate Notch during cell fate specification. These Brd gain-of-function phenotypes are specifically enhanced in flies heterozygous for loss of function mutations in *neuralized* or *Notch* , and are suppressed in flies heterozygous for loss of function mutations in *hairless* or in flies with an extra wild type Notch gene. Other genes on the Notch signaling pathway or regulating the Notch signaling pathway, such as Delta, Su(H), E(spl)-C, mastermind, big brain, shaggy, scabrous and the Achaete/Scute complex, do not show strong dosage sensitive interactions with Brd^1 or Brd^3 alleles.

In addition, loss of function mutations in *hairless* result in fewer SOP cells being formed. Simultaneous Hairless loss of function and Bearded overexpression result in fewer SOP cells. These data suggest that Bearded may function in a biochemical pathway upstream of Hairless to regulate Notch (Notch-Su(H)) signaling. The primary amino acid sequence of Bearded is not yet published but is reported to be very similar to the sequence of E(spl)m4 (LEVITEN and POSAKONY 1996). m4 is a 152 amino acid protein of novel structure which contains a potential tyrosine phosphorylation site at residue 96 (KLAMBT et al. 1989). Null alleles of *bearded* do not alter *Drosophila* development. It is believed that this is due to a functional redundancy between Bearded and E(spl)m4. Perhaps *bearded* and *E(spl)m4* must be mutated together in order to determine the developmental/signaling function of these related proteins (LEVITEN and POSAKONY 1996).

5.8 Neuralized

As noted above, loss of function mutations in the *neuralized* gene enhance the phenotypes associated with overexpression of Bearded (LEVITEN and POSAKONY 1996). *neuralized* is one of the classical *neurogenic* genes which when mutated in the *Drosophila* embryo results in overrepresentation of neural tissues (LEHMANN et al. 1983). The Neuralized protein is therefore required (together with Delta, Notch and the other neurogenics) for cell fate specification at this time. Overproduction of neural tissues in flies with mutations in *neuralized* still occurs in the presence of some "activated" alleles of Notch (activated Notch alleles with cysteine to serine mutations in the either of the two conserved cysteine residues C-terminal to the LNR repeats) but this neural overproduction is blocked by other activated alleles (the intracellular domain of Notch, or Notch with an LNR domain deletion) (LIEBER et al. 1993). Interestingly, the former alleles but not the latter are still dependent on Delta for signaling. *Neuralized* encodes a basic protein of 754 amino acids which contains a C-terminal C3HC4 ring finger motif (BOULIANNE et al. 1991; PRICE et al. 1993). Neuralized related sequences have been identified in *D. virilis*, *C.elegans* and mammals (ZHOU and BOULIANNE 1994) (see also public DNA sequence databases). In each case the encoded protein is homologous throughout its length (note: only partial clones of mammalian Neuralized have been analyzed to date). Each protein is also basic and contains a C-terminal C3HC4 ring finger (FREEMONT 1993). The ring finger motif of Neuralized proteins is most similar to the ring finger domain present in the cytoplasmic Inhibitors of Apoptosis Proteins (IAPs) (HAY et al. 1995; ROTHE et al. 1995) as opposed to the C3HC4 ring finger domain in the nuclear BRCA1, PML1 or BMI-1 proteins. What does Neuralized do? It is not yet clear how Neuralized controls neuronal differentiation during *Drosophila* development or if it directly interacts with components of the Notch signaling pathway. What is the subcellular localization of this protein? Why does *neuralized* show strong dosage sensitive interactions with *Notch* and *Brd*? Biochemical analysis of Neuralized should help to answer these questions.

6 Notch Receptor Signaling: Future Directions

In the preceding sections we have highlighted several of the extracellular and intracellular proteins which are required for Notch signaling or which regulate the efficiency of Notch signaling. From this analysis several themes emerge. Below we summarize some of the questions which remain, the approaches which are likely to yield insight on the molecular connections through which Notch functions, and a Notch-centric model of development.

6.1 Notch Receptor Signaling: Genetics and Molecular Biology

Only a fraction of the proteins which participate in this pathway have been identified. Over the next few years, we are likely to see a large number of additional proteins implicated in this pathway. For example, genetic screens have defined the existence of numerous other loci which enhance or suppress phenotypes associated with gain-of-function or loss-of-function Notch mutations (Maine and Kimble 1993; Verheyen et al. 1996). Proteins which bind to the extracellular domain of Notch receptors are eagerly anticipated; as are additional proteins which bind to the Notch/LIN-12/GLP-1 cytoplasmic domains. Proteins like Hairless, which bind to CSL proteins, will likely be important regulators of Notch signaling. Perhaps additional DNA-binding proteins complex with the Notch cytoplasmic domain in order to inhibit differentiation of myoblasts (a CSL-independent process). Several of the protein products which are discussed in the above sections contain domains which are *not* responsible for interacting with Notch receptors or with other components of the Notch pathway. The function of these domains has not been determined. For example, what partner proteins bind to the putative SH3-binding domain of Deltex, the Ring H2 finger of Deltex, the putative SH3-binding domain of Numb, the Ring C3HC4 finger of Neuralized? Two hybrid screens in yeast and affinity purification experiments should yield new components shortly. There are also several genes which produce neurogenic phenotypes for which no connection to the Notch pathway(s) have been made. For example, the Big Brain MIP-channel protein or the large transmembrane Pecanex protein are very important and highly conserved neurogenic proteins (Gilbert et al. 1992; LaBonne et al. 1989; Perrimon et al. 1989; Rao et al. 1990; Shields and Bassnett 1996). Are they involved in Notch signaling? Heterozygous mutations in Big Brain do not enhance or suppress phenotypes associated with altered Notch signaling but this criterion is insufficient reason for its exclusion as a candidate regulator of the Notch pathway. In addition, *strawberry notch* interacts with Notch genetically in several tissues (Coyle-Thompson and Banerjee 1993), suggesting that the product of this locus may be a critical component or regulator of a Notch signaling pathway. Recently, Strawberry Notch has been cloned and found to encode a nuclear protein which is required for Su(H) function during inductive Notch signaling in several tissues but is not required for lateral specification during neurogenesis (Majumdar et al. 1997). This work has highlighted the tissue specific nature of Notch signaling pathways.

6.2 Notch Receptor Signaling: Biochemistry and Cell Biology

Once a cell decides to transcribe Notch gene sequences into mRNA, a long and highly regulated journey is undertaken. Every aspect of Notch gene, mRNA and protein function is likely the target of specific regulatory mechanisms. The regulated steps in this journey may include mRNA translation, glycosylation, transport, localization, cleavage, ligand activation, partner binding, nuclear translocation, protein turnover, transcriptional activation and transcriptional autoregulation. Each of these processes may respond to distinct input signals from other cellular signaling pathways. Each process can be studied in isolation by defining the sequences within Notch which are required for regulation. In addition, the genetic screens in *Drosophila* and *C. elegans* with specific mutant Notch receptors are likely to identify enhancers and suppressors which function at distinct regulatory steps. Another fruitful approach will be to identify the phosphorylation sites on Notch, CSL proteins, Groucho proteins and other phosphorylated signaling proteins, then to define the kinases and the effect of phosphorylation on distinct functions of Notch signaling. It would be surprising if second messengers including cAMP, Ca^{2+}, PIP2 and its varied derivatives did not affect some aspect of Notch processing, activation, signaling or turnover. Indeed, the GSK3 kinases are thought to affect Notch function (RUEL et al. 1993).

6.3 Notch Receptor Signaling: Developmental Biology

Recent analysis in *Drosophila* has given profound insight into what may be a universal function of the Notch receptor system (BAKER et al. 1996; BAKER and YU 1997). During eye development, Notch is used to block cellular response to the secreted protein Hedgehog and to maintain expression of the Atonal bHLH transcription factor (BAKER and YU 1997). Paradoxically, Notch is later used to inhibit expression of Atonal in all of these cells except the future R8 photoreceptor. Differentiation of an R8 cell is activated by spatially restricted EGF receptor activation (FREEMAN 1996; GOLEMBO et al. 1996). Baker and Yu have speculated that the same receptor, *Notch*, is used to both activate and repress Atonal expression for two reasons (BAKER and YU 1997): (a) in order to ensure that signals to induce R8 cell fate and to block R8 cell fate do not compete and (b) to ensure that induction of R8 potential in "intermediate group" cells occurs prior to lateral inhibition which limits R8 fate to a single cell within the intermediate group. This situation, where Notch links one developmental event to another and where Notch induces two opposite effects on the same genes, may be a fundamental feature of this complex receptor system. Perhaps the Notch signaling pathways are employed every time two or more cell fates are required during development in order to ensure that each cell type is generated. Notch may have a "positive" pathway and a "negative" pathway, depending on the presence or absence of other signals. In the developing fruit fly eye, the other signal may be a tyrosine kinase signal (FREEMAN 1996). Biochemical connections between Notch and tyrosine kinase pathways can be envisioned at multiple levels, including: (a) Ras proteins may control the subcellular localization of Notch (via Canoe) (MIYAMOTO et al. 1995), (b) tyrosine kinase signals may alter the accessibility of SH3 containing protein partners of Deltex (BUSSEAU et al. 1994), (c) Map kinase may phosphorylate CSL proteins and thus alter their activity (CHRISTENSEN et al. 1996), (d) phosphorylation of bHLH proteins may be altered in response to tyrosine kinase signals (LI et al. 1992) and (e) phosphorylation of Groucho homologues may respond to tyrosine kinase pathways (HUSAIN et al. 1996).

In the developing wing margin, the "other signal" may be a Wingless signal. Notch activates expression of Wingless, yet Wingless functions to limit its own expression and to induce bristles at the margin by inhibiting Notch (RULIFSON et al. 1996). An interesting feature of Wingless/Notch pathway competition is that the cytoplasmic adapter protein Dishevelled is required for Wingless signaling yet it binds to Notch. Perhaps the Wingless receptor pathway cannot function in the wing unless Notch binds to Dishevelled. In this way, a cytoplasmic adapter protein could be used to set up a positive and negative pathway which both require Notch. The positive pathway would be Notch-mediated coupling of Dishevelled to the Wingless pathway. The negative pathway would be Delta/Notch-mediated lateral inhibition of Achaete/Scute expression, in order to block Wingless induction of bristles.

There are many cytoplasmic adapter proteins which have been implicated in Notch signaling during specific developmental processes in the fly. Perhaps each adapter protein links Notch to a distinct signaling pathway with both positive and negative effects on tissue induction in each case. For example, Canoe may connect Notch to a Scabrous pathway (MIYAMOTO et al. 1995). *Scabrous* and *canoe* interact genetically with *Notch*, and the secreted Scabrous protein was even thought to be a Notch ligand. Recent experiments have shown, however, that Scabrous does not bind with high affinity to Notch (LEE and BAKER 1996). The Canoe adapter protein may couple a Scabrous Receptor or components of a Scabrous signaling pathway ($p21^{Ras}$?) to Notch. Do the adapter proteins Deltex, Numb and Neuralized also couple Notch to distinct signaling pathways? Genetic analysis has demonstrated a positive role for Deltex and Neuralized in Notch dependent development. In contrast, Numb is known to be an inhibitor of Notch. Perhaps these adapter proteins have opposite functions in a connected signaling pathway.

Crosstalk between Notch and other signaling pathways may therefore be a universal theme which is essential for development of multiple complementary cell fate decisions during inductive patterning.

6.4 Notch Receptor Signaling: Pathology

The Notch family proteins have been implicated in the genesis of tumors in numerous species from worms to humans (BERRY et al. 1997; ELLISEN et al. 1991). Other Notch signaling pathway components have also been implicated in tumorigenesis. For example, the EBNA2 oncoprotein from Epsteine-Barr Virus binds to CBF1 in order to alter its transcriptional regulatory functions (HENKEL et al. 1994;

HONJO 1996). It is believed that EBNA2 functions to mimic the activated nuclear form of Notch. In addition, Notch proteins and Deltex proteins are dramatically overexpressed in some tumors, which suggests that oncogenic lesions alter regulation of this signaling pathway (ARTAVANIS-TSAKONAS et al. 1995; PAMPENO and MERUELO 1996). While the Notch and Deltex genomic loci may not be a direct target of carcinogenic mutations in these tumors, the effect of oncogenic lesions on increased Notch signaling may be critical for tumor development.

As mentioned above in Sect. 5.3, the Presenilin proteins have been implicated strongly in both Notch signaling and in early onset Alzheimer's disease. Mutations in Notch3 have also been associated with an inherited stroke and dementia syndrome (CADASIL) (JOUTEL et al. 1996). While the majority of cancer, stroke and Alzheimer's disease cases in humans are unlikely to result from mutations in genes encoding Notch family receptors, signaling partners or regulators, inappropriate Notch signaling may be a driving force underlying these devastating pathological conditions.

Acknowledgments. The authors wish to apologize for the omission of all appropriate citations. In many cases, several researchers, studying different model organisms have discovered similar concepts. In an effort to simply text, we have chosen to highlight concepts only as discovered in one system. We have also used phrases such as Notch family proteins or Notch-related receptors throughout the text in place of Notch *and* LIN-12/GLP-1. We apologize to researchers studying *C. elegans* for this simplification. The authors are grateful to Gabrielle Boulianne, Stefano Stifani, Jeanette Goguen and Christine Campbell for valuable comments on the manuscript. Sean Egan is supported by an MRC of Canada Scholarship. Benoit St-Pierre is supported by an MRC of Canada Studentship and Ching Ching Leow is supported by a grant from the Canadian Breast Cancer Research Initiative.

Note Added in Proof. It has recently been demonstrated that Fringe can alter the sensitivity of Notch for Delta and Serrate through a cell autonomous mechanism (Panin et al. 1997). In addition, a second mammalian Numb homologue has been identified (ZHONG et al. 1997), and a peptide ligand for the *Drosophila* Numb PTB domain defined (LI et al. 1997).

References

Ahmad I, Zaquoras P, Artavanis-Tsakonas S (1995) Involvement of Notch-1 in mammalian retinal neurogenesis: association of Notch-1 activity with both immature and terminally differentiated cells. Mech Dev 53:73–85

Artavanis-Tsakonas S, Matsuno K, Fortini ME (1995) Notch signaling. Science 268:225–232

Asai N, Iwashita T, Matsuyama M, Takahashi M (1995) Mechanism of activation of the ret proto-oncogene by multiple endocrine neoplasia 2A mutations. Mol Cell Biol 15:1613–1619

Aster J, Pear W, Hasserjian R, Erba H, Davi F, Luo B, Scott M, Baltimore D, Sklar J (1994) Functional analysis of the TAN-1 gene, a human homolog of Drosophila notch. Cold Spring Harb Symp Quant Biol LIX:125–136

Austin J, Kimble J (1987) glp-1 is required in the germ line for regulation of the decision between mitosis and meiosis in C. elegans. Cell 51:589–599

Axelrod JD, Matsuno K, Artavanis-Tsakonas S, Perrimon N (1996) Interaction between wingless and notch signaling pathways mediated by dishevelled. Science 271:1826–1832

Bailey AM, Posakony JW (1995) Suppressor of hairless directly activates transcription of enhancer of split complex genes in response to notch receptor activity. Genes Dev 9:2609–2622
Baker NE, Yu S-Y (1997) Proneural function of neurogenic genes in the developing Drosophila eye. Curr Biol 7:122–132
Baker NE, Yu S, Han D (1996) Evolution of proneural atonal expression during distinct regulatory phases in the development Drosophila eye. Curr Biol 6:1290–1301
Bang AG, Posakony JW (1992) The Drosophila gene hairless encodes a novel basic protein that controls alternative cell fates in adult sensory organ development. Genes Dev 6:1752–1769
Bang AG, Bailey AM, Posakony JW (1995) Hairless promotes stable commitment to the sensory organ precursor cell fate by negatively regulating the activity of the Notch signaling pathway. Dev Biol 172:479–494
Berry LW, Westlund B, Schedl T (1997) Germ-line tumor formation caused by activation of glp-1, a Caenorhabditis elegans member of the notch family of receptors. Development 124:925–936
Bettenhausen B, Hrabe de Angelis M, Simon D, Guenet J-L, Gossler A (1995) Transient and restricted expression during mouse embryogenesis of Dll1, a murine gene closely related to Drosophila delta. Development 121:2407–2418
Bettler D, Pearson S, Yedvobnick B (1996) The nuclear protein encoded by the Drosophila neurogenic gene mastermind is widely expressed and associates with specific chromosomal regions. Genetics 143:859–875
Bhanot P, Brink M, Samos CH, Hsieh J-C, Wang Y, Macke JP, Andrew D, Nathans J, Nusse R (1996) A new member of the frizzled family from Drosophila functions as a Wingless receptor. Nature 382: 225–230
Bier E, Vassin H, Younger-Shepherd S, Jan LY, Jan YN (1992) Deadpan, an essential pan-neural gene in Drosophila, encodes a helix-loop-helix protein similar to the hairy gene product. Genes Dev 6: 2137–2151
Bierkamp C, Campos-Ortega JA (1993) A Zebrafish homologue of the Drosophila neurogenic gene Notch and its pattern of transcription during early embryogenesis. Mech Dev 43:87–100
Blank V, Kourilsky P, Israel A (1992) NF-kB and related proteins: Rel/dorsal homologies meet ankyrin-like repeats. Trends Biochem Sci 17:135–140
Blaumueller CM, Qi H, Zagouras P, Artavanis-Tsakonis S (1997) Intracellular cleavage of Notch leads to a heterodimeric receptor on the plasma membrane. Cell 90:281–291
Borchelt DR, Thinakaran G, Eckman CB, Lee MK, Davenport F, Ratovitsky T, Prada C-M, Kim G, Seekins S, Yager D, Slunt HH, Wang R, Seeger M, Levey AI, Gandy SE, Copeland NG, Jenkins NA, Price DL, Younkin SG, Sisodia SS (1996) Familial Alzheimer's disease-linked presenilin 1 variants elevate AB1-42/1–40 ratio in vitro and in vivo. Neuron 17:1005–1013
Bortvin A, Winston F (1996) Evidence that Spt6p controls chromatin structure by a direct interaction with histones. Science 272:1473–1476
Boulianne GL, de la Concha A, Campos-Ortega JA, Jan LY, Jan YN (1991) The Drosophila Neurogenic gene neuralized encodes a novel protein and is expressed in precursors of larval and adult neurons. EMBO J 10:2975–2983 (see also 12:2586)
Boulliane GL, Livne-Bar I, Humphreys JM, Liang Y, Lin C, Rogaev E, St George Hyslop P (1997) Cloning and characterization of the Drosophila presenilin homologue. Neuroreport 8:1–5
Breeden I, Nasmyth K (1987) Similarity between cell-cycle genes of budding yeast and fision yeast and the Notch gene of Drosophila. Nature 329:651–654
Brou C, Logeat F, Lecourtois M, Vandekerckhove J, Kourrilsky P, Schweisguth F, Isreal A (1994) Inhibition of the DNA-binding activity of Drosophila Suppressor of Hairless and its human homolog, KBF2/RBP-Jk, by direct protein-protein interaction with Drosophila hairless. Genes Dev 8:2491–2503
Busseau I, Diederich RJ, Xu T, Artavanis-Tsakonas S (1994) A member of the Notch group of interacting loci, deltex encodes a cytoplasmic basic protein. Genetics 136:585–596
Bycroft M, Hubbard TJP, Proctor M, Freund SMV, Murzin AG (1997) Solution structure of the S1 RNA binding domain: a member of an ancient nucleic acid-binding fold. Cell 88:235–242
Cagan RL, Kramer H, Hart A, Zipursky SL (1992) The bride of sevenless and sevenless interaction: internalization of a transmembrane ligand. Cell 69:393–399
Campbell ID, Bork P (1993) Epidermal growth factor-like modules. Curr Opin Struct Biol 3:385–392
Campos-Ortega JA (1988) Cellular interactions during early neurogenesis of Drosophila melanogaster. Trends Neurosci 11:400–405
Campos-Ortega JA (1991) Mechanisms of early neurogenesis in Drosophila melanogaster. Comments Dev Neurobiol 1:251–266
Campos-Ortega JA (1996) Numb diverts Notch pathway off the tramtrack. Neuron 17:1–4

Campuzano S, Modolell J (1992) Patterning of the Drosophila nervous system: the achaete-scute gene complex. Trends Genet 8:202–208
Chenn A, McConnell B (1995) Cleavage orientation and the asymmetric inheritance of Notch1 Immunoreactivity in Mammalian Neurogenesis. Cell 82:631–641
Chiang PW, Wang S, Smithivas P, Song WJ, Ramamoorthy S, Hillman J, Puett S, Van Keuren ML, Crombez E, Kumar A, Glover TW, Miller DE, Tsai CH, Blackburn CC, Chen XN, Sun Z, Cheng JF, Korenberg JR, Kurnit DM (1996) Identification and analysis of the human and murine putative chromatin structure regulator SUPT6H and Supt6h. Genomics 34:328–333
Chitnis A, Henrique D, Lewis J, Ish-Horowicz D, Kintner C (1995) Primary neurogenesis in Xenopus embryos regulated by a homologue of the Drosophila neurogenic gene Delta. Nature 375:761–766
Christensen S, Kodoyianni V, Bosenberg M, Friedman L, Kimble J (1996) Lag-1, a gene required for lin-12 and glp-1 signaling in Caenorhabditis elegans, is homologous to human CBF1 and Drosophila Su(H). Development 122:1373–1383
Chung CN, Hamaguchi Y, Honjo T, Kawaichi M (1994) Site-directed mutagenesis study on DNA binding regions of the mouse homologue of suppressor of hairless, RBP-Jk. Nucleic Acids Res 22:2938–2944
Coffman C, Harris W, Kintner C (1990) Xotch, the Xenopus homologue of Drosophila notch. Science 249:1438–1441
Cohen B, Bashirullah A, Dagnino L, Campbell C, Fisher B, Leow CC, Whiting E, Ryan D, Zinyk D, Boulianne G, Hui CC, Gallie B, Phillips RA, Lipshitz HD, Egan SE (1997) Fringe boundaries coincide with Notch-dependent patterning centers in mammals and alter Notch-dependent development in Drosophila Nature Genetics 16:283–288
Cook DG, Sung JC, Golde TE, Felsenstein KM, Wojczyk BS, Tanzi RE, Trojanowski JQ, Lee VM-Y, Doms RW (1996) Expression and analysis of presenilin 1 in a human neuronal system: localization in cell bodies and dendrites. Proc Natl Acad Sci USA 93:9223–9228
Couso JP, Knust E, Martinez Arias A (1995) Serrate and wingless cooperate to induce vestigal gene expression and wing formation in Drosophila. Curr Biol 5:1437–1448
Coyle-Thompson CA, Banerjee U (1993) The strawberry notch gene functions with notch in common developmental pathways. Development 119:377–395
Crittenden SL, Troemel ER, Evans TC, Kimble J (1994) GLP-1 is localized to the mitotic region of the *C. elegans* germ line. Development 120:2901–2911
Dawson SR, Turner DL, Weintraub H, Parkhurst SM (1995) Specificity for the Hairy/Enhancer of split Basic Helix-Loop-Helix (bHLH) proteins maps outside the bHLH domain and suggests two seperable modes of transcriptional repression. Mol Cell Biol 15:6923–6931
De Celis JF, Barrio R, Del Arco A, Garcia-Bellido A (1993) Genetic and molecular characterization of a Notch mutation in its Delta- and Serrate-binding domain in Drosophila. Proc Natl Acad Sci USA 90:4037–4041
De Celis JF, Garcia-Bellido A, Bray SJ (1996) Activation and function of Notch at the dorsal-ventral boundary of the wing imaginal disc. Development 122:359–369
de la Pompa JL, Wakeham A, Correia KM, Samper E, Brown S, Aguilera RJ, Nakano T, Honjo T, Mak TW, Rossant J, Conlon RA (1997) Conservation of the Notch signalling pathway in mammalian neurogenesis. Development 124:1139–1148
Delidakis C, Artavanis-Tsakonas S (1992) The Enhancer of split [E(spl)] locus of Drosophila encodes seven independent helix-loop-helix proteins. Proc Natl Acad Sci USA 89:8731–8735
Diaz-Benjumea FJ, Cohen SM (1995) Serrate signals through Notch to establish a Wingless-dependent organizer at the dorsal/ventral compartment boundary of the Drosophila wing. Development 121:4215–4225
Diederich RJ, Matsuno K, Hing H, Artavanis-Tsakonas S (1994) Cytosolic interaction between deltex and Notch ankyrin repeats implicates deltex in the Notch signaling pathway. Development 120:473–481
Doan A, Thinakaran G, Borchelt DR, Slunt HH, Ratovitsky T, Podlisny M, Selkoe DJ, Seeger M, Gandy SE, Price DL, Sisodia SS (1996) Protein topology of presenilin 1. Neuron 17:1023–1030
Doe CQ, Goodman CS (1985) Early events in insect neurogenesis. II. Role of cell interactions and cell lineage in the determination of neuronal precursor cells. Dev Biol 111:206–219
Doherty D, Feger G, Younger-Shepherd S, Jan LY, Jan YN (1996) Delta is a ventral to dorsal signal complementary to Serrate, another Notch ligand, in Drosophila wing formation. Genes Dev 10:421–434
Dorsky RI, Chang WS, Rapaport DH, Harris WA (1997) Regulation of neuronal diversity in the Xenopus retina by Delta signalling. Nature 385:67–70

Duff K, Eckman C, Zehr C, Yu X, Prada C-M, Perez-tur J, Hutton M, Buee L, Harigaya Y, Yager D, Morgan D, Gordon MN, Holcomb L, Refolo L, Zenk B, Hardy J, Younkin S (1996) Increased amyloid-B42(43) in brains of mice expressing mutant presenilin 1. Nature 383:710–713

Ellisen LW, Bird J, West DC, Soreng AL, Reynolds TC, Smith SD, Sklar J (1991) TAN-1, the human homologue of the Drosophila Notch gene, is broken by chromosomal translocations in T lymphoblastic neoplasms. Cell 66:649–661

Fanning AS, Anderson JM (1996) Protein-protein interactions: PDZ domain networks. Curr Biol 6: 1385–1388

Feder JN, Jan LY, Jan YN (1993) A rat gene with sequence homology to the Drosophila gene hairy is rapidly induced by growth factors known to influence neuronal differentiation. Mol Cell Biol 13:105–113

Fehon RG, Kooh PJ, Rebay I, Regan CL, Xu T, Muskavitch MAT, Artavanis-Tsakonas S (1990) Molecular interactions between the protein products of the neurogenic loci notch and delta, two egf-homologous genes in Drosophila. Cell 61:523–534

Fehon RG, Johansen K, Rebay I, Artavanis-Tsakonas S (1991) Complex cellular and subcellular regulation of notch expression during embryonic and imaginal development of *Drosophila:* implications for notch function. J Cell Biol 113:657–669

Fitzgerald K, Greenwald I (1995) Interchangeability of Caenorhabditis elegans DSL proteins and intrinsic signalling activity of their extracellular domains in vivo. Development 121:4275–4282

Fleming RJ, Scottgale TN, Diederich RJ, Artavanis-Tsakonas S (1990) The gene Serrate encodes a putative EGF-like transmembrane protein essential for proper ectodermal development in Drosophila melanogaster. Genes Dev 4:2188–2201

Fleming RJ, Gu Y, Hukriede NA (1997) Serrate-mediated activation of Notch is specifically blocked by the product of the gene fringe in the dorsal compartment of the Drosophila wing imaginal disc. Development 124:2973–2981

Fortini ME, Artavanis-Tsakonas S (1993) Notch: neurogenesis is only part of the picture. Cell 75: 1245–1247

Fortini ME, Artavanis-Tsakonas S (1994) The suppressor of hairless protein participates in Notch receptor signaling. Cell 79:273–282

Franco Del Amo F, Smith DE, Swiatek PJ, Gendron-Meguire M, Greenspan RJ, McMahon AP, Gridley T (1992) Expression pattern of Motch, a mouse homolog of Drosophila Notch, suggests an important role in early postimplantation mouse development. Development 115:737–744

Freeman M (1996) Reiterative use of the EGF receptor triggers differentiation of all cell types in the Drosophila eye. Cell 87:651–660

Freemont PS (1993) The RING finger: a novel protein sequence motif related to the zinc finger. Ann N Y Acad Sci 684:174–192

Frise E, Knoblich JA, Younger-Shepherd S, Jan LY, Jan YN (1996) The Drosophila Numb protein inhibits signaling of the Notch receptor during cell-cell interaction in sensory organ lineage. Proc Natl Acad Sci USA 93:11925–11932

Furukawa T, Maruyama S, Kawaichi M, Honjo T (1992) The Drosophila homolog of the immunoglobulin recombination signal-binding protein regulates peripheral nervous system development. Cell 69:1191–1197

Furukawa T, Kobayakawa Y, Tamura K, Kimura K-I, Kawaichi M, Tanimura T, Honjo T (1995) Suppressor of hairless, the Drosophila homologue of RBPJk, transactivates the neurogenic gene E(spl)m8. Jpn J Genet 70:505–524

Gallahan D, Callahan R (1997) The mouse mammary tumor associated gene INT3 is a unique member of the NOTCH gene family (NOTCH4). Oncogene 14:1883–1890

Gho M, Lecourtois M, Geraud G, Posakony JW, Schweisguth F (1996) Subcellular localization of suppressor of hairless in Drosophila sense organ cells during Notch signalling. Development 122:1673–1682

Gilbert TL, Haldeman BA, Mulvihill E, O'Hara PJ (1992) A mammalian homologue of a transcript from the Drosophila pecanex locus. J Neurogenet 8:181–187

Golembo M, Schweitzer R, Freeman M, Shilo BZ (1996) argos transcription is induced by the Drosophila EGF receptor pathway to form an inhibitory feedback loop. Development 122:223–230

Gomez-Skarmeta, J-L, del Corral RD, de la Calle-Mustienes E, Ferres-Marco D, Modolell J (1996) Araucan and caupolican, two members of the novel Iroquois complex, encode homeoproteins that control proneural and vein-forming genes. Cell 85:95–105

Goode S, Melnick M, Chou T-B, Perrimon N (1996a) The neurogenic genes egghead and brainiac define a novel signaling pathway essential for epithelial morphogenesis during Drosophila oogenesis. Development 122:3863–3879

Goode S, Morgan M, Liang Y-P, Mahowald AP (1996b) Brainiac encodes a novel, putative secreted protein that cooperates with grk TGFalpha to produce the follicular epithelium. Dev Biol 178: 35–50
Grant B, Greenwald I (1996) The Caenorhabditis elegans sel-1 gene, a negative regulator of lin-12 and glp-1, encodes a predicted extracellular protein. Genetics 143:237–247
Grant B, Greenwald I (1997) Structure, function, and expression of SEL-1, a negative regulator of LIN-12 and GLP-1 in C. elegans. Development 124:637–644
Grbavec D, Stifani S (1996) Molecular interaction between TLE1 and the carboxyl-terminal domain of HES-1 containing the WRPW motif. Biochem Biophys Res Commun 223:701–705
Greenwald I (1994) Structure/function studies of lin-12/Notch proteins. Curr Opin Genet Dev 4:556–562
Greenwald I, Rubin GM (1992) Making a difference: the role of cell-cell interactions in establishing separate identities for equivalent cells. Cell 68:271–281
Greenwald I, Seydoux G (1990) Analysis of gain-of-function mutations of the lin-12 gene of C. elegans. Nature 346:197–199
Greenwald IS, Sternberg PW, Horvitz RH (1983) The lin-12 locus specifies cell fates in Caenorhabditis elegans. Cell 34:435–444
Gu Y, Hukriede NA, Fleming RJ (1995) Serrate expression can functionally replace Delta activity during neuroblast segregation in the Drosophila embryo. Development 121:855–865
Guan E, Wang J, Laborda J, Norcross M, Baeuerle PA, Hoffman T (1996) T cell leukemia-associated human notch/translocation-associated notch homologue has IkB-like activity and physically interacts with nuclear factor-kB proteins in T cells. J Exp Med 183:2025–2032
Guo M, Jan LY, Jan YN (1996) Control of daughter cell fates during asymmetric division: interaction of numb and notch. Neuron 17:27–41
Hampton RY, Gardner RG, Rine J (1996) Role of the 26S proteasome and HRD genes in the degradation of 3-hydroxy-3-methylglutaryl-CoA reductase, an integral endoplasmic reticulum membrane protein. Mol Biol Cell 7:2029–2044
Harrison SC (1996) Peptide-surface association: the case of PDZ and PTB domains. Cell 86:341–343
Hartenstein AY, Rugendorff A, Tepass U, Hartenstein V (1992) The function of the neurogenic genes during epithelial development in the Drosophila embryo. Development 116:1203–1220
Hartley DA, Preiss A, Artavani-Tsakonas S (1988) A deduced gene product from the Drosophila neurogenic locus, enhancer of split, shows homology to mammalian G-protein B subunits. Cell 55: 785–795
Hay BA, Wassarman DA, Rubin GM (1995) Drosophila homologs of Baculovirus Inhibitor of Apoptosis proteins function to block cell death. Cell 83:1253–1262
Hayashi H, Mochii M, Kodama R, Hamada Y, Mizuno N, Eguchi G, Tachi C (1996) Isolation of a novel chick homologue of serrate and its coexpression with C-notch-1 in chick development. Int J Dev Biol 40:1089–1096
Heitzler P, Simpson P (1991) The choice of cell fate in the epidermis of Drosophila. Cell 64:1083–1092
Heitzler P, Bourouis M, Ruel L, Carteret C, Simpson P (1996) Genes of the enhancer of split and achaete-scute complexes are required for a regulatory loop between Notch and Delta during lateral signalling in Drosophila. Development 122:161–171
Henderson ST, Gao D, Lambie EJ, Kimble J (1994) Lag-2 may encode a signaling ligand for the GLP-1 and LIN-12 receptors of C. elegans. Development 120:2913–2924
Henkel T, Ling PD, Hayward SD, Peterson MG (1994) Mediation of Epstein-Barr virus EBNA2 transactivation by recombination signal-binding protein Jk. Science 265:92–95
Henrique D, Adam J, Myat A, Chitnis A, Lewis J, Ish-Horowicz D (1995) Expression of a Delta homologue in prospective neurons in the chick. Nature 375:787–790
Honjo T (1996) The shortest path from the surface to the nucleus: RBPJk/Su(H) transcription factor. Genes to Cells 1:1–9
Hsieh JJ-D, Henkel T, Salmon P, Robey E, Peterson MG, Hayward SD (1996) Truncated mammalian Notch1 activates CBF1/RBPJk-repressed genes by a mechanism resembling that of Eptstein-Barr Virus EBNA2. Mol Cell Biol 16:952–959
Hua X, Nohturfft A, Goldstein JL, Brown MS (1996) Sterol resistance in CHO cells traced to point mutation in SREBP cleavage-activating protein. Cell 87:415–426
Hubbard EJA, Dong Q, Greenwald I (1996) Evidence for physical and functional association between EMB-5 and LIN-12 in C. elegans. Science 273:112–115
Husain J, Lo R, Grbavec D, Stifani S (1996) Affinity for the nuclear compartment and expression during cell differentiation implicate phosphorylated Groucho/TLE1 forms of higher molecular mass in nuclear functions. Biochem J 317:523–531

Irvine KD, Wieschaus E (1994) Fringe, a boundary-specific molecule, mediates interactions between dorsal and ventral cells during Drosophila wing development. Cell 79:595–606

Ishibashi M, Sasai Y, Nakanishi S, Kageyama R (1993) Molecular characterization of HES-2, a mammalian helix-loop-helix factor structurally related to Drosophila hairy and enhancer of split. Eur J Biochem 215:645–652

Ishibashi M, Moriyoshi K, Sasai Y, Shiota K, Nakanishi S, Kageyama R (1994) Persistent expression of helix-loop-helix factor HES-1 prevents mammalian neural differentiation in the central nervous system. EMBO J 13:1799–1805

Iyengar R (1997) Signal transduction: there are GAPS and there are GAPS. Science 275:42–43

Jarriault S, Brou C, Logeat F, Schroeter EH, Kopan R, Isreal A (1995) Signalling downstream of an activated mammalian Notch. Nature 377:355–358

Jen W-C, Wettstein D, Turner D, Chitnis A, Kintner C (1997) The Notch ligand, X-Delta-2, mediates segmentation of the paraxial mesoderm in Xenopus embryos. Development 124:1169–1178

Jennings B, Preiss A, Delidakis C, Bray S (1994) The Notch signalling pathway is required for Enhancer of split bHLH protein expression during neurogenesis in the Drosophila embryo. Development 120:3537–3548

Jennings B, de Celis J, Delidakis C, Preiss A, Bray S (1995) Role of Notch and achaete-scute complex in the expression of enhancer of split bHLH proteins. Development 121:3745–3752

Johansen KM, Fehon RG, Artavanis-Tsakonas S (1989) The Notch gene product is a glycoprotein expressed on the cell surface of both epidermal and neuronal precursor cells during Drosophila development. J Cell Biol 109:2427–2440

Johnston SH, Rauskolb C, Wilson R, Prabhakaran B, Irvine KD, Vogt TF (1997) A family of mammalian fringe genes implicated in boundary determination and the notch pathway. Development 124:2245–2254

Jonsson F, Knust E (1996) Distinct functions of the Drosophila genes Serrate and Delta revealed by ectopic expression during wing development. Dev Genes Evol 206:91–101

Joutel A, Corpechot C, Ducros A, Vahedi K, Chabriat H, Mouton P, Alamowitch S, Domenga V, Cecillion M, Marechal E, Maciazek J, Vayssiere C, Cruaud C, Cabanis E-A, Ruchoux MM, Weissenbach J, Bach JF, Bousser MG, Tournier-Lasserve E (1996) Notch3 mutations in CADASIL, a hereditary adult-onset condition causing stroke and dementia. Nature 383:707–710

Kannabiran C, Zeng X, Vales LD (1997) The mammalian transcriptional repressor RBP (CBF1) regulates interleukin-6 gene expression. Mol Cell Biol 17:1–9

Kelley MR, Kidd S, Deutsch WA, Young MW (1987) Mutations altering the structure of epidermal growth factor-like coding sequences at the Drosophila Notch locus. Cell 51:539–548

Kidd S, Kelley MR, Young MW (1986) Sequence of the Notch locus of Drosophila: relationship of the encoded protein to mammalian clotting and growth factors. Mol Cell Biol 6:3094–3108

Kidd S, Baylies MK, Gasic GP, Young MW (1989) Structure and distribution of the Notch protein in developing Drosophila. Genes Dev 3:1113–1129

Kim J, Irvine KD, Carrol SB (1995) Cell recognition, signal induction, and symmetrical gene activation at the Dorsal-Ventral boundary of the developing Drosophila wing. Cell 82:795–802

Kim J, Sebring A, Esch JJ, Kraus ME, Vorwerk K, Magee J, Carroll SB (1996) Integration of positional signals and regulation of wing formation and identity by Drosophila vestigal gene. Nature 382: 133–138

Kimble J (1981) Alterations in cell lineage following laser ablation of cells in the somatic gonad of Caenorhabditis elegans. Dev Biol 87:286–300

Kimble J, Hirsh D (1979) Postembryonic cell lineages of the hermaphrodite and male gonads in Caenorhabditis elegans. Dev Biol 81:208–221

Klambt C, Knust E, Tietze K, Campos-Ortega JA (1989) Closely related transcripts encoded by the neurogenic gene complex enhancer of split of Drosophila melanogaster. EMBO J 8:203–210

Klingensmith J, Nusse R (1994) Signaling by wingless in Drosophila. Dev Biol 166:396–414

Klingensmith J, Nusse R, Perrimon N (1994) The Drosophila segment polarity gene dishevelled encodes a novel protein required for responses to the wingless signal. Genes Dev 8:118–130

Knoblich JA, Jan LY, Jan YN (1995) Asymmetric segregation of Numb and Prospero during cell division. Nature 377:624–627

Knust E, Tietze K, Campos-Ortega JA (1987) Molecular analysis of the neurogenic locus Enhancer of split of Drosophila melanogaster. EMBO J 6:4113–4123

Knust E, Schrons H, Grawe F, Campos-Ortega JA (1992) Seven genes of the Enhancer of split complex of Drosophila melanogaster encode helix-loop-helix proteins. Genetics 132:505–518

Kooh PJ, Fehon RG, Muskavitch MAT (1993) Implications of dynamic patterns of Delta and Notch expression for cellular interactions during Drosophila development. Development 117:493–507

Kopan R, Turner DL (1996) The Notch pathway: democracy and aristocracy in the selection of cell fate. Curr Opin Neurobiol 6:594–601

Kopan R, Weintraub H (1993) Mouse Notch: expression in Hair Follicles correlates with cell fate determination. J Cell Biol 121:631–641

Kopan R, Nye JS, Weintraub H (1994) The intracellular domain of mouse Notch: a constitutively activated repressor of myogenesis directed at the basic helix-loop-helix region of MyoD. Development 120:2385–2396

Kopan R, Schroeter EH, Weintraub H, Nye JS (1996) Signal transduction by activated mNotch: importance of proteolytic processing and its regulation by the extracellular domain. Proc Natl Acad Sci USA 93:1683–1688

Kopczynski CC, Alton AK, Fechtel K, Kooh PJ, Muskavitch MAT (1988) Delta, a Drosophila neurogenic gene, is transcriptionally complex and encodes a protein related to blood coagulation factors and epidermal growth factor of vertebrates. Genes Dev 2:1723–1735

Kovacs DM, Fausett HJ, Page KJ, Kim T-W, Moir RD, Merriam DE, Hollister RD, Hallmark OG, Mancini R, Felsenstein KM, Hyman BT, Tanzi RE, Wasco W (1996) Alzheimer-associated presenilins 1 and 2: neuronal expression in brain and localization to intracellular membranes in mammalian cells. Nat Med 2:224–229

Krasnow RE, Wong LL, Adler PN (1995) dishevelled is a component of the frizzled signaling pathway in Drosophila. Development 121:4095–4102

Kraut R, Campos-Ortega JA (1996) Inscuteable, a neural precursor gene of Drosophila, encodes a candidate for a cytoskeleton adaptor protein. Dev Biol 174:65–81

Kraut R, Chia W, Jan LY, Jan YN, Knoblich JA (1996) Role of inscuteable in orienting asymmetric cell division in Drosophila. Nature 383:50–55

Kunisch M, Haenlin M, Campos-Ortega JA (1994) Lateral inhibition mediated by the Drosophila neurogenic gene Delta is enhanced by proneural proteins. Proc Natl Acad Sci USA 91:10139–10143

Kuriyama M, Harada N, Kuroda S, Yamamoto T, Nakafuku M, Iwamatsu A, Yamamoto D, Prasad R, Croce C, Canaani E, Kaibuchi K (1996) Identification of AF-6 and Canoe as putative targets for Ras. J Biol Chem 271:607–610

LaBonne SG, Sunitha I, Mahowald AP (1989) Molecular genetics of pecanex, a maternal-effect neurogenic locus of Drosophila melanogaster that potentially encodes a large transmembrane protein. Dev Biol 136:1–16

Lambie EJ (1996) Cell-cell communication: receptor function at the junction. Curr Biol 6:1089–1091

Lambie EJ, Kimble J (1991) Two homologous regulatory genes, lin-12 and glp-1, have overlapping functions. Development 112:231–240

Lardelli M, Dahlstrand, J, Lendahl U (1994) The novel Notch homologue mouse Notch3 lacks specific epidermal growth factor-repeats and is expressed in proliferating neuroepithelium. Mech Dev 46: 123–136

Lardelli M, Williams R, Lendahl U (1995) Notch-related genes in animal development. Int J Dev Biol 39:769–780

Lecourtois M, Schweisguth F (1995) The neurogenic suppressor of hairless DNA binding protein mediates the transcriptional activation of the enhancer of split complex genes triggered by Notch signaling. Genes Dev 9:2598–2608

Lee E-C, Baker NE (1996) GP300SCA is not a high affinity Notch ligand. Biochem Biophys Res Commun 225:720–725

Lehmann R, Jimenez F, Dietrich U, Campos-Ortega JA (1983) On the phenotype and development of mutants of early neurogenesis in Drosophila melanogaster. Rouxs Arch Dev Biol 192:62–74

Lemere CA, Lopera F, Kosik KS, Lendon CL, Ossa J, Saido TC, Yamaguchi H, Ruiz A, Martinez A, Madrigal L, Hincapie L, Arango LJC, Anthony DC, Koo EH, Goate AM, Selkoe DJ, Arango VJC (1996) The E280A presenilin 1 Alzheimer mutation produces increased AB42 deposition and severe cerebellar pathology. Nat Med 2:1146–1150

Levitan D, Greenwald I (1995) Facilitation of lin-12-mediated signalling by sel-12, a Caenorhabditis elegans S182 Alzheimer's disease gene. Nature 377:351–354

Levitan D, Doyle TG, Brousseau D, Lee MK, Thinakaran G, Slunt HH, Sisodia SS, Greenwald I (1996) Assessment of normal and mutant human presenilin function in Caenorhabditis elegans. Proc Natl Acad Sci USA 93:14940–14944

Leviten MW, Posakony JW (1996) Gain-of-function alleles of bearded interfere with alternative cell fate decisions in Drosophila adult sensory organ development. Dev Biol 176:264–283

Li L, Zhou J, James G, Heller-Harrison R, Czech MP, Olson EN (1992) FGF inactivates myogenic helix-loop-helix proteins through phosphorylation of a conserved protein kinase C site in their DNA-binding domains. Cell 71:1181–1194
Li S-C, Songyang Z, Vincent SJF, Zwaheen C, Wiley S, Cantley L, Kay LE, Forman-Kay J, Pawson T (1997) High-affinity binding of the Drosophila Numb phosphotyrosine-binding domain to peptides containing a Gly-Pro-(p)Tyr motif. Proc Natl Acad Sci (USA) 94:7204–7209
Li X, Greenwald I (1996) Membrane topology of the C. elegans SEL-12 Presenilin. Neuron 17:1015–1021
Lieber T, Wesley CS, Alcamo E, Hassel B, Krane JF, Campos-Ortega JA, Young MW (1992) Single amino acid substitutions in EGF-like elements of notch and delta modify drosophila development and effect cell adhesion in vitro. Neuron 9:847–859
Lieber T, Kidd S, Alcamo E, Corbin V, Young MW (1993) Antineurogenic phenotypes induced by truncated Notch proteins indicate a role in signal transduction and may point to a novel function for Notch in nuclei. Genes Dev 7:1949–1965
Lin H, Schagat T (1997) Neuroblasts: a model for the asymmetric division of stem cells. TIG 13:33–39
Lindsell CE, Shawber CJ, Boulter J, Weinmaster G (1995) Jagged: a mammalian ligand that activates Notch1. Cell 80:909–917
Lu FM, Lux SE (1996) Constitutively active human Notch1 binds to the transcription factor CBF1 and stimulates transcription through a promoter containing a CBF1-responsive element. Proc Natl Acad Sci USA 93:5663–5667
Lyman D, Young MW (1993) Further evidence for function of the Drosophila Notch protein as a transmembrane receptor. Proc Natl Acad Sci USA 90:10395–10399
Maclennan AJ, Shaw G (1993) A yeast SH2 domain. Trends Biochem 18:464–465
Maier D, Stumm G, Kuhn K, Preiss A (1992) Hairless, a Drosophila gene involved in neural development, encodes a novel, serine rich protein. Mech Dev 38:143–156
Maine EM, Kimble J (1993) Suppressors of glp-1, a gene required for cell communication during development in C. elegans, define a set of interacting genes. Genetics 135:1011–1022
Majumdar A, Nagaraj R, Banerjee U (1997) Strawberry notch encodes a conserved nuclear protein that functions downstream of Notch and regulates gene expression along the developing wing margin of Drosophila. Genes & Development 11:1341–1353
Mango SE, Maine EM, Kimble J (1991) Carboxy-terminal truncation activates glp-1 protein to specify vulval fates in Caenorhabditis elegans. Nature 352:811–815
Maratos-Flier E, Yang Kao CY, Verdin EM, King GL (1987) Receptor-mediated vectorial transcytosis of epidermal growth factor by madin-darby canine kidney cells. J Cell Biol 105:1595–1601
Matsunami N, Hamaguchi Y, Yamamoto Y, Kuze K, Kangawa K, Matsuo H, Kawaichi M, Honjo T (1989) A protein binding to the Jk recombination sequence of immunoglobulin genes contains a sequence related to the integrase motif. Nature 342:934–937
Matsuno K, Diederich RJ, Go MJ, Blaumueller CM, Artavanis-Tsakonas S (1995) Deltex acts as a positive regulator of Notch signaling through interactions with the Notch ankyrin repeats. Development 121:2633–2644
Matsuo T, Takahashi K, Kondo S, Kaibuchi K, Yamamoto D (1997) Regulation of cone cell formation by Canoe and Ras in the developing eye. Development 124:2671–2680
Mello CC, Draper BW, Priess JR (1994) The maternal genes apx-1 and glp-1 and establishment of dorsal-ventral polarity in the early C. elegans embryo. Cell 77:95–106
Minoguchi S, Taniguchi Y, Kato H, Okazaki T, Strobl LJ, Zimber-Strobl U, Bornkamm GW, Honjo T (1997) RBP-L, a transcription factor related to RBP-J_K. Mol Cell Biol 17:2679–2687
Miyamoto H, Nihonmatsu I, Kondo S, Ueda R, Togashi S, Hirata K, Ikegami Y, Yamamoto D (1995) canoe encodes a novel protein containing a GLGF/DHR motif and functions with Notch and scabrous in common developmental pathways in Drosophila. Genes Dev 9:612–625
Mulligan LM, Kwok JBJ, Healey CS, Elsdon MJ, Eng C, Gardner E, Love DR, Mole SE, Moore JK, Papi L, Ponder MA, Telenius H, Tunnacliffe A, Ponder BAJ (1993) Germ-line mutations of the Ret proto-oncogene in multiple endocrine neoplasia type 2A. Nature 363:458–460
Muskavitch MAT (1994) Delta-Notch signaling and Drosophila cell fate choice. Dev Biol 166:415–430
Myat A, Henrique D, Ish-Horowicz D, Lewis J (1996) A chick homologue of Serrate and its relationship with Notch and Delta homologues during central neurogenesis. Dev Biol 174:233–247
Newfeld SJ, Tachida H, Yedvobnick B (1994) Drive-selection equilibrium: homopolymer evolution in the Drosophila gene mastermind. J Mol Evol 38:637–641
Nishiwaki K, Sano T, Miwa J (1993) Emb-5, a gene required for the correct timing of gut precursor cell division in Caenorhabditis elegans gastrulation, encodes a protein similar to the yeast nuclear protein SPT6. Mol Gen Genet 239:313–322

Noordermeer J, Klingensmith J, Perrimon N, Nusse R (1994) dishevelled and armadillo act in the Wingless signalling pathway in Drosophila. Nature 367:80–83

Nye JS, Kopan R (1995) Vertebrate ligands for Notch. Curr Biol 5:966–969

Ohsako S, Hyer J, Panganiban G, Oliver I, Caudy M (1994) Hairy function as a DNA-binding helix-loop-helix repressor of Drosophila sensory organ formation. Genes Dev 8:2743–2755

Olson EN, Klein WH (1994) bHLH factors in muscle development: dead lines and commitments, what to leave in and what to leave out. Genes Dev 8:1–8

Pahl HL, Baeuerle PA (1996) Control of gene expression by proteolysis. Curr Opin Cell Biol 8:340–347

Pampeno CL, Meruelo D (1996) A novel cDNA transcript expressed in fractionated X-irradiation-induced murine thymomas. Cell Growth Differ 7:1113–1123

Pan D, Rubin GM (1997) Kuzbanian controls proteolytic processing of Notch and mediates lateral inhibition during *Drosphila* and vertebrate neurogenesis. Cell 90:271–280

Panin VM, Papayannopoulos V, Wilson R, Irvine KD (1997) fringe modulates Notch-ligand interactions. Nature 387:908–912

Paroush Z, Finley RL Jr, Kidd T, Wainwright SM, Ingham PW, Brent R, Ish-Horowicz D (1994) Groucho is required for Drosophila neurogenesis, segmentation, and sex determination and interacts directly with Hairy-related bHLH proteins. Cell 79:805–815

Perrimon N, Engstrom L, Mahhowald AP (1989) Zygotic lethals with specific maternal effect phenotypes in Drosophila melanogasterI Loci on the X. Chromos Genet 121:333–352

Pinto M, Lobe CG (1996) Products of the *grg* (*Groucho-related Gene*) family can dimerize through the amino-terminal Q domain. Journal of Biological Chemistry 271:33026–33031

Posakony JW (1994) Nature versus Nurture: asymmetric cell divisions in Drosophila bristle development. Cell 76:415–418

Poulson DF (1937) Chromosomal deficiencies and embryonic development of Drosophila melanogaster. Proc Natl Acad Sci USA 23:133–137

Prasad R, Gu Y, Alder H, Nakamura T, Canaani O, Saito H, Huebner K, Gale RP, Nowell PC, Kuriyama K, Miyazaki Y, Croce CM, Canaani E (1993) Cloning of the ALL–1 fusion partner, the AF-6 gene, involved in acute myeloid leukemias with the t(6:11) chromosomal translocation. Cancer Res 53:5624–5628

Price BD, Chang Z, Smith R, Bockheim S, Laughon A (1993) The Drosophila neuralized gene encodes a C3HC4 zinc finger. EMBO J 12:2411–2418

Rao Y, Jan LY, Jan YN (1990) Similarity of the product of the Drosophila neurogenic gene big brain to transmembrane channel proteins. Nature 345:163–167

Rao Z, Handford P, Mayhew M, Knott V, Brownlee GG, Stuart D (1995) The structure of a Ca^{2+}-binding Epidermal growth factor-like domain: its role in protein-protein interactions. Cell 82:131–141

Reaume AG, Conlon RA, Zirngibl R, Yamaguchi TP, Rossant J (1992) Expression of a Notch homologue in the mouse embryo. Dev Biol 154:377–387

Rebay I, Fleming RJ, Fehon RG, Cherbas L, Cherbas P, Artivanis-Tsakonas S (1991) Specific EGF repeats of Notch mediate interactions with Delta and Serrate: implications for Notch as a multifunctional receptor. Cell 67:687–699

Rebay I, Fehon RG, Artavanis-Tsakonas S (1993) Specific truncations of Drosophila Notch define dominant activated and dominant negative forms of the receptor. Cell 74:319–329

Ren R, Mayer BJ, Cicchetti P, Baltimore D (1993) Identification of a ten-amino acid proline-rich SH3 binding site. Science 259:1157–1161

Rhyu MS, Jan LY, Jan YN (1994) Asymmetric distribution of Numb protein during division of the sensory organ precursor cell confers distinct fates to daughter cells. Cell 76:477–491

Robbins J, Blondel BJ, Gallahan D, Callahan R (1992) Mouse mammary tumor gene int-3: a member of the notch gene family transforms mammary epithelial cells. J Virol 66:2594–2599

Robey E, Chang D, Itano A, Cado D, Alexander H, Lans D, Weinmaster G, Salmon P (1996) An activated form of Notch influences the choice between CD4 and CD8 T cell lineages. Cell 87:483–492

Roehl H, Kimble J (1993) Control of cell fate in C. elegans by a GLP-1 peptide consisting primarily of ankyrin repeats. Nature 364:632–635

Roehl H, Bosenberg M, Blelloch R, Kimble J (1996) Roles of the RAM and ANK domains in signaling by the C. elegans GLP-1 receptor. EMBO 15:7002–7012

Rogaev EI, Sherrington R, Rogaeva EA, Levesque G, Ikeda M, Liang Y, Chi H, Lin C, Holman K, Tsuda T, Mar L, Sorbi S, Nacmias B, Piacentini S, Amaducci L, Chumakov I, Cohen D, Lannfelt L, Fraser PE, Rommens JM, St George-Hyslop PH (1995) Familial Alzheimer's disease in kindreds with missense mutations in a gene on chromosome 1 related to the Alzheimer's disease type 3 gene. Nature 376:775–778

Rogers S, Wells R, Rechsteiner M (1986) Amino acid sequences common to rapidly degraded proteins: the PEST hypothesis. Science 234:364–368

Rohn JL, Lauring AS, Linenberger ML, Overbaugh J (1996) Transduction of Notch2 in feline leukemia virus-induced thymic lymphoma. J Virol 70:8071–8080

Rothe M, Pan M-G, Henzel WJ, Ayres TM, Goeddel DV (1995) The TNFR2-TRAF signaling complex contains two novel proteins related to baculoviral inhibitor of apoptosis proteins. Cell 83: 1243–1252

Ruel L, Bourouis M, Heitzler P, Pantesco V, Simpson P (1993) Drosophila shaggy kinase and rat glycogen synthase kinase-3 have conserved activities and act downstream of Notch. Nature 362: 557–560

Rulifson EJ, Micchelli CA, Axelrod JD, Perrimon N, Blair SS (1996) wingless refines its own expression domain on the Drosophila wing margin. Nature 384:72–74 (see also p 597)

Rushlow CA, Hogan A, Pinchin SM, Howe KM, Lardelli MT, Ish-Horowicz D (1989) The Drosophila hairy protein acts in both segmentation and bristle patterning and shows homology to N-myc. EMBO J 8:3095–3103

Sasai Y, Kageyama R, Tagawa Y, Shigemoto R, Nakanishi S (1992) Two mammalian helix-loop-helix factors structurally related to Drosophila hairy and enhancer of split. Genes Dev 6:2620–2634

Scheuner D, Eckman C, Jensen M, Song X, Citron M, Suzuki N, Bird TD, Hardy J, Hutton M, Kukull W, Larson E, Levy-Lahad E, Viitanen M, Peskind E, Poorkaj P, Schellenberg G, Tanzi R, Wasco W, Lannfelt L, Selkoe D, Younkin S (1996) Secreted amyloid B-protein similar to that in the senile plaques of Alzheimer's disease is increased in vivo by the presenilin 1 and 2 and APP mutations linked to familial Alzheimer's disease. Nat Med 2:864–870

Schmidt CJ, Sladek TE (1994) A rat homologue of the Drosophila enhancer of split (groucho) locus lacking WD40 repeats. J Biol Chem 268:25681–25686

Schweisguth F, Posakony JW (1992) Suppressor of hairless, the Drosophila homolog of the mouse recombination signal-binding protein gene, controls sensory organ fates. Cell 69:1199–1212

Shawber C, Boulter J, Lindsell CE, Weinmaster G (1996a) Jagged2: a serrate-like gene expressed during rat embryogenesis. Dev Biol 180:370–376

Shawber C, Nofziger D, Hsieh J-DJ, Lindsell C, Bogler O, Hayward D, Weinmaster G (1996b) Notch signaling inhibits muscle cell differentiation through a CBF1-independent pathway. Development 122:3765–3773

Sherrington R, Rogaev EI, Liang Y, Rogaeva EA, Levesque G, Ikeda M, Chi H, Lin C, Li G, Holman K, Tsuda T, Mar L, Foncin J-F, Bruni AC, Montesi MP, Sorbi S, Rainero I, Pinessi L, Nee L, Chumakov I, Pollen D, Brookes A, Sanseau P, Polinsky RJ, Wasco W, Da Silva HAR, Haines JL, Pericak-Vance MA, Tanzi RE, Roses AD, Fraser PE, Rommens JM, St George-Hyslop PH (1995) Cloning of a gene bearing missense mutations in early-onset familial Alzheimer's disease. Nature 375:754–760

Shields A, Bassnett S (1996) Mutations in the founder of the MIP gene underlie cataract development in the mouse. Nat Genet 12:212–215

Siegel PM, Dankort DL, Hardy WR, Muller WJ (1994) Novel activating mutations in the neu proto-oncogene involved in induction of mammary tumors. Mol Cell Biol 14:7068–7077

Siegfried E, Wilder EL, Perrimon N (1994) Components of wingless signalling in Drosophila. Nature 367:76–80

Simske JS, Kaech SM, Harp SA (1996) LET-23 receptor localization by the cell junction protein LIN-7 during C. elegans vulval induction. Cell 85:195–204

Smith GH, Gallahan D, Diella F, Jhappan C, Merlino G, Callahan R (1995) Constitutive expression of a truncated int3 gene in mouse mammary epithelium impairs differentiation and functional development. Cell Growth Differ 6:563–577

Smoller D, Friedel C, Schmid A, Bettler D, Lam L, Yedvobnick B (1990) The Drosophila neurogenic locus mastermind encodes a nuclear protein unusually rich in amino acid polymers. Genes Dev 4:1688–1700

Songyang Z, Fanning AS, Fu C, Xu J, Marfatia SM, Chishti AH, Crompton A, Chan AC, Anderson JM, Cantley LC (1997) Recognition of unique carboxy-terminal motifs by distinct PDZ domains. Science 275:73–77

Spana EP, Doe CQ (1996) Numb antagonizes Notch signaling to specify sibling neuron cell fates. Neuron 17:21–26

Speicher SA, Thomas U, Hinz U, Knust E (1994) The serrate locus of Drosophila and its role in morphogenesis of the wing imaginal discs: control of cell proliferation. Development 120:535–544

Sternberg PW (1993) Falling off the knife edge. Curr Biol 3:763–765

Stifani S, Blaumueller CM, Redhead NJ, Hill RE, Atavanis-Tsakonas S (1992) Human homologues of a Drosophila Enhancer of Split product define a novel family of nuclear proteins. Nat Genet 2:119–127

Struhl G, Fitzgerald K, Greenwald I (1993) Intrinsic activity of the Lin-12 and Notch intracellular domains in vivo. Cell 74:331–345

Sun X, Artavanis-Tsakonas S (1996) The intracellular deletions of Delta and Serrate define dominant negative forms of the Drosophila Notch ligands. Development 122:2465–2474

Sundaram M, Greenwald I (1993a) Genetic and phenotypic analysis of hypomorphic lin-12 mutants in Caenorhabditis elegans. Genetics 135:755–763

Sundaram M, Greenwald I (1993b) Suppressors of a lin-12 hypomorph define genes that interact with both lin-12 and glp-1 in Caenorhabditis elegans. Genetics 135:765–783

Sussman DJ, Klingensmith J, Salinas P, Adams PS, Nusse R, Perrimon N (1994) Isolation and characterization of a mouse homolog of the Drosophila segment polarity gene dishevelled. Dev Biol 166:73–86

Swanson MS, Carlson M, Winson F (1990) SPT6, an essential gene that affects transcription in Saccharomyces cerevisiae, encodes a nuclear protein with an extremely acidic amino terminus. Mol Cell Biol 10:4935–4941

Takebayashi K, Sasai Y, Sakai Y, Watanabe T, Nakanishi S, Kageyama R (1994) Structure, chromosomal locus and promoter analysis of the gene encoding the mouse helix-loop-helix factor HES-1 Negative autoregulation through the multiple N-box elements. J Biol Chem 269:5150–5156

Tamura K, Taniguchi Y, Minoguchi S, Sakai T, Tun T, Furukawa T, Honjo T (1995) Physical interaction between a novel domain of the receptor Notch and the transcription factor RBPJk/Su(H). Curr Biol 5:1416–1423

Tanzi RE, Kovacs DM, Kim T-W, Moir RD, Guenette SY, Wasco W (1996) The Presenilin genes and their role in early-onset familial Alzheimer's disease. Alzheimers Dis Rev 1:90–98

Tax FE, Yeargers JJ, Thomas JH (1994) Sequence of C. elegans lag-2 reveals a cell signaling domain shared with Delta and Serrate of Drosophila. Nature 368:150–154

Tepass U, Hartenstein V (1995) Neurogenic and proneural genes control cell fate specification in the Drosophila endoderm. Development 121:393–405

Theisen H, Purcell J, Bennet M, Kansagara D, Syed A, Marsh JL (1994) dishevelled is required during wingless signaling to establish both cell polarity and cell identity. Development 120:347–360

Thinakaran G, Borchelt DR, Lee MK, Slunt HH, Spitzer L, Kim G, Ratovitsky T, Davenport F, Nordstedt C, Seeger M, Hardy J, Levey AI, Gandy SE, Jenkins NA, Copeland NG, Price DL, Sisodia SS (1996) Endoproteolysis of presenilin 1 and accumulation of processed derivatives in vivo. Neuron 17:181–190

Thomas U, Speicher SA, Knust E (1991) The Drosophila gene Serrate encodes an EGF-like transmembrane protein with a complex expression pattern in embryos and wing discs. Development 111:749–761

Tietze K, Oellers N, Knust E (1992) Enhancer of split D, a dominant mutation of Drosophila, and its use in the study of functional domains of a helix-loop-helix protein. Proc Natl Acad Sci USA 89: 6152–6156

Uemura T, Shepard S, Ackerman L, Jan LY, Jan YN (1989) Numb, a gene required for determination of cell fate during sensory organ formation in Drosophila embryos. Cell 58:349–360

Uyttendaele H, Marazzi G, Wu G, Yan Q, Sassoon D, Kitajewski J (1996) Notch4/int-3, a mammary proto-oncogene, is an endothelial cell-specific mammalian Notch gene. Development 122:2251–2259

Van Doren M, Bailey AM, Esnayra J, Ede K, Posakony JW (1994) Negative regulation of proneural gene activity: hairy is a direct transcriptional repressor of achaete. Genes Dev 8:2729–2742

Vassin H, Bremer KA, Knust E, Campos-Ortega JA (1987) The neurogenic gene Delta of Drosophila melanogaster is expressed in neurogenic territories and encodes a putative transmembrane protein with EGF-like repeats. EMBO J 6:3431–3440

Verdi JM, Schmandt R, Bashirullah A, Jacob S, Salvino R, Craig CG, Program AE, Lipshitz HD, McGlade CJ (1996) Mammalian NUMB is an evolutionarily conserved signaling adapter protein that specifies cell fate. Curr Biol 6:1134–1145

Verheyen EM, Purcell KJ, Fortini ME, Artavanis-Tsakonas S (1996) Analysis of dominant enhancers and suppressors of activated Notch in Drosophila. Genetics 144:1127–1141

Wang X, Sato R, Brown MS, Hua X, Goldstein JL (1994) SREBP-1, a membrane-bound transcription factor released by sterol-regulated proteolysis. Cell 77:53–62

Wang Y, Macke JP, Abella BS, Andreasson K, Worley P, Gilbert DJ, Copeland NG, Jenkins NA, Nathans J (1996) A large family of putative transmembrane receptors homologous to the product of the Drosophila tissue polarity gene frizzled. J Biol Chem 271:4468–4476

Weidemann A, Paliga K, Durrwang U, Czech C, Evin G, Masters CL, Beyreuther K (1997) Formation of stable complexes between two Alzheimer's disease gene products: presenilin-2 and beta-amyloid precursor protein. Nat Med 3:328–332
Weinmaster G, Roberts VJ, Lemke G (1991) A homologue of Drosophila notch expressed during mammalian development. Development 113:199–205
Weinmaster G, Roberts VJ, Lemke G (1992) Notch 2: a second mammalian Notch gene. Development 116:931–941
Weintraub H (1993) The MyoD family and myogenesis: redundancy, networks, and thresholds. Cell 75:1241–1244
Welshons WJ (1956) Dosage experiments with split mutants in the presence of an enhancer of split. Drosophila Inform Serv 30:157–158
Wettstein DA, Turner DL, Kintner C (1997) The Xenopus homologue of Drosophila suppressor of hairless mediates Notch signaling during primary neurogenesis. Development 124:693–702
Wharton KA, Johansen KM, Xu T, Artavanis-Tsakonas S (1985) Nucleotide sequence from the neurogenic locus Notch implies a gene product that shares homology with proteins containing EGF-repeats. Cell 43:567–581
Wilkinson HA, Fitzgerald K, Greenwald I (1994) Reciprocal changes in expression of the receptor Lin-12 and its ligand Lag-2 prior to commitment in a C. elegans cell fate decision. Cell 79:1187–1198
Williams JA, Paddock SW, Vorwerk K, Carroll SB (1994) Organization of wing formation and induction of a wing-patterning gene at the dorsal/ventral compartment boundary. Nature 368:299–305
Wu JY, Wen L, Zhang W-J, Rao Y (1996) The secreted product of Xenopus gene lunatic Fringe, a vertebrate signaling molecule. Science 273:355–358
Yanagawa S-I, van Leeuwen F, Wodarz A, Klingensmith J, Nusse R (1995) The dishevelled protein is modified by Wingless signaling in Drosophila. Genes Dev 9:1087–1097
Yanker BA (1996) New clues to Alzheimer's disease: unraveling the roles of amyloid and tau. Nat Med 2:850–852
Yochem J, Greenwald I (1989) glp-1 and lin-12, genes implicated in distinct cell-cell interactions in C. elegans, encode similar transmembrane proteins. Cell 58:553–563
Yochem J, Weston K, Greenwald I (1988) The Caenorhabditis elegans lin-12 gene encodes a transmembrane protein with overall similarity to Drosophila Notch. Nature 335:547–550
Yuan YP, Schultz J, Mlodzik M, Bork P (1997) Secreted fringe-like signaling molecules may be glycosyltransferases. Cell 88:9–11
Zagouras P, Stifani S, Blaumueller CM, Carcangiu ML, Artavanis-Tsakonas S (1995) Alterations in Notch signaling in neoplastic lesions of the human cervix. Proc Natl Acad Sci USA 92:6414–6418
Zhong W, Feder JN, Jiang M-M, Jan LY, Jan YN (1996) Asymmetric localization of a mammalian Numb homolog during mouse cortical neurogenesis. Neuron 17:43–53
Zhong W, Jiang M-M, Weinmaster G, Jan LY, Jan YN (1997) Differential expression of mammalian Numb, Numblike and Notch 1 suggests distinct roles during mouse cortical neurogenesis. Development 124:1887–1897
Zhou L, Boulianne GL (1994) Comparison of the neuralized genes of Drosophila virilis and D. melanogaster. Genome 37:840–847
Zimrin AB, Pepper MS, McMahon GA, Nguyen F, Montesano R, Maciag T (1996) An antisense oligonucleotide to the Notch ligand Jagged enhances Fibroblast growth factor-induced angiogenesis in vitro. J Biol Chem 271:32499–32502

Signaling Through Grb2/Ash-Control of the Ras Pathway and Cytoskeleton

T. TAKENAWA[1], H. MIKI[1] and K. MATUOKA[2]

1 Introduction

Src homology regions (SH) are recognized as a conserved sequence among non-receptor protein tyrosine kinases (PTK), including v-Src, v-Abl, and v-Fps (SADOWSKI et al. 1986). Besides the kinase domain (termed SH1), SH comprises two independent modular units, SH2 and SH3, and has been found in a variety of proteins that quite often contain both SH2 and SH3 domains (COHEN et al. 1995; PAWSON 1995). These proteins can be categorized broadly into two groups: (1) molecules that themselves have enzymatic activity, such as phospholipase Cγ, Src family tyrosine kinases and ras GTPase-activating protein(GAP); and (2) proteins

[1] Department of Biochemistry, Institute of Medical Science, University of Tokyo, Shirokanedai, Minato-ku, Tokyo 108 Japan
[2] Department of Biosignaling, Tokyo Metropolitan Institute of Gerontology, Sakae-cho, Itabashi-ku, Tokyo 173 Japan

that lack a catalytic domain but contain only the SH domains and other signaling modules, such as Nck, c-Crk, Shc and the phosphoinositide 3-kinase 85-kDa subunit (PI3 K p85). Members of the latter group are often referred to as "adaptors." The interaction between SH2 domains and tyrosine-phosphorylated proteins is important in the assembly of signal transduction complexes. On the other hand, SH3 domains recognize proline-rich sequences, thereby signaling downstream targets. Adaptors containing SH2 and SH3 domains form signaling machines in themselves. Grb2/Ash is one such adaptor. This molecule has already been established to be important in the intracellular signal transduction system, as reviewed by DOWNWARD (1994) and CHARDIN et al. (1995), but many laboratories are still accumulating evidence for its pivotal and indispensable roles in a variety of cell regulatory processes.

2 Grb2/Ash and Its Signaling to Ras

2.1 Link Between Receptor/Protein Tyrosine Kinases and Ras

A molecule mediating the signal from a receptor tyrosine kinase to the Ras guanine nucleotide-binding protein was first identified genetically in the nematode worm *Caenorhabditis elegans*. Vulval formation in *C. elegans* is initiated by a signal from a gonadal cell to vulval precursor cells. Many genes including *let-23* and *let-60* are required for this differentiation event (HORVITZ and STERNBERG 1991). The *let-23* and *let-60* genes encode an EGF receptor-like tyrosine kinase (AROIAN et al. 1990) and a Ras protein (HAN and STERNBERG 1990), respectively. Thus signal transduction in vulval formation has been implicated in Ras signaling. A novel gene isolated by CLARK et al. (1992) is a key molecule in this signaling. The gene, *sem-5*, is essential for vulval formation and encodes the Sem-5 protein that is composed almost exclusively of a single SH2 domain and two SH3 domains. Since SH2 domains recognize phosphotyrosine residues in tyrosine kinase receptors, Sem-5 became the most promising candidate for coupling the tyrosine kinase receptor encoded by *let-23* to Ras signaling.

2.2 Discovery of Grb2/Ash

Soon after the discovery of Sem-5, its mammalian counterpart was identified independently by two groups using different strategies. MATUOKA et al. (1992) cloned the rat and human cDNAs by hybridization with mixed oligonucleotides encoding a consensus sequence to the SH2 domain. A cDNA clone encoding 217 amino acids with a molecular mass of 25 kDa was isolated and termed Ash (Abundant Src homology). On the other hand, LOWENSTEIN et al. (1992) utilized the CORT (cloning of receptor targets) method for the human cDNA in which phosphorylated tyrosine-containing fragments in EGF receptors were used as a probe and λgt11 phages were screened for the proteins binding to the probe. They isolated the same Ash cDNA and named it Grb2 (Growth Factor Receptor Bound Protein 2). Grb2/Ash showed a striking structural similarity to Sem-5. They are similar in size (217 amino acids for Grb2/Ash and 228 amino acids for Sem-5), show the same SH orientation(SH3-SH2-SH3), and are highly homologous (58.5% identity with an additional 9.2% amino acid similarity). Moreover, they both lack functional domains other than the SH domains, indicating that they belong to the same protein class as v-Crk and Nck. It was also confirmed that Grb2/Ash binds to the activated EGF receptor and transmits signals downstream, resulting in the enhancement of DNA synthesis (MATUOKA et al. 1992; LOWENSTEIN et al. 1992). Homologs of Grb2/Ash and Sem-5 have been identified to date in fruit fly (termed Drk; OLIVER et al. 1993; SIMON et al. 1993), mouse (SUEN et al. 1993), and chicken (WASENIUS et al. 1993). In addition, subtypes and related molecules, including Grb3-3 (FATH et al. 1994), Ash-m, Ash-s (WATANABE et al. 1995), and Grap (FENG et al. 1996), have been reported.

2.3 Grb2/Ash Interacts with Sos, the Ras Activator

Analogous to Sem-5 functioning in nematodes, Grb2/Ash is believed to play critical roles in the signaling between tyrosine kinase receptors and Ras in mammals. Therefore, the search for a missing link between Grb2/Ash and Ras was started immediately after the discovery of Grb2/Ash. Soon, Sos (guanine nucleotide exchange factor for Ras) was found as a candidate molecule linking Grb2/Ash to Ras by several groups (see DOWNWARD 1994). Studies demonstrated that Sos is brought from the cytoplasm into the vicinity of its target, plasma membrane-associated Ras, by the adaptor Grb2/Ash. Thus, it is clear that Grb2/Ash is an adaptor molecule that associates physically with Sos, thereby allowing ligand-activated tyrosine kinase receptors to modulate Ras activity. Today, the mechanism for the recruitment of the Grb2/Ash-Sos complex to the plasma membrane is thought to involve two types of intermolecular interactions: a ligand-dependent association of the SH2 domain of Grb2/Ash with specific phosphotyrosine residues, pYXNX motifs and the binding of the SH3 domains of Grb2/Ash to proline-rich motifs in the C-terminal domain of Sos proteins.

2.4 Signaling Toward Grb2/Ash

The receptors thus far identified as direct binders for Grb2/Ash include receptor/PTKs EGF receptor (BATZER et al. 1994; OKUTANI et al. 1994), RET (LIU et al. 1996), MET (PELICCI et al. 1995), and erbB-2 (or HER-2/neu; JANES et al. 1994). A receptor protein tyrosine phosphatase alpha also binds to Grb2/Ash directly (DEN HERTOG and HUNTER 1996).

The other type of interaction has been shown to employ one more linker to Grb2/Ash, in many cases the adaptor molecules Syp and Shc. Other docking proteins that function only in specific signaling events have also been identified, for example, p36 for T-cell receptor (BUDAY et al. 1994) and p89 for fibroblast growth factor (FGF) receptor 2 (KLINT et al. 1995). Insulin receptor substrate 1 and 2 (IRS-1, IRS-2) and Gab-1 have been shown to be docking elements specific for insulin and IGF-I receptors, but it now appears likely that they also function with cytokines and many other receptors (TOBE et al. 1993; SKOLNIK et al. 1993; SUN et al. 1995; HOLGADO-MADRUGA et al. 1996).

Syp (also known as SHP-2, SH-PTP2, and PTP-1D) is a protein tyrosine phosphatase bearing SH2, through which it interacts with activated (i.e., tyrosine-phosphorylated) receptors and is then tyrosine-phosphorylated. The phosphorylated Syp subsequently interacts with the SH2 domain of Grb2/Ash. Thus, Syp is involved in the signaling for platelet-derived growth factor (PDGF) receptor β (LI et al. 1994) and the proto-oncogene product c-Kit (TAUCHI et al. 1994), both receptor/PTKs. Syp also plays a role in erythropoietin signaling (TAUCHI et al. 1996). Shc (or ShcA), comprising variants of 46, 52 and 66 kDa, is a ubiquitously expressed protein composed of a C-terminal SH2 domain, a central collagen homology domain and an N-terminal phosphotyrosine binding (PTB) domain (PELICCI et al. 1992). Two more Shc-like genes ShcB and ShcC, have been cloned recently (O'BRYAN et al. 1996). Shc is in itself equipped with dual pY binding modules and a tyrosine phosphorylation sequence, and thus is a suitable adaptor for pY-mediated signaling. Similar to Syp, Shc binds to activated receptor/PTKs such as nerve growth factor receptor (OHMICHI et al. 1994; HASHIMOTO et al. 1994) and FGF receptor 1 (MOHAMMADI et al. 1996), becomes phosphorylated, and recruits Grb2/Ash. Recent reports, however, have established its critical roles in signaling of the other type of receptors, including T-cell receptor ζ (RAVICHANDRAN et al. 1993), GM-CSF receptor (JUCKER and FELDMAN 1996), FcγRI (PARK et al. 1996), FcεRI (JABRIL-CUENOD et al. 1996), IgM receptor (KUMAR et al. 1995), and growth hormone receptor (VANDERKUUR et al. 1995). Also, heterotrimeric G-protein-coupled receptors, e.g., those for endothelin (CAZAUBON et al. 1994), thrombin (CHEN et al. 1996), and angiotensin II (SADOSHIMA and IZUMI 1996), employ Shc for their interaction with Grb2/Ash. These pathways are mediated by the βγ-subunit of heterotrimeric G-proteins and stimulated by the calcium dependent activation of Pyk2, a tyrosine kinase homologous to Fak (LEV et al. 1995). Recently, DIKIC et al. (1996) demonstrated that the activation of G-protein-coupled receptors, such as lysophosphatidic acid (LPA) and bradykinin, induces the tyrosine phosphorylation of Pyk2 and complex formation of Pyk2 and Src, resulting in Ras activation via Shc-Grb2/Ash-Sos. These results support the idea that G-protein-coupled receptors mediate the same signaling events as receptor tyrosine kinases through cellular tyrosine kinases such as Pyk2 and Src. In the case of receptors lacking intrinsic PTK activity, Shc needs to be phosphorylated by the action of an accompanying PTK. Some PTKs other than Pyk2 have been identified, including JAK2 for GM-CSF (JUCKER and FELDMAN 1996) and growth hormone receptor (VANDERKUUR et al. 1995) and Syk for FcεRI (JABRIL-CUENOD et al.

1996). Although there appear to be various types of intermolecular interactions involving Grb2/Ash, it is possible to reduce them essentially to the scheme proposed by EGAN et al. (1993): ligand stimulation of its receptor elicits the tyrosine phosphorylation of a protein, the receptor itself, or an adaptor through the action of a PTK, thereby producing a binding site for Grb2/Ash. This scheme can apply to the interaction of Grb2/Ash with other molecules such as focal adhesion-associated kinase (FAK), which becomes phosphorylated upon integrin engagement with fibronectin (SCHLAEPFER et al. 1994), and Bcr-Abl, which occurs by the fusion of c-Abl and breakpoint cluster region (Bcr) genes and is constitutively phosphorylated (PENDERGAST et al. 1993). Both phosphorylations result in an association with Grb2/Ash and subsequent Ras activation. Recently, a proto-oncogene product, c-Cbl, which is an early tyrosine kinase substrate following T cell and B cell activation, has been shown to associate with various adaptors, Grb2/Ash, c-Crk and Crk-L, suggesting its involvement in various signaling events (SMIT et al. 1996; BUDAY et al. 1996).

2.5 Structural Basis for Understanding Grb2/Ash Specificity

It is now clear that the four C-terminal amino acids of pY are the most critical for the interaction with a specific SH2 and that some other residues around pY also affect affinity and selectivity (Songyang et al. 1993, 1994). The SH2 domain of Grb2/Ash prefers phosphotyrosyl peptides with the consensus sequence pYXNX, in which the specificity determining residue is N+2 (where the position of pY is 0). The site surrounding Tyr 1068 (EpYINQ) of the EGF receptor is a specific, high-affinity binding site. X-ray crystallographic studies of Grb2/Ash SH2 domains in complex with the high affinity phosphotyrosyl peptide ligand help provide an understanding of the structural basis for specificity (RAHUEL et al. 1996). The general fold of the SH2 domain is the same as reported for the SH2 domains of other proteins and consists of a central antiparallel β-sheet flanked by α-helices. However, substantial differences appear in the conformation of the loops forming the binding site of the target peptide. The loop that links βC and βD is five residues shorter in the Grb2/Ash SH2 domain than in the Src and Lck SH2 domains and more closely resembles that of the SypN, p85, SykC, and Zap 70 SH2 domains. Analysis of the conformation of the specificity-determining EF loop showed that Trp-121 in the SH2 domain is located in the immediate environment of the N+2 binding pocket. The bulky side chain of Trp-121 closes the binding cleft C-terminal to pY of the ligand and consequently forces the ligand to adapt a β-turn configuration (RAHUEL et al. 1996). This would lead to the expectation that substitutions of either Trp in the SH2 domain or Asn in the ligand would affect the interaction. The Grb2/Ash SH2 domain containing the substitution W121T is unable to bind to its natural target, the pYVNV motif of Shc. The Src SH2 domain does not naturally bind to the pYVNV motif, but if Thr, corresponding to residue 121 of Grb2/Ash, is replaced by Trp, the modification enables Src SH2 to bind to pYVNV (MARENGERE et al. 1994). In addition, the preference for pYXNX appears to be unique to Grb2/Ash. Molecules bearing this motif include Shc, EGF receptor, Met, Syp, FAK, and Bcr-Abl, and all exhibit a high binding affinity for Grb2/Ash.

Analyses of the steric structure have revealed that the SH3 domain constitutes a beta barrel equipped with a shallow hydrophobic pocket, where the interaction with its ligand, a left-handed polyproline type II helix (PPII) occurs (Cohen et al. 1995; Pawson 1995). Similar to SH2, the affinity and specificity of the acceptor-ligand interaction of SH3 are determined by specific amino acids in the ligand sequence and SH3 pocket (Yu et al. 1994; Sparks et al. 1996). Although the binding capacity is greatly affected by a single substitution of some critical amino acids in the binding region, the preference for specific SH3 domains seems not to change dramatically (Erpel et al. 1995). Interestingly, however, a structural detail of the franking sequences, such as the RT loop, is likely to play a critical role in determining the specificity of SH3-PPII interactions (Lee et al. 1995; Feng et al. 1995). One striking aspect of Grb2/Ash is that its SH3 domains bind to PPII in an opposite orientation to that observed for the SH3 domain of Abl, Fyn, and PI3 K p85 (Feng et al. 1994; Lim et al. 1994). The structural basis for the bidirectional interaction has now been reported (see Mayer and Eck 1995). Grb2/Ash has been found to be not alone in binding to the ligand in the "class II" or "minus" orientation, although it belongs to the minority. Crystal structure analysis of full length Grb2/Ash shows three distinct domains (Maignan et al. 1995). Within a molecule of Grb2/Ash, the SH3 domains are not in contact with the SH2 domain from which they are separated by an interlaced junction with few contacts that resembles a pair of suspenders. This type of construct enables these domains to access to their ligands independently.

2.6 Regulation of Grb2/Ash Signaling to Ras

Mammalian cells contain two Sos protein subtypes, termed Sos1 and Sos2, both of which interact with the SH3 domains (mainly the N-terminal SH3) of Grb2/Ash through their proline-rich C-terminal regions (Sastry et al. 1995). The apparent binding capacity of Sos2 for Grb2/Ash is significantly higher than that of Sos1 both in vitro and in vivo (Yang et al. 1995). Similarly, two isoforms of hSos1 that differ only by an insertion of 15 amino acids exhibit differences in Grb2/Ash-binding affinity and tissue/stage expression (Rojas et al. 1996). These molecules may contribute to Ras activation differentially and, therefore, this divergence may play a regulatory role in the mechanisms controlling Ras function.

Although Sos was once thought to be in constitutive association with Grb2/Ash, it is now known that the Grb2/Ash-Sos interaction is rather dynamic. For example, T-cell antigen receptor stimulation enhances the association between Grb2/Ash and Sos, which is promoted by the binding of Grb2/Ash to Shc through a function of the collagen homology region of Shc (Ravichandran et al. 1995). This would be beneficial in stabilizing signal transduction toward Ras. On the contrary, another regulatory mechanism is employed to deactivate Grb2/Ash-mediated signaling after the signal has been transduced successfully to the nucleus.

This negative feedback should be important, because continuous activation of Ras results in the deregulation of cell proliferation as seen in malignant cell transformation (MARSHALL 1995). Stimulation by mitogens such as EGF or insulin induce Ser/Thr phosphorylation of the Sos C-terminal proline-rich region (PORFIRI and MCCORMICK 1996). Since this is also the binding sequence for Grb2/Ash SH3, it is conceivable that such modification would affect Grb2/Ash binding. In fact, stimulation by insulin or phorbol esters elicits the phosphorylation of Sos and its dissociation from Grb2/Ash. It has been suggested that MEK (HOLT et al. 1996a) or ERK (BUDAY et al. 1995; DONG et al. 1996) is responsible for this phosphorylation event. In contrast, however, EGF stimulation leads to the dissociation of the EGF receptor or Shc from the Grb2/Ash/-Sos complex rather than the dissociation of Sos and Grb2/Ash (HOLT et al. 1996b; PORFIRI and MCCORMICK 1996). Another study indicates that EGF induces both of the above (HU and BOWTELL 1996). One possible explanation for the discrepancy is that the Grb2/Ash and Sos pools utilized upon stimulation by EGF and insulin are distinct (WATERS et al. 1996), since only a small portion of Grb2/Ash associates with only a small portion of Sos. Further studies are needed to specify the mode and differences in the above events controlling signal transduction.

There is one more point related to the mechanisms of Ras regulation. As mentioned earlier, molecules related to Grb2/Ash have been found, three deletion mutants and one homolog. Grb3-3 lacks 41 amino acids of the SH2 domain (FATH et al. 1994). Since it fails to bind to activated EGF receptor or Shc (CAZAUBON et al. 1994), Grb3-3 may function as a dominant negative protein over Grb2/Ash and a trigger for apoptosis. Ash-m lacks 14 amino acids from the C-SH3 domain and Ash-s consists of only the N-SH3 domain (WATANABE et al. 1995). These are also considered to function as dominant negative regulators, since they exhibit a different binding spectrum for the SH3-binding proteins than Grb2/Ash and, moreover, their injection suppresses mitogenesis. Grap is a protein highly homologous to Grb2/Ash but differs in the size of its transcript and in expression (FENG et al. 1996). Whereas Grb2/Ash is ubiquitous, Grap is predominantly expressed in thymus and spleen. With these naturally occurring Grb2/Ash derivatives it is possible to believe that the signal transduction from receptor and cytoplasmic PTKs to Ras is under refined control.

3 Grb2/Ash Signals to Cytoskeletal Regulation

3.1 Signaling Pathways Other than Ras

Unexpectedly, a recent study has shown that the injection of an antibody against Grb2/Ash into cells abolishes the growth factor-induced membrane ruffling (MATUOKA et al. 1993) controlled by Rac, a Rho family small GTPase (RIDLEY et al. 1992). In addition, dynamin, a GTPase required for synaptic vesicle endocytosis, has been found to bind to the Grb2/Ash SH3 domain through its carboxy-terminal proline-rich region (GOUT et al. 1993; HERSKOVITS et al. 1993; MIKI et al. 1994). These findings suggest the possibility that Grb2/Ash functions not only in Ras signaling but also in other pathways involving cytoskeletal regulatory signals. When Grb2/Ash binding proteins in rat brain were investigated, two other brain specific proteins, synapsin I and p145, were found to bind to the Grb2/Ash SH3 domain (MCPHERSON et al. 1994a). Synapsin I is a synaptic vesicle-associated nerve terminal protein that interacts with actin and is thought to mediate the interactions of synaptic vesicles with the presynaptic cytomatrix. On the other hand, the fourth Grb2/Ash binding protein in brain, p145, is present exclusively in neurons and co-localizes with dynamin at nerve-endings (MCPHERSON et al. 1994b). Like dynamin, p145 has been shown to be rapidly depohosphorylated in response to depolarization as dynamin (MCPHERSON et al. 1994b), suggesting that p145 also participates in endocytosis. Recently, p145 was found to be a novel inositolphosphate 5-phosphatase and was named as synaptojanin (MCPHERSON et al. 1996).

3.2 Isolation of Grb2/Ash-Binding Proteins in Bovine Brain

To clarify the downstream signalings of Grb2/Ash, proteins that bind specifically to Grb2/Ash were examined in bovine brain. GST-fused Grb2/Ash was used to prepare the affinity-column. Several proteins were found to bind to the column; especially, 180-kDa, 150-kDa, 110-kDa, 100-kDa, and 65-kDa proteins were the major Grb2/Ash-binding proteins in bovine brain cytosol (MIKI et al. 1994; MIURA et al. 1996). However, it is uncertain whether these proteins also form complexes with Grb2/Ash in vivo. Thus, to clarify this point, an Ash SH2 binding motif-containing peptide (EGF receptor Grb2/Ash SH2 binding site, PVPE-pYINQSVPK) was used to affinity-purify Grb2/Ash and the in vivo binding proteins. This method is thought to reflect more precisely the physiological status of Grb2/Ash-containing protein complexes than the method using GST-Grb2/Ash, since only proteins that have formed complexes with Grb2/Ash in vivo can bind to this peptide; on the other hand, GST-Grb2/Ash only binds to proteins free from Grb2/Ash. Various protein bands were detected (MIURA et al. 1996; ITOH et al. 1996) more clearly than with GST-Grb2/Ash, with the major bands at 180 kDa, 150 kDa, 110 kDa, 100 kDa, 65 kDa, 55 kDa, and 28 kDa . Among them, the 28-kDa protein was found to be Grb2/Ash by Western blotting. Amino acid sequencing analyses revealed the 180-kDa,110-kDa, 100-kDa, and 55-kDa proteins to be Sos, c-Cbl, dynamin, and β-tubulin, respectively. None of the amino acid sequences derived from the 150-kDa (p150) or 65-kDa (p65) proteins showed any homology to other proteins in the protein sequence database.

3.3 β-Tubulin Binds to SH2 Domains Through a Region Different from the Tyrosine-Phosphorylated Protein-Recognition Site

It is believed that Grb2/Ash binds to its binding proteins via SH2 or SH3 domains because Grb2/Ash is composed of only SH2 and SH3 domains. However, β-tubulin is not tyrosine phosphorylated and has no proline-rich motif that is likely to bind to SH3 domains. Therefore, it is of great interest to identify the β-tubulin binding domain within Grb2/Ash. In experiments to evaluate the binding of β-tubulin to GST-fusion proteins containing each domain of Grb2/Ash, it was found that the SH2 domain binds strongly to β-tubulin (ITOH et al. 1996). SH2 domains of the 85-kDa subunit of PI 3-kinase (p85) and PLCγ1 could also bind to β-tubulin, suggesting that the association with β-tubulin is a common characteristic conserved among SH2 domains. Furthermore, β-tubulin co-immunoprecipitates with Grb2/Ash, PLCγ1, and p85, indicating their in vivo association.

Tubulin is a heterodimer composed of α-and β-tubulin subunits. These subunits polymerize to generate microtubules that provide a cytoskeletal network and also associate with various proteins designated microtubule-associated proteins. KAPELLER et al. (1993) showed that PI 3-kinase localizes at microtubules and moves to the perinuclear area when cells are stimulated by PDGF, suggesting that PDGF receptor-PI 3-kinase complexes internalize and transit in association with the microtubule cytoskeleton. More recently, the same authors (KAPELLER et al. 1995) reported that tubulin binds to the inter-SH2 domain of PI 3-kinase between the N-terminal and C-terminal SH2 domains, but not to an SH2 domain. However, they did not succeed in determining the precise site of β-tubulin binding. To identify the β-tubulin binding site within the Grb2/Ash SH2 domain, a series of GST-SH2 deletion mutants was constructed and used for β-tubulin binding assays. As a result, a region (amino acids 119–135) was shown to be required and sufficient for β-tubulin binding and it was concluded that the SH2 region including amino acids 119–135 is the β-tubulin binding site (ITOH et al. 1996). In addition, a synthetic peptide (FLWVVKFNSLNELVDYH) corresponding to the same sequence of the SH2 domain (including β-strands E and F, and α-helix B) inhibited the association of β-tubulin with the Grb2/Ash SH2 domain. It also inhibited β-tubulin-binding to the SH2 domains of p85 and PLCγ1.

This mode of interaction with SH2 domains is clearly different from that of usual tyrosine-phosphorylated proteins, leading to a very simple question: Is the association of tyrosine-phosphorylated proteins with SH2 domains affected by the tubulin-binding or vice versa? The answer is "no." Tubulin-binding has been shown to have no effect on the binding of tyrosine-phosphorylated EGF receptor to the SH2 domain of Grb2/Ash. Thus, SH2 domains can associate with both tubulin and tyrosine-phosphorylated proteins at the same time.

In the tyrosine kinase signaling system, it is clear that SH2 and SH3 domains are crucial for the assembly of signaling molecules at the plasma membrane to generate downstream signals. After signal generation by receptor activation, clustering and internalization of the ligand-receptor complex are rapidly induced in most cases. The relationship between the initial recruitment of cytoplasmic proteins to the plasma membrane and the rapid response of receptor endocytosis is unknown. Tubulin may play important roles in the ligand binding-induced endocytosis of receptors by binding to SH2-containing proteins. It is also possible that microtubules regulate the assembly and disassembly of signaling molecules containing SH2 domains.

3.4 N-WASP Regulates the Cortical Actin Cytoskeleton Downstream of Grb2/Ash

3.4.1 Sequence of N-WASP

Since p65, a Grb2/Ash binding protein in bovine brain (Miura et al. 1996), seems to be a novel protein, its cDNA was cloned from a bovine brain cDNA library (Miki et al. 1996). The predicted amino acid sequence revealed a protein composed of 505 amino acid residues showing approximately 50% homology to the Wiskott-Aldrich syndrome protein (WASP) over its full length (Fig. 1). The p65 was termed neural WASP (N-WASP). N-WASP possesses many functional sequence motifs including a PH domain, IQ-motif, GBD/CRIB motif, proline-rich region, and verprolin- and cofilin-homologous regions, most of which are also conserved in WASP.

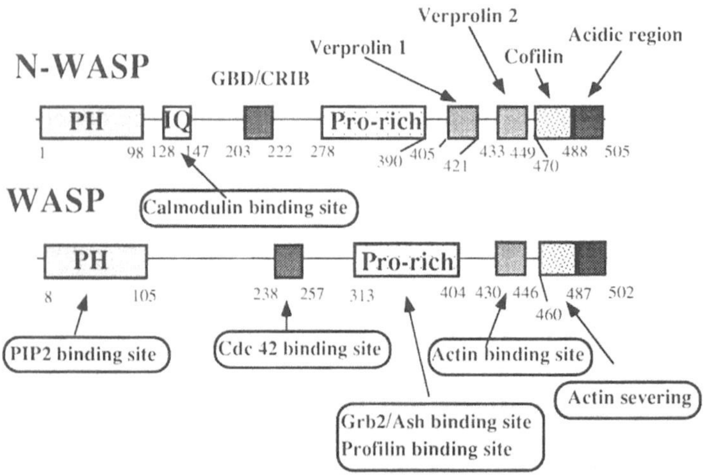

Fig. 1. A variety of functional motifs and domains in N-WASP and WASP. N-WASP contains one PH domain, one IQ motif, one GBD/CRIB motif, several proline-rich sequences, two verprolin-like regions, and one cofilin-like region. On the other hand, WASP contains one PH domain, one GBD/CRIB motif, several proline-rich sequences, one verprolin-like region and one cofilin-like region. WASP does not contain a consensus IQ motif

3.4.2 N-WASP Regulates Cortical Actin Assembly In Vivo

A study of the ectopic expression of N-WASP in COS cells revealed that N-WASP regulates the actin cytoskeleton (Miki et al. 1996). N-WASP expression induces the destruction of the thick fibers of actin filaments and the accumulation of actin filaments in cortical areas under serum-containing culture conditions. The expressed N-WASP localizes predominantly in the nucleus and in cortical areas that overlap actin filaments, suggesting that N-WASP participates directly in the regulation and/or maintenance of the actin cytoskeleton. In contrast, the mutant ΔVCA, which lacks the verproline- and cofilin-homologous regions, localizes only in the nucleus. Further, the mutant C38 W, in which the cysteine residue at position 38 in the PH domain of the original N-WASP is replaced by a tryptophan residue, also localizes in the nucleus. The mutation that corresponds to this residue in WASP has been reported to occur in one WAS patient (Kwan et al. 1995) and so it is supposed that the mutation would also inactivate N-WASP. Interestingly, this point mutation significantly reduces the PIP2-binding activity of the PH domain, suggesting that PIP2-binding through the PH domain may be required for retaining N-WASP in the membrane areas (Miki et al. 1996).

In serum-starved cells, N-WASP localizes predominantly at membranes, not in the nucleus (Miki et al. 1996). When stimulated by EGF, the formation of many microspikes is induced. This result suggests that N-WASP is a key molecule in the transmission of signals from tyrosine kinases to the actin cytoskeleton. In support of this possibility, it has been shown that N-WASP associates constitutively with Grb2/Ash and that the Grb2/Ash-N-WASP protein complexes translocate to the tyrosine-phosphorylated EGF receptor.

The formation of microspikes is known to be regulated by Cdc42 (Kozma et al. 1995; Nobes and Hall 1995). In N-WASP, there is a putative Cdc42-binding motif, the GBD/CRIB motif, suggesting that N-WASP is an effector of Cdc42 for microspike formation.

3.4.3 N-WASP Has Actin Depolymerizing Activity

It remains unclear how N-WASP regulates the actin cytoskeleton. In vitro binding assays have shown that the cofilin-homologous region is critical for binding to actin. Cofilin is known to be an actin-severing protein. In fact, the G-VCA protein, which contains verprolin- and cofilin-homologous regions, can sever actin filaments very quickly (Miki et al. 1996).

3.4.4 N-WASP Generates Signals Leading to Actin Polymerization

At first, the fact that N-WASP severs actin filaments seems contradictory to the result of the overexpression study in which N-WASP causes the accumulation of actin filaments and then co-localizes with them. However, it is known that the overexpression of cofilin stimulates the polymerization and bundling of actin

(Aizawa et al. 1996), possibly by digesting the filamentous network of actin and stimulating its reorganization.

N-WASP has several poly-proline sequences that are putative profilin binding sites. Profilin is known to promote actin polymerization to uncapped barbed ends in the presence of ATP. Therefore, after N-WASP severs actin filaments resulting in providing new barbed ends, profilin may organize the new actin filaments. Indeed, it has been found that profilin binds to and co-immunoprecipitates with N-WASP. This suggests that N-WASP plays a role not only in inducing the reorganization of the actin cytoskeleton by filament severing but also in the actin polymerization process itself.

3.4.5 Putative Model of N-WASP Function in Microspike Formation

The putative microspike formation mechanism is summarized in Fig. 2. In the resting state, actin filaments are stable with most barbed ends capped. When some signals (for example, tyrosine kinase activator) recruit Grb2/Ash-N-WASP complexes to the tyrosine-phosphorylated sites at the plasma membranes and stimulate GDP/GTP exchange on Cdc42 simultaneously, the GTP-bound Cdc42 acquires the ability to bind to N-WASP through its GBD/CRIB motif. A GTP-form Cdc 42 binding stimulates the actin severing activity of N-WASP. The actin filaments are partially severed and the uncapped filament ends become exposed. At the uncapped site, profilin, which exists in association with N-WASP, that is, in close proximity to the newly created barbed ends, promotes the polymerization of actin at plasma membranes. The newly formed actin filaments are assembled by actin bundling proteins and this enforces the protrusion of the plasma membranes.

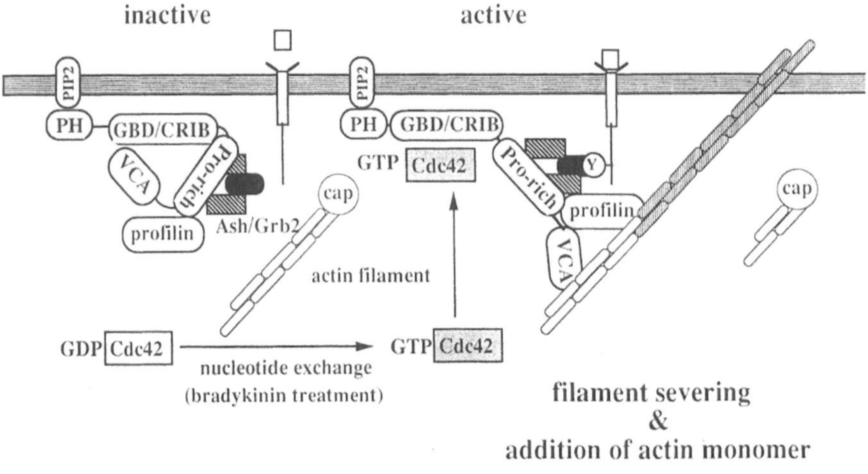

Fig. 2. The putative mechanism of N-WASP action on filopodium formation

3.5 WASP Forms Clusters Enriched in Actin Filaments Downstream of Tyrosine Kinases

WASP has been identified as the gene product whose mutation is responsible for the human hereditary disease Wiskott-Aldrich syndrome (WAS) (Derry et al. 1994). As mentioned above, WASP also contains many functional motifs including a PH domain, GBD/CRIB motif, proline-rich regions, and verprolin-like and cofilin-like domains. The existence of the PH domain is very hard to detect by looking at the sequence, but by comparison with the sequence of N-WASP, its existence has been confirmed (Miki et al. 1996). Therefore, WASP, like N-WASP, is expected to bind tightly to PIP2 through its PH domain.

N-WASP expression is confined to hematopoietic lineages and the cellular defects in WAS patients are also limited to these tissues. N-WASP, on the other hand, is expressed dominantly in the brain. Recently, it was found that the ectopic expression of WASP leads to the clustering of actin filaments in a process inhibited by the co-expression of the dominant negative form of Cdc42 (Symons et al. 1996). Further, WASP has been shown to associate with activated (GTP-bound) Cdc42 through the GBD/CRIB motif. These findings suggest that WASP regulates the actin cytoskeleton under the control of Cdc42, although the mechanism is unclear. Since WASP contains a cofilin-homologous domain like N-WASP (Miki et al. 1996), it is expected to sever actin filaments. However, its regulatory effect on actin filaments appears to be quite different from that of N-WASP despite the structural similarities between WASP and N-WASP. WASP induces the microvesicle formation that is enriched in actin filaments, while N-WASP forms microspikes on plasma membranes, suggesting the different physiological roles played by these two proteins.

WASP also contains several proline-rich regions that are binding targets for SH3 domain-containing proteins. Indeed, Nck, an adaptor protein, has been shown to bind to WASP through its SH3 domains (Rivero-Lezcano et al. 1995). Taken together, the data suggest that WASP is transducer of signals from tyrosine kinases to the cytoskeleton. In MEG-01 megakaryoblastic cells, microvesicle formation is also regulated by tyrosine kinases through Grb2/Ash (Miki et al. 1997). In parallel with the differentiation to megakaryocytes induced by TPA treatment, WASP becomes co-localized with actin filaments at microvesicles and associates with tyrosine-phosphorylated Shc, whose phosphorylation is induced, through Grb2/Ash. Microvesicle formation is specifically inhibited by a tyrosine kinase inhibitor, herbimycin A. Furthermore, treatment withWASP antisense oligonucleotides abolishes both microvesicle formation and the gathering of actin filaments (Miki et al. 1997). All these data suggest that WASP controls the actin filament assembly and/or polymerization required for microvesicle formation downstream of tyrosine kinases.

A comparison of the expression pattern, physiological function, and regulatory mechanism of WASP and N-WASP is presented in Table 1. WASP and N-WASP are very similar to one another in sequence and in that they both transmit signals from tyrosine kinases to the actin cytoskeleton. However, the resultant phenotypes

Table 1 Comparison of WASP and N-WASP

	WASP	N-WASP
Tissue distribution	Hematopoietic cells	Dominantly in brain, but ubiquitously distributed
Intracellular distribution		
Resting state	Cytoplasm	Plasma membranes,
Activating state	Clusters in cytoplasm	Filopodia, nuclei
Motifs and domains	PH,GBD/CRIB, verproline-, cofilin-like polyproline	PH, GBD/CRIB, IQ verproline-, cofilin-like polyproline
Upstream signals	Shc-Grb2/Ash, Cdc42 Nck?	Grb2/Ash, Cdc42 Ca^{2+}-calmodulin?
Downstream signals	Actin, profilin	Actin, profilin
Function	Microvesicle and cluster formation of actin filaments	Filopodium formation of actin filaments

are quite different, that is, WASP induces microvesicles, while N-WASP induces microspikes. Thus, WASP and N-WASP are not homologous proteins performing the same functions in different cells or tissues, but are functionally different proteins.

References

Aizawa H, Sutoh K, Yahara I (1996) Overexpression of cofilin stimulates bundling of actin filaments, membrane ruffling, and cell movement in Dictyostelium. J Cell Biol 132:335–344

Aroian RV, Koga M, Mendel JE, Ohshima Y, Sternberg PW (1990) The let-23 gene necessary for Caenorhabditis elegans vulval induction encodes a tyrosine kinase of the EGF receptor subfamily. Nature 348:693–699

Batzer AG, Rotin D, Urena JM, Skolnik EY, Schlessinger J (1994) Hierarchy of binding sites for Grb2 and Shc on the epidermal growth factor receptor. Mol Cell Biol 14:5192–5201

Buday L, Egan SE, Rodriguez VP, Cantrell DA, Downward J (1994) A complex of Grb2 adaptor protein, Sos exchange factor, and a 36-kDa membrane-bound tyrosine phosphoprotein is implicated in ras activation in T cells. J Biol Chem 269:9019–9023

Buday L, Warne PH, Downward J (1995) Downregulation of the Ras activation pathway by MAP kinase phosphorylation of Sos. Oncogene 11:1327–1331

Buday L, Khwaja A, Sipeki S, Farago A, Downward J (1996) Interactions of Cbl with two adapter protein, Grb2 and Crk, upon T cell activation. J Biol Chem 271:6159–6163

Cazaubon SM, Ramos MF, Fischer S, Schweighoffer F, Strosberg AD, Couraud PO (1994) Endothelin induces tyrosine phosphorylation and GRB2 association of Shc in astrocytes. J Biol Chem 269:24805–24809

Chardin P, Cussac D, Maignan S, Ducruix A (1995) The Grb2 adaptor. FEBS Lett 369:47–51

Chen Y, Grall D, Salcini AE, Pelicci PG, Pouyssegur J, Van O, Schilling E (1996) Shc adaptor proteins are key transducers of mitogenic signaling mediated by the G protein-coupled thrombin receptor. EMBO J 15:1037–1044

Clark SG, Stern MJ, Horvitz HR (1992) C. elegans cell-signalling gene sem-5 encodes a protein with SH2 and SH3 domains. Nature 356:340–344

Cohen GB, Ren R, Baltimore D (1995) Modular binding domains in signal transduction proteins. Cell 80:237–248

den Hertog J, Hunter T (1996) Tight association of GRB2 with receptor protein-tyrosine phosphatase a is mediated by the SH2 and C-terminal SH3 domains. EMBO J 15:3016–3027
Derry JMJ, Ochs HD, Francke U (1944) Isolation of a novel gene mutated in Wiskott-Aldrich syndrome. Cell 78:635–644
Dikic I, Tokiwa G, Lev S, Courtneidge SA, Schlessinger J (1996) A role for Pyk2 and Src in linking G-protein-coupled receptors with MAP kinase activation. Nature 383:547–550
Dong C, Waters SB, Holt KH, Pessin JE (1996) SOS phosphorylation and disassociation of the Grb2-SOS complex by the ERK and JNK signaling pathways. J Biol Chem 271:6328–6332
Downward J (1994) The GRB2/Sem-5 adaptor protein. FEBS Lett 338:113–117
Egan SE, Giddings BW, Brooks MW, Buday L, Sizeland AM, Weinberg RA (1993) Association of Sos Ras exchange protein with Grb2 is implicated in tyrosine kinase signal transduction and transformation. Nature 363:45–51
Erpel T, Superti FG, Courtneidge SA (1995) Mutational analysis of the Src SH3 domain: the same residues of the ligand binding surface are important for intra- and intermolecular interactions. EMBO J 14:963–975
Fath I, Schweighoffer F, Rey I, Multon MC, Boiziau J, Duchesne M, Tocque B (1994) Cloning of a Grb2 isoform with apoptotic properties. Science 264:971–974
Feng GS, Ouyang YB, Hu DP, Shi ZQ, Gentz R, Ni J (1996) Grap is a novel SH3-SH2-SH3 adaptor protein that couples tyrosine kinases to the Ras pathway. J Biol Chem 271:12129–12132
Feng S, Chen JK, Yu H, Simon JA, Schreiber SL (1994) Two binding orientations for peptides to the Src SH3 domain: development of a general model for SH3-ligand interactions. Science 266:1241–1247
Feng S, Kasahara C, Rickles RJ, Schreiber SL (1995) Specific interactions outside the proline-rich core of two classes of Src homology 3 ligands. Proc Natl Acad Sci USA 92:12408–12415
Gout I, Dhand R, Hiles ID, Fry MJ, Panayotou G, Das P, Truong O, Totty NF, Hsuan J, Booker GW, Campbell ID, Waterfield MD (1993) The GTPase dynamin binds to and is activated by a subset of SH3 domains. Cell 75:25–36
Han M, Sternberg PW (1990) Let-60, a gene that specifies cell fates during C. elegans vulval induction, encodes a ras protein. Cell 63:921–931
Hashimoto Y, Matuoka K, Takenawa T, Muroya K, Hattori S, Nakamura S (1994) Different interactions of Grb2/Ash molecule with the NGF and EGF receptors in rat pheochromocytoma PC12 cells. Oncogene 9:869–875
Herskovits JS, Shpetner HS, Burgess CC, Vallee RB (1993) Microtubules and Src homology 3 domains stimulate the dynamin GTPase via its C-terminal domain. Proc Natl Acad Sci USA 90:11468–11472
Holgado-Madruga MM, Emlet DR, Moscatello DK, Godwin AK, Wong AJ (1996) A Grb2-associated docking protein in EGF- and insulin-receptor signalling. Nature 379:560–564
Holt KH, Kasson BG, Pessin JE (1996a) Insulin stimulation of a MEK-dependent but ERK-independent SOS protein kinase. Mol Cell Biol 16:577–583
Holt KH, Waters SB, Okada S, Yamauchi K, Decker SJ, Saltiel AR, Motto DG, Koretzky GA, Pessin JE (1996b) Epidermal growth factor receptor targeting prevents uncoupling of the Grb2-SOS complex. J Biol Chem 271:8300–8306
Horvitz HR, Sternberg PW (1991) Multiple intercellular signalling systems control the development of the Caenorhabditis elegans vulva. Nature 351:535–541
Hu Y, Bowtell DD (1996) Sos1 rapidly associates with Grb2 and is hypophosphorylated when complexed with the EGF receptor after EGF stimulation. Oncogene 12:1865–1872
Ito T, Miura K, Miki H, Takenawa T (1996) β-Tubulin binds Src homology 2 domains through a region different from the tyrosine-phosphorylated protein-recognizing site. J Biol Chem 271:27931–27935
Jabril-Cuenod CB, Zhang C, Scharenberg AM, Paolini R, Numerof R, Beaven MA, Kinet JP (1996) Syk-dependent phosphorylation of Shc A potential link between FcεRI and the Ras/mitogen-activated protein kinase signaling pathway through SOS and Grb2. J Biol Chem 271:16268–16272
Janes PW, Daly RJ, deFazio A, Sutherland RL (1994) Activation of the Ras signalling pathway in human breast cancer cells overexpressing erbB-2. Oncogene 9:3601–3608
Jucker M, Feldman RA (1996) Novel adapter proteins that link the human GM-CSF receptor to the phosphatidylinositol 3-kinase and Shc/Grb2/ras signaling pathways. Curr Top Microbiol Immunol 211:67–75
Kapeller R, Chakrabarti R, Cantley L, Fay F, Corvera S (1993) Internalization of activated platelet-derived growth factor receptor-phosphatidylinositol-3′ kinase complexes: potential interactions with the microtubule cytoskeleton. Mol Cell Biol 13:6052–6063
Kapeller R, Toker A, Cantley LC, Carpenter CL (1995) Phosphoinositide 3-kinase binds constitutively to alpha/beta-tubulin and binds to gamma-tubulin in response to insulin. J Biol Chem 270:25985–25991

Klint P, Kanda S, Claesson WL (1995) Shc and a novel 89-kDa component couple to the Grb2-Sos complex in fibroblast growth factor-2-stimulated cells. J Biol Chem 270:23337–23344
Kozma R Ahmed S, Best A, Lim L (1995) The Ras-related protein Cdc42Hs and bradykinin promote formation of peripheral actin microspikes and filopodia in Swiss 3T3 fibroblasts. Mol Cell Biol 15:1942–1952
Kumar G, Wang S, Gupta S, Nel A (1995) The membrane immunoglobulin receptor utilizes a Shc/Grb2/hSOS complex for activation of the mitogen-activated protein kinase cascade in a B-cell line. Biochem J 307:215–223
Kwan S-P, Hagemann TL, Radtke BE, Blease RM, Rosen FS (1995) Identification of mutations in the Wiskott-Aldrich syndrome gene and characterization of a polymorphic dinucleotide repeat at DXS6940, adjacent to the disease gene. Proc Natl Acad Sci USA 92:4706–4710
Lee CH, Leung B, Lemmon MA, Zheng J, Cowburn D, Kuriyan J, Saksela K (1995) A single amino acid in the SH3 domain of Hck determines its high affinity and specificity in binding to HIV-1 Nef protein. EMBO J 14:5006–5015
Lev S, Moreno H, Martinez R, Canoll P, Peles E, Musacchio JM, Plowman GD, Rudy B, Schlessinger J (1995) Protein tyrosine kinase PYK2 involved in Ca(2+)-induced regulation of ion channel and MAP kinase functions. Nature 376:737–745
Li W, Nishimura R, Kashishian A, Batzer AG, Kim WJ, Cooper JA, Schlessinger J (1994) A new function for a phosphotyrosine phosphatase: linking GRB2-Sos to a receptor tyrosine kinase. Mol Cell Biol 14:509–517
Lim WA, Richards FM, Fox RO (1994) Structural determinants of peptide-binding orientation and of sequence specificity in SH3 domains. Nature 372:375–379
Liu X, Vega QC, Decker RA, Pandey A, Worby CA, Dixon JE (1996) Oncogenic RET receptors display different autophosphorylation sites and substrate binding specificities. J Biol Chem 271:5309–5312
Lowenstein EJ, Daly RJ, Batzer AG, Li W, Margolis B, Lammers R, Ullrich A, Skolnik EY, Bar SD, Schlessinger J (1992) The SH2 and SH3 domain-containing protein GRB2 links receptor tyrosine kinases to ras signaling. Cell 70:431–442
Maignan S, Guilloteau JP, Fromage N, Arnoux B, Becquart J, Ducruix A (1995) Crystal structure of the mammalian Grb2 adaptor. Science 268:291–293
Marengere LE, Songyang Z, Gish GD, Schaller MD, Parsons JT, Stern MJ, Cantley LC, Pawson T (1994) SH2 domain specificity and activity modified by a single residue. Nature 369:502–505
Marshall MS (1995) Ras target proteins in eukaryotic cells. FASEB J 9:1311–1318
Matuoka K, Shibata M, Yamakawa A, Takenawa T (1992) Cloning of ASH, a ubiquitous protein composed of one Src homology region (SH) 2 and two SH3 domains, from human and rat cDNA libraries. Proc Natl Acad Sci USA 89:9015–9019
Matuoka K, Shibasaki F, Shibata M, Takenawa T (1993) Ash/Grb-2, a SH2/SH3-containing protein, couples to signaling for mitogenesis and cytoskeletal reorganization by EGF and PDGF. EMBO J 12:3467–3473
Mayer BJ, Eck MJ (1995) SH3 domains Minding your p's and q's. Curr Biol 5:364–367
McPherson PS, Czernik AJ, Chilcote TJ, Onofri F, Benfenati F, Greengard P, Schlessinger J, Camilli PD (1994a) Interaction of Grb2 via its Src homology 3 domains with synaptic proteins including synapsin I. Proc Natl Acad Sci USA 91:6486–6490
McPherson PS, Takei K, Schmid SL, Camilli PD (1994b) p145, a major Grb2-binding protein in brain, is co-localized with dynamin in nerve terminals where it undergoes activity-dependent dephosphorylation. J Biol Chem 269:30132–30139
McPherson PS, Garcia EP, Slepnev VI, David C, Zhang X, Grabs D, Sossin WS, Bauerfeind R, Nemoto Y, De Camilli P (1996) A presynaptic inositol-5-phosphatase. Nature 379:353–357
Miki H, Miura K, Matuoka K, Nakata T, Hirokawa N, Orita S, Kaibuchi K, Takai Y,Takenawa T (1994) Association of Ash/Grb-2 with dynamin through the Src homology 3 domain. J Biol Chem 269:5489–5492
Miki H, Miura K, Takenawa T (1996) N-WASP, a novel actin-depolymerizing protein, regulates the cortical cytoskeletal rearrangement in a PIP2-dependent manner downstream of tyrosine kinases. EMBO J 15:5326–5335
Miki H, Nonoyama S, Zhu Q, Aruffo A, Ochs HD, Takenawa T (1997) A tyrosine kinase signalling regulates WASP function, which is essential for megakaryocyte differentiation. Cell Growth Differ 8: 195–202
Miura K, Miki H, Shimazaki K, Kawai N, Takenawa T (1996) Interaction of Ash/Grb2 via its SH3 domains with neuron-specific p150 and p65. Biochem J 316:639–645

Mohammadi M, Dikic I, Sorokin A, Burgess WH, Jaye M, Schlessinger J (1996) Identification of six novel autophosphorylation sites on fibroblast growth factor receptor 1 and elucidation of their importance in receptor activation and signal transduction. Mol Cell Biol 16:977–989

Mohammadi M, Dikic I, Sorokin A, Burgess WH, Jaye M, Schlessinger J (1996) Identification of six novel autophosphorylation sites on fibroblast growth factor receptor 1 and elucidation of their importance in receptor activation and signal transduction. Mol Cell Biol 16:977–989

Nobes CD, Hall A (1995) Rho, rac, and cdc42 GTPases regulate the assembly of multimolecular focal complexes associated with actin stress fibers, lamellipodia, and filopodia. Cell 81:53–62

O'Bryan JP, Songyang Z, Cantley L, Der CJ, Pawson T (1996) A mammalian adaptor protein with conserved Src homology 2 and phosphotyrosine-binding domains is related to Shc and is specifically expressed in the brain. Proc Natl Acad Sci USA 93:2729–2734

Ohmichi M, Matuoka K, Takenawa T, Saltiel AR (1994) Growth factors differentially stimulate the phosphorylation of Shc proteins and their association with Grb2 in PC-12 pheochromocytoma cells. J Biol Chem 269:1143–1148

Okutani T, Okabayashi Y, Kido Y, Sugimoto Y, Sakaguchi K, Matuoka K, Takenawa T, Kasuga M (1994) Grb2/Ash binds directly to tyrosines 1068 and 1086 and indirectly to tyrosine 1148 of activated human epidermal growth factor receptors in intact cells. J Biol Chem 269:31310–31314

Olivier JP, Raabe T, Henkemeyer M, Dickson B, Mbamalu G, Margolis B, Schlessinger J, Hafen E, Pawson T (1993) A Drosophila SH2-SH3 adaptor protein implicated in coupling the sevenless tyrosine kinase to an activator of Ras guanine nucleotide exchange, Sos. Cell 73:179–191

Park RK, Liu Y, Durden DL (1996) A role for Shc, Grb2, and Raf-1 in FcγRI signal relay. J Biol Chem 271:13342–13348

Pawson T (1995) Protein modules and signalling networks. Nature 373:573–580

Pelicci G, Lanfrancone L, Grignani F, McGlade J, Cavallo F, Forni G, Nicoletti I, Grignani F, Pawson T, Pelicci PG (1992) A novel transforming protein (SHC) with an SH2 domain is implicated in mitogenic signal transduction. Cell 70:93–104

Pelicci G, Giordano S, Zhen Z, Salcini AE, Lanfrancone L, Bardelli A, Panayotou G, Waterfield MD, Ponzetto C, Pelicci PG (1995) The motogenic and mitogenic responses to HGF are amplified by the Shc adaptor protein. Oncogene 10:1631–1638

Pendergast AM, Quilliam LA, Cripe LD, Bassing CH, Dai Z, Li N, Batzer A, Rabun KM, Der CJ, Schlessinger J (1993) BCR-ABL-induced oncogenesis is mediated by direct interaction with the SH2 domain of the GRB-2 adaptor protein. Cell 75:175–185

Porfiri E, McCormick F (1996) Regulation of epidermal growth factor receptor signaling by phosphorylation of the ras exchange factor hSOS1. J Biol Chem 271:5871–5877

Rahuel J, Gay B, Erdmann D, Strauss A, Garcia-Echeverria C, Furet P, Caravatti G, Fretz H, Schoepfer J, Grutter MG (1996) Structural basis for specificity of Grb2-SH2 revealed by a novel ligand binding mode. Nat Struct Biol 3:586–589

Ravichandran KS, Lee KK, Songyang Z, Cantley LC, Burn P, Burakoff SJ (1993) Interaction of Shc with the zeta chain of the T cell receptor upon T cell activation. Science 262:902–905

Ravichandran KS, Lorenz U, Shoelson SE, Burakoff SJ (1995) Interaction of Shc with Grb2 regulates association of Grb2 with mSOS. Mol Cell Biol 15:593–600

Ridley AJ, Paterson HF, Johnston CL, Diekman D, Hall A (1992) The small GTP-binding protein rac regulates growth factor-induced membrane ruffling. Cell 70:401–410

Rivero-Lezcano OM, Marcilla A, Sameshima JH, Robbins KC (1995) Wiskott-Aldrich syndrome protein physically associated with Nck through Src homology 3 domains. Mol Cell Biol 15:5725–5731

Rojas JM, Coque JJ, Guerrero C, Aroca P, de Mora JF, de la Cruz X, Lorenzi MV, Esteban LM, Santos E (1996) A 15 amino acid stretch close to the Grb2-binding domain defines two differentially expressed hSos1 isoforms with markedly different Grb2 binding affinity and biological activity. Oncogene 12:2291–2300

Sadoshima J, Izumo S (1996) The heterotrimeric G q protein-coupled angiotensin II receptor activates p21 ras via the tyrosine kinase-Shc-Grb2-Sos pathway in cardiac myocytes. EMBO J 15:775–787

Sadowski I, Stone JC, Pawson T (1986) A noncatalytic domain conserved among cytoplasmic protein-tyrosine kinases modifies the kinase function and transforming activity of Fujinami sarcoma virus P130gag-fps. Mol Cell Biol 6:4396–4408

Sastry L, Lin W, Wong WT, Di Fiore PP, Scoppa CA, King CR (1995) Quantitative analysis of Grb2-Sos1 interaction: the N-terminal SH3 domain of Grb2 mediates affinity. Oncogene 11:1107–1112

Schlaepfer DD, Hanks SK, Hunter T, van der Geer P (1994) Integrin-mediated signal transduction linked to Ras pathway by GRB2 binding to focal adhesion kinase. Nature 372:786–791

Simon MA, Dodson GS, Rubin GM (1993) An SH3-SH2-SH3 protein is required for p21Ras1 activation and binds to sevenless and Sos proteins in vitro. Cell 73:169–177

Skolnik EY, Batzer A, Li N, Lee CH, Lowenstein E, Mohammadi M, Margolis B, Schlessinger J (1993) The function of GRB2 in linking the insulin receptor to Ras signaling pathways. Science 260:1953–1955

Smit L, van der Horst G, Borst J (1996) Formation of Shc/Grb2- and Crk adaptor complexes containing tyrosine phosphorylated Cbl upon stimulation of the B-cell antigen receptor. Oncogene 13:381–389

Songyang Z, Shoelson SE, Chaudhuri M, Gish G, Pawson T, Haser WG, King F, Roberts T, Ratnofsky S, Lechleider RJ, Neel BG, Birge RB, Fajardo JE, Chou MM, Hanafusa H, Schffhausen B, Cantley LC (1993) SH2 domains recognize specific phosphopeptide sequences. Cell 72:767–778

Songyang Z, Shoelson SE, McGlade J, Olivier P, Pawson T, Bustelo XR, Barbacid M, Sabe H, Hanafusa H, Yi T (1994) Specific motifs recognized by the SH2 domains of Csk, 3BP2, fps/fes, GRB-2, HCP, SHC, Syk, and Vav. Mol Cell Biol 14:2777–2785

Sparks AB, Rider JE, Hoffman NG, Fowlkes DM, Quillam LA, Kay BK (1996) Distinct ligand preferences of Src homology 3 domains from Src, Yes, Abl, Cortactin, p53bp2, PLCγ, Crk, and Grb2. Proc Natl Acad Sci USA 93:1540–1544

Suen KL, Bustelo XR, Pawson T, Barbacid M (1993) Molecular cloning of the mouse grb2 gene: differential interaction of the Grb2 adaptor protein with epidermal growth factor and nerve growth factor receptors. Mol Cell Biol 13:5500–5512

Sun XJ, Wang LM, Zhang Y, Yenush L, Myers MJ, Glasheen E, Lane WS, Pierce JH, White MF (1995) Role of IRS-2 in insulin and cytokine signalling. Nature 377:173–177

Symons M, Derry JMJ, Karlak B, Jiang S, Lemahieu V, McCormick F, Francke U, Abo A (1996) Wiskott-Aldrich syndrome protein, a novel effector for the GTPase CDC42Hs, is implicated in actin polymerization. Cell 84:723–734

Tauchi T, Feng GS, Marshall MS, Shen R, Mantel C, Pawson T, Broxmeyer HE (1994) The ubiquitously expressed Syp phosphatase interacts with c-kit and Grb2 in hematopoietic cells. J Biol Chem 269:25206–25211

Tauchi T, Damen JE, Toyama K, Feng GS, Broxmeyer HE, Krystal G (1996) Tyrosine 425 within the activated erythropoietin receptor binds Syp, reduces the erythropoietin required for Syp tyrosine phosphorylation, and promotes mitogenesis. Blood 87:4495–4501

Tobe K, Tamemoto H, Yamauchi T, Aizawa S, Yazaki Y, Kadowaki T (1995) Identification of a 190-kDa protein as a novel substrate for the insulin receptor kinase functionally similar to insulin receptor substrate-1. J Biol Chem 270:5698–5701

VanderKuur J, Allevato G, Billestrup N, Norstedt G, Carter SC (1995) Growth hormone-promoted tyrosyl phosphorylation of SHC proteins and SHC association with Grb2. J Biol Chem 270: 7587–7593

Wasenius VM, Merilainen J, Lehto VP (1993) Sequence of a chicken cDNA encoding a GRB2 protein. Gene 134:299–300

Watanabe K, Fukuchi T, Hosoya H, Shirasawa T, Matuoka K, Miki H, Takenawa T (1995) Splicing isoforms of rat Ash/Grb2 Isolation and characterization of the cDNA and genomic DNA clones and implications for the physiological roles of the isoforms. J Biol Chem 270:13733–13739

Waters SB, Chen D, Kao AW, Okada S, Holt KH, Pessin JE (1996) Insulin and epidermal growth factor receptors regulate distinct pools of Grb2-SOS in the control of Ras activation. J Biol Chem 271:18224–18230

Yang SS, Van AL, Bar SD (1995) Differential interactions of human Sos1 and Sos2 with Grb2. J Biol Chem 270:18212–18215

Yu H, Chen JK, Feng S, Dalgarno DC, Brauer AW, Schreiber SL (1994) Structural basis for the binding of proline-rich peptides to SH3 domains. Cell 76:933–945

Genetic Analysis of Sevenless Tyrosine Kinase Signaling in *Drosophila*

T. RAABE

1 Introduction

In a multicellular organism, pattern formation and cell fate determination depend on the ability of cells to communicate with their environment. Inductive cellular interactions play an important role in regulating developmental decisions which finally lead to cellular diversity. One of the requirements in this process is the temporally and spatially controlled release of signaling molecules by one cell which can be received and interpreted by neighboring cells. One model system which has been extensively used to study inductive cellular interactions and to elucidate the underlying machinery is the developing compound eye of *Drosophila*. This review will focus on the genetic and biochemical approaches that have identified signaling molecules involved in the determination of a single photoreceptor cell type in the eye, namely the R7 cell.

Theodor Boveri Institut für Biowissenschaften, Lehrstuhl für Genetik, Universität Würzburg, D-97074 Würzburg, Germany

2 Cell Fate Specification in the *Drosophila* Eye: an Overview

The *Drosophila* eye is composed of approximately 800 single eye units or ommatidia. Each ommatidium comprises 20 cells. These are the eight photoreceptor cells (R1–R8), four lens secreting cone cells and eight accessory cells. The identity of each photoreceptor cell is defined by its position, morphology, spectral sensitivity, axonal projection and its expression of cell-type specific markers. During late larval and pupal stages, the fly eye develops progressively from a monolayer epithelium, the eye-antennal imaginal disc. An indentation, the morphogenetic furrow, traverses the eye disc from posterior to anterior. Ahead of the morphogenetic furrow, the cells of the eye imaginal disc are undifferentiated. The recruitment and differentiation of the different cell types of each ommatidium starts at the posterior edge of the morphogenetic furrow and follows a highly stereotypic temporal sequence. R8 is the first cell to differentiate, subsequently R2/R5 and R3/R4 are added pairwise. After the remaining undifferentiated cells have undergone a last round of mitosis, photoreceptors R1/R6, then finally R7, are recruited. Addition of the non-neuronal cone and pigment cells completes ommatidial development. As a consequence of furrow movement and sequential recruitment of the different cell types, a single third instar eye imaginal disc displays all steps of ommatidial maturation as a graded series starting in the morphogenetic furrow (for a detailed description see Wolff and Ready 1993).

The lack of any cell lineage relationship between the different photoreceptor cells in an ommatidium (Ready et al. 1976; Lawrence and Green 1979), the defined order of cell differentiation and the invariant position of the different cells in the ommatidial cluster led to the proposal that cell fate determination within the ommatidium depends on a series of local cell-cell interactions (Tomlinson and Ready 1987b). In this model, cells that have already differentiated instruct neighboring, yet undifferentiated, cells to adopt a particular cell fate. The most thoroughly studied cell choice event in this hierarchy is the recruitment of the R7 photoreceptor cell. Several features make the R7 cell an attractive system to study inductive cellular interactions. Firstly, only two cells are involved, namely the already differentiated R8 cell as the sender of the inductive signal and the presumptive R7 cell as the recipient. Secondly, depending on the presence or absence of the signal, the R7 precursor cell can only choose between two alternative cell fates, the neuronal R7 or the non-neuronal cone cell fate, respectively. Thirdly, the presence of the R7 cell can be easily assayed in living animals. This allows extensive genetic screens for mutations which specifically interfere with the formation of the R7 cell and therefore might disrupt reception, transduction or interpretation of the inductive signal in the R7 precursor cell. Finally, by expression of transgenes in the eye, the function of potential signaling molecules can be analyzed in vivo.

3 R8 Induces R7 Development: The Signal and the Receptor

If the induction model is correct one might expect to be able to isolate mutations that specifically block R7 cell development either because the inductive signal is missing or the R7 precursor cell is unable to receive and interpret the signal. The isolation of two mutations, *sevenless (sev)* and *bride of sevenless (boss)*, was decisive in defining the inductive signaling pathway critical for R7 cell development (HARRIS et al. 1976; REINKE and ZIPURSKY 1988). Loss-of-function mutations in *boss* and *sev* are homozygous viable and produce an identical mutant phenotype. Although the R7 precursor cell is found in its wild type location, making appropriate contacts with R8, R1 and R6, it fails to initiate neuronal development and instead develops as a non-neuronal cone cell (TOMLINSON and READY 1986; CAGAN et al. 1993). Subsequently, the first distinction between the functions of SEV and BOSS came by genetic mosaic analysis. Whereas SEV activity is required in the R7 precursor cell itself (CAMPOS-ORTEGA et al. 1979; TOMLINSON and READY 1987a), BOSS is required in the R8 cell for proper development of the R7 cell (REINKE and ZIPURSKY 1988). The *boss* gene encodes a 100-kDa protein with seven putative transmembrane domains (HART et al. 1990) and consistent with the mosaic analysis, BOSS expression is restricted to the R8 cell at the time of R7 cell specification (KRÄMER et al. 1991). In contrast, SEV expression is less restricted. The SEV protein has all the characteristic features of a receptor tyrosine kinase (RTK) and is expressed in a highly dynamic manner in a subpopulation of ommatidial cells including R3/R4, R7 and cone cells (HAFEN et al. 1987; TOMLINSON et al. 1987; BASLER and HAFEN 1988). A direct interaction between SEV and BOSS was demonstrated on expression of both proteins in S2 cell lines (KRÄMER et al. 1991). The immediate consequence of this interaction is the stimulation of the SEV kinase activity and autophosphorylation of the receptor (HART et al. 1993). In summary, these experiments established a specific role for the R8 cell in inducing the development of the R7 cell and, secondly, showed that SEV and BOSS are the direct mediators of this cell-cell interaction. However, these data do not readily explain whether other SEV expressing cells have the potential to develop as an R7 cell and if so why there is only a single R7 cell in each ommatidium. To address this, the *boss* and *sev* cDNAs were ubiquitously expressed in the developing eye. Whilst a single R7 cell was observed when wild type SEV was expressed (BASLER and HAFEN 1989; BOWTELL et al. 1989), ubiquitous expression of BOSS resulted in transformation of cone cells into additional R7 cells (VAN VACTOR et al. 1991). The same cell fate transformation event was also obtained by expression of constitutively activated versions of the *sev* gene in the cone cell precursors (BASLER et al. 1991; DICKSON et al. 1992a). Hence, the cone cell precursors are competent to respond to the inductive signal but in the wild type situation where they do not contact the signal-providing R8 cell they are unable to adopt the R7 cell fate. Due to their similar developmental potential, the R7 and cone cells are often referred to as the R7 equivalence group (GREENWALD and RUBIN 1992). In contrast, R3/R4 express SEV and contact R8, but even expression of a constitutively activated version of

SEV in these cells docs not result in a transformation to the R7 cell fate. As discussed later, prior commitment to the R3 and R4 cell fates obviously prevents them adopting the R7 cell fate.

4 Genetic Dissection of SEV Receptor Tyrosine Kinase Mediated Signaling

Tremendous progress has been made recently in elucidating the principles of RTK-mediated signaling from the activated receptor to the nuclear effector proteins. Beside SEV, genetic analyses in *Drosophila* have focused mainly on two RTKs: TORSO (TOR), which is required for specification of terminal structures in the embryo (reviewed in DUFFY and PERRIMON 1994) and the EGF receptor homologue, DER, which fulfills multiple roles during development, such as specification of dorsoventral polarity in the egg, formation of wing veins and development of photoreceptor cells (PRICE et al. 1989; SCHEJTER and SHILO 1989; DIAZ-BENJUMEA and GARCIA-BELLIDO 1990; CLIFFORD and SCHÜPBACH 1992; XU and RUBIN 1993; FREEMAN 1996). For all three RTKs, genetic systems have been established to identify proteins involved in signaling downstream of the corresponding RTK (SIMON et al. 1991; OLIVIER et al. 1993; DOYLE and BISHOP 1993; DIAZ-BENJUMEA and HAFEN 1994; HOU et al. 1995). This has enabled novel components identified in any of these pathways to be readily tested for their functional relevance in the other pathways. In fact, our current understanding of the cytoplasmic events involved in SEV signaling results from the exceptional convergence of molecular, biochemical and genetic data from a number of distinct RTKs in a variety of different organisms, including *C. elegans* and mammals.

As mentioned above, the BOSS-SEV interaction acts as a switch between two different cell fates in the R7 equivalence group. It is this simple cell choice that has been used in large scale screens to identify mutations in genes whose products are required for signaling downstream of the SEV RTK. Two classes of mutations are anticipated. The first class are homozygous viable mutations like *sev* or *boss* which specifically affect R7 cell development. The second class of mutations is more difficult to isolate. If the SEV cascade shares components with other RTKs such as TOR and DER which have functions essential for viability, animals homozygous for loss-of-function mutations in the corresponding genes would presumably die prior to R7 cell commitment. Two strategies have been employed to circumvent this problem. SIMON et al. (1991) used a temperature sensitive allele of *sev* (sev^{ts}) which provides just sufficient activity to allow R7 cell development. The second approach, pioneered by BASLER et al. (1991), involved expression of a constitutively activated version of *sev* (sev^{S11}) in the cells of the R7 equivalence group, which produced a multiple-R7 cell phenotype and resulted in a roughening of the eye surface. In both approaches, R7 cell specification became sensitive to the gene dosage of rate-limiting components acting downstream of SEV. Hence, for a recessive lethal mutation, reducing the gene dose by half was sufficient to produce an enhancement of the *sev*^{ts} phenotype or a suppression of the *sev*^{S11} phenotype, which could be scored in living animals. Large scale screens conducted with both systems unraveled the first steps in the SEV signaling cascade. Some of the signaling molecules identified (e.g., RAS1, RAF) were in turn used as additional entry points for further genetic screens (Dickson et al. 1996; Karim et al. 1996).

5 Downstream of SEV: The Cytoplasmic Signaling Network

The genes identified so far by the various genetic screens encode proteins that function in, or regulate, a highly conserved signaling cascade which is used by RTKs in a variety of different organisms and which is commonly referred to as the RAS GTPase/mitogen activated protein kinase (MAPK) pathway (reviewed by Marshall 1994, 1995; McCormick 1994). In general, stimulation of the RTK results in an increased level of active, GTP-bound RAS protein. The level of GTP-RAS is determined by the ratio between the activity of the RAS guanine nucleotide exchange factor SOS and the intrinsic GTPase activity of RAS which is enhanced by GTPase activating proteins (GAPs). GTP-bound RAS recruits and stimulates the RAF serine threonine kinase. RAF phosphorylates and thereby activates the dual specificity MAPK-kinase (MAPKK), which in turn activates the extracellular signal-regulated kinase (ERK) subgroup of the MAPK family by phosphorylation on threonine and tyrosine residues. *Drosophila* homologues of these proteins are encoded by the *Sos, Gap1, ras1, raf, Dsor1* (Dsor1/MAPKK) and *rolled* (rl/MAPK) genes, respectively (see below for a detailed discussion of each locus). The defined order of protein–protein interactions implies a single, linear pathway necessary and sufficient to transduce all information from the RTK to the nuclear effector proteins. However, the abundance of proteins identified in vertebrate systems as targets of RTKs (Kazlauskas 1994) and, more recently, genetic analyses in *Drosophila* (see below) strongly suggest that the RAS/MAPK pathway is only part of a more complex network of cytoplasmic signaling molecules.

5.1 RAS1 Activation by SEV: DRK and SOS

Binding of BOSS to the SEV receptor results in stimulation of SEV tyrosine kinase activity, resulting in autophosphorylation of the receptor (Hart et al. 1993). In mammalian systems it has been shown that the sequence context of the phosphotyrosine (pTyr) residue provides specificity for the binding of SH2 or PTB domain containing proteins (reviewed in Pawson 1995). For SEV, the binding specificity and in vivo functional role of only one pTyr has been characterized so far. Tyrosine 2546, within the motif YXN, functions as a binding site for the SH2/SH3 domain containing protein DRK (Olivier et al. 1993; Simon et al. 1993), the *Drosophila* homologue of the vertebrate adaptor protein GRB2 (Lowenstein et al. 1992) and the *C. elegans* protein SEM-5 (Clark et al. 1992). Binding of DRK to SEV is abolished by mutation of either Tyr 2546 or of conserved residues in the pTyr binding pocket of the DRK SH2 domain (Olivier et al. 1993; Raabe et al. 1995). The N-and C-terminally located SH3 domains represent a second class of protein–protein interaction domains in DRK; these mediate binding to proline rich sequences located in the C-terminal tail of the RAS1 guanine nucleotide exchange factor SOS (Simon et al. 1991, 1993; Bonfini et al. 1992; Olivier et al. 1993; Raabe et al. 1995). Therefore, DRK appears to provide a link between the guanine nucleotide exchange factor SOS and the activated SEV receptor. However, a detailed structure-function analysis of the SOS protein has revealed a more complex picture (Karlovich et al. 1995; McCollam et al. 1995). The SOS catalytic domain is flanked at the N-terminus by Pleckstrin (PH) and Dbl homology domains whereas the proline rich DRK binding sites are located at the C-terminus. Surprisingly, a mutant SOS construct lacking the PH and Dbl domains dominantly inhibited R7 cell development whereas a construct lacking the DRK binding sites promoted R7 cell development. Complementary experiments using the *Drosophila* SOS protein expressed in mammalian cell lines demonstrated a requirement for the PH and Dbl domains to stimulate the exchange of GDP bound to p21RAS, for GTP, by the SOS catalytic domain. In contrast, the C-terminal DRK/GRB2 binding domain was not sufficient to confer p21RAS exchange activity to the SOS catalytic domain in the absence of the PH and Dbl domains. Hence, in addition to its adaptor function, DRK may be required to alleviate inhibition by the C-terminal regions of SOS on SOS activity. The role of the SOS PH domain may be that of a membrane localization module which brings SOS in close proximity to its membrane-bound substrate RAS1.

Whereas the DRK/SOS/RAS1 cassette is a key target for all RTKs investigated so far in *Drosophila* and removal of any of these components results in lethality, loss-of-function mutations in the GTPase activating protein GAP1 are homozygous viable (Gaul et al. 1992; Buckles et al. 1992; Rogge et al. 1992). The most prominent mutant phenotype of *Gap1* flies is the presence of additional R7 cells, even in the absence of a functional SEV protein. The sequence homology of GAP1 to the mammalian GTPase activating proteins suggests that GAP1 fulfills a similar function by enhancing the intrinsic GTPase activity of RAS1 in the R7 precursor cell. It remains to be verified whether SEV associates with and thereby regulates the activity of GAP1 despite the lack of an SH2 domain in the GAP1 protein.

5.2 Different Routes to RAS1 Activation?

Two important questions regarding receptor signaling to RAS1 are whether DRK and SOS are the only mediators between SEV and RAS1 and, secondly, whether RAS1 activation can account for all of the effects of SEV activation. The ability of a constitutively activated version of RAS1 ($RAS1^{V12}$) to bypass the requirement of

SEV activation in R7 cell development has led to the conclusion that RAS1 activation is necessary and sufficient for R7 cell development (FORTINI et al. 1992). However, overexpression of RAS1^{V12} could potentially mask the requirement for RAS1-independent signaling pathways for proper R7 cell development in the wild type. This would parallel the situation described for mammalian RTKs, where multiple receptor targets, including phospholipase Cγ (PLCγ), phosphatidylinositol (PI) 3-kinase and the GRB2/SOS/RAS cassette may contribute to efficient signaling (reviewed in KAZLAUSKAS 1994; PAWSON 1995). The presence of multiple potential autophosphorylation sites on the SEV protein suggests that DRK might not be the only SH2 domain containing protein bound to SEV. Furthermore, genetic evidence for the existence of parallel pathways downstream of SEV was provided by mutation of Tyr 2546, the DRK binding site on SEV (RAABE et al. 1995). This mutation abolished detectable binding of DRK to SEV, but the mutant receptor was still able to induce R7 cell development, albeit with a reduced efficiency. Strikingly, signaling from the mutant receptor was still sensitive to the gene dosage of *drk, Sos* and *ras1* (RAABE et al. 1996). A solution to this apparent contradiction was provided by the identification of a potential multiadaptor protein (Daughter of Sevenless, DOS; HERBST et al. 1996; RAABE et al. 1996). Genetic interaction experiments indicated a requirement for DOS function upstream or in parallel to RAS1 for signaling from the SEV, DER and TOR receptors. The DOS protein sequence predicts a N-terminal PH domain, a potential SH3 domain binding site and ten tyrosine residues within consensus binding sites for the SH2 domains of DRK, the CORKSCREW (CSW) tyrosine phosphatase, the PI-3-kinase regulatory subunit and PLCγ. DOS might therefore be a functional homologue of the vertebrate receptor substrates IRS-1, IRS-2 and GAB1, which bind a variety of SH2 domain containing proteins upon RTK induced tyrosine phosphorylation (SUN et al. 1993, 1995; SKOLNIK et al. 1993; HOLGADO-MADRUGA et al. 1996). Although biochemical data are still lacking, the presence of putative DRK SH2 domain binding sites on DOS might provide an indirect route to activate SOS/RAS1 in the absence of a direct DRK binding site on SEV.

Further support for the existence of alternative routes for signaling has come from the genetic and biochemical characterization of CSW, the *Drosophila* homologue of the vertebrate protein tyrosine phosphatase SYP (PERKINS et al. 1992; FENG et al. 1993; VOGEL et al. 1993). CSW function is necessary for signaling from the TOR and SEV RTKs, but from genetic experiments with a number of mutant *csw* alleles it has been difficult to precisely assign the step at which CSW function is required (PERKINS et al. 1992; ALLARD et al. 1996). To gain further insight into CSW function, HERBST et al. (1996) applied a biochemical approach. Using a catalytically inactive *csw* mutant (csw^{CS}) as a substrate trap in *Drosophila* SL2 cells, the DOS protein was identified as the first physiological substrate of CSW. Further experiments suggested a link between SEV, CSW and DOS (HERBST et al. 1996). While transfected SEV and CSW can be coimmunoprecipitated, the CSW SH2 domains are critical for binding to tyrosine phosphorylated DOS. Furthermore, SEV activation leads to an enhanced tyrosine phosphorylation of DOS. Consistent with the biochemical analysis, on expression in the eye, the catalytically inactive *csw*^{CS} mutant behaved as a dominant negative allele and caused the frequent absence of photoreceptor cells. This phenotype was further enhanced by halving the gene dosage of *dos, Sos* or *ras1* (HERBST et al. 1996).

While the genetic and biochemical data are consistent, a number of questions remain unresolved. Most importantly, it will be necessary to verify the proposed function of DOS as a multiadaptor protein and to prove the functional relevance of potential DOS binding proteins such as PLCγ and PI-3-kinase in signaling from the SEV receptor. Other important issues include whether CSW activity is regulated by SEV and at which step CSW activity is required for DOS function.

5.3 Downstream of RAS1

The major route by which the signal is propagated from RAS1 to the nucleus in the TOR, DER and SEV pathways involves the sequential activation of the kinases RAF, DSOR1/MAPKK and RL/MAPK (reviewed in DUFFY and PERRIMON 1994; DOMINGUEZ and HAFEN 1996). The phenotypic analysis of various combinations of loss- and gain-of function alleles of these genes has complemented and confirmed the detailed biochemical characterization of the homologous proteins in vertebrates (reviewed in MARSHALL 1995) and suggested that the same chain of phosphorylation and protein–protein interaction events described for RAF, MAPKK and MAPK in mammalian cells also exists in *Drosophila.*

However, several aspects need to be discussed in more detail. One of them is the mechanism by which RAF becomes activated. Analogous to the situation found in mammalian cells (LEEVERS et al. 1994; STOKOE et al. 1994), one important step in RAF activation seems to be the recruitment to the cell membrane, probably via direct interaction with membrane localized RAS-GTP. Expression of a hybrid construct consisting of the RAF kinase domain fused to extracellular and transmembrane domains of a constitutively activated TOR protein (TOR^{4021}-RAF) in the cells of the R7 equivalence group is sufficient to allow R7 cell development, even in a *sev* mutant background (DICKSON et al. 1992b). This experiment raises two questions: firstly, is RAF the only effector of RAS1 in the SEV cascade and, secondly, does RAF activation depend solely on RAS1 function? In mammalian systems there is considerable evidence for effectors of RAS other than RAF (reviewed by WITTINGHOFER and NASSAR 1996) and it has also been shown that RAF activation is not only dependent on RAS binding but also requires additional signals (LEEVERS et al. 1994). The observation that, in *Drosophila*, complete removal of TOR or RAF function produces more severe mutant phenotypes in the embryo than removal of RAS1 function is consistent with this view (HOU et al. 1995).

Kinase suppressor of RAS (KSR), a protein kinase distantly related to RAF, could be a missing piece in this jigsaw. Mutations in the *ksr* locus were isolated as suppressors of the multiple R7 cell phenotype caused by constitutively activated RAS1^{V12} but they do not interfere with the TOR^{4021}-RAF phenotype in the eye (THERRIEN et al. 1995). This indicated a requirement of KSR function downstream or in parallel to RAS1 but upstream or in parallel to RAF. KSR is also required for signaling from the TOR RTK and, consistent with a general role for KSR in RTK mediated signaling, homologous proteins have been identified in *C. elegans*, mouse and human (KORNFELD et al. 1995; SUNDARAM and HAN 1995; THERRIEN et al. 1995). Assuming that the direct interaction of RAF with RAS1 is not sufficient to fully stimulate RAF, one attractive model for KSR function would be the direct or indirect regulation of RAF activity. Indeed, KSR and RAF interact in a RAS dependent manner at the plasma membrane but neither RAF nor KSR appears to be a substrate for the kinase activity of the other protein (THERRIEN et al. 1996). A structure-function analysis of mouse KSR expressed in *Xenopus* oocytes and in mammalian fibroblasts has revealed a more complex picture. It appears that KSR cooperates with RAS and facilitates signal propagation between RAF, MAPKK and MAPK in an as yet uncharacterized manner (THERRIEN et al. 1996). Although physiological substrates have not been identified so far, the identification and characterization of KSR points to a complex network of protein interactions required to mediate the biological effects of RAS1. This network may also include further pathways which cooperate and might be controlled by RAS1 dependent or independent mechanisms. Further complexity could be added through feedback loops. For instance, KSR and SOS possess potential MAPK phosphorylation sites which could play an important role in signal maintenance or attenuation. Finally, the role of protein phosphatases as regulators of signal propagation from RAS1 to MAPK is poorly understood. Recently, the serine/threonine phosphatase PP2A has been shown to negatively regulate signal propagation from RAS1 to RAF but also to stimulate signaling downstream of RAF in the R7 precursor cell (WASSARMAN et al. 1996).

Despite these complexities, the major convergence point of RTK signaling seems to be RL/MAPK (BIGGS and ZIPURSKY 1992; BIGGS et al. 1994). This conclusion can be drawn from the phenotypic and biochemical analysis of a dominant gain-of-function allele of *rl*, rl^{Sem}, isolated in a genetic screen for mutations that bypass the requirement for SEV activity in R7 cell development (BRUNNER et al. 1994a). The phenotypes of rl^{Sem} animals are similar to those described for constitutively activated versions of TOR, DER and SEV RTKs. The molecular defect in rl^{Sem} is a single amino acid substitution (D334N) in the catalytic domain which results in an increased kinase activity (OELLERS and HAFEN 1996). Hence, activation of RL/MAPK is not only necessary but also sufficient to mediate the signal from the receptor to the nuclear target proteins.

6 Nuclear Targets of RL/MAPK

Activation of RL/MAPK results in its translocation from the cytoplasm to the nucleus, where it regulates the activity of transcription factors through phosphorylation on serine and threonine residues. So far, three proteins have been identified as nuclear targets of RL/MAPK. Two of these proteins, YAN and POINTED, belong to the Ets domain family of transcription factors (LAI and RUBIN 1992; TEI et al. 1992; KLÄMBT 1993), whereas the third protein is the *Drosophila* homologue of one of the signal-responsive bZIP transcription factors of the AP-1 family in vertebrates, namely C-JUN (PERKINS et al. 1990; ZHANG et al. 1990).

Although no mutations in the *Drosophila jun* gene *(D-jun)* have been described so far, a number of observations indicate an important role for D-JUN in the development of all photoreceptor cells. Firstly, expression of D-JUN in the eye imaginal disc mirrors the sequential recruitment of the ommatidial cells but precedes the appearence of neuronal differentiation markers (BOHMANN et al. 1994). Secondly, D-JUN is a substrate for RL/MAPK. Phosphopeptide mapping identified Ser82, Thr92 and Thr107 as targets for RL/MAPK phosphorylation in vitro (PEVERALI et al. 1996). The in vivo relevance of these sites was tested by creating two mutant versions of D-JUN. Expression in the developing eye of a D-JUN mutant lacking the phosphoacceptor site acted as a dominant negative: it suppressed the differentiation of photoreceptor cells. Conversely, replacing the RL/MAPK phosphorylation sites for phosphorylation mimicking Asp residues promoted photoreceptor cell differentiation (PEVERALI et al. 1996). These experiments suggested that RL/MAPK induced phosphorylation of D-JUN in the R7 precursor cell is a crucial step for transcriptional activation of target genes which are required for execution of the R7 cell differentiation program.

The second target protein for RL/MAPK, POINTED (PNT) was originally identified as a mutation affecting the development of the embryonic nervous system and subsequently shown to be required for normal photoreceptor cell differentiation. Two alternatively spliced transcripts are expressed, PNT^{P1} and PNT^{P2}. Both proteins share the C-terminal sequences including the Ets domain but differ in their N-terminal sequence (KLÄMBT 1993; SCHOLZ et al. 1993). PNT^{P2} has a single MAPK phosphorylation consensus site (Thr 151) and, at least in vitro, PNT^{P2} is a substrate for RL/MAPK (BRUNNER et al. 1994b; O'NEILL et al. 1994). Using an Ets binding domain reporter construct, O'NEILL et al. (1994) demonstrated that PNT^{P2} becomes a strong transcriptional activator upon stimulation of the RAS/MAPK pathway. As anticipated, the crucial step in the activation of PNT^{P2} is phosphorylation of Thr 151. In the eye imaginal disc, PNT^{P2} is expressed in all nondifferentiated ommatidial precursor cells and removal of PNT^{P2} activity results in the absence of photoreceptors, including R7. This phenotype can be rescued by a transgene encoding the wild type PNT^{P2} protein whereas a mutant version lacking the unique MAPK phosphorylation site does not rescue (BRUNNER et al. 1994b; O'NEILL et al. 1994). Thus, activation of the MAPK pathway is necessary for PNT^{P2} to act as a positive regulator of neuronal development.

The second Ets domain protein that acts downstream of RL/MAPK is YAN. However, the mutant phenotype of *yan* hypomorphic alleles is the converse of that described for PNT^{P2}, namely the recruitment of extra photoreceptor cells, most prominently additional R7 cells (LAI and RUBIN 1992; TEI et al. 1992). In a series of studies, the role of YAN as an inhibitor of cell differentiation was demonstrated and the link between YAN and the MAPK pathway was made. YAN contains eight MAPK phosphorylation consensus sites and at least some of these sites are used in vitro (BRUNNER et al. 1994b; O'NEILL et al. 1994). Cotransfection of YAN and PNT^{P1} in tissue culture cells decreases PNT^{P1}-mediated transcription of a reporter construct containing multiple Ets binding sites. However, this repression is alleviated by coexpression of activated versions of either RAS ($RAS1^{V12}$) or MAPK (RL^{Sem}), indicating that activation of the MAPK pathway negatively regulates the ability of YAN to repress transcription, probably via phosphorylation of YAN (O'NEILL et al. 1994). The behavior of YAN in vitro reflects the in vivo situation. Wild type YAN is expressed in a number of tissues during development, including all undifferentiated cells behind the morphogenetic furrow. YAN disappears from the nuclei of ommatidial cells as soon as they start to differentiate. In contrast, a mutant version of YAN lacking all the predicted MAPK phosphorylation sites remains in the nuclei of ommatidial cells when expressed in the eye imaginal disc and differentiation of neuronal and non-neuronal cells is blocked (REBAY and RUBIN 1995). One conclusion drawn from these studies is that dephosphorylated YAN keeps cells in an undifferentiated state and therefore maintains the competence of cells to respond to the appropriate signal, in the case of the R7 precursor cell, activation of the RAS/MAPK pathway by SEV.

7 Cooperativity

One important question not been addressed so far is whether D-JUN, PNT^{P2} and YAN act in concert to regulate the transcription of target genes in the process of R7 cell development. Although there is only very limited information on potential target genes (see below) and no regulatory elements have been characterized so far, the available genetic and biochemical data suggest cooperativity. However, this does not exclude alternative mechanisms for different target genes.

To examine of the role of cooperativity, TREIER et al. (1995) used a reporter construct carrying justaposed AP-1 and Ets binding sites and demonstrated a synergistic effect of various PNT and D-JUN isoforms in activating transcription of the reporter gene. Furthermore they found that also cotransfection of mutant PNT^{P2} and D-JUN constructs lacking predicted MAPK phosphorylation sites activated transcription significantly compared to either protein alone. In this assay, YAN acts as an antagonist of D-JUN/PNT^{P2} mediated transcriptional activation, supporting the concept that PNT^{P2} and YAN compete for the same Ets domain binding site. The cooperativity between dephosphorylated JUN and PNT^{P2} in vitro might also explain why ectopic R7 cells arise in the absence of YAN function, even in a *sev* mutant background. Further evidence that tight coupling of PNT^{P2} and YAN activities occurs during R7 cell development was provided by BRUNNER et al. (1994b). Neither overexpression of PNT^{P2} nor halving the gene dosis of *yan* is on its own sufficient to transform cone cells into R7 cells; only together do these changes lead to the formation of multiple R7 cells. It appears, therefore, that the sum of the activities of PNT^{P2}, D-JUN and YAN determines whether the R7 precursor cell is committed to the neuronal R7 cell fate or to the non-neuronal cone cell fate. RL/MAPK induced phosphorylation has a dual function: it relieves the inhibitory influence of YAN and it enhances the transcriptional activity of PNT^{P2} and D-JUN in order to induce the expression of target genes that ultimately lead to R7 cell differentiation.

8 Making a Difference

Genetic and biochemical analyses of the different proteins involved in signaling from the SEV receptor has suggested that the same signaling cascade is required for the proper development of all photoreceptor cells, via stimulation of the DER receptor tyrosine kinase (Xu and Rubin 1993; Freeman 1996). If the RAS/MAPK pathway is a common inducer of photoreceptor cell development, then the information for photoreceptor cell identity must be integrated. In the case of the R7 cell, this could happen at different levels. At the level of receptor activation, the ligand BOSS and the receptor SEV are required only for R7 cell development. It could thus be envisaged that SEV might be able to recruit and activate a unique set of proteins in addition to the RAS/MAPK-pathway responsible for R7 cell determination. However, SEV can be replaced by activated versions of other RTKs which activate the RAS/MAPK pathway. For example, expression of activated versions of the DER and TOR RTKs in cells of the R7 equivalence group result in the same cone to R7 cell transformation produced by expression of activated versions of SEV (Basler et al. 1991; Freeman 1996; Dickson and Hafen, unpublished). Furthermore, dependent on the stage of ommatidial assembly, activated SEV can produce photoreceptors other than R7 (Dickson et al. 1992a). These experiments imply that it is not the type of RTK but the time point of signal reception which is critical for cell fate determination. In this model, the developmental potential of the cells of the R7 equivalence group is determined independently of the SEV signal; stimulation of the RAS/MAPK pathway at the right time by SEV only allows execution of the R7 developmental program. Further evidence for this model has been provided by altering the developmental program of the R7 precursor cell. Ectopic expression of ROUGH, a homeo domain containing protein expressed in R2/R5 and R3/R4 photoreceptors, and required in R2/R5 for correct specification of the R3/R4 cell fates, transforms the R7 cell into an R1-R6 like photoreceptor cell. Most strikingly, neuronal differentiation of this cell still depends on SEV activity (Basler et al. 1990; Kimmel et al. 1990). Similar results were obtained with SEVEN-UP (SVP), a protein which belongs to the steroid receptor family and is expressed in the R3/R4/R1/R6 photoreceptors. In *svp* mutant ommatidia, these cells differentiate as R7 photoreceptors whereas ectopic expression of *svp* specifies the presumptive R7 cell as an R1-R6 photoreceptor cell (Mlodzik et al. 1990; Hiromi et al. 1993). Thus, from a mechanistic point of view R7, and probably photoreceptor cell development in general, can be viewed as a process involving at least two steps: specification of the photoreceptor identity and execution of the developmental program triggered by activation of the RAS/MAPK pathway. This does not exclude the possibility that the RAS/MAPK cascade is additionally required in the process of subtype specification.

An important question arising from this model is the issue of signal integration: how is the RAS/MAPK pathway in the R7 precursor cell linked to the cell type specific program? Although this question is difficult to answer at the moment, the identification of three genes, *phyllopod (phyl)*, *seven in absentia (sina)* and *prospero (pros)*, has provided the first insights.

9 PHYL and SINA

The best candidate for a RAS/MAPK inducible target gene in the nucleus of the R7 precursor cell is *phyl* (CHANG et al. 1995; DICKSON et al. 1995). Genetically PHYL has been placed downstream of RL/MAPK, YAN and D-JUN in the SEV pathway. However, whereas D-JUN, PNT and YAN are expressed and required in all photoreceptor cells, PHYL expression is restricted to R1, R6 and R7. In the absence of PHYL these three cells adopt the cone cell fate. The fact that ectopic expression of PHYL is sufficient to transform cone cells into R7 cells, together with the observations that activation of the RAS/MAPK pathway in the cone cells is accompanied by PHYL expression whereas in *sev* mutant flies only R1 and R6 express PHYL, indicate that PHYL regulates R7 cell development and distinguishes R7 and cone cell precursors at the level of gene expression. On the other hand, the apparent contradiction between the general requirement for the RAS/MAPK pathway and the restricted expression of PHYL as a RAS/MAPK inducible target gene emphasizes the need for additional subtype specific information in R1, R6 and R7. Interestingly, the expression pattern of *phyl* reflects the distinct ontogenesis of the different photoreceptor cells. Photoreceptors R8, R2/R5, and R3/R4 arise from cells born anterior to the morphogenetic furrow, whereas R1/R6 and R7 are derived from a second wave of mitosis posterior to the furrow (see WOLFF and READY 1993).

Characteristic features of the PHYL protein are a highly basic domain followed by a strongly acidic domain but no significant homology to other proteins has been identified so far. Recently, KAUFFMANN et al. (1996) employed the yeast two-hybrid system to show that PHYL interacts with another nuclear protein of unknown function, SINA. Beside *sev* and *boss*, *sina* is the third homozygous viable mutation known to prevent formation of the R7 cell (CARTHEW and RUBIN 1990). Remarkably, there is an absolute requirement for SINA in R7 cell development, even when the RAS/MAPK pathway is constitutively activated (BASLER et al. 1991; FORTINI et al. 1992; DICKSON et al. 1992b; BRUNNER et al. 1994a) or when PHYL is ectopically expressed in the cone cell precursors (CHANG et al. 1995). SINA is expressed in a pattern similar to that of SEV, but SINA expression does not depend on a functional SEV protein (CARTHEW and RUBIN 1990). One very attractive working model to account for the roles of SINA and PHYL in the R7 cell is that SINA acts as a transcriptional activator of R7 cell-specific differentiation genes but only after complex formation with PHYL that has been activated by the RAS/MAPK pathway. In photoreceptors R1 and R6, which also express SINA and PHYL, R7 cell formation is prevented by the presence of SVP (see above). Consistent with this model, expression of PROSPERO, a putative transcription factor required for differentiation of the embryonic nervous system and proper connectivity of the R7 cell axons to their synaptic targets in the brain, is upregulated in the R7 cell in a PHYL and SINA dependent manner (DOE et al. 1991; VAESSIN et al. 1991; KAUFFMANN et al. 1996). Thus, for the first time, a potential link between R7 cell specification, induction of the RAS/MAPK pathway and one aspect of R7 cell differentiation has been made.

10 Concluding Remarks

I have described the molecular nature and functional roles of each of the known components of the SEV signaling pathway. Since only a fraction of loci identified in the various genetic screens have been characterized at the molecular level, the model shown in Fig. 1 represents only a snapshot of our current understanding of RAS/

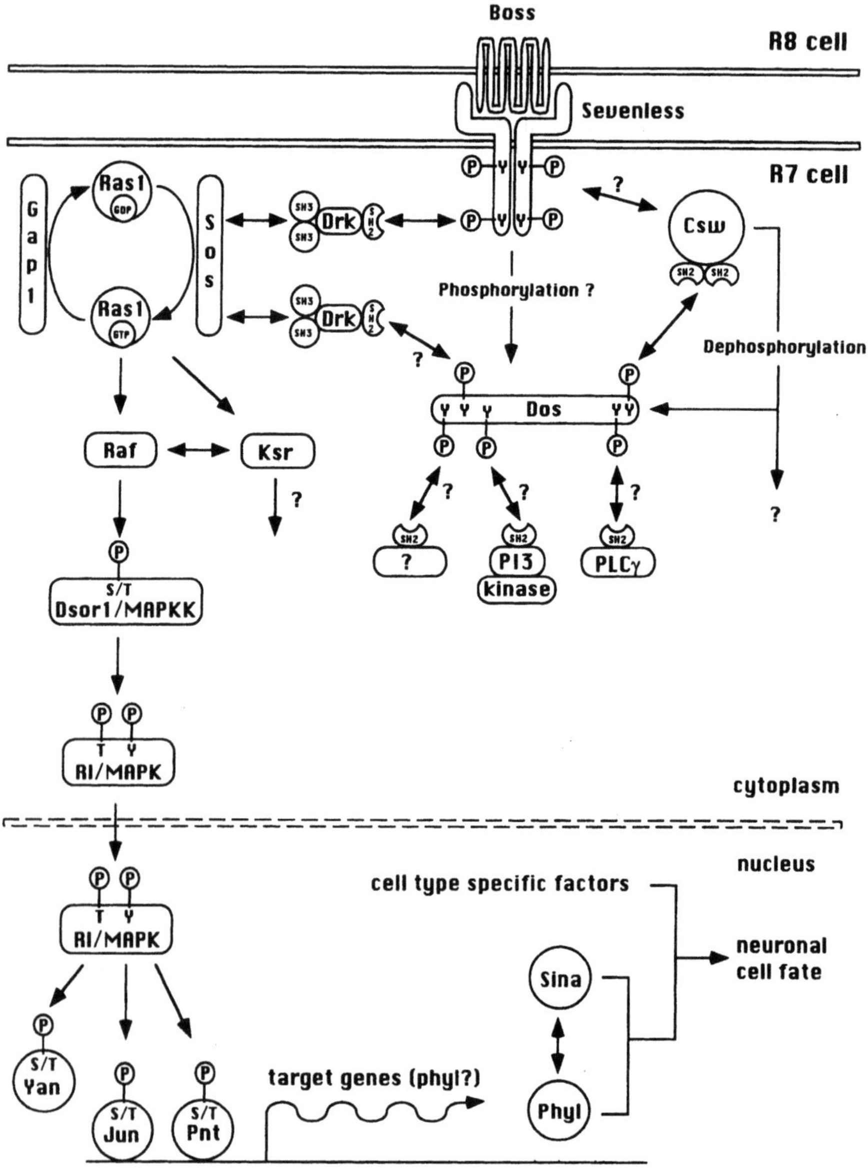

Fig. 1. Model for the Sevenless signaling pathway. See text for a detailed description

MAPK mediated signal transduction in the R7 cell. A detailed understanding of the SEV signaling cascade will require not only the identification but also the functional characterization of the individual signaling proteins. Although the genetic approach has been extremely useful to dissect SEV signaling, it has to be complemented by a biochemical approach to describe the growing complexity of the RAS/MAPK pathway and its regulation, to investigate the nature of alternative routes of signaling and to understand the integration process taking place in the nucleus.

Ten years have now elapsed since the cloning of the *sev* gene. At that time, it was anticipated that the further analysis of this gene would provide insights into the fundamental mechanisms of cell determination – an expectation that has only partially been fulfilled in the intervening decade. While great progress has been made in unraveling the signal transduction cascade from SEV to the nucleus, the basic question of what actually determines the R7 cell fate remains unanswered. What is presently clear is that the activation of the SEV RTK is the final step in a series of events that steer an undetermined cell towards the R7 fate. Along with the genetic and biochemical characterization of additional components of the SEV signal transduction cascade, these earlier events in the determination of the R7 cell fate will also pose a challenge for future research.

Acknowledgements. As any attempt to review a broad research field will lead to omission of important citations, I have to apologize to those investigators whose work was not fully discussed or acknowleged. I would like to thank Ernst Hafen and Martin Heisenberg for continuous support and Barry Dickson, Ernst Hafen and Lindsay MacDougall for critical reading of the manuscript. I am supported by the University of Würzburg and the Deutsche Forschungsgemeinschaft (DFG).

References

Allard JD, Chang HC, Herbst R, McNeill H, Simon MA (1996) The SH2-containing tyrosine phosphatase corksrew is required during signaling by sevenless, Ras1 and Raf. Development 122:1137–1146

Basler K, Hafen E (1988) Control of photoreceptor cell fate by the sevenless protein requires a functional tyrosine kinase domain. Cell 54:299–311

Basler K, Hafen E (1989) Ubiquitous expression of sevenless: position dependent specification of cell fate. Science 243:931–934

Basler K, Yen D, Tomlinson A, Hafen E (1990) Reprogramming cell fate in the developing Drosophila retina: transformation of R7 cells by ectopic expression of rough. Genes Dev 4:728–739

Basler K, Christen B, Hafen E (1991) Ligand-independent activation of the sevenless receptor tyrosine kinase changes the fate of cells in the developing Drosophila eye. Cell 64:1069–1082

Biggs WH, Zipursky SL (1992) Primary structure, expression, and signal-dependent tyrosine phosphorylation of a Drosophila homolog of the extracellular signal-regulated kinase. Proc Natl Acad Sci USA 89:6295–6299

Biggs WH, Zavitz KH, Dickson B, van der Straten A, Brunner D, Hafen E, Zipursky SL (1994) The Drosophila rolled locus encodes a MAP kinase required in the sevenless signal transduction pathway. EMBO J 13:1628–1635

Bohmann D, Ellis MC, Staszewski LM, Mlodzik M (1994) Drosophila Jun mediates Ras-dependent photoreceptor determination. Cell 78:973–986

Bonfini L, Karlovich CA, Dasgupta C, Banerjee U (1992) The son of sevenless gene product: a putative activator of Ras. Science 255:603–606

Bowtell DDL, Simon MA, Rubin GM (1989) Ommatidia in the developing Drosophila eye require and can respond to sevenless for only a restricted period. Cell 56:931–936
Brunner D, Oellers N, Szabad J, Biggs WH, Zipursky SL, Hafen E (1994a) A gain-of-function mutation in Drosophila MAP kinase activates multiple receptor tyrosine kinase pathways. Cell 76:875–888
Brunner D, Dücker K, Oellers N, Hafen E, Scholz H, Klämbt C (1994b) The ETS domain protein pointed-P2 is a target of MAP kinase in the sevenless signal transduction pathway. Nature 370:386–389
Buckles GR, Smith ZDJ, Katz FN (1992) mip causes hyperinnervation of a retinotopic map in Drosophila by excessive recruitment of R7 photoreceptor cells. Neuron 8:1015–1029
Cagan RL, Thomas BJ, Zipursky SL (1993) The role of induction in cell choice and cell cycle in the developing Drosophila retina. In: Bernstein M (ed) Molecular basis of morphogenesis. 51st annual symposium of the Society for Developmental Biology. Wiley Liss, New York, pp 109–133
Campos-Ortega JA, Juergens G, Hofbauer A (1979) Cell clones and pattern formation: studies on sevenless, a mutant of Drosophila melanogaster. Rouxs Arch Dev Biol 186:27–50
Carthew RW, Rubin GM (1990) seven in absentia, a gene required for specification of R7 cell fate in the Drosophila eye. Cell 63:561–577
Chang HC, Solomon NM, Wassarman DA, Karim FD, Therrien M, Rubin GM, Wolff T (1995) Phyllopod functions in the fate determination of a subset of photoreceptors in Drosophila. Cell 80:463–472
Clark SG, Stern MJ, Horvitz HR (1992) C. elegans cell signaling gene sem-5 encodes a protein with SH2 and SH3 domains. Nature 356:340–344
Clifford R, Schüpbach T (1992) The torpedo (DER) receptor tyrosine kinase is required at multiple times during Drosophila embryogenesis. Development 115:853–872
Diaz-Benjumea FJ, Garcia-Bellido A (1990) Behaviour of cells mutant for an EGF receptor homologue of Drosophila in genetic mosaics. Proc R Soc Lond B 242:36–44
Diaz-Benjumea FJ, Hafen E (1994) The sevenless signalling cassette mediates Drosophila EGF receptor function during epidermal development. Development 120:569–578
Dickson B, Sprenger F, Hafen E (1992a) Prepattern in the developing Drosophila eye revealed by an activated torso-sevenless chimeric receptor. Genes Dev 6:2327–2339
Dickson B, Sprenger F, Morrison D, Hafen E (1992b) Raf functions downstream of Ras1 in the Sevenless signal transduction pathway. Nature 360:600–603
Dickson BJ, Dominguez M, van der Straten A, Hafen E (1995) Control of Drosophila photoreceptor cell fates by Phyllopod, a novel nuclear protein acting downstream of the Raf kinase. Cell 80:453–462
Dickson BJ, van der Straten A, Dominguez M, Hafen E (1996) Mutations modulating Raf signaling in Drosophila eye development. Genetics 142:163–171
Doe CQ, Cu-LaGraff Q, Wright DM, Scott MP (1991) The prospero gene specifies cell fates in the Drosophila central nervous system. Cell 65:451–464
Dominguez M, Hafen E (1996) Genetic dissection of cell fate specification in the developing eye of Drosophila. Semin Cell Dev Biol 7:219–226
Doyle HJ, Bishop JM (1993) Torso, a receptor tyrosine kinase required for embryonic pattern formation, shares substrates with the sevenless and EGF-R pathways in Drosophila. Genes Dev 7:633–646
Duffy JB, Perrimon N (1994) The torso pathway in Drosophila: lessons on receptor tyrosine signaling and pattern formation. Dev Biol 166:380–395
Feng G-S, Hui C-C, Pawson T (1993) SH2-containing phosphotyrosine phosphatase as a target of protein-tyrosine kinases. Science 259:1607–1611
Fortini ME, Simon MA, Rubin GM (1992) Signaling by the sevenless protein tyrosine kinase is mimicked by Ras1 activation. Nature 355:559–561
Freeman M (1996) Reiterative use of the EGF receptor triggers differentiation of all cell types in the Drosophila eye. Cell 87:651–660
Gaul U, Mardon G, Rubin GM (1992) A putative Ras GTPase activating protein acts as a negative regulator of signaling by the sevenless receptor tyrosine kinase. Cell 68:1007–1019
Greenwald I, Rubin GM (1992) Making a difference: the role of cell-cell interactions in establishing separate identities for equivalent cells. Cell 68:271–281
Hafen E, Basler K, Edstroem J-E, Rubin GM (1987) Sevenless, a cell specific homeotic gene of Drosophila, encodes a putative transmembrane receptor with a tyrosine kinase. Science 236:55–63
Harris WA, Stark WS, Walker JA (1976) Genetic dissection of the photoreceptor system in the compound eye of Drosophila melanogaster. J Physiol (Lond) 256:415–439

Hart AC, Krämer H, Van Vactor DL, Paidhungat M, Zipursky SL (1990) Induction of cell fate in the Drosophila retina: the bride of sevenless protein is predicted to contain a large extracellular domain and seven transmembrane segments. Genes Dev 4:1835–1847
Hart AC, Krämer H, Zipursky SL (1993) Extracellular domain of the boss transmembrane ligand acts as an antagonist of the sev receptor. Nature 361:732–736
Herbst R, Carroll PM, Allard JD, Schilling J, Raabe T, Simon MA (1996) Daughter of sevenless is a substrate of the phosphotyrosine phosphatase corkscrew and functions during sevenless signaling. Cell 85:899–909
Hiromi Y, Mlodzik M, West SR, Rubin GM, Goodman CS (1993) Ectopic expression of seven-up causes cell fate changes during ommatidial assembly. Development 118:1123–1135
Holgado-Madruga M, Emlet RD, Moscatello DK, Godwin AK, Wong AJ (1996) A Grb2-associated docking protein in EGF- and insulin-receptor signalling. Nature 379:560–564
Hou XS, Chou T-B, Melnick MB, Perrimon N (1995) The torso receptor tyrosine kinase can activate Raf in a Ras-independent pathway. Cell 81:63–71
Karim FD, Chang HC, Therrien M, Wassarman DA, Laverty T, Rubin GM (1996) A screen for genes that function downstream of Ras1 during Drosophila eye development. Genetics 143:315–329
Karlovich CA, Bonfini L, McCollam L, Rogge RD, Daga A, Czech MP, Banerjee U (1995) In vivo functional analysis of the Ras exchange factor son of sevenless. Science 268:576–579
Kauffmann RC, Li S, Gallagher PA, Zhang J, Carthew RW (1996) Ras1 signaling and transcriptional competence in the R7 cell of Drosophila. Genes Dev 10:2167–2178
Kazlauskas A (1994) Receptor tyrosine kinases and their targets. Curr Opin Gen Dev 4:5–14
Kimmel BE, Heberlein U, Rubin GM (1990) The homeo domain protein rough is expressed in a subset of cells in the developing Drosophila eye where it can specify photoreceptor cell subtype. Genes Dev 4:712–727
Klämbt C (1993) The Drosophila gene pointed encodes two ETS-like proteins which are involved in the development of the midline glial cells. Development 117:163–176
Kornfeld K, Hom DB, Horvitz HR (1995) The ksr-1 gene encodes a novel protein kinase involved in Ras-mediated signaling in C. elegans. Cell 83:903–913
Krämer H, Cagan RL, Zipursky SL (1991) Interaction of bride of sevenless membrane-bound ligand and the sevenless tyrosine-kinase receptor. Nature 352:207–212
Lai Z-C, Rubin GM (1992) Negative control of photoreceptor cell development in Drosophila by the product of the yan gene, an ETS domain protein. Cell 70:609–620
Lawrence PA, Green SM (1979) Cell lineage in the developing retina of Drosophila. Dev Biol 71:142–152
Leevers SJ, Paterson HF, Marshall CJ (1994) Requirement for Ras in Raf activation is overcome by targeting Raf to the plasma membrane. Nature 369:411–414
Lowenstein EJ, Daly RJ, Batzer AG, Li W, Margolis B, Lammers R, Ullrich A, Skolnik EY, Bar-Sagi D, Schlessinger J (1992) The SH2 and SH3 domain-containing protein GRB2 links receptor tyrosine kinases to ras signaling. Cell 70:431–442
Marshall CJ (1994) MAP kinase kinase kinase, MAP kinase kinase and MAP kinase. Curr Opin Gen Dev 4:82–89
Marshall CJ (1995) Specificity of receptor tyrosine kinase signaling: transient versus sustained extracellular signal-regulated kinase activation. Cell 80:179–185
McCollam L, Bonfini L, Karlovich CA, Conway BR, Kozma LM, Banerjee U, Czech MP (1995) Functional roles for the Pleckstrin and Dbl homology regions in the Ras exchange factor Son-of-sevenless. J Biol Chem 270:15954–15957
McCormick F (1994) Activators and effectors of ras p21 proteins. Curr Opin Gen Dev 4:71–76
Mlodzik M, Hiromi Y, Weber U, Goodman CS, Rubin GM (1990) The Drosophila seven-up gene, a member of the steroid receptor gene superfamily, controls photoreceptor cell fates. Cell 60:211–224
Oellers N, Hafen E (1996) Biochemical characterization of Rolled^{Sem}, an activated form of Drosophila mitogen-activated protein kinase. J Biol Chem 271:24939–24944
Olivier JP, Raabe T, Henkemeyer M, Dickson B, Mbamalu G, Margolis B, Schlessinger J, Hafen E, Pawson T (1993) A Drosophila SH2-SH3 adaptor protein implicated in coupling the sevenless tyrosine kinase to an activator of Ras guanine nucleotide exchange, Sos. Cell 73:179–191
O'Neill EM, Rebay I, Tjian R, Rubin GM (1994) The activities of two Ets-related transcription factors required for Drosophila eye development are modulated by the Ras/MAPK pathway. Cell 78:137–147
Pawson T (1995) Protein modules and signalling networks. Nature 373:573–579
Perkins KK, Admon A, Patel N, Tjian R (1990) The Drosophila fos-related AP-1 protein is a developmentally regulated transcription factor. Genes Dev 4:822–834

Perkins LA, Larson I, Perrimon N (1992) corkscrew encodes a putative protein tyrosine phosphatase that functions to transduce the terminal signal from the receptor tyrosine kinase torso. Cell 70:225–236
Peverali FA, Isaksson A, Papavassiliou AG, Plastina P, Staszewski LM, Mlodzik M, Bohmann D (1996) Phosphorylation of Drosophila Jun by the MAP kinase rolled regulates photoreceptor differentiation. EMBO J 15:3943–3950
Price JV, Clifford RJ, Schüpbach T (1989) The maternal ventralizing locus torpedo is allelic to faint little ball, an embryonic lethal, and encodes the Drosophila EGF receptor homolog. Cell 56:1085–1092
Raabe T, Olivier JP, Dickson B, Liu X, Gish GD, Pawson T, Hafen E (1995) Biochemical and genetic analysis of the Drk SH2/SH3 adaptor protein of Drosophila. EMBO J 14:2509–2518
Raabe T, Riesgo-Escovar J, Liu X, Bausenwein BS, Deak P, Maröy P, Hafen E (1996) Dos, a novel pleckstrin homology domain-containing protein required for signal transduction between sevenless and Ras1 in Drosophila. Cell 85:911–920
Ready DF, Hanson TE, Benzer S (1976) Development of the Drosophila retina a neurocrystalline lattice. Dev Biol 53:217–240
Rebay I, Rubin GM (1995) Yan functions as a general inhibitor of differentiation and is negatively regulated by activation of the Ras1/MAPK pathway. Cell 81:857–866
Reinke R, Zipursky SL (1988) Cell-cell interaction in the Drosophila retina: the bride of sevenless gene is required in photoreceptor cell R8 for R7 cell development. Cell 55:321–330
Rogge R, Cagan R, Majumdar A, Dulaney T, Banerjee U (1992) Neuronal development in the Drosophila retina: the sextra gene defines an inhibitory component in the developmental pathway of R7 photoreceptor cells. Proc Natl Acad Sci USA 89:5271–5275
Schejter ED, Shilo BZ (1989) The Drosophila EGF receptor homolog (DER) gene is allelic to faint little ball, a locus essential for embryonic development. Cell 56:1093–1104
Scholz H, Deatrick J, Klaes A, Klämbt C (1993) Genetic dissection of pointed, a Drosophila gene encoding two ETS-related proteins. Genetics 135:455–468
Simon MA, Bowtell DL, Dodson GS, Laverty TR, Rubin GM (1991) Ras1 and a putative guanine nucleotide exchange factor perform crucial steps in signaling by the Sevenless protein tyrosine kinase. Cell 67:701–716
Simon MA, Dodson GS, Rubin GM (1993) An SH3-SH2-SH3 protein is required for $p21^{Ras1}$ activation and binds to sevenless and Sos proteins in vitro. Cell 73:169–177
Skolnik EY, Batzer A, Li N, Lee C-H, Lowenstein E, Mohammadi M, Margolis B, Schlessinger J (1993) The function of GRB2 in linking the insulin receptor to ras signaling pathways. Science 260:1953–1955
Stokoe D, Macdonald SG, Cadwallader K, Symons M, Hancook JF (1994) Activation of Raf as a result of recruitment to the plasma membrane. Science 264:1463–1467
Sun XJ, Crimmins DL, Myers MG Jr, Miralpeix M, White MF (1993) Pleiotropic insulin signals are engaged by multisite phosphorylation of IRS-1. Mol Cell Biol 13:7418–7428
Sun XJ, Wang L-M, Zhang Y, Yenush L, Myers MG Jr, Glasheen E, Lane WS, Pierce JH, White MF (1995) Role of IRS-2 in insulin and cytokine signalling. Nature 377:173–177
Sundaram M, Han M (1995) The C. elegans ksr-1 gene encodes a novel Raf-related kinase involved in Ras-mediated signal transduction. Cell 83:889–901
Tei H, Nihonmatsu I, Yokokura T, Udea R, Sano Y, Okuda T, Sato K, Hirata K, Fujita SC, Yamamoto D (1992) pokkuri, a Drosophila gene encoding an E-26-specific (Ets) domain protein, prevents overproduction of the R7 photoreceptor. Proc Natl Acad Sci USA 89:6856–6860
Therrien M, Chang HC, Solomon NM, Karim FD, Wassarman DA, Rubin GM (1995) KSR, a novel protein kinase required for RAS signal transduction. Cell 83:879–888
Therrien M, Michaud NR, Rubin GM, Morrison DK (1996) KSR modulates signal propagation within the MAPK cascade. Genes Dev 10:2684–2695
Tomlinson A, Ready DF (1986) sevenless: a cell-specific homeotic mutation of the Drosophila eye. Science 231:400–402
Tomlinson A, Ready DF (1987a) Cell fate in the Drosophila ommatidium. Dev Biol 123:264–275
Tomlinson A, Ready DF (1987b) Neuronal differentiation in the Drosophila ommatidium. Dev Biol 120:366–376
Tomlinson A, Bowtell DD, Hafen E, Rubin GM (1987) Localization of the sevenless protein, a putative receptor for positional information, in the eye imaginal disc of Drosophila. Cell 51:143–150
Treier M, Bohmann D, Mlodzik M (1995) JUN cooperates with the ETS domain protein pointed to induce photoreceptor R7 fate in the Drosophila eye. Cell 83:753–760

Vaessin H, Grell E, Wolff E, Bier E, Jan LY, Jan YN (1991) prospero is expressed in neuronal precursors and encodes a nuclear protein that is involved in the control of axonal outgrowth in Drosophila. Cell 67:941–953

Van Vactor DL, Cagan RL, Krämer H, Zipursky SL (1991) Induction in the developing compound eye of Drosophila: multiple mechanisms restrict R7 induction to a single retinal precursor cell. Cell 67:1145–1155

Vogel W, Lammers R, Huang J, Ullrich A (1993) Activation of a phosphotyrosine phosphatase by tyrosine phosphorylation. Science 259:1611–1614

Wassarman DA, Solomon NM, Chang HC, Karim FD, Therrien M, Rubin GM (1996) Protein phosphatase 2A positively and negatively regulates Ras1-mediated photoreceptor development in Drosophila. Genes Dev 10:272–278

Wittinghofer A, Nassar N (1996) How Ras-related proteins talk to their effectors. TIBS 21:488–491

Wolff T, Ready DF (1993) Pattern formation in the Drosophila retina. In: Bate M, Martinez-Arias A (eds) The development of Drosophila melanogaster. Cold Spring Harbor Laboratory Press, Cold Spring Harbor, pp 1277–1326

Xu T, Rubin GM (1993) Analysis of genetic mosaics in developing and adult Drosophila tissues. Development 117:1223–1237

Zhang K, Chaillet R, Perkins LA, Halazonetis TD, Perrimon N (1990) Drosophila homolog of the mammalian jun oncogene is expressed during embryonic development and activates transcription in mammalian cells. Proc Natl Acad Sci USA 87:6281–6285

Subject Index

Current Topics in Microbiology and Immunology

Volumes published since 1989 (and still available)

Vol. 188: **Letvin, Norman L.; Desrosiers, Ronald C. (Eds.):** Simian Immunodeficiency Virus. 1994. 37 figs. X, 240 pp. ISBN 3-540-57274-0

Vol. 189: **Oldstone, Michael B. A. (Ed.):** Cytotoxic T-Lymphocytes in Human Viral and Malaria Infections. 1994. 37 figs. IX, 210 pp. ISBN 3-540-57259-7

Vol. 190: **Koprowski, Hilary; Lipkin, W. Ian (Eds.):** Borna Disease. 1995. 33 figs. IX, 134 pp. ISBN 3-540-57388-7

Vol. 191: **ter Meulen, Volker; Billeter, Martin A. (Eds.):** Measles Virus. 1995. 23 figs. IX, 196 pp. ISBN 3-540-57389-5

Vol. 192: **Dangl, Jeffrey L. (Ed.):** Bacterial Pathogenesis of Plants and Animals. 1994. 41 figs. IX, 343 pp. ISBN 3-540-57391-7

Vol. 193: **Chen, Irvin S. Y.; Koprowski, Hilary; Srinivasan, Alagarsamy; Vogt, Peter K. (Eds.):** Transacting Functions of Human Retroviruses. 1995. 49 figs. IX, 240 pp. ISBN 3-540-57901-X

Vol. 194: **Potter, Michael; Melchers, Fritz (Eds.):** Mechanisms in B-cell Neoplasia. 1995. 152 figs. XXV, 458 pp. ISBN 3-540-58447-1

Vol. 195: **Montecucco, Cesare (Ed.):** Clostridial Neurotoxins. 1995. 28 figs. XI., 278 pp. ISBN 3-540-58452-8

Vol. 196: **Koprowski, Hilary; Maeda, Hiroshi (Eds.):** The Role of Nitric Oxide in Physiology and Pathophysiology. 1995. 21 figs. IX, 90 pp. ISBN 3-540-58214-2

Vol. 197: **Meyer, Peter (Ed.):** Gene Silencing in Higher Plants and Related Phenomena in Other Eukaryotes. 1995. 17 figs. IX, 232 pp. ISBN 3-540-58236-3

Vol. 198: **Griffiths, Gillian M.; Tschopp, Jürg (Eds.):** Pathways for Cytolysis. 1995. 45 figs. IX, 224 pp. ISBN 3-540-58725-X

Vol. 199/I: **Doerfler, Walter; Böhm, Petra (Eds.):** The Molecular Repertoire of Adenoviruses I. 1995. 51 figs. XIII, 280 pp. ISBN 3-540-58828-0

Vol. 199/II: **Doerfler, Walter; Böhm, Petra (Eds.):** The Molecular Repertoire of Adenoviruses II. 1995. 36 figs. XIII, 278 pp. ISBN 3-540-58829-9

Vol. 199/III: **Doerfler, Walter; Böhm, Petra (Eds.):** The Molecular Repertoire of Adenoviruses III. 1995. 51 figs. XIII, 310 pp. ISBN 3-540-58987-2

Vol. 200: **Kroemer, Guido; Martinez-A., Carlos (Eds.):** Apoptosis in Immunology. 1995. 14 figs. XI, 242 pp. ISBN 3-540-58756-X

Vol. 201: **Kosco-Vilbois, Marie H. (Ed.):** An Antigen Depository of the Immune System: Follicular Dendritic Cells. 1995. 39 figs. IX, 209 pp. ISBN 3-540-59013-7

Vol. 202: **Oldstone, Michael B. A.; Vitković, Ljubiša (Eds.):** HIV and Dementia. 1995. 40 figs. XIII, 279 pp. ISBN 3-540-59117-6

Vol. 203: **Sarnow, Peter (Ed.):** Cap-Independent Translation. 1995. 31 figs. XI, 183 pp. ISBN 3-540-59121-4

Vol. 204: **Saedler, Heinz; Gierl, Alfons (Eds.):** Transposable Elements. 1995. 42 figs. IX, 234 pp. ISBN 3-540-59342-X

Vol. 205: **Littman, Dan R. (Ed.):** The CD4 Molecule. 1995. 29 figs. XIII, 182 pp. ISBN 3-540-59344-6

Vol. 206: **Chisari, Francis V.; Oldstone, Michael B. A. (Eds.):** Transgenic Models of

Human Viral and Immunological Disease. 1995. 53 figs. XI, 345 pp. ISBN 3-540-59341-1

Vol. 207: **Prusiner, Stanley B. (Ed.):** Prions Prions Prions. 1995. 42 figs. VII, 163 pp. ISBN 3-540-59343-8

Vol. 208: **Farnham, Peggy J. (Ed.):** Transcriptional Control of Cell Growth. 1995. 17 figs. IX, 141 pp. ISBN 3-540-60113-9

Vol. 209: **Miller, Virginia L. (Ed.):** Bacterial Invasiveness. 1996. 16 figs. IX, 115 pp. ISBN 3-540-60065-5

Vol. 210: **Potter, Michael; Rose, Noel R. (Eds.):** Immunology of Silicones. 1996. 136 figs. XX, 430 pp. ISBN 3-540-60272-0

Vol. 211: **Wolff, Linda; Perkins, Archibald S. (Eds.):** Molecular Aspects of Myeloid Stem Cell Development. 1996. 98 figs. XIV, 298 pp. ISBN 3-540-60414-6

Vol. 212: **Vainio, Olli; Imhof, Beat A. (Eds.):** Immunology and Developmental Biology of the Chicken. 1996. 43 figs. IX, 281 pp. ISBN 3-540-60585-1

Vol. 213/I: **Günthert, Ursula; Birchmeier, Walter (Eds.):** Attempts to Understand Metastasis Formation I. 1996. 35 figs. XV, 293 pp. ISBN 3-540-60680-7

Vol. 213/II: **Günthert, Ursula; Birchmeier, Walter (Eds.):** Attempts to Understand Metastasis Formation II. 1996. 33 figs. XV, 288 pp. ISBN 3-540-60681-5

Vol. 213/III: **Günthert, Ursula; Schlag, Peter M.; Birchmeier, Walter (Eds.):** Attempts to Understand Metastasis Formation III. 1996. 14 figs. XV, 262 pp. ISBN 3-540-60682-3

Vol. 214: **Kräusslich, Hans-Georg (Ed.):** Morphogenesis and Maturation of Retroviruses. 1996. 34 figs. XI, 344 pp. ISBN 3-540-60928-8

Vol. 215: **Shinnick, Thomas M. (Ed.):** Tuberculosis. 1996. 46 figs. XI, 307 pp. ISBN 3-540-60985-7

Vol. 216: **Rietschel, Ernst Th.; Wagner, Hermann (Eds.):** Pathology of Septic Shock. 1996. 34 figs. X, 321 pp. ISBN 3-540-61026-X

Vol. 217: **Jessberger, Rolf; Lieber, Michael R. (Eds.):** Molecular Analysis of DNA Rearrangements in the Immune System. 1996. 43 figs. IX, 224 pp. ISBN 3-540-61037-5

Vol. 218: **Berns, Kenneth I.; Giraud, Catherine (Eds.):** Adeno-Associated Virus (AAV) Vectors in Gene Therapy. 1996. 38 figs. IX,173 pp. ISBN 3-540-61076-6

Vol. 219: **Gross, Uwe (Ed.):** Toxoplasma gondii. 1996. 31 figs. XI, 274 pp. ISBN 3-540-61300-5

Vol. 220: **Rauscher, Frank J. III; Vogt, Peter K. (Eds.):** Chromosomal Translocations and Oncogenic Transcription Factors. 1997. 28 figs. XI, 166 pp. ISBN 3-540-61402-8

Vol. 221: **Kastan, Michael B. (Ed.):** Genetic Instability and Tumorigenesis. 1997. 12 figs.VII, 180 pp. ISBN 3-540-61518-0

Vol. 222: **Olding, Lars B. (Ed.):** Reproductive Immunology. 1997. 17 figs. XII, 219 pp. ISBN 3-540-61888-0

Vol. 223: **Tracy, S.; Chapman, N. M.; Mahy, B. W. J. (Eds.):** The Coxsackie B Viruses. 1997. 37 figs. VIII, 336 pp. ISBN 3-540-62390-6

Vol. 224: **Potter, Michael; Melchers, Fritz (Eds.):** C-Myc in B-Cell Neoplasia. 1997. 94 figs. XII, 291 pp. ISBN 3-540-62892-4

Vol. 225: **Vogt, Peter K.; Mahan, Michael J. (Eds.):** Bacterial Infection: Close Encounters at the Host Pathogen Interface. 1998. 15 figs. IX, 169 pp. ISBN 3-540-63260-3

Vol. 226: **Koprowski, Hilary; Weiner, David B. (Eds.):** DNA Vaccination/Genetic Vaccination. 1998. 31 figs. approx. IX, 200 pp. ISBN 3-540-63392-8

Vol. 227: **Vogt, Peter K.; Reed, Steven I. (Eds.):** Cyclin Dependent Kinase (CDK) Inhibitors. 1998. 15 figs. approx. IX, 175 pp. ISBN 3-540-63429-0

Springer and the environment

At Springer we firmly believe that an international science publisher has a special obligation to the environment, and our corporate policies consistently reflect this conviction.
We also expect our business partners – paper mills, printers, packaging manufacturers, etc. – to commit themselves to using materials and production processes that do not harm the environment. The paper in this book is made from low- or no-chlorine pulp and is acid free, in conformance with international standards for paper permanency.

MIX
Papier aus verantwortungsvollen Quellen
Paper from responsible sources
FSC® C105338

If you have any concerns about our products,
you can contact us on
ProductSafety@springernature.com

In case Publisher is established outside the EU,
the EU authorized representative is:
Springer Nature Customer Service Center GmbH
Europaplatz 3, 69115 Heidelberg, Germany

Printed by Libri Plureos GmbH
in Hamburg, Germany